A TUTORIAL GUIDE TO
AutoCAD®
RELEASE 14

A TUTORIAL GUIDE TO
AutoCAD®
RELEASE 14

Shawna D. Lockhart

with Marc Finseth

 **ADDISON-WESLEY**

Addison-Wesley is an imprint of Addison Wesley Longman, Inc.

Reading, Massachusetts • Harlow, England • Menlo Park, California
Berkeley, California • Don Mills, Ontario • Sydney
Bonn • Amsterdam • Tokyo • Mexico City

The Publishing Team

Senior Producer: Denise Descoteaux
Executive Editor: Robert Woodbury
Manager, Educational Technology: Janet Drumm
Producer's Assistant: Morgan Baker
Project Manager: Cindy Johnson
Text Designer: Jean Hammond

Copyeditor: Jerrold Moore
Proofreader: Sarah Corey
Technical Validator: Tomislav Mutak
Page Layout: Publishing Services
Cover Designer: Alwyn Velásquez
Cover Illustration: John Sanderson, Horizon Design
Senior Manufacturing Coordinator: Janet Weaver

Autodesk Trademarks

The following are registered trademarks of Autodesk, Inc.: ADI, Advanced Modeling Extension, AME, Animator Pro, ATC, AutoCAD, AutoCAD Development System, Autodesk, Autodesk Animator, the Autodesk logo, AutoLISP, AutoShade, AutoSketch, AutoSnap, AutoSolid, AutoSurf, Geodyssey, HOOPS, Multimedia Explorer, Office Series, TinkerTech, World-Creating Toolkit, 3D Plan, and 3D Studio.

The following are trademarks of Autodesk, Inc.: ACAD, Advanced User Interface, AEMULUS, AEMULUS(mf), AME Link, Animation Partner, Animation Player, Animation Pro Player, A Studio in Every Computer, ATLAST, AUI, AutoCAD Data Extension, AutoCAD Simulator, AutoCAD SQL Extension, AutoCAD SQL Interface, Autodesk Animator Clips, Autodesk Animator Theatre, Autodesk Device Interface, Autodesk Software Developer's Kit, Autodesk WorkCenter, AutoCDM, AutoEDM, AutoFlix, AutoLathe, AutoVision, DXF, FLI, Flic, Generic 3D, SketchTools, SmartCursor, Supportdesk, Texture Universe, Transforms ideas Into Reality, and Visual Link.

The following are service marks of Autodesk, Inc.: Autodesk Strategic Developer, Autodesk Strategic Developer logo, Autodesk Registered Developer, Autodesk Registered Developer logo, and the Autodesk University logo.

Third Party Trademarks

Renderman is a registered trademark of Pixar used by Autodesk under license from Pixar.
Microsoft, Windows, MS-DOS, Windows NT, and the Windows logo are either registered trademarks or trademarks of Microsoft Corporation.
All other trademarks are trademarks of their respective holders.

ISBN 0-201-82371-3
1 2 3 4 5 6 7 8 9 10—CRS—02 01 00 99 98

AutoCAD® is the most widely used design and drafting software for desktop computers in the world. AutoCAD® Release 14 provides you with the capability to create complex and accurate drawings. Its position as the industry standard makes it an essential tool for anyone preparing for a career in engineering, design, or technology.

Because it is the industry standard, AutoCAD is the ideal cornerstone of your design and drafting skill set. With a knowledge of AutoCAD, you will find it easy to add any number of a wide range of applications to create a complete design environment suited to your needs.

This tutorial-based manual will teach you step-by-step to use AutoCAD Release 14 running under a Windows® or Windows NT® operating system to create 2D and 3D models and the engineering drawings that describe them. Written for the novice AutoCAD user, the manual uses a proven tutorial approach that guides you through the creation of actual drawings and models. Information about AutoCAD is presented in a need-to-know fashion that makes it easy to remember. Tips and shortcuts are included where appropriate to help you become an efficient and proficient AutoCAD user. A Command Summary and Glossary make it easy to review what you've learned and to use the manual as a reference.

This manual may be used in conjunction with a basic engineering graphics course, introductory engineering, introductory architecture and/or design courses, or independently for self-study. Completing this entire set of tutorials provides a solid groundwork enabling you to apply AutoCAD in a professional setting including creating standard template drawings; using blocks, libraries, attributes, and Xrefs; managing dimension styles; printing and plotting standard drawing views; generating and rendering 3D solid models; applying other Windows software along with AutoCAD to create parts lists; customizing your AutoCAD toolbars, using geometric dimensioning and tolerancing symbols, and even publishing your drawings on the World Wide Web.

Features

To facilitate your study of AutoCAD Release 14, this tutorial guide includes:

- Step-by-step tutorials written for the novice user
- A complete chapter on configuring AutoCAD for performance and use with these tutorials
- Tutorials organized to parallel an introductory engineering graphics course
- "TIP" boxes that offer suggestions and warnings to students as they progress through the tutorials
- Challenging end-of-tutorial drawing exercises, with applications in mechanical, civil, and electrical engineering, as well as architecture
- Key Terms and Key Commands summaries to recap important topics and commands learned in each tutorial

Organization

The tutorials proceed in a logical fashion to guide you from drawing basic shapes to building three-dimensional models. Later tutorials introduce techniques for automatically generating 2D drawing views from your solid models and for advanced dimensioning including geometric dimensioning and tolerancing symbols. The final tutorial shows you how to create photo-realistic renderings from your models.

Part 1, Getting Started, first helps you configure AutoCAD's environment and menus for the step-by-step instructions in this manual, then takes you on a guided tour of the AutoCAD Release 14 screen display, help facility, and keyboard/mouse usage conventions. Both of the Getting Started chapters address AutoCAD operations under Windows 95 and Windows NT.

Part 2, Tutorials, introduces AutoCAD Release 14 in the context of technical drawing.

- Tutorials 1 and 2 introduce AutoCAD's basic drawing commands and build proficiency with the menus, toolbars, and drawing aids.

- Tutorials 3 and 4 introduce editing commands and geometric constructions and provide instruction on plotting a drawing.

- Tutorial 5 focuses on good drawing management with template drawings, layers, and plotting from paper space.

- Tutorial 6 is devoted to the concepts of drawing orthographic views.

- Tutorial 7 introduces basic dimensioning and the use of dimension styles.

- Tutorial 8 shows how to use blocks and to customize toolbars.

- Tutorials 9, 10, and 11 teach 3D solid modeling from a single part to an assembly drawing. Tutorial 9 introduces solid modeling techniques, Tutorial 10 shows you how to edit solid models and create 2D drawings

from them, and in Tutorial 11, you use the solid models to check for interference and create an assembly drawing.

- Tutorials 12 and 13 introduce section and auxiliary views and show how to create them in 2D and from a 3D model. These tutorials may be undertaken after solid modeling, Tutorials 9 and 10.

- Tutorial 14 is devoted to advanced dimensioning topics such as geometric tolerances. It may be used immediately after Tutorial 7, if desired.

- Tutorial 15 shows how to render and shade 3D models and export them to 3D Studio. This tutorial may be completed any time after solid modeling (Tutorial 9) if you wish to render models created in later tutorials.

The Glossary defines key terms used in the tutorials. The Command Summary lists the key commands used in the tutorials, defines them, and recaps their toolbar icon, menu location, and keyboard equivalent.

Acknowledgments

I wish to acknowledge the individuals who contributed to the conceptualization and implementation of these tutorials. First, and foremost, thanks to Marc Finseth of Montana State University, who was a major contributor to this Release 14 version of the text. Thanks also to the many teaching colleagues who responded to inquiries and helped to shape the tutorials in this book. We cannot list all the individuals with whom we spoke in the course of our research, but we are grateful to the reviewers listed below for their help with this edition.

James G. Raschka
Delaware County Community College
Reuben Aronovitz
Delaware County Community College

John Matson
Arizona State University

Craig L. Miller
Purdue University

Michael H. Gregg
Virginia Polytechnic Institute and State University

Our thanks go to Doug Baese, Steve Brackman, Craig Bradley, Tom Bryson, Karen Coen-Brown, Joe Evers, Mary Ann Koen, D. Krall, Kim Manner, Shawn Murphy, Torian Roesch, Kyle Tage, John Walker, and Wendy Warren for submitting exercises and creating art files for us; to Shannon Kyles and James Bethune, authors of previous AutoCAD tutorial guides, from which some material was adapted for this manual; and to James Earle, for permission to reprint several exercises from his text.

I wish to thank Jim Purcell, Director of Education; Mark Sturges, Education Business Development Manager; Jimm Meloy, Worldwide Learning and Training Manager; Carrie Bustillos, Manager ATC Programs; Maureen Barrow, Contracts Administrator; and Daniel Vinson, Consultant, Education Department, who lent their talents to the partnership between Addison-Wesley and Autodesk, Inc. that supported the development of this manual.

Editing, testing, and producing step-by-step tutorials requires a tremendous amount of publishing expertise, and I would be remiss if I did not acknowledge those who worked to make this book complete, technically accurate, and lovely to look at, especially Denise Descoteaux, Morgan Baker, Alwyn Velásquez, Karen Wernholm, Phoebe Ling, Janet Drumm, and Cindy Johnson for their enthusiasm, efficiency, and professionalism throughout this project. Last but not least, I wish to thank the many great students I have had the pleasure to teach at Montana State University and from whom I have learned a great deal.

Shawna D. Lockhart

CONTENTS

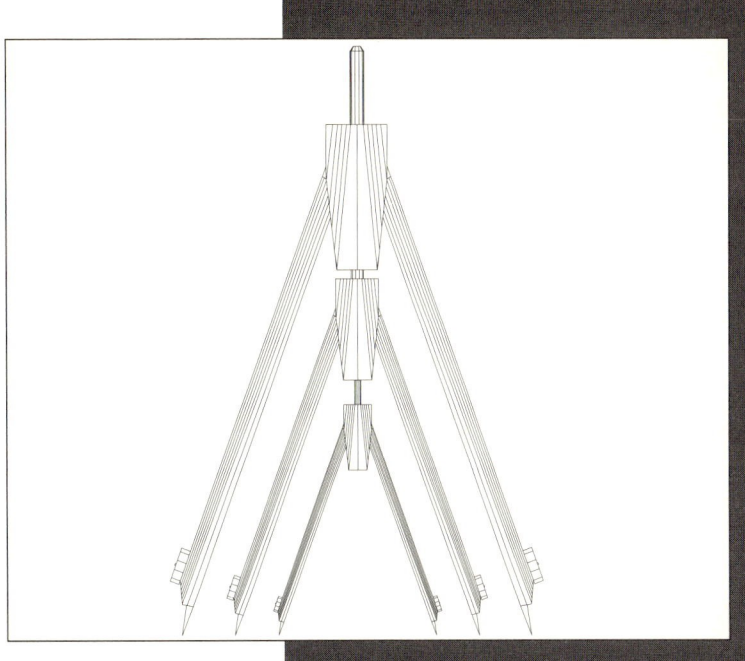

PART ONE

Getting Started

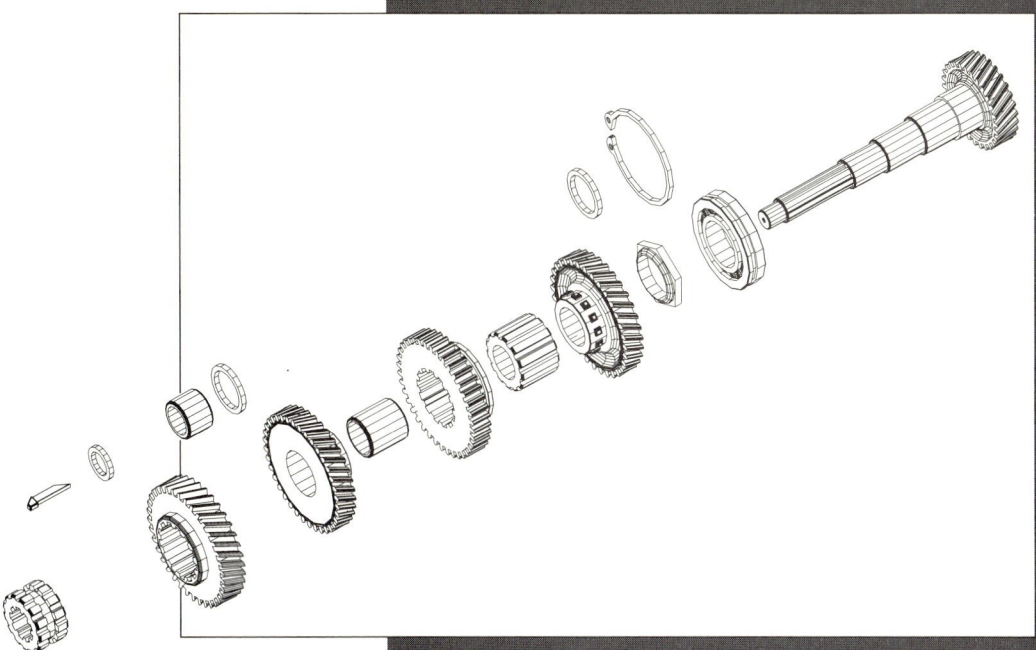

CHAPTER 1

Preparing AutoCAD for the Tutorials

Objectives

This chapter describes how to prepare your computer system and *AutoCAD Release 14 for Windows 95/NT* for use with the tutorials in this manual. As you read the chapter, you will

1. Understand how to use a mouse to complete these tutorials.

2. Recognize the typographical conventions used in this book.

3. Configure Windows for maximum AutoCAD performance and support.

4. Create a separate AutoCAD shortcut to use with these tutorials.

5. Create and prepare datafile and working directories.

6. Configure your copy of AutoCAD for use with these tutorials.

Introduction to the Tutorials

The fifteen tutorials in this manual will teach you to use *AutoCAD Release 14 for Windows 95/NT* through a series of step-by-step exercises. You will learn the fundamental operating procedures and how to use the drawing tools of each application. As you progress through the tutorials, you will employ some of the more advanced features of AutoCAD to create, dimension, shade, and print drawings.

The tutorials in this manual are based on the assumption that you are using AutoCAD's default settings. If your system has been customized for other uses, some of the settings may not match those used in the step-by-step instruction in the tutorials. Please check your system against the configuration in this chapter so that you can work through the tutorials as instructed.

If you are using AutoCAD on a network at your school, ask your professor or system administrator about how the software is configured. You should not try to reconfigure AutoCAD unless instructed to do so by network personnel. Read about mouse techniques, typographical conventions, and end-of-tutorial exercises in this chapter, and then go on to Chapter 2, *AutoCAD Basics*.

To complete these tutorials, your system must be running Microsoft Windows 95 or Windows NT.

Basic Mouse Techniques

The tutorials in this manual are based on the assumption that you will be using a mouse or pointing device to work with AutoCAD Release 14 for Windows 95/NT. The following terms will be used to streamline the instructions for use of the mouse.

Term	Meaning
Pick or Click	To press quickly and then release the left mouse button
Right click	To quickly press and release the right mouse button
Double-click	To click the left mouse button twice in rapid succession
Drag	To press and hold down the left mouse button while you move the mouse
Point	To move the mouse until the mouse pointer on the screen is positioned above the item you want

> ■ *TIP* Your mouse may have more than one button. To click (or pick) in Windows and in AutoCAD, use the left mouse button. This button is also referred to as the *pick button*. To return, pick the right mouse button, also referred to as the *return button*. ■

Recognizing Typographical Conventions

When you work with AutoCAD, you'll use your keyboard and your pointing device to input information. As you read this manual, you'll encounter special type styles that will help you determine the commands required and the information you must input.

Special symbols illustrate certain computer keys. For example, the Enter or Return key is represented by the symbol ⏎.

> ■ *TIP* The right button on your mouse performs the same function as the ⏎ key. This button is also referred to as the *return button*. ■

The manual also employs special typefaces when it presents instructions for performing computer operations. Instructions are set off from the main text to indicate a series of actions or AutoCAD command prompts. Boldface type in all capital letters is used for the letters and numbers to be input by you. For example,

Command: **LINE**

instructs you to type the Line command.

From point: **(pick point A)**

instructs you to select point A on your screen by clicking it with the mouse. The words "Command" and "From point" represent the AutoCAD prompts that you would see on your screen. The AutoCAD prompt line is always in regular type.

Instruction words, such as "pick," "type," "select," and "press," appear in italic type. For example, "type" instructs you to strike or press several keys in sequence. For the following instruction,

Type: **4,4**

you would type *4,4* in that order on your keyboard. Remember to type exactly what you see, including spaces, if any.

"Pick" tells you to pick an object or an icon or to choose commands from the menu. For example,

Pick: **Line icon**

instructs you to use your pointing device to select the Line icon from the toolbar on your screen. To select an icon, move the pointer until the cursor is over the command and press the pick button.

"Press" means to strike a key once. For example,

Press: ⏎

means strike the Enter or Return key once. Sometimes "press" is followed by two keys, such as

Press: Alt – F1

In this case, press and hold down the first key, press the second key once and release it, and then release the first key.

Default values that are displayed as part of a command or prompt are represented in angle brackets: < >. For example,

Trace width<0.0500>:

is the command prompt when the Trace command has a default value of 0.05.

New terms are in italics in the text when they are introduced and are defined in the Glossary at the end of the manual.

End-of-Tutorial Exercises

Exercises at the end of the tutorials are divided into four types, designated by the following icons.

Mechanical Engineering—Exercises in the design of machines and tools.

Electrical Engineering—Exercises in the design of electrical systems.

Civil Engineering—Exercises in the design of roads, bridges, and other public and private works.

Architecture—Exercises in the design of buildings.

Configuring Windows for AutoCAD

AutoCAD Release 14 for Windows 95/NT takes advantage of common Windows file operations and allows you to run AutoCAD while running other applications.

Windows 95 and Windows NT

AutoCAD Release 14 is a 32-bit program, designed to take full advantage of today's computers. AutoCAD can be installed with Windows NT 3.51, Windows NT 4.0, or Windows 95. The figures in the tutorials were captured in Windows 95. The AutoCAD screens in Windows NT 4.0 are similar to the screens that you will see in this manual. File operations and some screens may be slightly different running under Windows NT 3.51, but the steps are described in sufficient detail that you should be able to follow them on your own.

Screen Resolution

In Windows 95, you can change your screen resolution in the Control Panel. From Windows 95,

> Pick: **Start, Settings, Control Panel** *(from the bottom left of your screen)*
>
> Pick: **Display** *(from the Control Panel)*
>
> Pick: **Settings** *(in the dialog box that appears)*

The Display Properties dialog box appears on your screen, as shown in Figure G1.1. You can change the color selection and the pixel selection with the scroll bars. Set your display to 256 Color palette and 800 x 600 pixels. Windows 95 will look for and load the appropriate driver for your video display. (Refer to your system documentation if you have trouble with your screen resolution settings.) If you have a monitor that is capable of higher resolution and want to use a higher setting, your screens will look slightly

different from the ones captured for this book, but it should not affect the overall operation of the commands as described in the tutorials.

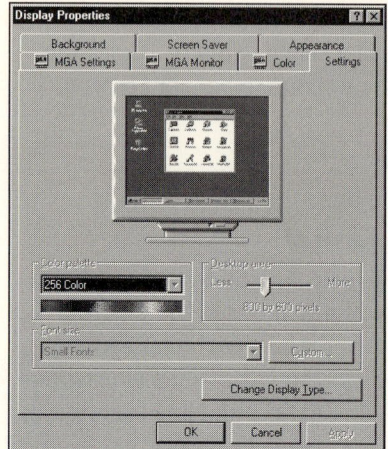

Figure G1.1

> Pick: **OK** *(to exit the Display Properties dialog box)*

You will be prompted to restart Windows so that the change takes effect. Restart Windows now.

Backing Up AutoCAD Defaults

Before you begin to configure and customize AutoCAD for the purposes of these tutorials, you will want to back up your configuration and menu files to save the previous configuration. You can do this by copying the customizable files listed below from the *c:\AutoCAD R14* and *c:\AutoCAD R14\support* directories to a floppy disk or to a safe location on your hard disk or a remote network drive where they will not be overwritten.

If you want to restore these AutoCAD defaults, you can copy them back to the *c:\AutoCAD R14* and *c:\AutoCAD R14\support* directories or your working directory and overwrite the files there. (This manual is based on the assumption that you are working from the *c:* drive and that AutoCAD is installed on that drive in a directory called AutoCAD R14. If you are not, substitute your drive name and path as needed in these instructions.) If you are on a networked system, contact your network administrator about backing up these files.

Customizable support files

File	Description
*.ahp	AutoCAD Help files. Associated help index files have the extension .hdx.
acad.ase	Names and locations of ASE database drivers.
acad14.cfg	AutoCAD device configuration file.
*.ccp	CalComp color palette files. For use with CalComp printers and plotters.
*.cus	Custom dictionary files.
*.dcl	AutoCAD Dialog Control Language (DCL) descriptions of dialog boxes.
acad.dcl	Describes the AutoCAD standard dialog boxes.
*.dxt	DXFIX translator file.
*.lin	AutoCAD linetype definition files.
acad.lin	The standard AutoCAD linetype library file.
*.lsp	AutoLISP program files.
acad.lsp	A user-defined AutoLISP routine that loads each time you start a drawing.
*.mln	A multiline library file.
*.mnd	A special type of uncompiled menu file that contains macros. The MC executable file is used to compile this file.
*.mnl	AutoLISP routines used by AutoCAD menus. A .mnl file must have the same filename as the .mnu file it supports.
acad.mnl	AutoLISP routines used by the standard AutoCAD menu.
*.mns	AutoCAD generated menu source files. Contains the command strings and macro syntax that define AutoCAD menus.
acad.mns	Source file for the standard AutoCAD menu.
*.mnu	AutoCAD menu source files. Contain the command strings and macro syntax that define AutoCAD menus.
acadfull.mnu	Source file for the standard AutoCAD menu. Source file for the DOS-like menu.
*.pat	AutoCAD hatch pattern definition files.
acad.pat	The standard AutoCAD hatch pattern library file.
*.pcp	AutoCAD plot configuration parameters files. Each .pcp file stores configuration information for a specific plotter.
acad.pgp	The AutoCAD program parameters file. Contains definitions for external commands and command aliases.
fontmap.ps	The AutoCAD Font Map file. Used by PSIN; the catalog (or font map) of all fonts known to the AutoCAD PostScript interpreter.
acad.psf	AutoCAD PostScript Support file; the master support file for the PSOUT and PSFILL commands.

*.rpf	Raster-pattern fill definition file. For use with Hewlett-Packard printer and plotter drivers.
acad.rx	Lists ARX applications that load when you start AutoCAD.
*.scr	AutoCAD script files. A script file contains a set of AutoCAD commands processed as a batch.
*.shp	AutoCAD shape/font definition files. Compiled shape/font files have the extension .shx.
acad.unt	AutoCAD unit definition file. Contains data that lets you convert from one set of units to another.

Creating a Working Directory

To make saving and opening easier as you complete the tutorials, you will create and specify a new working directory for AutoCAD. The working directory is the default directory used by AutoCAD to open and save files. By setting a working directory, when asked to save a drawing, you will not have to search through the computer's directories every time.

This is a good practice in general for helping organize your files. You should not save your drawing files and other work files in the same directory as the AutoCAD application because doing so makes it likely that they will be over-written or lost when you install software upgrades. It is also easier to organize and backup your project files and drawings if they are in a separate directory. Making subdirectories within this directory for different types of projects helps you organize your drawings even further.

You will create a new shortcut for launching AutoCAD that will use the directory *c:\work* as the starting directory. The *work* directory will be the default choice when you save your AutoCAD files after launching AutoCAD with this shortcut.

Before continuing, use your Windows Explorer to create a new directory, *c:\work*.

When you are finished, return to the Windows 95 desktop.

Creating Multiple Shortcuts

In Windows 95, you will start AutoCAD from the Start menu. Items on the Start menu are shortcuts to the executable file in the AutoCAD R14 directory. You can create several shortcuts to the same file, each with different properties. You will add a new shortcut using the Settings option on the Start menu.

> *Pick:* **Start, Settings, Taskbar**
> *Pick:* **(the Start Menu programs tab)**
> *Pick:* **Advanced**

Locate the AutoCAD R14 directory in the list of directories on the left. It may be in the Programs directory on your system. When you have located it, double-click it so its contents are displayed in the list on the right. Highlight the AutoCAD R14 shortcut by picking it once. With the shortcut highlighted,

> *Pick:* **File, Create Shortcut**

These actions will create a new copy of your AutoCAD R14 shortcut, AutoCAD R14(2). You will rename this copy of the shortcut *AutoCAD R14 Tutorials*.

Windows 95 allows you to rename a file by picking on its name with the right mouse button and selecting Rename from the list in the resulting pop-up menu. (You can also rename the file by picking it twice, but NOT double-clicking. Pick on the shortcut once and then after a few seconds pick on it again; a cursor will start blinking to the right of the name allowing you to change it.)

Rename the shortcut on your own now.

You will now set *c:\work* as your working directory.

Setting the Working Directory

The AutoCAD session you start with this shortcut will save your drawings to *c:\work* by default if you set it as your working directory.

To set *c:\work* as your working directory, you will modify the settings stored with the new AutoCAD R14 Tutorials shortcut.

> Pick: **(the new AutoCAD R14 Tutorials shortcut with the right mouse button)**
>
> Pick: **Properties (from the pull-down menu)**
>
> Pick: **(the Shortcut tab at the top of the Properties dialog box)**
>
> Pick: **(the text box labeled Start in and highlight the existing text)**
>
> Type: **C:\WORK**

Your dialog box should look like Figure G1.2. If you have created your *work* directory on another drive, substitute the correct drive letter.

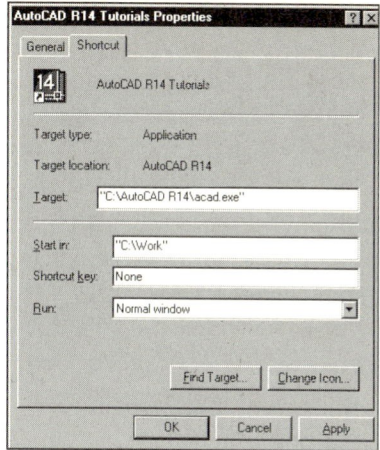

Figure G1.2

> Pick: **OK (to exit the Properties dialog box)**

When you start AutoCAD from this shortcut on the Start menu, the drawings you save will be saved to *c:\work*. The settings necessary for running AutoCAD will not be affected by the change in the working directory. Those drive paths are laid out within AutoCAD's Preferences.

Installing Data Files for the Tutorials

The next step is to create a directory for the data files used in the tutorials and install them. Throughout the tutorials, you will be instructed to work with files already prepared for you. You will not be able to complete the tutorials without these files. You can download these data files via FTP or with a Web browser such as Netscape Navigator. You will create a directory for them and then copy them into that directory.

To download via FTP, connect to *ftp.awl.com/cseng/authors/lockhart/r14* and look for the file called *r14data.exe*. If you have trouble, connect to *ftp.awl.com* and change to the *cseng/authors/lockhart/r14* directory. Use *anonymous* as your user name and your email address as your password when you log on.

To access the files from a web browser, use the URL *http://www.awl.com/cseng/authors/lockhart/r14* and look for the file *r14data.exe*.

If you do not have access to the Internet, you may request a copy of the files from Addison Wesley Longman by calling (781) 944-3700, ext. 2294, or by sending e-mail to cad-info@awl.com.

After you have retrieved the files, use Windows to create a new directory, *c:\datafile*. Copy *r14data.exe* into the directory and then run the file to decompress it. From Windows 95, you may double-click the file name in the Windows Explorer, or use Start, Run and enter the path and file name in the Run dialog box. When the file is decompressed, you will see an assortment of files in the directory, most of which have the *.dwg* extension.

Now you are ready to customize your AutoCAD menus.

AutoCAD Configuration for Tutorials

The tutorials in this manual are based on the assumption that you are using AutoCAD's default settings. If your system has been customized for other uses, some of the settings may not match those in the step-by-step instructions in the tutorials. You will now set up your system so that you can work through the tutorials as instructed.

■ *Warning:* If you are working on a networked system or on a workstation in a classroom or lab, your system has already been configured for you. Please skip to Chapter 2, *AutoCAD Basics.* ■

To start AutoCAD from Windows 95, pick Start, Programs, AutoCAD R14, AutoCAD R14 Tutorials.

The AutoCAD Start Up dialog box appears, as shown in Figure G1.3.

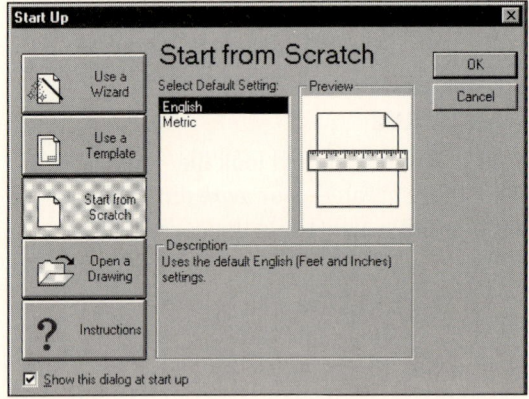

Figure G1.3

The following options are available: Use a Wizard, Use a Template, Start from Scratch, Open a Drawing, and Instructions.

Pick: **Instructions**

A brief description of each button appears. The Use a Wizard button will guide you through setting up a drawing before you begin one. However, to be able to use it, you need a clear understanding of Units, Paper Space, Model Space, and Limits (to name a few), which you will learn how to use and manipulate later in this guide. The Use a Template button allows you to start a drawing based on an existing template (either in **.dwt* or **.dwg* format), without changing the template drawing. The other buttons are self-explanatory, and for now you will choose Start from Scratch with English units.

Pick: **Start from Scratch**

Pick: **OK**

The AutoCAD graphics window appears on your screen, as shown in Figure G1.4, with the following menus: File, Edit, View, Insert, Format, Tools, Draw, Dimension, Modify, Bonus, and Help. The following toolbars are also showing: Draw and Modify (docked on the left side of the screen) and Object Properties and Standard (docked on the upper part of the screen).

An Internet Utilities toolbar may be shown, docked on the right-hand of the screen, depending on which Internet setup you have on your system. If AutoCAD does not directly detect an Internet service on your computer, the Internet Utilities toolbar will not appear.

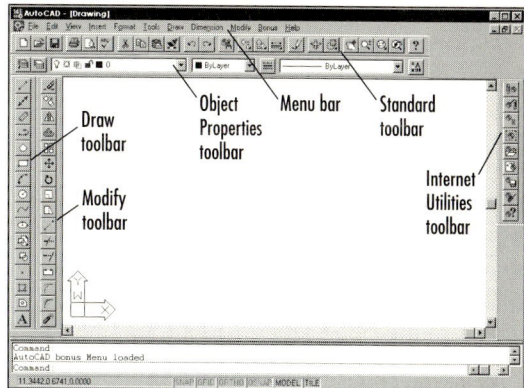

Figure G1.4

You will use the Tools menu to open the Preferences dialog box.

Pick: **Tools, Preferences**

The Preferences dialog box appears on your screen, as shown in Figure G1.5.

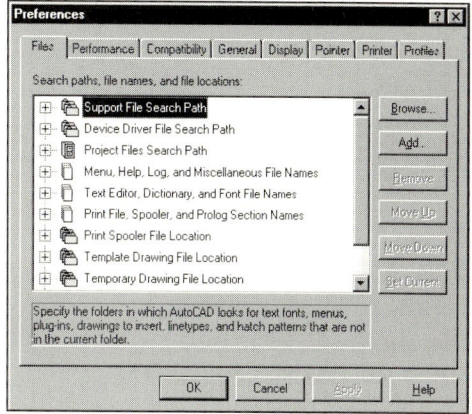

Figure G1.5

The index cards are titled Files, Performance, Compatibility, General, Display, Pointer, Printer, and Profiles. The Preferences dialog box uses the Windows interface to set up AutoCAD's search paths, performance, display, pointer, and printer configurations, and various general items. When you make changes to the Preferences dialog box and pick Apply, AutoCAD will update the system registry.

To change from one card to the next, you will pick on the name at the top of the card. The card that you select will then appear on top of the stack.

The Files card shows the search paths that AutoCAD follows during particular operations. They are what AutoCAD uses to find the configuration, menu, and toolbar files. The Performance card allows for changes to the display of an object, including solid, arc, and circle smoothness. The Compatibility card shows priorities that AutoCAD will follow when responding to other application objects and other compatible functions. The General card changes the AutoCAD automatic saving times, file extensions for temporary files, and other general functions. The Display card allows for changes to the AutoCAD display, including fonts and colors. The items under Pointer change the current pointing device and its size relative to the screen. The Printer card is similar to the Pointer card in that it allows you to change the current printer and modify its settings.

The Profiles card changes the AutoCAD environment to user-specified customizations. You will use this card to create your own Profile with the configuration settings used for this manual.

■ **Warning:** Changes to AutoCAD's configuration can affect its performance greatly. If you are working on a networked system, you must check with the network administrator before creating your own profile. Whatever changes you make must be made only to *your* user Profile setting. ■

Now you will create the user Profile for these tutorials.

User Profiles

AutoCAD Release 14 for Windows 95/NT allows you to save multiple configuration profiles, making it possible to configure AutoCAD for different uses and peripherals.

If you are working on a shared machine, you can create a special profile to use for these tutorials so that another user doesn't change your settings and defaults between sessions. You will use AutoCAD's Preferences to save these settings.

If you're working on a networked computer, your professor or system administrator has probably configured the system already. Ask which profile you should use when in AutoCAD.

To create a user-defined profile or configuration, first

Pick: **Profiles index card**

The Profiles index card appears on top of the other index cards, as shown in Figure G1.6.

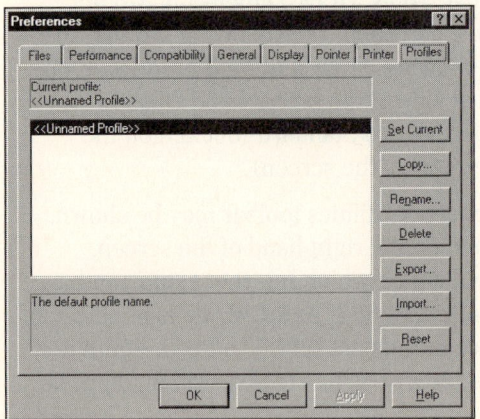

Figure G1.6

As you can see, the current profile is <<Unnamed Profile>>, which is what affects the *acad.pgp* file. Any changes made to the other cards will be stored in this Profile because it is selected as current. You will make a new profile for the tutorials, set it as current, and then make the necessary changes to the other cards.

Pick: **Copy**

The Copy Profile dialog box appears where you will type the name of your new Profile.

Type: **AutoCAD Tutorials** *(similar to Figure G1.7)*

Figure G1.7

A description is optional, but not necessary.

Pick: **OK**

You will see "AutoCAD Tutorials" appear in the space below Profile name. Highlight it, if necessary, and

Pick: **Set Current**

AutoCAD should not change in the background because the configuration settings for this Profile are the same as for the Unnamed Profile. Now you will make changes to the other cards, as necessary, to follow the tutorials in this manual.

Changing the Configuration

Now you will change the Performance card, if necessary.

Pick: **Performance**

The Performance card should appear on top of the other cards. On your own, be sure that the settings on your computer are the same as those shown in Figure G1.8.

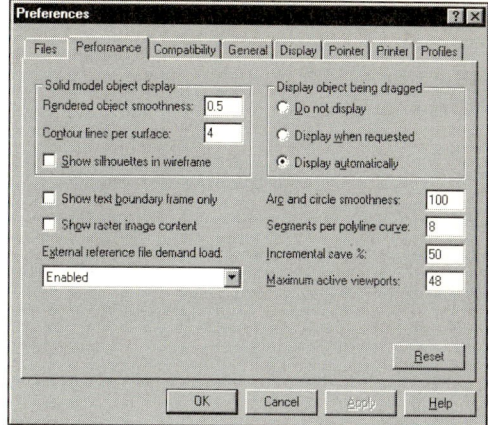

Figure G1.8

When you have finished, choose the next card, Compatibility.

Pick: **Compatibility**

The Compatibility card should appear on top of the other cards. On your own, be sure that the settings on your computer are the same as those shown in Figure G1.9.

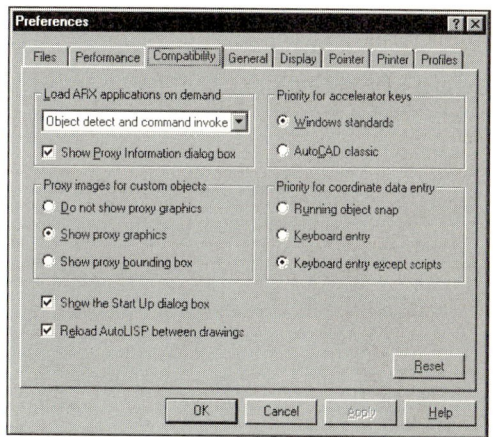

Figure G1.9

When you have finished, choose the next card, General.

Pick: **General**

The General card should appear on top of the other cards. On your own, be sure that the settings on your computer are the same as those shown in Figure G1.10.

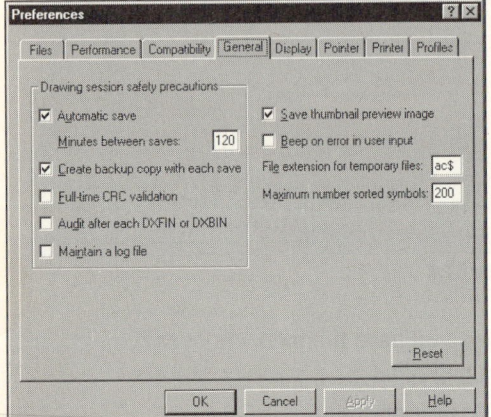

Figure G1.10

When you have finished, choose the next card, Display.

Pick: **Display**

The Display card should appear on top of the other cards. On your own, be sure that the settings on your computer are the same as those shown in Figure G1.11.

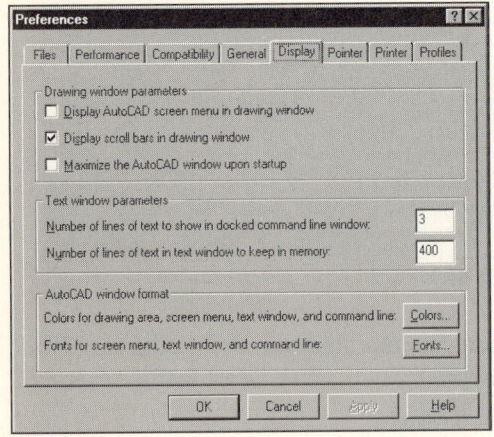

Figure G1.11

The background screen color for this Tutorial is set to white. However, you do not have to change your background color to white. To see the available options, you can choose the color button on the lower part of the dialog box.

When you have finished, choose the next card, Pointer.

Pick: **Pointer**

The Pointer card should appear on top of the other cards. On your own, be sure that the settings on your computer are the same as those shown in Figure G1.12.

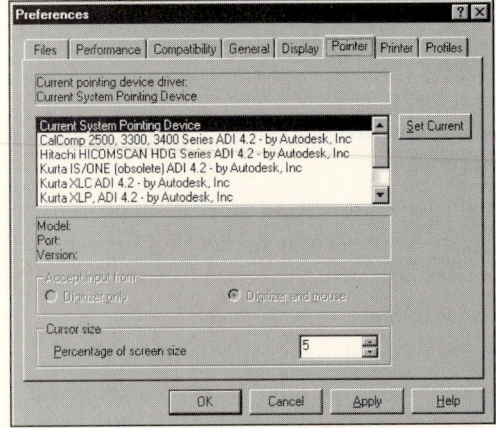

Figure G1.12

The default digitizer is the Current System Pointing Device (your mouse). AutoCAD can also support a tablet. To configure AutoCAD to use a tablet, refer to your AutoCAD documentation.

When you have finished, choose the next card, Printer.

Pick: **Printer**

The Printer card should appear on top of the other cards, as shown in Figure G1.13.

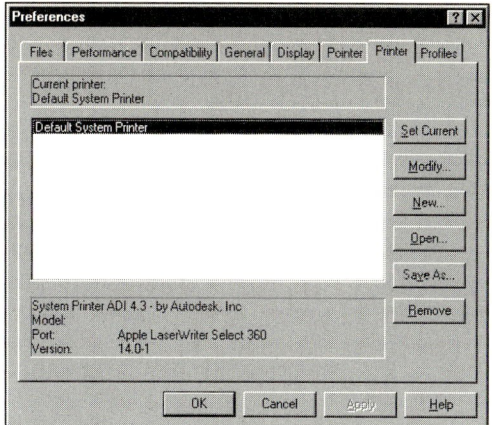

Figure G1.13

If you did not select a printer when you installed AutoCAD, the simplest way to print your AutoCAD drawings is to pick New, select the System Printer ADI 4.3—by Autodesk, Inc. (Version 14.0-1), and accept the Control Panel default settings. Doing so allows AutoCAD to use the printer that you have already set up to work with Windows. If you want to choose one of the other plotter options, refer to your AutoCAD documentation.

If you are using a laser printer, the default orientation is probably portrait. In these tutorials you will be plotting to a landscape orientation. You may want to change the default orientation of your system printer to landscape so that you do not have to rotate your plots in AutoCAD. In Windows 95 you can make this change from the Start, Settings, Printers option; select your printer and choose the File, Properties, Paper options.

If you do not want to change your system printer to landscape, you can do a detailed plotter configuration in AutoCAD to rotate your

plot 90° so that your drawings are plotted to landscape. To do so from the Printer tab in Preferences, select Modify, Reconfigure, OK. In the text window that appears, you will be asked whether you want to change anything. Type *Y* and accept the default at each prompt except the Rotate Plot default. Rotate your plot 90°.

If you do not change your printer orientation or configuration, you will need to rotate your plot each time you print in a tutorial.

Accepting Configuration Changes

Now you will go back to the Profiles card.

Pick: **Profiles**

To accept the changes that you have just made (if any), you will pick the Apply button in the lower part of the dialog box. However, be sure that the Current profile is set to AutoCAD Tutorials (the new profile name you created previously). Then,

Pick: **Apply**

Pick: **OK**

The AutoCAD screen appears again, with no significant changes. Now whenever you begin AutoCAD, you can go into Preferences and select the AutoCAD Tutorials Profile (if it is not already selected) to use this manual. You can also export your profile and save it in a separate directory where it will not be changed by other users. To do this, pick Export from the Profiles tab and use the dialog box to save the file ending in .*arg* to a secure folder. When you need to restore it, pick Import from the Profiles tab and select your file ending in .*arg* to import.

Dialog Box Settings

Three AutoCAD system variables control whether dialog boxes appear when certain commands are issued. Filedia controls the display of dialog boxes for file operations, such as New, Open, and Save. Cmddia controls whether dialog boxes are displayed for Plot and external database commands. Attdia controls whether dialog boxes are displayed with the Insert command. The default value for each of these variables (displayed in the angle brackets) should be 1.

Command: **FILEDIA** ⏎

New value for FILEDIA <1>: *(if the default value is 1, press ⏎ ; if the default value is 0, type 1 ⏎)*

Command: **CMDDIA** ⏎

New value for CMDDIA <1>: *(if the default value is 1, press ⏎ ; if the default value is 0, type 1 ⏎)*

Command: **ATTDIA** ⏎

New value for ATTDIA <0>: *(if the default value is 1, press ⏎ ; if the default value is 0, type 1 ⏎)*

As you follow the instructions in the tutorials, a dialog box should appear on your screen every time one is used in the tutorials.

You have completed the configuration and setup necessary to do the tutorials in this manual. Proceed to Chapter 2 to start to learn to use AutoCAD.

CHAPTER 2

AutoCAD Basics

Objectives

When you have completed this chapter, you will be able to

1. Minimize your AutoCAD drawing session and switch between AutoCAD and other Windows applications.

2. Start a new AutoCAD drawing.

3. Recognize the icons, menus, and commands used in AutoCAD Release 14 for Windows 95/NT.

4. Use the mouse to pick commands, menu options, and objects.

5. Work with a dialog box.

6. Access AutoCAD's on-line help facility.

7. Save a drawing file.

8. Exit from the graphics window and return to the Windows operating system.

Introduction

In this chapter, you will learn the basics of AutoCAD Release 14's screen display, menus, and on-line help. The chapter also describes the type of instructions you'll encounter in the tutorials.

The tutorials in this manual are based on the assumption that you are using AutoCAD Release 14 for Windows 95/NT. AutoCAD should be installed and configured as described in Chapter 1, *Preparing AutoCAD for the Tutorials*. The tutorials are based on the assumption that you are using AutoCAD's standard configuration with the menus described in Chapter 1.

■ *Warning:* If you are using AutoCAD on a network, ask your technical support person about the software configuration. Changes from AutoCAD's standard configuration might have been made; for example, the program might be under a different directory name or require a special command or password to launch. ■

Loading AutoCAD Release 14

Load AutoCAD Release 14 to display the AutoCAD Start Up dialog box, as in Chapter 1. On your own, choose Start from Scratch with English units and pick OK.

The graphics window is displayed on your screen, as shown in Figure G2.1.

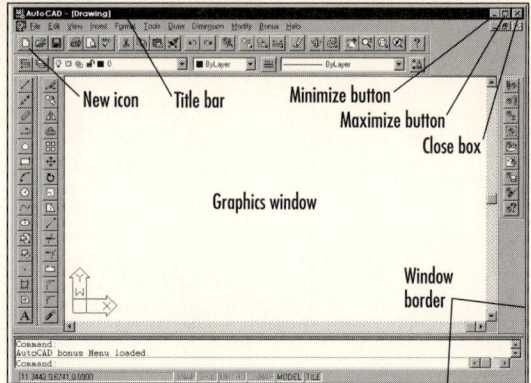

Figure G2.1

Microsoft Windows and AutoCAD

AutoCAD Release 14 for Windows uses many of the same conventions as other applications that run in Windows. This section will identify some of the techniques that you will be using to complete the tutorials in this manual.

The tutorials are based on the assumption that you will be using a mouse or pointing device to work with AutoCAD Release 14. If you are unfamiliar with mouse operations, refer to Chapter 1 or your on-line Microsoft Windows tutorial for basic mouse techniques.

Navigating in Windows

Using a mouse is usually easier than using the keyboard, although a combination of mouse use and keyboard shortcuts is the most efficient way to navigate. For information on keyboard shortcuts in Windows, see your on-line Microsoft Windows tutorial.

Basic Elements of an AutoCAD Graphics Window

The AutoCAD *graphics window* is the main workspace. It has elements that are common to all applications written for the Windows environment. These elements are labeled in Figure G2.1 and described as follows.

- The *close box* is a button in the upper right corner of each window. Picking this button closes the window.
- The *title bar* shows the name of the application (in this case, AutoCAD) followed by the document name (in this case, *Drawing*).
- The *window border* is the outside edge of a window. You can change the window size by moving the cursor over the border until it becomes a double-ended arrow. Holding the mouse button down while you move the mouse (dragging) resizes the window.
- The *maximize and minimize buttons* are in the upper right corner of the window. Clicking the maximize button with the mouse enlarges the active window so that it fills the entire screen; this condition is the default for AutoCAD. You will learn to use the minimize button in the next section.

Minimizing and Restoring AutoCAD for Windows

At times you may want to leave AutoCAD temporarily while you are in the middle of a work session, perhaps to access another application. Minimizing AutoCAD allows you to leave the application and return to it more quickly than you could if you had to exit and start AutoCAD all over again.

Picking the minimize button reduces the AutoCAD window to a title button on the taskbar and makes the Windows desktop accessible.

Pick: (the minimize button)

AutoCAD is reduced to a button at the bottom of your screen. (In a working AutoCAD session, you should always save your work before minimizing AutoCAD.)

AutoCAD is still running in the background, but other applications are accessible to you. When you are ready to return to AutoCAD, click the minimized AutoCAD program button.

Click: (the minimized AutoCAD button)

Microsoft Windows Multitasking Options

Microsoft Windows allows you to have several applications running and to switch between them. This feature is called *multitasking*. You will use multitasking in some of the tutorials in this manual to access your spreadsheet or word processing application while you are still running AutoCAD. To switch between active applications, you press [Alt]-[TAB]. You will minimize AutoCAD and open another application to see how this works.

Pick: (the minimize button)

Pick: (another Program group icon from Start in Windows 95)

Pick: (an application name in Windows 95)

Press: [Alt]-[TAB] (twice)

A small window appears on your screen with the name of an application in it each time you press [Alt]-[TAB]. (When you stop typing, your AutoCAD drawing session should be on your screen. If it is not, continue pressing [Alt]-[TAB] until it appears.)

Press: [Alt]-[TAB]

You should have returned to your other application. (If not, continue typing Alt-TAB until you do.) Quit this application now. You will return to your AutoCAD drawing session automatically if it was the last application you were in. (If you do not see AutoCAD on your screen, press Alt-TAB until you return to AutoCAD.)

Pointing Techniques in AutoCAD

A drawing is made up of separate elements, called *objects*, that consist of lines, arcs, circles, text, and other elements that you access and draw by using AutoCAD's commands and menu options.

The pointing device (usually a mouse) is the most common means of picking commands and menu options, selecting objects, or locating points in AutoCAD.

Picking Commands and Menu Options

The tutorials in this manual assume that you will be using a mouse or pointing device to work with AutoCAD Release 14. The left mouse button is referred to as the *pick button*. The AutoCAD menus let you enter a command simply by pointing to the command and pressing the pick button to choose it. In this way you are instructed to *Pick:* specific commands. Recall that the right mouse button is referred to as the *return button*; it duplicates the action of the ↵ key when you are selecting objects or repeating commands.

Entering Points

You can specify points in a drawing either by typing in the coordinates from the keyboard or by using your mouse to locate the desired points in the graphics window.

When you move the mouse around on the mouse pad or table surface, "crosshairs" (the small intersecting vertical and horizontal lines on the screen) follow the motion of the mouse. These crosshairs form the AutoCAD cursor in the graphics window. When you are selecting points during the execution of some commands, the location of the intersection of the crosshairs will be the selected point when you press the pick button. This is how you pick a point. In some AutoCAD modes the crosshairs change to arrows or boxes with target areas during the execution of certain commands. You will learn about these modes in the tutorials. The cursor on the screen always echoes the motion of the mouse.

Dragging

Many AutoCAD commands permit dynamic specification, or *dragging,* of an image of the object on the screen. You can use your mouse to move an object, rotate it, or scale it graphically. You will learn about drag mode in Tutorials 1 and 2. Once you have selected an object in drag mode, AutoCAD draws tentative images as you move your pointing device. When you are satisfied with the appearance of the object, press the pick button to confirm it.

Object Selection

Many of AutoCAD's editing commands ask you to select one or more objects for processing. This collection of objects is called the *selection set*. You can use your mouse to add objects to the selection set or remove objects from it. AutoCAD has various tools and commands that you can work with when selecting objects; you use the cursor to point to objects in response to specific prompts. AutoCAD highlights the objects as you select them. You will learn about the various ways to select objects in the tutorials.

AutoCAD Commands and Menu Options

AutoCAD gives you several options for entering commands. You can select them from the toolbars, pick them from the pull-down menus, or type them at the command prompt. These elements are labeled in Figure G2.2. Again, depending on your computer's Internet setup, you may or may not see the Internet Utilities toolbar on the right side of the screen.

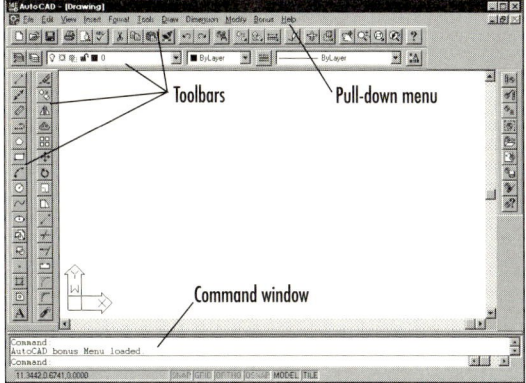

Figure G2.2

How you invoke a command may affect the order and wording of the prompts displayed on the screen; these differences offer you options for more efficient use of AutoCAD. Most AutoCAD commands can be typed at the command prompt, but picking them from a toolbar or a menu usually is easier. For the purposes of the tutorials in this manual, you should pick or type commands *exactly as instructed.* The command selection location you should use will be indicated after the command. When you are supposed to type a command it will appear in all capital letters and bold; for example, **NEW**. You must hit the ⏎ key to activate the command. When you are supposed to invoke a

command by picking its icon, the complete icon name, including the word *icon*, will appear in the instruction line; for example, **New icon**. When you are supposed to invoke a command by selecting it from the pull-down menu, you will see the list of menu names at the instruction line; for example, **File, New**.

As you progress through the tutorials, you will learn about the benefits of choosing commands in different ways in different situations. Remember, for the purposes of the tutorials in this manual, you should pick or type commands exactly as instructed.

Typing Commands

The command window at the bottom of your screen is one way of interacting with AutoCAD. All commands that you select by any means are echoed there, and AutoCAD responds with additional prompts that tell you what to do next. *Command:* in this window is a signal that AutoCAD is ready for a command. For commands with text output, you might need a larger command window display area. You can use the scroll arrows at the right of the command window to review the lines that may have been passed. You can also enlarge the window by positioning the cursor at the border of the window and dragging to enlarge the window so that it shows more lines.

Position the cursor at the top border of the command window. Your cursor will change to a double-headed arrow. Hold down the cursor and drag the mouse up to enlarge the command window to show two more lines. Your screen should look like Figure G2.3.

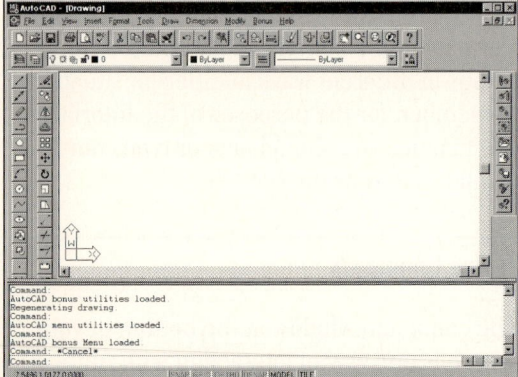

Figure G2.3

You can also review more lines of the command prompt by viewing an AutoCAD text window, as shown in Figure G2.4. You can use the F2 key to toggle to a text window. Doing so will enlarge the command window and move it in front of your AutoCAD graphics window. To return to your graphics window, use the F2 key again or close the text window with the Windows close box.

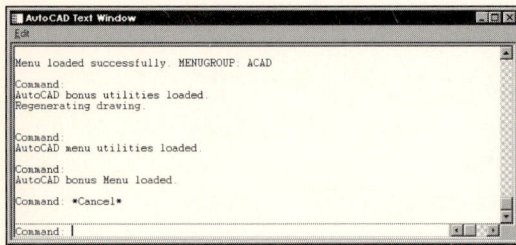

Figure G2.4

When you are supposed to type a command, it will appear in all capital letters; for example,

LINE. Watch the command line in the window as you type.

   Command: **LINE** ⏎

The Line command is activated when you press ⏎, and the Line command prompt *From point:* is in the command window. This prompt tells you to enter the first point from which a line will be drawn. You will learn about the Line command in Tutorial 1. For now,

   Press: Esc

to cancel the command and return to the command prompt.

Using the Toolbars

Toolbars contain icons that represent commands. The Standard, Object Properties, Draw, and Modify toolbars are visible by default, as shown in Figure G2.5 (and perhaps the Internet Utilities, depending on your system). The Standard toolbar contains frequently used commands, such as Redraw, Undo, and Zoom. The advantage of using icons to begin commands is that they are "heads-up"; that is, you do not need to look down at the keyboard to enter the command.

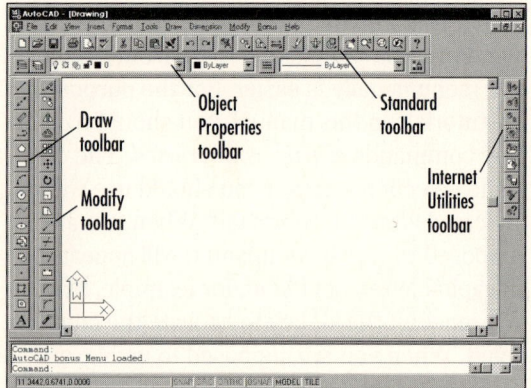

Figure G2.5

You can change the contents of toolbars, resize them, and *dock* or *float* them. A docked toolbar, such as the toolbars shown in Figure G2.5, attaches to any edge of the graphics window. A floating toolbar can lie anywhere on the application screen and can be resized. A docked toolbar cannot be resized and doesn't overlap the graphics window. You will learn to click and drag toolbars to reposition them in Tutorial 1.

When pointing to an icon, the cursor changes to an arrow that points up and to the left. You use this arrow to select command icons. When you move the pointing device over an icon, the name of the icon appears below the cursor, as shown in Figure G2.6. This feature is called a *tool tip* and is an easy way to identify icons.

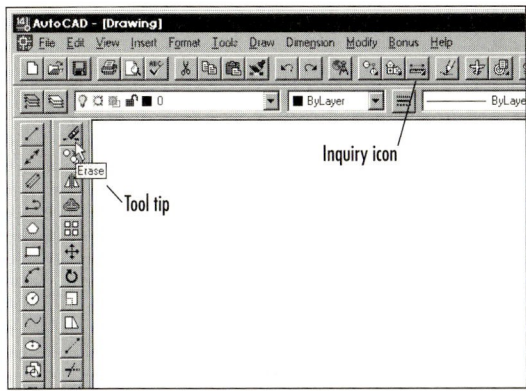

Figure G2.6

Icons with a small black triangle have *flyouts* that contain subcommands, such as the Inquiry flyout on the Standard toolbar at the top of the screen. Hold the pick button down with the cursor over the Inquiry icon on the Standard toolbar until the flyout appears, as shown in Figure G2.7.

Figure G2.7

As new commands are introduced in the tutorials in this manual, their toolbar icons will appear at the beginning of the paragraph, as the Circle icon does here. Now pick the Circle icon from the Draw toolbar to see the options that appear. The icon looks like a circle with a radius drawn in it. As you move your mouse over the icons (you may need to pause), the command names are confirmed by the tool tips that appear below your cursor, and the icon's function appears on the status bar at the bottom of the graphics window; in this case, *Creates a circle: circle.* (Sometimes the tool tip command name is slightly different from the name echoed at the command prompt or the command that you would type at the command prompt. If you are confused about a specific command name, check the AutoCAD Command Summary at the end of the manual.)

Pick: **Circle icon**

The following options will appear at the command line.

Command: 3P/2P/TTR/<Center point>:

The letters that appear at the command prompt, 3P/2P/TTR, are command options for the Circle command. Subcommands and command options work properly only when entered in response to the appropriate prompts on the command line. When a subcommand or option includes one or more uppercase letters in its name, it is a signal that you can type those letters at the prompt as a shortcut for the option name; for example, *X* for eXit. If a number appears in the option name— for example, 2P for Circle 2 Points, which creates a circle using two endpoints of the diameter— you need to type both the number and the capital letter(s).

Backing Up and Backing Out of Commands

When you type a command name or any data in response to a prompt on the command line, the typed characters "wait" until you press the spacebar or ⏎ to instruct AutoCAD to perform the command or enter and act on the entered data.

If you have not already pressed ⏎ or the spacebar, use the [BACKSPACE] key (generally located above and to the right of the ⏎ key on standard IBM PC keyboards and represented by a long back arrow) to delete one character at a time from the command/prompt line. Pressing [Ctrl]-*H* has the same effect as pressing [BACKSPACE].

Pressing [Esc] terminates the currently active command (if any) and reissues the *Command:* prompt. You can cancel a command at any time: while typing the command name, during command execution, or during any time-consuming process. A short delay may occur

before the cancellation takes effect and the prompt *Cancel* confirms the cancellation.

Pressing [Esc] is also useful if you want to cancel a selection process. If you are in the middle of selecting an object, press [Esc] to cancel the selection process and discard the selection set. Any item that was highlighted because you selected it will return to normal.

If you complete a command and the result is not what you had expected or wanted, use the Undo command or type *U* to reverse the effect. You will learn more about Undo in Tutorial 1.

You will press [Esc] now to cancel the Circle command.

Command: 3P/2P/TTR/<Center point>: [Esc]

> ■ *TIP* The tutorials in this manual will instruct you to cancel operations with [Esc]; however, if you have used AutoCAD Release 12 and are used to typing [Ctrl]-*C* to cancel a command, you can set your AutoCAD preferences to use AutoCAD Classic Keystrokes. To do so, select Tools, Preferences from the pull-down menu to bring up the Preferences dialog box. The Compatibility tab of the Preferences dialog box contains a Priority for accelerator keys section. Pick the AutoCAD Classic radio button. Pick OK to exit the Preferences dialog box and save this setting. ■

Repeating Commands

You can press the spacebar or ↵ at the command prompt to repeat the previous command, regardless of the method you used to enter that command. You can also press the return button on your pointing device to repeat the command. Some commands, especially those that prompt you for settings when first invoked, assume default settings when repeated in this manner. You will try this now, and you will use the [Esc] key to cancel the command again.

Press: (SPACEBAR)

The Circle command is invoked again, and the prompt appears in the command window. You will cancel it with the [Esc] key.

Command: 3P/2P/TTR/<Center point>: [Esc]

Using Pull-Down Menus

The menu bar and its associated pull-down menus provide another way of executing AutoCAD commands. The menu bar in AutoCAD for Windows operates in the same way as menus in other applications for Windows. When in the menu area, the cursor changes to an arrow. You use this arrow to select a menu.

You open a menu in the menu bar by picking its name. To practice, you will open the Tools menu.

Pick: **Tools (from the menu bar)**

Picking a menu item executes the commands associated with it, opens a submenu of options, or opens a dialog box of options to be used to control the command. A triangle to the right of a menu item indicates that a submenu is associated with it. You will pick Inquiry to see its submenu.

Pick: **Inquiry (from the Tools menu)**

The Inquiry submenu shown in Figure G2.8 appears next to the Tools menu.

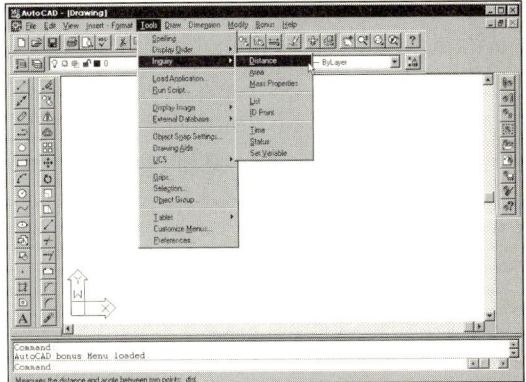

Figure G2.8

A pulled-down menu remains displayed until you

- pick an item from it;
- pick on the menu name again;
- pull down another menu by picking it from the menu bar,
- remove the menu by picking an unused area of the menu bar,
- pick a point on the graphics window;
- type a character on the keyboard;
- pick an item from a tablet or button menu; or
- move your pointing device into the regular screen menu area and click.

To remove the Tools and Inquiry menus from the screen,

Pick: **(a point in the graphics window)**

Working with Dialog Boxes

An ellipsis (...) after a pull-down menu item indicates that it will open a dialog box. Several commands let you set AutoCAD modes or perform operations by checking boxes or filling in fields in a dialog box.

When a dialog box is displayed, the cursor changes to an arrow. You use this arrow to select items from the dialog box.

You will open the Dimension Styles dialog box by picking Dimension Style from the Format menu.

Pick: **Format, Dimension Style**

The dialog box shown in Figure G2.9 appears on your screen.

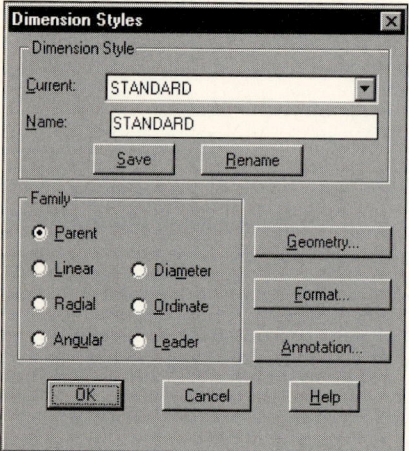

Figure G2.9

Dimension Styles is a typical dialog box. Most dialog boxes have an OK button that confirms the settings or options that you have selected in the dialog box. Clicking it is analogous to pressing ← to send a command to AutoCAD. Clicking the Cancel button disregards all changes made in the dialog box and closes the

dialog box; it has the same effect as pressing Esc. You can use the Help button to get more information about a particular command.

■ *Warning:* The term "button" used in conjunction with a dialog box refers to these selection options and should not be confused with the buttons on your mouse or digitizer. ■

Some dialog boxes have subboxes that pop up in front of them. When this occurs, you must respond to the "top" dialog box and close it before the underlying one can continue. The Geometry, Format, and Annotation buttons at the right of the dialog box bring up such subboxes.

Pick: **Geometry**

The Geometry dialog box shown in Figure G2.10 appears on your screen. Any item in a dialog box that is "grayed" cannot be picked, such as the Extension option in the Dimension Line section.

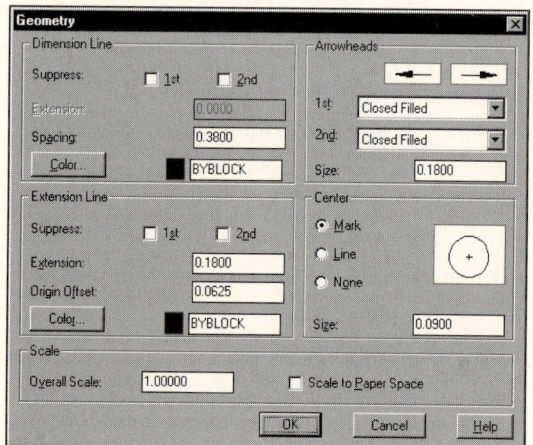

Figure G2.10

Most dialog boxes have several types of buttons that control values or commands in AutoCAD.

Check buttons: A check button is a small rectangle that is either blank or shows a check mark. Check buttons control an on/off switch—for example, turning Dimension Extension Line Suppression on or off—or control a choice from a set of alternatives—for example, determining which modes are on. A blank check button is off.

Radio buttons: A radio button also turns an option on or off. The Dimension Center options are indicated by radio buttons labeled Mark, Line, and None in this dialog box. A filled-in radio button is on. You can select only one radio button at a time; picking one button automatically turns any other button off.

Action buttons: An action button doesn't control a value but causes an action. The OK action button causes the dialog box to close and all the selected options to go into effect. When an action button is highlighted (outlined with a heavy rule), you can press ← to activate it.

Text boxes: A text box allows you to specify a value, such as Dimension Line Spacing in Figure G2.10. Picking a text box puts the cursor in it and lets you type values in it or alter values already there. If you enter an invalid value, the OK button has no effect; you must highlight the value, correct it, and select OK again.

Input buttons: An input button chooses among preset options, such as the Arrowheads 1st button in Figure G2.10. Input buttons have a small arrow at the right end. Picking the arrow causes the value area to expand into a menu of options that you can use to select the value for the button.

Image tiles: An image tile displays choices as graphic images (image tiles) rather than words. The two arrowheads shown in the Arrowheads section are image tiles. Picking the arrowhead style image tile causes the next option in the image tile menu to appear.

You can also use the keyboard to move around in dialog boxes. Pressing TAB moves among the options in the dialog box. Try this by pressing TAB several times now. Once you have highlighted an input button, you can use the arrow keys to move the cursor in text boxes. The SPACEBAR toggles options on and off.

Another common feature of dialog boxes is the scroll bar. A dialog box may contain more entries than can be displayed at one time. You use the scroll bars to move (scroll) the items up or down. You will use a scroll bar in the section Accessing On-line Help.

Most dialog boxes save changes to the current drawing only. Two exceptions to this rule are the Preferences dialog box and the Plot dialog box; these dialog boxes change the way AutoCAD operates, not the appearance of the current drawing itself.

The Cancel button ignores all the selections that you made while this dialog box was open and returns you to the most recent settings. To return to the graphics window without making any changes,

Pick: **Cancel *(twice)***

You are returned to the graphics window.

> ■ *TIP* A feature of many Windows dialog boxes allows you to double-click on the desired selection to select it and exit the dialog box in one action without having to pick OK. ■

AutoCAD On-Line Help

AutoCAD has a *context-sensitive* on-line help facility. The tutorials in this manual explain the most efficient ways in which you can access a command. The AutoCAD Command Summary at the end of this manual also lists the various locations for a command, as well as the command names and aliases that you can enter at the command prompt. In addition, you can choose AutoCAD's on-line help to find other ways of selecting a command.

Accessing On-Line Help

You can get help in a number of ways. You can pick the Help icon, type *HELP* or *?* at the command prompt, or press [F1] to bring up the Help window. You will use the Help icon, which looks like a question mark (?), from the Standard toolbar to get AutoCAD's on-line help.

Pick: **Help icon**

The AutoCAD Help window is shown in Figure G2.11.

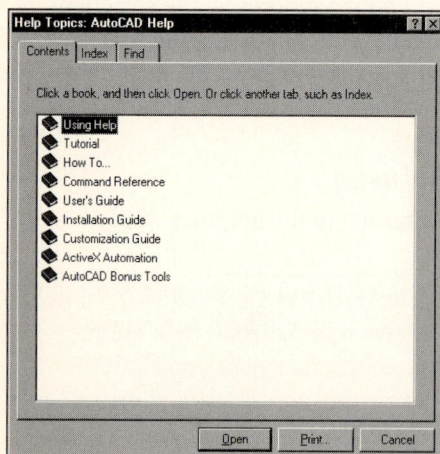

Figure G2.11

The window has three main tabs: Contents, Index, and Find. You may look under Contents at general topics that AutoCAD has set up for you, search through an Index of topics, or use Find to search for help by referencing a single word. For now you will pick the Index tab.

Pick: **Index**

The Index portion of the dialog box appears on your screen, as shown in Figure G2.12.

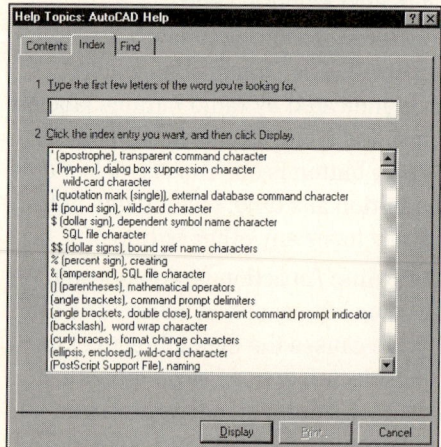

Figure G2.12

You will now type in an AutoCAD command instructing it to search for a particular topic in the Index.

Type: **ERASE**

The Erase command should become highlighted in the Index list.

Pick: **Display *(at the bottom of the dialog box)***

A Topics Found dialog box similar to Figure G2.13 now appears on your screen.

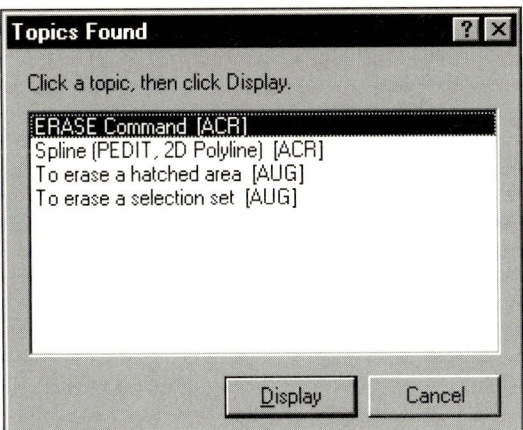

Figure G2.13

Pick: **Display**

The help screen for the Erase command appears on your screen, as shown in Figure G2.14.

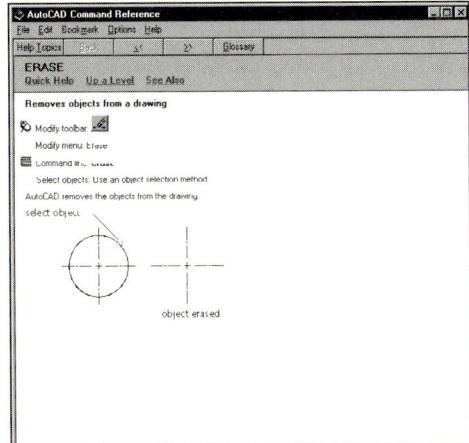

Figure G2.14

To return to the previous screen (if there was a previous screen), you can select Back from the menu bar. To return to the Help Topics screen of the AutoCAD Help window, select Help Topics, as shown in Figure G2.14.

Pick: **Help Topics**

Next, you will Pick the Find tab.

Pick: **Find**

A dialog box with an alphabetical list of anything having to do with AutoCAD appears in the Help window, similar to Figure G2.15.

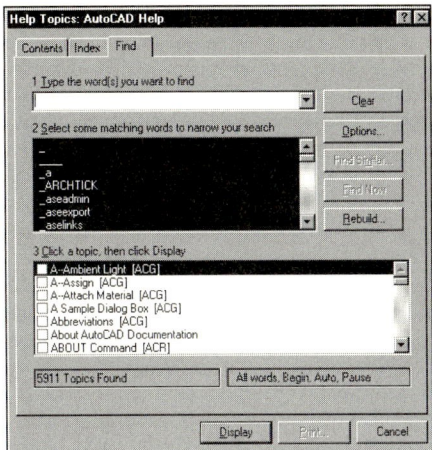

Figure G2.15

You can type in a word that AutoCAD will try to find by accessing all the help functions available. This method is helpful when you do not know the exact function or help topic to search for.

Using a Scroll Bar

A scroll bar allows you to access all items on a list that are too long to be fully displayed in the box. The up or down triangles are used to scroll in the desired direction. A slider box is located between the up and down triangles on the scroll bar. You can pick the slider box and drag it to move the entries up and down; with long lists this action is often faster than using the arrows.

Use the up and down triangles now to move through the list titled "(3) Click a topic" in the lower part of the dialog box. Then place the cursor on the slider box and click and drag it up and down to move quickly through the list.

To get information about the Circle command, you can pick it from the list or type the command in the box above the list of topics.

Type: **Circle** *(in the section titled "(1) Type the word(s) you want to find")*

Note that, as you type the word, the selector moves down both lists displayed.

Pick: **Display**

AutoCAD displays a text screen of information about the Circle command. If there is additional information that will not fit on the screen, use the scroll bar to move through it.

You will use the Windows close box to close the Help window and return to the AutoCAD graphics window.

Click: **(the Windows close box in the top right corner of the Help window)**

Transparent Help

You can easily get help during execution of any command by using AutoCAD's transparent Help command. A transparent command is one that you can execute during another command. In AutoCAD you can pick transparent commands by picking their toolbar icon. If you type these commands, you must insert an apostrophe (') in front of them. If you pick the Help icon from the Standard toolbar, type '? or 'HELP, or press [F1] during a command, you will enter the Help window, which shows help for the command that you are currently using; doing so eliminates the need for searching a topic or navigating through the pages of the Help menu. You will begin the Line command and then type the transparent Help command to bring up the Help window for the Line command.

Command: **LINE** ⏎
From point: **'?** ⏎

The AutoCAD on-line Help window for the Line command opens. Click the Windows close box to close the Help window.

Note the message in the Command window: *Resuming LINE command.* The command is still active. Press [Esc] to cancel the Line command.

AutoCAD's on-line help is a great resource for learning how to use this powerful software. You should use help whenever you want more information about commands and options as you work through the tutorials.

■ **TIP** For information on what's new in AutoCAD Release 14, you can select What's New from the Help menu. ■

Working with Documents

All the documents that you will be instructed to use are held in the directories *c:\datafile* and *c:\work*, which you created in Chapter 1. The best procedure is to save all files on your hard drive because it has more storage space and is faster than accessing a floppy disk drive. Saving directly from AutoCAD to your floppy disk can result in unrecoverable, corrupted drawing files.

AutoCAD Release 14 allows only one drawing file to be open at a time. If you have a drawing open when you select the Open or New commands, AutoCAD automatically closes the current drawing file before opening the new one. (If you have not saved your drawing, AutoCAD will prompt you to save changes before you close the document.)

Using Dialog Boxes to Locate and Manage Files

AutoCAD offers many ways to manage files without exiting the program. You can use the dialog box that appears at the Open command to find a file if you cannot remember its directory or the drive that it's on. In the Open dialog box, the Find File button will bring up the Browse/Search dialog box. You use this dialog box to select the drives and directories you want to search for your file name or file type. All AutoCAD drawing files have the file suffix *.dwg*. You will use other file types in this manual; some are text files (*.txt*), template files (*.dwt*), and some are bitmap files (*.bmp*). You can also use the wildcard characters asterisk (*) and question mark (?) to find and manage files. The asterisk wildcard character matches any string, including a null string. A search including **.dwg*, for example, would call up any drawing files in the directory being searched, as shown in Figure G2.16. The question mark, when used as a wildcard, matches any single character.

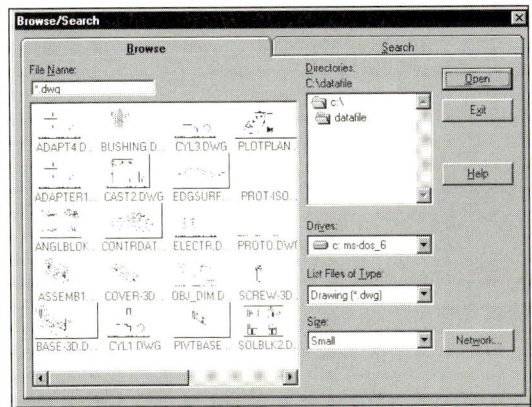

Figure G2.16

You can use the File Utilities dialog box to list the contents of a directory or delete, rename, or copy files. To access the File Utilities dialog box, select File, Management, Utilities.

Exiting AutoCAD

When you have finished an AutoCAD session, you will exit the program by choosing File, Exit. Before you exit this session, return your command window to its original three-line display so that it will match the figures in the tutorials.

Pick: **File, Exit** *(from the pull-down menu)*

Because quitting AutoCAD unintentionally could cause the loss of a lengthy editing session, AutoCAD prompts you to save the changes if you want to. An AutoCAD dialog box appears that contains a *Save Changes to DRAWING?* message and gives you three options: *Yes, No,* and *Cancel.*

Pick: **Yes**

The Save Drawing As dialog box appears, as shown in Figure G2.17. (Your dialog box will not list any files because you have not saved anything to your work directory yet.)

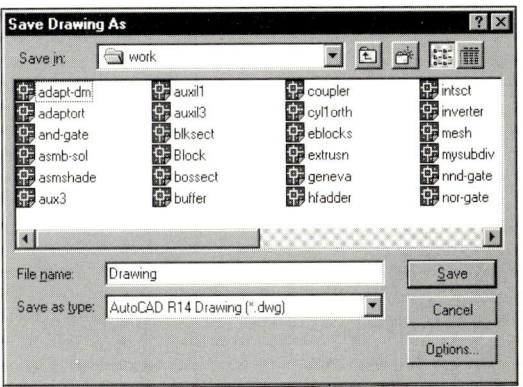

Figure G2.17

This dialog box is similar to other Windows dialog boxes for managing files. The Save in: box contains the directories available on the drives attached to your system; pressing the down arrow will display a list of these drives and directories. By double-clicking on a directory, you can open and select it as the place to store your file.

When you set up AutoCAD in Chapter 1, you created a working directory called *c:\work* in which to save your files. This directory should be open now. If it is not, refer to the Setting the Working Directory section of Chapter 1 to make *c:\work* your working directory. This change will eliminate the need to select the correct directory every time you create, open, or save a drawing file.

If you want to save your file to a disk, use the list that appears when you press the down arrow next to Save in: to select the floppy disk drive to be used for your drawing. Saving directly from AutoCAD to your floppy disk isn't a good idea. However, if you have named your drawing so that it is on a floppy drive, such as drive *a:*, be sure to exit the AutoCAD program before you remove your disk from the drive. Always leave the same disk in the drive the entire time that you are in AutoCAD. Otherwise, you can end up with part of your drawing not saved, which will result in an unrecoverable, corrupted drawing file. The best procedure is to create files and save them on your hard drive.

Now you need to give your file a name. AutoCAD supports long file names; if you do not have long file names enabled on your system, however, your file name cannot be longer than eight characters and cannot contain any spaces. (AutoCAD will automatically add the drawing file suffix *.dwg* to your file name, so you need not type it. All file names in this manual will contain the appropriate suffix for clarity.) To continue practicing AutoCAD basics, name the file *basics.dwg*. The file name *drawing* should be highlighted in the File Name text box. (If it is not, click in the File Name text box and drag to highlight all the text.)

Type: **BASICS**

Pick: **Save**

Your file is saved, and AutoCAD automatically returns you to Windows.

> ■ *TIP* If you did not want to save your drawing, you would have picked No. If you have made some irreversible error and want to discard all changes made in an AutoCAD session, you can choose Exit from the File menu (or type *QUIT* at the command prompt) and select No so that the changes are not saved to your file. ■

You are now ready to complete the tutorials in this manual.

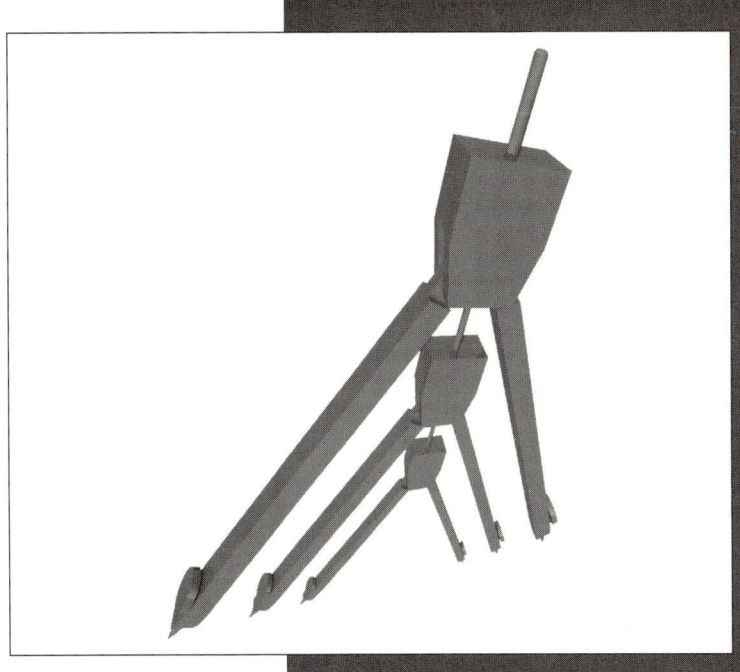

PART TWO

Tutorials

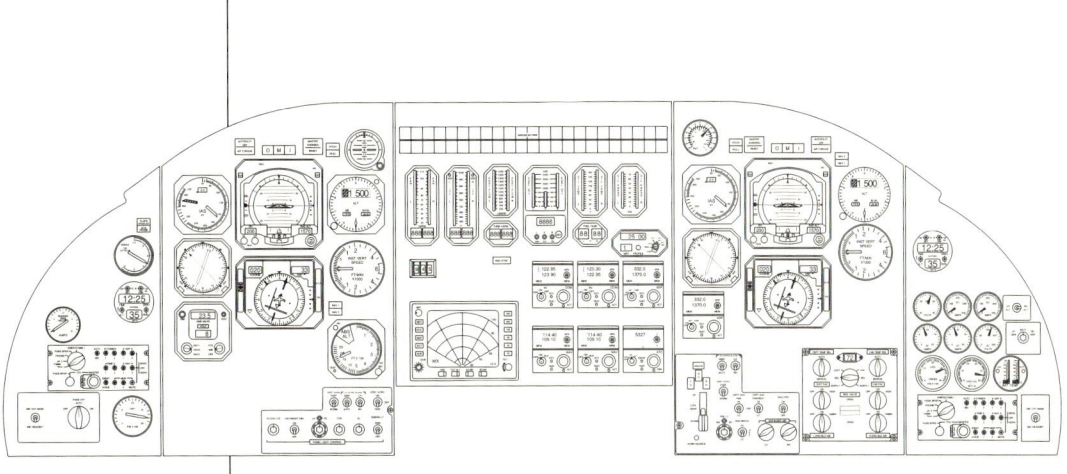

Introduction to AutoCAD

Objectives

When you have completed this tutorial, you will be able to

1. Select commands from the AutoCAD menus.

2. Use AutoCAD's toolbars.

3. Create a drawing file, using AutoCAD.

4. Use AutoCAD's Help command.

5. Enter coordinates.

6. Draw lines, circles, and rectangles.

7. Erase drawing objects.

8. Select drawing objects, using implied Window and Crossing.

9. Add text to a drawing and edit it.

10. Save a drawing and transfer it from one drive to another.

Introduction

This tutorial introduces the fundamental operating procedures and drawing tools of AutoCAD. It explains how to create a new drawing, draw lines, circles, and rectangles, and save a drawing to a new name. You will learn how to erase items and select groups of objects. This tutorial also explains how to add text to a drawing.

Starting

Before you begin, launch AutoCAD Release 14 from your Windows Desktop. If you need assistance in loading AutoCAD, refer to the Getting Started section of this manual.

The Start Up dialog box should appear when you begin a new drawing.

Pick: **Start from Scratch**

Pick: **OK**

For more information on this dialog box, refer to the AutoCAD Basics section of this manual.

The AutoCAD Screen

Your computer display screen should look something like Figure 1.1, which shows AutoCAD's drawing editor. When you first start, be sure to familiarize yourself with the main screen areas. (Depending on your system's configuration, the Internet Utilities toolbar may not appear on your screen.)

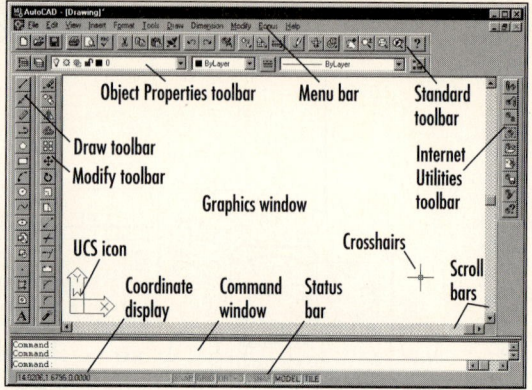

Figure 1.1

■ *Warning:* If your screen does not look something like Figure 1.1, be sure that you have configured AutoCAD properly for your computer system. Refer to the Getting Started portion of this manual. ■

The Graphics Window: The *graphics window* is the central part of the screen, which you use to create and display drawings.

The Graphics Cursor, or Crosshairs: The *graphics cursor*, or *crosshairs*, shows the location of your pointing device in the graphics window. You use the graphics cursor to draw, select, or pick menu items. The cursor's appearance depends on the command or option you select.

The Command Window: The command window is located at the bottom of the screen. It is called a floating window because you can move it anywhere on the screen. You can also resize it to show more or fewer command lines. Note that the command window also has scroll bars so that you can scroll to see commands that are not visible in the active area. Pay close attention to the command window because that's where AutoCAD prompts you when you need to enter information or make selections.

The User Coordinate System (UCS) Icon:
The *UCS icon* helps you keep track of the current X-, Y-, Z-coordinate system that you are using and the direction from which the coordinates are being viewed in 3D drawings.

The Toolbars: A *toolbar* is a group of *buttons* that allow you to pick commands quickly, without having to search through a menu structure. This feature usually increases the speed with which you can select commands. When the cursor is positioned over a button, its tool tip appears. A *tool tip* is text that describes the command invoked by the button. It is useful if you are unsure about what the icon on the button represents. A help line for the button also appears at the bottom of the command window.

Toolbars can be *floating* or *docked* to the edge of the graphics window. You can float a toolbar around on your screen by picking on its title bar or picking within the toolbar border (but not on a button), and holding the pick button down as you *drag* the toolbar to its new location. If you drag a toolbar near the edge of the window, the toolbar docks to the edge of the graphics window. The *title bar* at the top of a floating toolbar helps identify it. When a toolbar is docked, its title bar does not appear. You can also change the shape of a toolbar. To do so, position the arrow cursor near the edge of the toolbar until the cursor changes to a double-headed arrow. While the cursor has this appearance, pick on the edge of the toolbar and, keeping the pick button depressed, drag the edge of the toolbar to reshape it. You can customize the toolbars to show the commands you use most frequently and create new toolbars of your own. You will learn to do so in Tutorial 8.

Items on the toolbar can also be flyouts, where additional buttons are located. Figure 1.2 shows the docked Standard toolbar and the Object Snap flyout. A *flyout* is identified by

a small triangle in the button's lower right corner. To see the buttons that are located on a flyout, position the arrow cursor over the button that shows the flyout triangle and hold the pick button down. After you have held it down for a moment, you will see the additional buttons fly out. When you select a command from a flyout, that selection becomes the top button shown on the flyout. At first you may find this method confusing. However, once you become familiar with the flyouts, you'll find that it is a helpful feature because the most recently used command appears as the top button on the flyout, making it easy to select again.

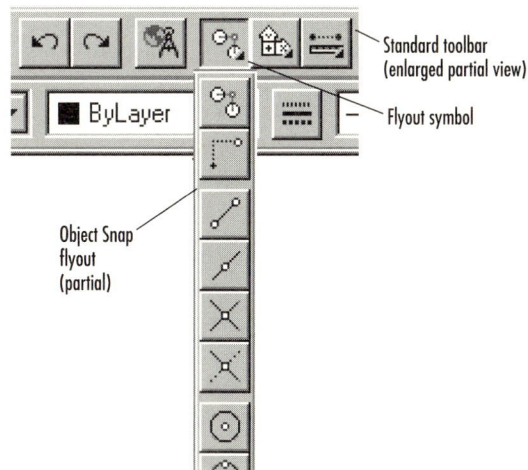

Figure 1.2

The Standard Toolbar: The top of the screen displays the *Standard toolbar*, which shows buttons for frequently used commands. Move the graphics cursor up into the Standard toolbar area. Note that the graphics cursor changes from the crosshairs to an arrow when it moves out of the graphics window. Position the arrow cursor over an item on the Standard toolbar, but do not press the pick button. After the arrow cursor remains over a button for a few seconds, the button's tool tip appears below it. Move the arrow cursor over each button in turn to familiarize yourself with the buttons on the Standard toolbar. If you are unsure about which command an icon represents, use the help line at the bottom of the command window to help identify it. The Standard toolbar can float anywhere on the screen. Its default location is docked near the top of the screen.

The Object Properties Toolbar: The *Object Properties toolbar* contains tools to help you manipulate the properties, such as color, linetype, and layer, of the graphical objects you create. On the Object Properties toolbar, from left to right, the icons are: Make Object's Layer Current, Layers, Layer Control, Color Control, Linetype, Linetype Control, and Properties. The layer name 0 is shown in the Layer Control display, and the word BYLAYER is shown in the Linetype Control and Color Control display. At this point, the properties of newly created drawing objects will be determined by the current layer 0. These tools will be covered more thoroughly in Tutorial 2.

The Status Bar: The *status bar* at the bottom of the screen shows important settings and modes that may be in effect. The coordinate reference of the crosshairs' location appears in a box at the left of the status bar as three numbers with the general form X.XXXX, Y.YYYY, and Z.ZZZZ. The specific numbers displayed on your screen tell you the location of your crosshairs. The words SNAP, GRID, ORTHO, OSNAP, MODEL, and TILE appearing to the right of the coordinates are buttons that you can double-click to turn these special modes on and off. When the SNAP, GRID, ORTHO, OSNAP, or TILE modes are in effect, their buttons are *highlighted*; when they are turned off, their buttons are grayed out. The remaining button, MODEL, lets you quickly switch between model space and paper space, which you will learn to use in

Tutorial 5. When you select paper space, the word PAPER will replace the word MODEL on this button.

The Menu Bar: The row of words at the top of the screen is called the *menu bar*. The menu bar allows you to select commands. You pick commands by moving the cursor to the desired menu heading and pressing the pick button on your pointing device to pop down the available choices. To select an item from the selections that appear, position the cursor over the item and press the pick button. The menu bar shows the following major headings.

File	Used to open existing files, create new files, save, export, print, recover, and perform other file operations
Edit	Contains editing commands for cutting and pasting operations similar to other Windows applications
View	Lets you select commands that manipulate the view of your drawing on the screen, as well as turn on and off and customize the toolbars
Insert	Used to insert blocks, 3D Studio files, ACIS solids, and other object formats
Format	Allows you to select commands to set color, linetype, text style, and other properties for new *objects* you create
Tools	Contains the spell checker, menu customization, and preferences, among other tools
Draw	Allows you to draw lines and circles, insert hatch patterns, and input text
Dimension	Allows you to dimension distances, lengths, and diameters and use many other dimensioning tools

Modify	Contains commands allowing you to erase, copy, and change properties, and make other object modifications
Bonus	Contains bonus commands not included in the regular menus for drawing, layer manipulation, text and object modifications
Help	Lets you find help for commands, toolbars, and variables

Canceling Commands

You can easily cancel commands by pressing [Esc]. If you make a selection by mistake, press [Esc] to cancel it. Sometimes you may have to press [Esc] twice to cancel a command, depending on where you are in the command sequence.

Picking Commands from the Menu Bar

Use your mouse to move the cursor to the menu bar.

Pick: **Tools**

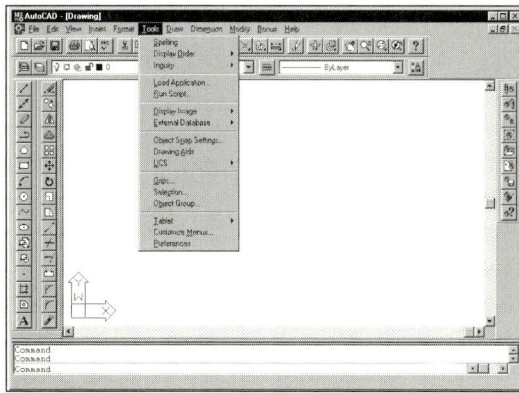

Figure 1.3

Figure 1.3 shows the Tools selection pulled down from the menu bar. Menu bar items with a triangle after the name activate another menu

when picked. Items on the menu with three dots after them cause a dialog box to appear on the screen when picked. Note that, as you move the cursor along the selections of the menu, the status bar at the bottom of your screen shows a help line describing the menu selection that is currently highlighted.

> ■ *TIP* If you are unsure about which menu bar selection contains the command you want to use, move the cursor to the menu bar area and press the pick button. Move from selection to selection as you look for the command. ■

Press: (the pick button and hold it down while moving down the Tools menu)

To remove the menu from the screen,

Pick: (in the graphics window)

Using the Preferences Dialog Box

You can control many things about AutoCAD's setup and general appearance using the Preferences dialog box. You will select Tools from the menu bar and then Preferences from the menu that appears, as shown in Figure 1.4.

Pick: **Tools, Preferences**

Figure 1.4

The Preferences dialog box appears on your screen, as shown in Figure 1.5.

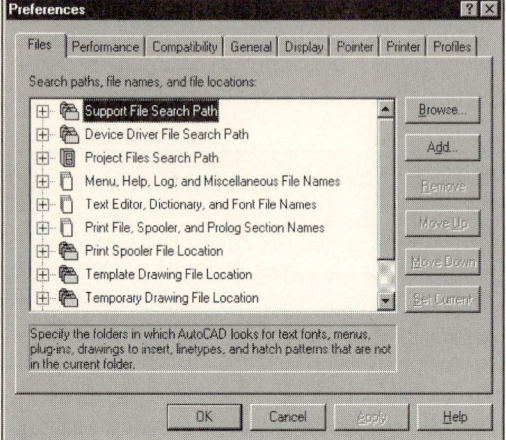

Figure 1.5

The Preferences dialog box uses the Windows interface to set up AutoCAD's search paths, performance, display, pointer, and printer configurations, and various general items. When you make changes to the Preferences dialog box and pick Apply, the system registry will be updated. If you pick OK, the system registry will be updated and the Preferences dialog box will close. For now, you will choose Cancel when exiting the Preferences dialog box, avoiding any system registry changes. However, keep track of changes if you do make any so that you can change them back if you are not happy with the result.

The Preferences dialog box looks like a stack of eight index cards, labeled Files, Performance, Compatibility, General, Display, Pointer, Printer, and Profiles. The Files card appears to be at the top of the stack. Next, you will look through the stack of cards to familiarize yourself with the selections you can make. To change from one card to the next, you will pick on the index name at the top of the card. The card that you select will then appear at the top of the stack.

Pick: **Performance**

The Performance card appears at the top of the stack, as shown in Figure 1.6.

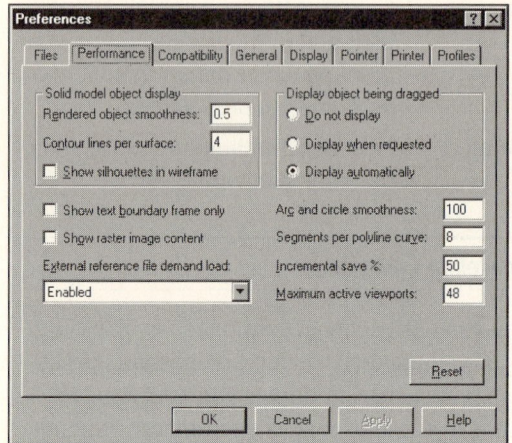

Figure 1.6

The Performance card allows you to change the display of an object, including solid, arc, and circle smoothness. Use of the Reset button, as shown in Figure 1.6, changes the values to the AutoCAD default options. At this time, do not change anything.

Pick: **Compatibility**

The Compatibility card shows priorities that AutoCAD will follow when responding to other application objects, as well as other compatible functions. This card also contains a Reset button, which you can use to restore defaults. It is unnecessary to change anything at this time.

Pick: **General**

The General card allows you to change the AutoCAD automatic saving times, *file extensions* for temp files, and various other general functions. No changes should be made at this time, however, you can use the Reset button to restore any unintended changes.

Pick: **Display**

The Display card allows you to change of the AutoCAD display, including fonts and colors. Again, don't change anything at this time.

Pick: **Pointer**

The items under Pointer let you change the current pointing device and its size relative to the screen. Don't make any changes for now.

Pick: **Printer**

The Printer card is similar to the Pointer card, allowing you to change the current printer and modify its settings. Once again you will not make any changes.

Pick: **Profiles**

The Profiles card changes the AutoCAD environment to user-specified customizations. Again, a Reset button is available, allowing you to restore the default functions. Don't make any changes for now.

Pick: **Files**

The Files card should again be at the top of the pile. This card shows the search paths that AutoCAD follows during particular operations. Don't make any changes for now.

Activating and Using the Screen Menu

A screen menu is available, similar to that used in older versions of AutoCAD. To use it, turn it on using the Preferences dialog box. Remaining in the Preferences dialog box, choose the Display card.

Pick: (the check box to the left of "Display AutoCAD screen menu in drawing window")

Pick: **OK** *(to exit the Preferences dialog box)*

The screen menu appears as a column of words along the right-hand side of the screen, as shown in Figure 1.7.

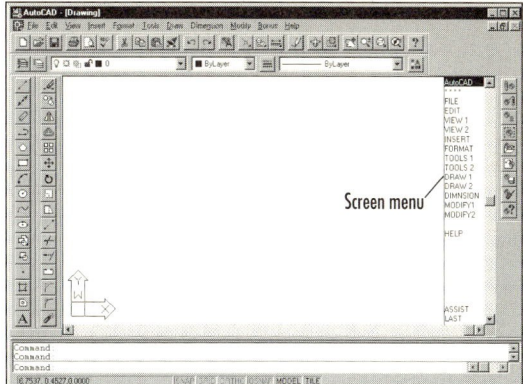

Figure 1.7

To select an item, position the cursor over the name of the item, as you did when using the menu bar. The items that you see on the screen in capital letters are the names of menus. When you select a menu name, a new list of items replaces the previous list of words in the right-hand column. You will select Modify1 from the screen menu.

Pick: **Modify1**

Note the new items that appear on the screen menu, as shown in Figure 1.8. These are the names of commands that modify objects in your drawing. Locate the Erase command on the screen menu.

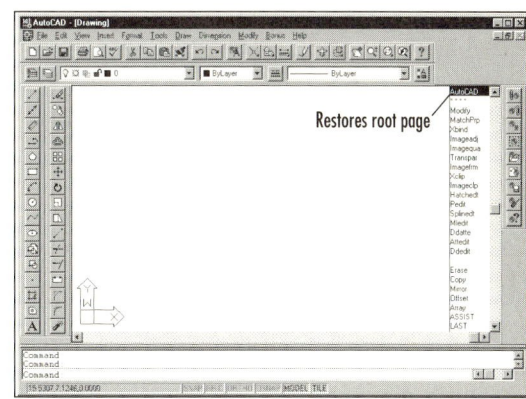

Figure 1.8

Next, you will return to the root page of the menu. The word AutoCAD at the top of the screen menu has a special function. You can pick it to bring back the original screen menu selections (called the *root page* of the menu and shown in Figure 1.7). Picking AutoCAD will also cancel any current command. This function is very helpful when you are first learning AutoCAD. If you are lost or confused, picking AutoCAD will allow you to start with the root page of the menu again. The items ASSIST and LAST always appear at the very bottom of the screen menu. Picking ASSIST displays *options* you can use within commands. Picking LAST returns the last screen menu page you were using. Next, try picking AutoCAD from the screen menu to return to the root page.

Pick: **AutoCAD**

Now you will turn off the screen menu by deselecting it within the Display card in the Preferences dialog box.

Pick: **Tools, Preferences**

Pick: **Display**

Pick: (the check box to the left of "Display AutoCAD screen menu...", so that the check disappears)

Pick: **OK**

The screen menu should have disappeared after you selected OK; however, if it didn't, try the preceding steps again. Because the screen menu is not the fastest way to select commands, only toolbars, menus, and typed commands will be used throughout the rest of the tutorials.

Typing Commands

You can type commands directly at the *command prompt* in the command window. To do so you must type the exact command name. Remember that many of the words on the menu bar are menu names, not commands.

Only the actual command name, not menu names, when typed, activate a command. You will type the command Line in the command prompt area. You may type all capitals, capitals and lowercase, or all lowercase letters; AutoCAD is not *case-sensitive*.

■ **TIP** Whenever you see the ⏎ key in an instruction line in this manual you must press ⏎, sometimes labeled Enter or Return on your keyboard, to enter the command or response. You can achieve the same result quickly by pressing the return button on your pointing device (the right button on a two-button mouse), except when you are entering text during a text command. ■

Command: **LINE** ⏎
From point: Esc

Pressing Esc cancels the command. You will learn more about the Line command later in this tutorial.

■ **TIP** If you are a fan of the AutoCAD Release 12 keystrokes, typing Ctrl-C to cancel, for example, you can use the Preferences dialog box Compatibility tab to turn on AutoCAD Classic keystrokes. ■

Command Aliasing

You can type commands quickly through *command aliasing*. You can give any command a shorter name, called an alias. Some sample commands already have aliases assigned to them to help you get started. They are stored in the file *acad.pgp*, which is part of the AutoCAD software and usually found in the directory *\r14\support*. You can create your own command aliases by editing the file *acad.pgp* with a text editor and adding lines that give the new, shorter names. (These lines take the form ALIAS, *COMMAND in the *acad.pgp* file.) After doing so, you can use the shortened name at the AutoCAD command prompt. The following commands are some for which aliases have already been created. When you need one of the following commands, you can type its alias at the command prompt instead of the entire command name.

Alias	Command
A	Arc
C	Circle
CP	Copy
DV	Dview
E	Erase
L	Line
LA	Layer
LT	Linetype
M	Move
MS	Mspace
P	Pan
PS	Pspace
PL	Pline
R	Redraw
T	Mtext
Z	Zoom

Bear in mind these are only a few of the aliases available. For a more complete list, please reference the Command Summary at the end of this manual. The Command Summary indicates where to find a command, its icon, the actual command name, which you can type to start the command, and its alias, if there is one.

AutoCAD provides many different ways to select any command. After you have worked through the tutorials in this book, decide which methods work best for you. While you are working through the tutorials, however, be sure to select the commands from the locations specified. The subsequent command prompts and options may differ, depending on how you selected the command.

Typing a Command Option Letter

Once you have selected a command, a number of options may appear at the command prompt. You can type the letter or letters that are capitalized and then press ↵ to select the option you want.

Starting a New Drawing

When you start AutoCAD, it opens a Start Up dialog box that allows various options, including opening an existing drawing and starting a new drawing from scratch. You will select the New command by picking File from the menu bar and then picking New from the pull-down list.

Pick: **File, New**

The message "Save Changes to Drawing.dwg" appears as AutoCAD recognizes that you have made changes since you started AutoCAD. Since you do not want to save what you have done so far,

Pick: **No**

The Create New Drawing dialog box appears on the screen as shown in Figure 1.9.

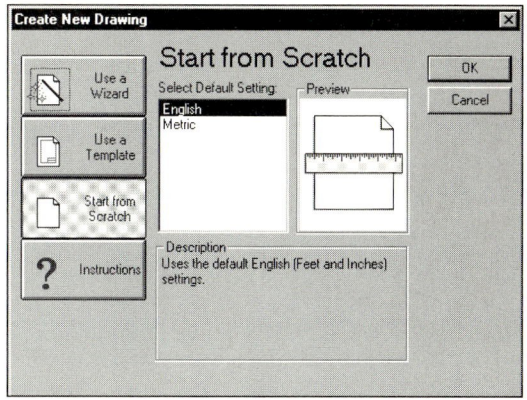

Figure 1.9

The Start from Scratch button allows you to start a new drawing automatically, without setting any preferences. The Use a Template button is similar to the Prototype button in Release 13, and you can use any drawing already created as a template. A template can contain graphical objects or preset AutoCAD settings and preferences. (You will learn more about creating and using templates in Tutorial 5.) The Use a Wizard button will take you through either quick or advanced steps to set up basic units and sizes in your new drawing. For your new drawing, select the Start from Scratch button, which will use an AutoCAD template named *acad.dwt*.

Naming Drawing Files

After the Create a New Drawing dialog box disappears, you will save the drawing under a new name. The drawing name *shapes.dwg* will serve as a good name for this sample drawing.

> *Pick:* **File, Save**
>
> *Type:* **SHAPES**

AutoCAD follows the Windows 95 rules for naming a file. Names may include as many characters and/or numbers as you choose, along with other characters: underscores (_), dashes (-), commas (,), and periods (.). However, forward slashes (/) and backward slashes (\) aren't allowed in the drawing name.

> ■ *TIP* It is not generally good practice to include periods (.) in your AutoCAD drawing names. Periods are generally used to divide the file name from the file extension (for example, *shapes.dwg*). Using a period within the name of the file may cause confusion with the extension and can make it difficult to use the file with other operating system versions, or inside other AutoCAD drawings. ■

All AutoCAD drawing files are automatically assigned the file extension *.dwg*, so typing the extension is unnecessary. The tutorials in this book use drawing names that are descriptive of the drawing being created. For example, the drawing that you will create for this first tutorial consists of basic shapes. Descriptive names make recognizing completed drawings easier. Drawing names can include a drive and directory specification. If you don't specify a drive, the drawing is saved on the computer's hard drive (usually drive C:) in the *default directory*. In the Getting Started section of this manual, you should have already created a default working directory for AutoCAD, called *C:\work*, where you will save your drawing files. If you have not done so, you may want to review Getting Started. The Save Drawing As displays the dialog box shown in Figure 1.10, which you can use to name your file and select the drive and directory.

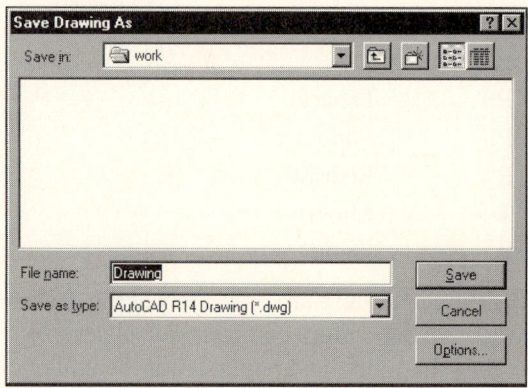

Figure 1.10

If you want to save a drawing on a disk in drive A:, the drawing name *A:\shapes.dwg* will automatically send the drawing file to drive A:. Saving a drawing directly from AutoCAD to your floppy disk is a poor practice. However, if you have named your drawing so that it is on a floppy drive, such as drive A:, be sure to exit the AutoCAD program before you remove your disk from the drive. Always leave the same disk in the drive the entire time you are in AutoCAD. Otherwise, you can end up with part of your drawing not saved, which will result in an unrecoverable, corrupted drawing file.

> ■ **TIP** If you cannot open your drawing for use with the AutoCAD program because it is corrupted, you may be able to recover it with the selection File, Drawing Utilities, Recover from the menu bar. ■

Pick: **Save**

This action returns you to the drawing editor, where you can begin work on your drawing, called *shapes.dwg*. Unless you are using a different drive or working directory on your system, the entire default name of your drawing is *C:\work\shapes.dwg*. Note that the name *shapes* now appears in the title bar as the current file name. Depending on your system, the *.dwg* extension may not show up in the title bar; however, AutoCAD recognizes it as such.

■ *Warning:* You should initially store all drawings on the hard drive (usually drive C:). As you create a drawing, more and more data are being stored in the drawing file. If the file size exceeds the remaining capacity of a disk (say, a floppy in drive A:), a FATAL ERROR may occur, resulting in the loss of the entire drawing file. To prevent problems with files that may become too large for a floppy, create all drawings on the hard drive and then transfer them to a floppy disk for storage after you exit AutoCAD. After completing the drawing for this tutorial, you will learn how to transfer drawings from the hard drive to a floppy disk. ■

Using Grid

Adding a *grid* to the graphics window to act as a reference for your drawing is often helpful. A grid is a background area in the graphics window covered with regularly spaced dots. The grid does not show up in your drawing when you print. You will add a grid background to the graphics window by typing the Grid command at the prompt.

Command: **GRID** ↵

The grid spacing prompt displays the options available for the command. The default option, in this case <0.5000>, appears within *angle brackets*. Pressing ↵ on your pointing device accepts the default. To set the size of a grid, you will type in the numerical value of the spacing and press ↵.

Grid spacing(X) or ON/OFF/Snap/Aspect <0.5000>: ↵

> ■ *TIP* If you choose a grid size that is too small—so that it would effectively fill your graphics window with a solid pattern of grid dots—the error message "Grid too dense to display" appears. If that happens, restart the Grid command and change to a larger value. (You can simply press ↵ to restart any command.) If the grid values are too large, nothing appears on the screen because the grid is beyond the visual limits of the screen. ■

Your screen should look like Figure 1.11, which shows a background grid drawn with 0.5-unit spacing. Note that GRID now shows clearly on the status bar to let you know that the grid is turned on.

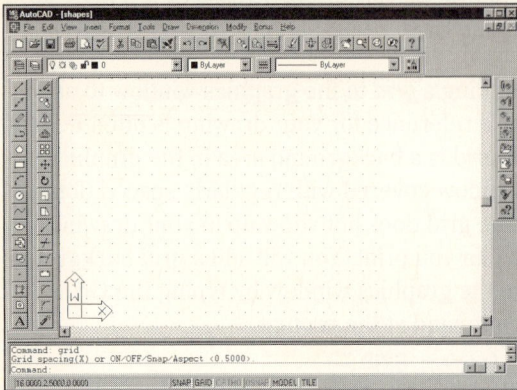

Figure 1.11

Note that SNAP is highlighted on the status bar to let you know that the Snap mode is in effect.

> ■ *TIP* If you set the GRID spacing to 0.0000, and then set the SNAP, the grid spacing will automatically follow the same spacing as SNAP. ■

Move the crosshairs around on the screen. Note that, instead of the smooth movement that you saw previously, the crosshairs jump or "snap" from point to point on the snap spacing.

Using Snap

The Snap command restricts the movement of the cursor to make selecting locations with a specified spacing easier. When Snap is on, your cursor jumps to the specified intervals. Snap is a very helpful feature when you are locating objects or distances in a drawing. If each jump on the snap interval is exactly 0.5 unit (horizontally or vertically only, not diagonally), you could draw a line 2.00 units long by moving the cursor four snap intervals. You will activate Snap by typing the command.

Command: **SNAP** ⏎

You should align the snap function with the grid size so that the snap spacing is the same as the grid spacing, at half the grid spacing, or some other even fraction of the grid spacing. The grid spacing is 0.5, you will accept the snap default value of <0.5000>.

Snap spacing or ON/OFF/Aspect/Rotate/Style <0.5000>: ⏎

Grid and Snap Toggles

You can quickly turn the grid and snap on and off by double-clicking their respective buttons on the status bar. If their buttons on the status bar appear in black type, they are on; if in gray type, they are off. Each button on the status bar acts as a toggle: that is, double-clicking it once turns the function on; double-clicking it a second time turns the function off. You can toggle grid and snap on and off during the execution of other commands.

Double-click: **SNAP button**

Move the crosshairs around on the screen. Note that they have been released from the snap constraint. To turn Snap on again,

Double-click: **SNAP button**

The crosshairs jump from snap location to snap location again.

Double-click: **GRID button**

The grid disappears from your screen, but the snap constraint remains. To show the grid again,

Double-click: **GRID button**

Note how the *command window* keeps a visual record of all the commands.

> ■ **TIP** Pressing ⌨-G also toggles the grid on and off; ⌨-B toggles the snap function. ■

Using Line

 The Line command draws straight lines between endpoints that you specify. The Line command is located on the Draw toolbar, docked on the left-hand side of the screen. (AutoCAD arranges toolbars and menus according to general spoken English syntax. This arrangement is helpful when you are trying to remember where commands are located. For example, you might say, "I want to draw a line.")

For the next sequence of commands, you will use the Draw toolbar to pick the Line command. When selecting the Line command, be sure to pick the icon that shows dots at both endpoints. This icon indicates that the command draws a line between the two points selected. You can continue drawing lines from point to point until you press ↵ to end the command.

> ■ **TIP** Using the toolbars and typing command aliases at the prompt are two of the fastest ways to select commands. ■

Pick: **Line icon**

AutoCAD prompts you for the starting point of the line. You can answer prompts requesting the input of a point in two different ways:

- By moving the cursor into the graphics window and choosing a screen location by pressing the pick button (grid and snap help to locate specifically defined points)
- By entering X-, Y-, and Z-coordinate values for the point

> ■ **TIP** If you make an error when using any of the menu commands, press Esc to return to the command prompt. ■

Getting Help

You can get help during any command by using AutoCAD's transparent Help command. A *transparent command* is one that you can use during execution of another command. In AutoCAD these commands are preceded by an apostrophe ('). If you type '? or 'Help or press F1 during a command, you will enter the Help dialog box, which shows help for the command that you are currently using. While you are still at the From point: prompt, you will use the Standard toolbar to pick the transparent Help command. Its icon looks like a question mark (?).

From point: *(pick Help icon)*

The AutoCAD Help dialog box with help for the Line command is shown in Figure 1.12. AutoCAD's Help feature is *context-sensitive*; it recognizes that you were using the Line command and therefore displays help for that command.

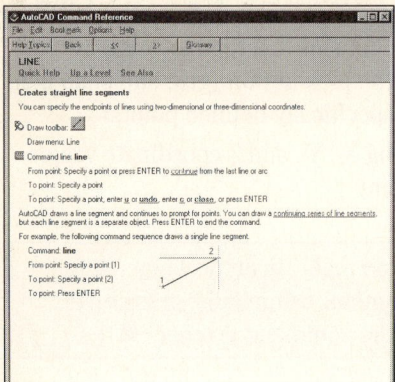

Figure 1.12

When you move the cursor over items displayed in a contrasting color, the cursor changes to a hand. You can then select these items, changing to the related help window. For now, you will use the buttons at the top of the AutoCAD Help window and pick to display the help topics.

Pick: **Help Topics**

The Help Topics window is divided into three different cards: Contents, Index, and Find. Select each card to familiarize yourself with the options available and then pick the Index card. As shown in Figure 1.13, you have the option of (1) typing the help topic you want or (2) scanning through all the help topics available.

Type: **Draw Toolbar**

Help Topics either matches exactly what is typed or locates the closest option on the list shown in box 2. In this case, Draw Toolbar is an exact option and appears as a selected entry in the window. If you pick the display button, press the return key, or double-click on the Draw Toolbar option, a new help window with information about the entry appears.

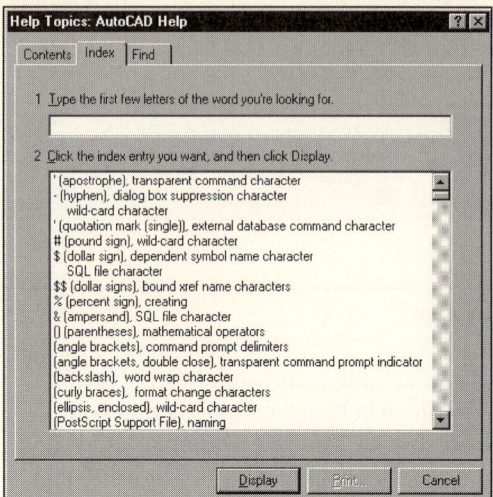

Figure 1.13

As you do with other Windows 95 applications, you may use the close box in the upper right-hand corner to close the Help window. If you are used to picking File, Close from the menu to close a window, you can do that instead.

Click: **(the X box in the top right corner of the AutoCAD Help window)**

Click: **(the X box in the Command reference dialog box)**

Entering Coordinates

AutoCAD stores your drawing geometry by using World Coordinates, which is AutoCAD's default *Cartesian coordinate system*, where X-, Y-, and *Z-coordinate values* specify locations in your drawing. The UCS icon near the bottom left corner of your screen shows the positive X- and positive Y-directions on the screen. You can think of the default orientation of the Z-axis as the location in front of the monitor for positive values and inside the monitor for negative values. AutoCAD functions according to the right-hand rule for coordinate systems. If you orient your right hand palm up and point your thumb in the positive X-direction and index finger in the positive Y-direction, the direction your other fingers curl will be the positive Z-direction. Use the right hand rule and the UCS icon to figure out the directions of the coordinate system. The letter W appearing in the UCS icon indicates that the *World Coordinate System (WCS)* is active.

You can specify a point explicitly by entering the X-, Y-, and Z-coordinates, separated by a comma. You can leave the Z-coordinate off when you are drawing in 2D, as you will be in this tutorial. If you do not specify the Z-coordinate, AutoCAD considers it to be the current elevation in the drawing, for which the default value is 0. For now type only the X- and Y-values, and the default value of 0 for Z will be assumed.

Using Absolute Coordinates

You often need to type the exact location of specific points to represent the geometry of the object you are creating. To do so, type the X-, Y-, and Z-coordinates to locate the point on the current coordinate system. Called the *absolute coordinates*, they specify a distance along the X-, Y-, Z-axes from the origin, or (point (0, 0, 0) of the coordinate system).

> ■ *TIP* In AutoCAD the spacebar acts the same way as the ⏎ key, to end the command. When typing in coordinates, enter only commas between them. Do not enter a space between the comma and the following coordinate of a coordinate pair because a space has the same effect as ⏎. Recall that you can press the spacebar to enter commands quickly. ■

Resume use of the Line command and use absolute coordinates to specify endpoints for the line you will draw. (Restart the Line com-

mand by picking the Line icon from the Draw toolbar if you do not see the From point: prompt.)

> From point: **5.26,5.37** ⏎

As you move your cursor, you will see a line rubberbanding from the point you typed to the current location of your cursor. The next prompt asks for the endpoint of the line.

> To point: **8.94,5.37** ⏎

Once you have entered the second point of a line, the line appears on the screen. The rubberband line continues to stretch from the last endpoint you selected to the current location of the cursor, and AutoCAD prompts you for another point. This feature allows lines to be drawn end to end. Continue as follows:

> To point: **8.94,8.62** ⏎
> To point: **5.26,8.62** ⏎

> ■ *TIP* If you enter a wrong point while still in the Line command, you can back up one endpoint at a time by selecting the Undo option by typing the letter U and pressing ⏎. ■

■ *Warning:* Undo functions differently if you pick it as a command. If you select Undo at the command prompt, you may undo entire command sequences. If necessary, you can use Redo to return something that has been undone; however, Redo only restores the last thing undone. ■

The Close option of the Line command joins the last point drawn to the first point drawn, thereby closing the lines and forming an area. To close the figure, type *c* in response to the prompt.

> To point: **C** ⏎

Using Grid and Snap

Use the grid and snap settings to draw another rectangle to the right of the one you just drew, as shown in Figure 1.14. (Be sure that Snap is highlighted on the status bar. If it is not, double-click SNAP on the status bar or press F9 to turn it on.)

Pick: **Line icon**

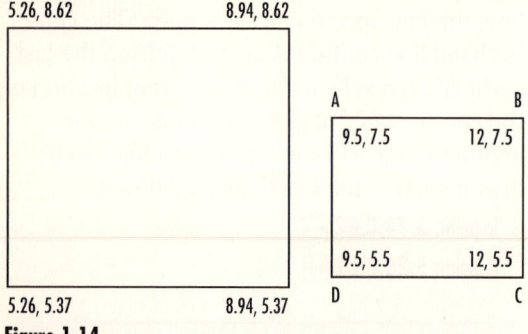

Figure 1.14

> ■ *TIP* You can restart the last command by pressing the ⏎ key, the spacebar, or the return button on your pointing device. Since Line was the last command used, you could have pressed ⏎ or the spacebar to restart the Line command. Using these shortcuts will help you produce drawings in less time. Keep in mind, however, that some commands selected from the toolbar do not always function the same when restarted at the command prompt. Restarting a command has the same effect as typing the command name at the prompt. Some toolbar and menu selections contain special programming to select certain options for you automatically. ■

On your own, choose the starting point on the grid by moving the cursor to point A, as shown in Figure 1.14 and pressing the pick button. Remember to use the coordinate display on the status bar to help you select the points.

Create the 2.5-unit long horizontal line by moving the cursor 5 snaps to the right and pressing the pick button to select point B.

Complete the rectangle by drawing a vertical line 4 spaces (2.00 units) long to point C and another horizontal line 5 spaces (2.5 units) long to point D. Then use the Close command option to complete the rectangle and end the Line command on your own.

Using Last Point

AutoCAD always remembers the last point that you specified. Often you will need to specify a point that is exactly the same as the preceding point. The @ symbol on your keyboard is AutoCAD's name for the last point entered. You will use last point entry with the Line command. This time you will restart the Line command by pressing ⏎ at the blank command prompt to restart the previous command.

Command: ⏎
From point: **2,5** ⏎
To point: **4,5** ⏎
To point: ⏎
Press: ⏎ *(or the return button on your mouse to restart the Line command)*
From point: **@** ⏎

> ■ *TIP* Once in the Line command, you can also press the spacebar, right mouse button, or ⏎ at the *From point* prompt to start your new line from the last point. ■

Your starting point is now the last point that you entered in the previous step (4,5). Note the line rubberbanding from that point to the current location of your cursor.

> ■ *TIP* When you use AutoCAD, moving the cursor away from the point you have selected is often helpful by letting you see the effect of the selection. You will not notice rubberband lines if you leave the cursor over the previously selected point. ■

Using Relative X- and Y-Coordinates

Relative coordinates allow you to select a point at a known distance from the last point specified. To do so, you must insert the @ symbol before your X- and Y-coordinate values. Now, continue the Line command.

> To point: **@0,3** ⏎

Note that your line is drawn 0 units in the X-direction and 3 units (6 spaces on the grid) in the positive Y-direction from the last point specified, (4,5). Complete the shape.

> To point: **@–2,0** ⏎
>
> To point: **@0,–3** ⏎
>
> To point: ⏎ *(or the return button on your mouse to end the command)*

Your drawing should look like Figure 1.15. You may need to scroll your screen to see all three rectangles.

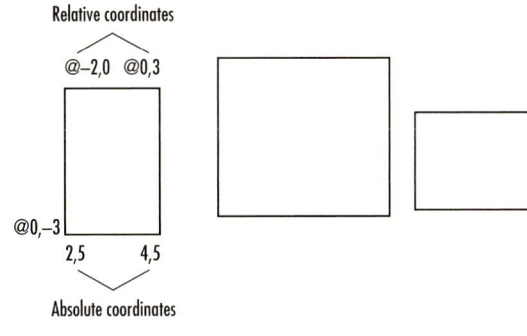

Figure 1.15

Using Polar Coordinates

When prompted to enter a point, you can specify the point by using *polar coordinates*. Polar coordinate values use the input format @DISTANCE<ANGLE. When using the default decimal units, you don't need to specify any units when typing the angle and length. Later in this tutorial you will learn how to select different units of measure for lengths and angles. When you are using polar coordinates, each new input is calculated relative to the last point entered. The default system for measuring angles in AutoCAD defines a horizontal line to the right of the current point as 0°. As shown in Figure 1.16, angular values are positive counterclockwise. Both distance and angle values may be negative.

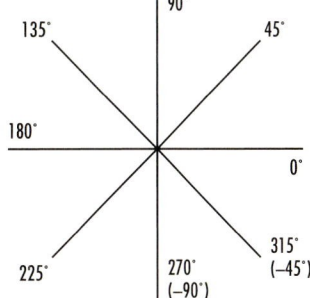

Figure 1.16

Figure 1.17 shows an example of lines drawn using relative coordinate entry. A line was drawn from the starting point to the distance and at the angle specified. In the figure, the line labeled 1 extends 3 units from the starting point at an angle of 30°. Remember that the default measures the angle from a horizontal line to the right of the starting point. Line 2 was begun at the last point of line 1 and extended 2 units at an angle of 135°, again measured from a horizontal line to the right of that line's starting point. Polar coordinates are very important for creating drawing geometry.

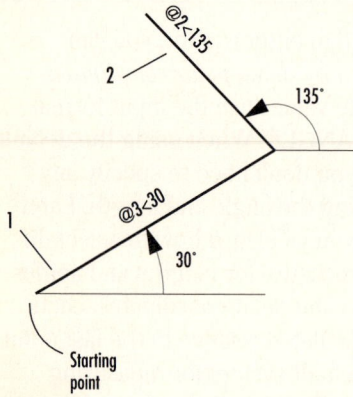

Figure 1.17

You will now draw a rectangle, using polar coordinate values. The starting point for this rectangle is specified as point A in Figure 1.18. Use the toolbar to select the Line command. You will draw a horizontal line 3.5 units to the right of the starting point by responding to the prompts.

Pick: **Line icon**

From point: *(pick point A, coordinates 1.0,4.0)*

To point: @3.5<0 ⏎

Complete the rectangle, using polar coordinate values.

To point: @2<–90 ⏎

To point: @3.5<180 ⏎

To point: C ⏎

Your drawing should look like that in Figure 1.18.

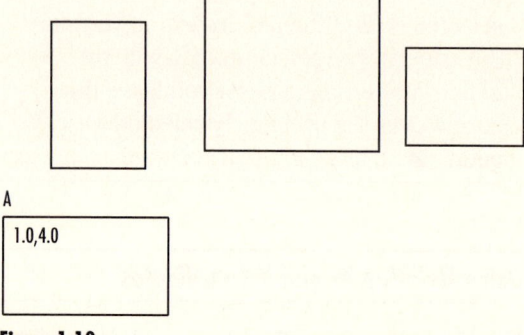

Figure 1.18

■ *TIP* Double-clicking on the coordinates displayed on the status bar toggles their display; that is, you can toggle the coordinate display off and on. When you toggle the coordinate display off, the coordinates do not change when you move the crosshairs and they appear grayed out on the status bar. When you toggle the coordinates back on, they once again display the X-, Y-, and Z-location of the crosshairs. During execution of the commands a third toggle is available for the coordinate display: In a command such as Line, when you double-click the coordinate display, AutoCAD displays the length and angle from the last point picked. This display can be very useful in helping you determine approximate distances and angles for polar coordinate entry. You can also toggle the coordinates by pressing F6 or Ctrl-D. ■

Using Save

 The Save command lets you save your work to the file name you have previously specified, in this case *shapes.dwg*. You should save your work frequently. If the power went off or your computer crashed, you would lose all the work that you had not previously saved. If you save every ten minutes or so, you will never lose more than that amount of work. Save your work at this time by using the Standard toolbar to pick the Save icon, which looks like a floppy disk.

Pick: **Save icon**

Your work will be saved to *shapes.dwg*. If you had not previously named your drawing, the Save Drawing As dialog box would appear, allowing you to specify a name for the drawing. Save your work periodically as you work through these tutorials.

Using Erase

 The Erase command removes objects from a drawing. The Erase command is located on the Modify toolbar.

> ■ *TIP* Typing E ⏎ (the command alias) will also start the Erase command. ■

Pick: **Erase icon**

A small square replaces the crosshairs, prompting you to select objects. To identify an object to be erased, place the cursor over the object and press the pick button. Each line that you have drawn is a separate drawing object, so now erase the objects A and B in Figure 1.19. If Snap is turned on, turning it off makes object selection easier.

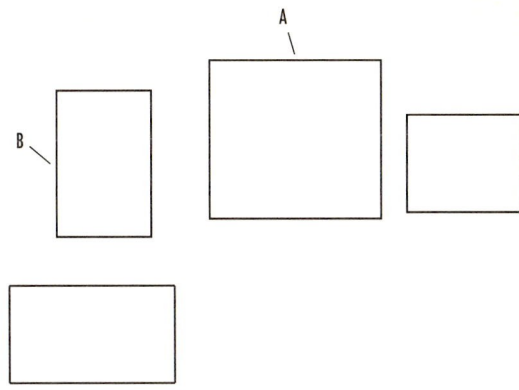

Figure 1.19

Select objects: *(pick on line A)*
Select objects: *(pick on line B)*

The lines you have selected become dashed but are not erased. When using AutoCAD, you can select more than one item to create a *selection set*. You will learn more ways to do so later in this tutorial. When you have finished selecting, you will press the return button on your pointing device or ⏎ to tell AutoCAD that you are done selecting. Pressing ⏎ or the return button completes the selection set and erases the selected objects.

Press: ⏎ *(or the return button)*

The selected lines are erased from the screen.

> ■ *TIP* Don't position the selection cursor at a point where two objects cross because you cannot be certain which will be selected. The entire object is selected when you point to any part of it, so select the object at a point that is not ambiguous. ■

Using Redraw All

Redraw All works in all viewports to remove the excess blip marks added to the drawing screen when you are drawing objects and restores objects partially erased while you are editing other objects. Redraw All is located on the Standard toolbar. Redraw All has no prompts; it is simply activated when you pick it.

Pick: **Redraw All icon**

Your screen is redrawn.

■ *TIP* You can also just type R ↵ at the command prompt; R is the command alias for Redraw. ■

If you erase an object by mistake, you may activate the OOPS command at the command prompt.

Type: **OOPS**

The lines last erased are restored on your screen. OOPS restores only the most recently erased objects. It won't restore objects beyond the last Erase command. However, it will work to restore erased objects even if other commands have been used in the meantime.

Erasing with Window or Crossing

You need to clear your drawing editor to make room for new shapes. You will erase all the rectangles on your screen. You can pick multiple objects by using the Window and Crossing selection modes.

To use AutoCAD's *implied Windowing mode*, at the Select objects: prompt, pick a point on the screen that is not on an object and move your pointing device from left to right. A window-type box will start to rubberband from the point you picked. In the implied Windowing

mode (a box drawn from left to right), the window formed selects everything that is entirely enclosed in it. However, drawing the box from right to left causes the *implied Crossing* mode to be used. Everything that either crosses the box drawn or is enclosed in the box is selected. You can use implied Windowing and implied Crossing to select objects during any command in which you are prompted to select objects.

You can turn implied Windowing on and off. Before using it, you will check to see that implied Windowing is turned on.

Pick: **Tools, Selection**

The dialog box shown in Figure 1.20 appears on your screen. For implied Windowing to work, it must be selected (a check must appear in the box to the left of Implied Windowing). On your own, select this mode if it is not currently selected; then pick OK to exit the dialog box. If it is already turned on, pick Cancel to exit without making any changes.

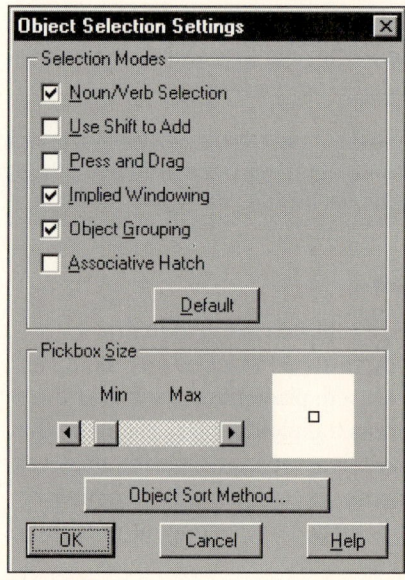

Figure 1.20

Pick: **Erase icon**

Select objects: ***(pick a point above and to the left of your upper three rectangles, identified as A in Figure 1.21)***

Other corner: ***(pick a point below and to the right of your upper three rectangles, identified as B)***

Figure 1.21

Note that a "window" box formed around the area specified by the upper left and lower right corners. Your screen should look like that in Figure 1.22.

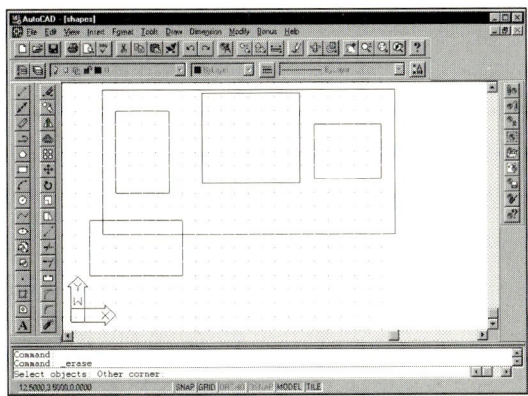

Figure 1.22

When you picked the second corner of the window (point B), the objects that were entirely enclosed in the window became highlighted. You will end the object selection by pressing ⏎ and erasing the selected objects.

Select objects: ⏎

Only objects that were entirely within the window were erased. The fourth rectangle (represented by four separate lines) was not completely enclosed and therefore was not erased.

Moving the Toolbars

The toolbars Draw, Modify, and Internet Utilities initially are docked on either side of the AutoCAD graphics window. (Depending on the Internet configuration of your system, you may not see the Internet Utilities toolbar.) You can float the toolbars or redock them to any side of the graphics window (top, bottom, left, or right). You will close the Internet Utilities toolbar until it is covered later in this manual. You will float the Modify toolbar, then redock it to the right edge of the screen. First, you will move the Modify toolbar by picking immediately to the right of any of the Modify buttons, keeping the pick button depressed, and dragging it into the center of the screen.

Pick: ***(immediately to the right of a Modify toolbar button and drag it to the center of the screen)***

The Modify toolbar should now appear as a box with a title bar and an X in the upper right corner. Next, perform the same task on the Internet Utilities toolbar, located on the right side of the screen.

Pick: ***(immediately to the left of an Internet Utilities toolbar button and drag it to the center of the screen)***

After the Internet Utilities toolbar has taken on an appearance similar to that the Modify toolbar, select the X in the upper right corner to close it. Next, you will redock the Modify toolbar to the right edge of the screen.

Pick: ***(on the title bar of the Modify toolbar and drag it to the right edge of the screen)***

When the toolbar is near the correct location, its outline will change to the docked shape. Release the pick button to leave the toolbar in its new docked location. When you have finished docking the toolbar, your screen should look like that in Figure 1.23, allowing for a more balanced drawing space.

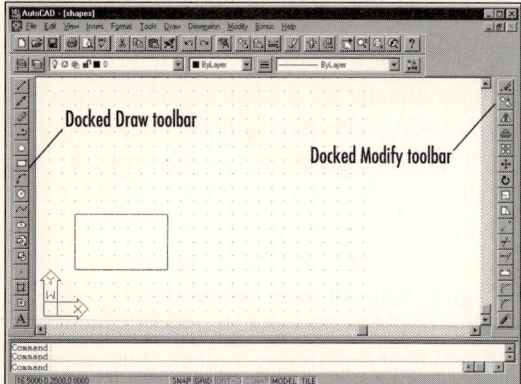

Figure 1.23

Next, you will use the Erase command and implied Crossing to erase the fourth rectangle.

    Pick: **Erase icon**

    Select objects: *(pick a point below and to the right of the remaining rectangle)*

    Other corner: *(pick a point above and to the left of the rectangle)*

    Select objects: ⏎

The remaining rectangle is erased from the drawing.

Using Undo

The Undo command removes the effect of previous commands. If you make a mistake, pick the Undo command from the Standard toolbar, and the effect of the previous command will disappear.

You will try this now to see the last rectangle you erased reappear.

    Pick: **Undo icon**

> ■ *TIP* If you undo the wrong object, a Redo command lets you restore the last item you undid. It is the next icon to the right of Undo on the Standard toolbar. ■

Line and other commands also include Undo as a command option that allows you to undo the last action within the command.

    Pick: **Line icon**

    From point: *(pick any point)*

    To point: *(pick any point)*

    To point: *(pick any point)*

    To point: *(pick any point)*

    To point: **U** ⏎

    To point: ⏎

One line segment disappeared while you remained within the Line command. Now, you will undo the entire Line command with the Undo command. You will pick Undo from the Standard toolbar.

    Pick: **Undo icon** ⏎

All the line segments drawn with the last *instance* of the command disappear.

Typing Undo at the command prompt offers more options for undoing your work. You will draw some lines to have some objects to undo.

On your own, use Erase from the toolbar and erase any remaining objects in your graphics window. Use Line on your own to draw six parallel lines anywhere on your drawing screen. Use Snap and Grid as needed.

In the next step, you will type the Undo command to remove the last three lines that you drew.

Type: **UNDO** ⏎

Auto/Control/Begin/End/Mark/Back/<number>: **3** ⏎

The last three lines drawn should disappear, corresponding to the last three instances of the Line command. You could have selected any number, depending on the number of command steps you wanted to undo.

> ■ *TIP* Backing up one step at a time by using Undo from the Standard toolbar (the U command) often is the easiest way to undo previous steps. Some commands do several steps in one sequence, so you may not be sure of the exact number you want to back up. ■

Using Redo

If the last command undid something you really did not want to undo, you can use the Redo command to reverse the effects. The Redo icon appears on the Standard toolbar to the right of Undo.

Pick: **Redo icon**

The three lines should reappear. Redo reverses only the last Undo; you must use it immediately after you use the Undo command.

Using Back

The Back option of the Undo command takes the drawing back to a mark that you set with the Mark option or to the beginning of the *drawing session* if you haven't set any marks. Be careful when selecting these options or you may undo too much.

Type: **UNDO** ⏎

Auto/Control/BEgin/End/Mark/Back/<number>: **B** ⏎

This will undo everything. OK?<Y>: ⏎

All operations back to the beginning of the drawing session (or mark, if you had previously set one with the Mark option of the Undo command) will be undone. (You can use Help to find out what the remaining options, Auto, Control, Begin, and End, are used for.)

Pick: **Redo icon**

The previous appearance of the drawing should return. On your own, erase all objects in this drawing session from the screen.

Next, you will create a new drawing showing a plot plan. A plot plan is a a plan view of a lot boundary and the location of utilities. You will set the units for this drawing, instead of using the defaults. To start a new file, you will pick the New command from the Standard toolbar. It is the left-most icon.

Pick: **New icon**

Because you have been working in a drawing already, AutoCAD will ask you whether you want to save your changes. You do not need to save *shapes*.

Select Start From Scratch, if it is not already selected, and pick OK from the Create New Drawing dialog box. On your own, save your drawing to a new name: *plotplan*. Be sure to verify that you're creating the new drawing in the *C:\work* directory.

Setting the Units

AutoCAD allows you to work in the type of units that are appropriate for your drawing. The *default units* are decimal. You can change the type of units for lengths to *architectural units* that appear in feet and fractional inches, *engineering units* that appear in feet and decimal inches, *scientific units* that appear in exponential format, or *fractional units* that are whole numbers and fractions. When you use architectural or engineering units, one drawing unit is equal to one inch; to specify feet you must type the feet mark after the numerical value (e.g., 50.5' or 20'2" or 35'-4"). You can think of the other types of units as representing any type of real-world measurement you want: decimal miles, furlongs, inches, millimeters, microns, or anything else. When the time comes to plot the drawing, you determine the final relationship between your drawing database, in which you create the object the actual size it is in the real world, and the paper plot.

Angular measurements can be given in decimal degrees; degrees, minutes, and seconds; gradians; radians; or *surveyor units*, such as the bearing N45d0'0"E.

You will use the Format selection from the menu bar to access the Units Control dialog box.

Pick: **Format, Units**

The Units Control dialog box appears on your screen, as shown in Figure 1.24.

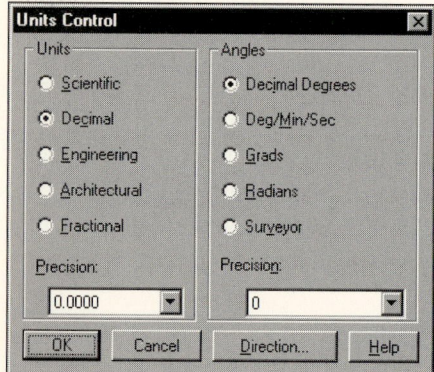

Figure 1.24

In the Units area, the radio buttons to the left let you select the type of unit for the drawing. Pick the button to the left of Architectural. The center of the button becomes a filled black circle to tell you that it has been selected. You can select only one button at a time. Note that the units displayed under the heading Precision change to architectural units. Pick the button for Decimal once again. Note that the units in the Precision area change back to decimal units. The dimensions for the plot plan that you

will create will be in decimal feet, so you will leave the units set to Decimal. Next, pick on the number 0.0000 displayed in the box below Precision. The choices for unit precision pull down below the pick box, as shown in Figure 1.25.

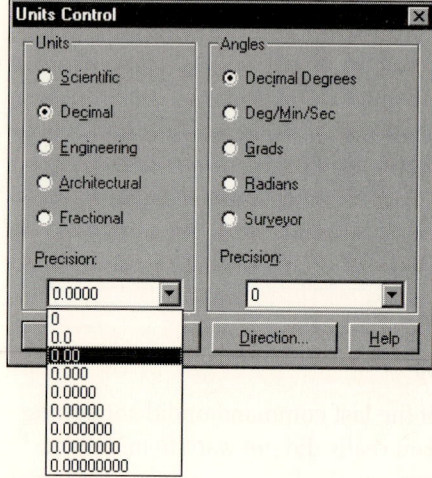

Figure 1.25

Pick on the selection 0.00 to set the display for two decimal places. When specifying coordinates and lengths, you can still type a value from the keyboard with more precision and AutoCAD's drawing database will keep track of your drawing with this accuracy. However, because you have selected this precision, only two decimal places will be displayed. Units precision determines the display of the units on your screen and in the prompts, not the accuracy internal to the drawing. Though AutoCAD keeps track internally to at least fourteen decimal places, only eight decimal places of accuracy will ever appear on the screen. You can change the type of unit and precision at any point during the drawing process.

The right side of the dialog box controls the type and precision of angular measurements. Remember that the default measurement of angles is counterclockwise, starting from a horizontal line to the right as 0°. You can change the default setting by picking on the Direction button and making a new selection. Leave the direction set at the default of 0° towards East.

Select the button for Surveyor angles. Note that the display in this Precision area changes to list the angle as a *bearing*. When this mode is active, you can type in a surveyor angle. AutoCAD will measure the angle from the specified direction, North or South, toward East or West, as specified. The default direction, North, is straight toward the top of the screen. If you want to view greater angle precision, pick on the box containing the precision, N 0d E. The list of the available precisions pops down, as shown in Figure 1.26, allowing you to select to display degrees, minutes, and seconds. On your own, pick the selection N0d00"00"E to display degrees, minutes, and seconds.

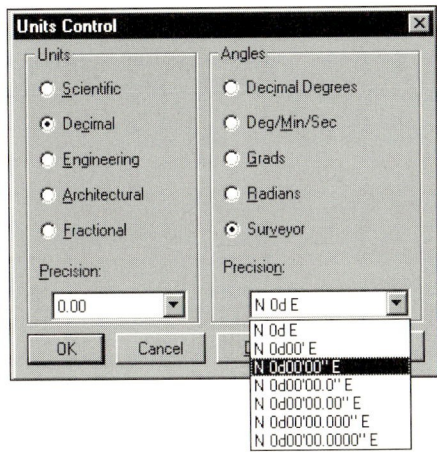

Figure 1.26

Pick: **OK** *(to exit the dialog box)*

Look at the status bar: The coordinate display has changed to show two instead of four places after the decimal.

Sizing Your Drawing

In AutoCAD you always create your drawing geometry in *real-world units*. In other words, if the object is 10' long, you make it exactly 10' long in the drawing database. If it is a few millimeters long, you create the drawing to those lengths. You can think of the decimal units as representing whatever decimal measurement system you are using. For this drawing, the units will represent decimal feet, so 10 units will equal 10 feet in the drawing. After you have created the drawing geometry, you can decide on the scale at which you want to plot your final hardcopy drawing on the sheet of paper. The ability to create exact drawings from which you can make accurate measurements and calculations on the computer is one of the powerful features of CAD. Also, you can plot the final drawing to any scale, saving a great deal of time because you don't have to remake drawings just to change the scale, as you would with paper drawings.

For your plot plan drawing, you need to create a larger graphics window to accommodate the site plan shown in Figure 1.27.

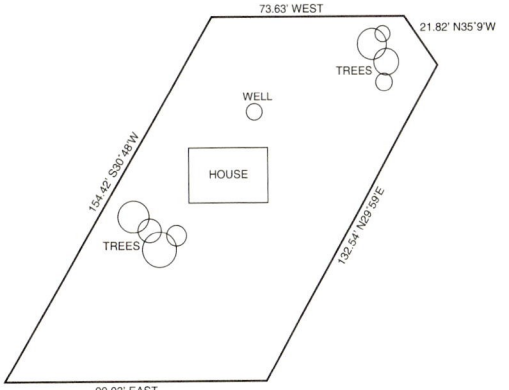

Figure 1.27

Using Limits

The Limits command sets the size of your drawing area. You can also turn off Limits if you do not want to preset the size of the drawing. Use the Drawing Limits selection from the Format menu to change the overall size of the drawing to reflect metric drawing units.

Pick: **Format, Drawing Limits**

Reset model space limits:

ON/OFF/<Lower left corner> <0.00,0.00>: ⏎
   (to accept default of 0,0)

Upper right corner <12.00,9.00>: **300,225** ⏎

Note the message that reads "Reset model space limits:". The default space in which you create your drawing geometry or model is called model space. It is where you accurately create real-world–size models. In Tutorial 5, you will learn to use paper space, where you lay out views, as you do on a sheet of paper. You can set different sizes for model space and paper space with the Limits command.

Using Zoom

The Zoom command enlarges or reduces areas of the drawing on your screen. It is different from the Scale command, which actually makes the selected items larger or smaller in your drawing database. You can use the Zoom command when you want to enlarge something on the screen so that you can more easily see details. You can also zoom out so that objects appear smaller on your display. The Zoom command options are located on the Standard toolbar and on the View menu.

Using Zoom All

 The Zoom All option displays the drawing limits or shows all the drawing objects on the screen, depending on which is larger. Zoom All is located on the Zoom flyout on the Standard toolbar, which we will cover in greater detail in Tutorial 2. You will select View, Zoom All from the menu bar to show this larger area on your display.

Pick: **View, Zoom, All**

Move the crosshairs to the upper right-hand corner of the screen. You will see that the status bar displaying the coordinate location of the crosshairs indicates that the size of the graphics window has changed to reflect the limits of the drawing.

> ■ *TIP* If the coordinate display does not show a larger size, check to see that you set the limits correctly and that you picked Zoom All. Be sure that the coordinates are turned on by double-clicking on the coordinates located on the status bar until you see them change as you move the mouse. ■

Now, you will set the grid spacing to a larger value so that you can see the grid on your screen.

Command: **GRID** ⏎

Grid spacing(x) or ON/OFF/Snap/Aspect <0.50>: **10** ⏎

To draw the site boundary, you will use absolute and polar coordinates as appropriate. You will be using surveyor angles to specify the directions for the lines. AutoCAD's surveyor angle selections, which you selected when setting the type of units, are given in bearings. Bearings give the angle to turn from the first direction stated toward the second direction stated. For example, a bearing of N29°59'E means to start at the direction North and turn an angle of 29°59' toward the East.

Now you are ready to start drawing the lines of the site boundary. Use the Draw toolbar.

Pick: **Line icon**

From point: **50,30** ⏎

To point: **@99.03<E** ⏎

To point: **@132.54<N29d59'E** ⏎

To point: **@21.82<N35d9'W** ⏎

To point: **@73.63<W** ⏎

To point: **c** ⏎

Your screen should look like Figure 1.28.

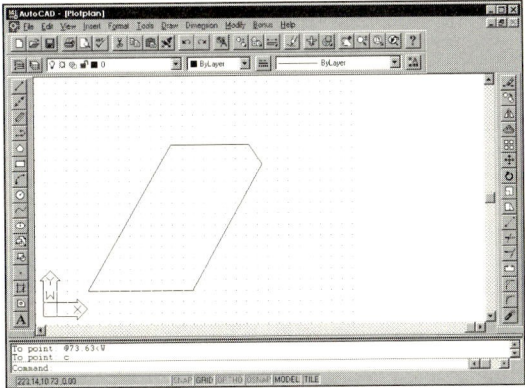

Figure 1.28

> ■ *TIP* Something to remember for later in this manual when drawing property boundaries, you can change the UCS X-axis position to match any boundary line. Thus, when drawing lines referenced to inside angles, you can draw that angle, based on the X-axis, if it is lined up with the boundary that the inside angle is referring to. ■

Using Rectangle

You are going to draw a rectangle to represent a house on the plot plan. Using coordinate values, you will place the rectangle roughly in the center of your plot plan shown in Figure 1.28. Pick Rectangle from the Draw toolbar.

Pick: **Rectangle icon**

AutoCAD gives you various options, as well as a prompt for the first corner of the rectangle.

Chamfer/Elevation/Fillet/Thickness/Width/<First Corner>:

These options allow you to manipulate the appearance and placement of the rectangle before it is drawn. Chamfer and Fillet are options that change the corners of the rectangle, which are covered in Tutorial 3. Elevation and Thickness are options best used in 3D modeling. The Width option allows you to specify a width for the lines making up the rectangle. You will type the command option letter W to change Width and enter a value of 5.

Chamfer/Elevation/Fillet/Thickness/Width/
    <First Corner>: **W** ⏎

Width for rectangles <0.00>: **5** ⏎

Chamfer/Elevation/Fillet/Thickness/Width/<First Corner>:

AutoCAD prompts you for the first corner again, and you will arbitrarily pick a point on the screen for the first corner. As you move the crosshairs away from the point, a box with a line width of 5 stretches from the first point to the current location of the crosshairs. Now press [Esc] to cancel the command or go through the necessary steps to erase the rectangle if you have selected a second point. On your own, choose the Rectangle icon again and change the Width back to the initial default value of 0. The Rectangle options line with a prompt for the First Corner should be showing again; if it isn't, pick the Rectangle icon again.

Now you are ready to draw a rectangle to represent a house. You will type the coordinates for the corners of the rectangle.

Chamfer/Elevation/Fillet/Thickness/Width/
    <First Corner>: **120,95** ⏎

Other corner: **150,115** ⏎

Your drawing should be similar to the one in Figure 1.29.

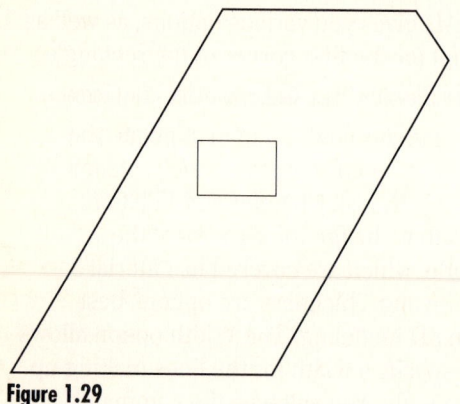

Figure 1.29

Saving Your Drawing

Having completed the site boundary, you should now save the drawing on your disk. You should do so when you have completed a major step or every 10 to 15 minutes. This way, if your computer crashes, you will not have lost more than a few minutes worth of work. Also, if you make a mistake that you don't know how to correct, you can pick File, Open, then discard your drawing changes, and open the saved version.

Pick: **Save icon**

AutoCAD saves your file to the name you assigned when you first saved the drawing.

Drawing Circles

 You will draw a circle to represent the location of the well on the plot plan. You draw circles using the Circle command located on the Draw toolbar.

Pick: **Circle Center Radius icon**

Command: _circle 3P/2P/TTR/<Center point>: **145,128** ⏎

As you move the cursor away from the center point, a circle continually reforms, using the distance from the center you selected to the cursor location as the radius value. You will type the value to specify the exact size for the circle's radius.

Diameter/<Radius>: **3** ⏎

Now, draw some circles to represent trees or shrubs in the plot plan. You will specify the locations for the trees by picking them from the screen.

Pick: **Circle Center Radius icon**

3P/2P/TTR/<Center point>: *(pick a point for the center of circle 1, as shown on Figure 1.30)* ⏎

Diameter/<Radius>: *(move the crosshairs away until the circle appears similar to that shown in Figure 1.30, then press the pick button)* ⏎

On your own, draw the remaining circles representing trees using this method.

Your drawing should look like that in Figure 1.30.

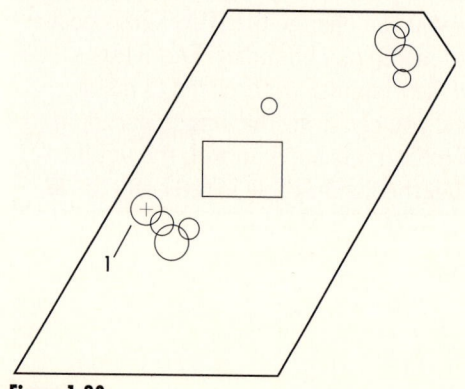

Figure 1.30

Adding Text

Use the Dtext (Dynamic Text) command to add single lines of text, words, or numbers to a drawing. Dtext, or Single Line Text is located in the Draw menu, under text.

Pick: **Draw, Text, Single Line Text**

Unless you specify otherwise, text is added to the right of a designated starting point. You can also center text about a point or add it to the left of a point if you use the Justify option. To specify the starting point for the text, you will pick from the screen. (You can also type the coordinates of the location for the starting point.) You will label the lengths and bearings of the lot lines shown in the plot plan. Start with the bottom lot line by picking a point below the lot line.

Command: _dtext Justify/Style/<Start Point>: **(select a point at about coordinates 90,25)**

The 0.20 value shown in the next prompt is the default text height. The current drawing must be scaled to fit on a piece of paper, so this height is much too small in proportion to the final drawing. In this case a height of 3 will create text of proportionate size.

■ *TIP* To determine the actual text height necessary, you must know the size of paper and scale you will use to plot the final drawing. In Tutorial 5 you will learn to use paper space to set up scaling and text for plotting. If you are unsure about what text height to use, move the crosshairs on the screen and read the coordinate display on the toolbar. Use the lengths shown to get an appropriate size for your text. ■

Height<0.20>: **3** ⏎

The default value of E in the Rotation angle <E> prompt generates horizontal (East) text. You can specify any angle. Text on technical drawings is usually drawn horizontally and is called unidirectional text. Accept the default rotation angle.

Rotation angle <E>: ⏎

AutoCAD is now ready to accept typed-in text.

Text: **99.03' EAST** ⏎

Text: ⏎

You can type the text just as you would with a word processing program. Dtext will not wrap text, so you must designate the end of each line by pressing ⏎. If you make a mistake before exiting Dtext, backspace to erase the text. After exiting the Dtext command, you can use Edit Text to make corrections to lines of text. The Text: prompt always appears after every line of text. Press ⏎ without entering any text (a null reply to the prompt) to end the Dtext command. (You must use the keyboard, not the return button on your pointing device.) When you exit Dtext, AutoCAD recreates the text as a permanent object. Your drawing should look like that in Figure 1.31.

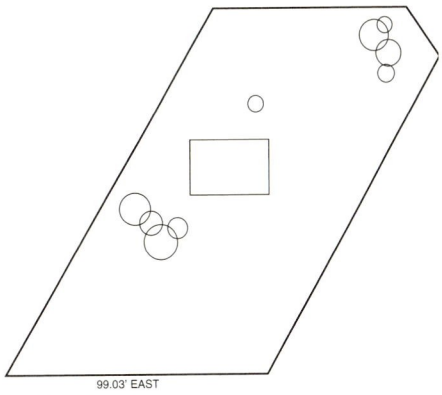

99.03' EAST

Figure 1.31

Using Special Text Characters

You will add the text labeling the length and angle for the right-hand lot line. When you are using the standard AutoCAD fonts, the special text item %%d creates the degree symbol. Type the line of text with %%d in place of the degree symbol. When you have finished typing in the text and press ⏎, AutoCAD will replace %%d with a degree symbol. (You will use more special text characters in Tutorial 7.) You will also specify a rotation angle for the text so that it is aligned with the lot line. To familiarize yourself with typing a command, as well as finding it within the Menu system, you will type Dtext to activate the Single Line Text or Dynamic Text command.

Type: **Dtext**

Justify/Style/<Start Point>: **(pick a point slightly to the right of the right-hand lot line)**

Height <3.00>: ⏎

Rotation angle <E>: **N29d59'E** ⏎

> ■ **TIP** You can also specify the rotation by picking when the line that rubberbands from the point you picked for the text location is oriented at the rotation angle you desire. If you do not want to type in bearings for the text angle, you can change the units at any time. ■

Text: **132.54' N29%%d59'E** ⏎

Text: ⏎

The text should align with the lot line in your drawing.

On your own, use the techniques that you have learned to add the bearings and distances for the remaining lot lines. To have the text on the left lot line be parallel to the line, but face the same direction as the text on the right lot line, reverse the direction of the bearings (for

example, if a bearing was SW, use NE), and keep the same degree and minute values. You can also specify the angle by picking two points that align with the lot line along which you are creating text.

> ■ **TIP** Each time you use the Dtext command, AutoCAD returns the height and rotation angle to the last height and rotation angle used. This action will help with labeling the left lot line because the rotation angle will already be set to the right lot line. ■

Add more text identifying the house and the well locations. Your drawing should look like that in Figure 1.32.

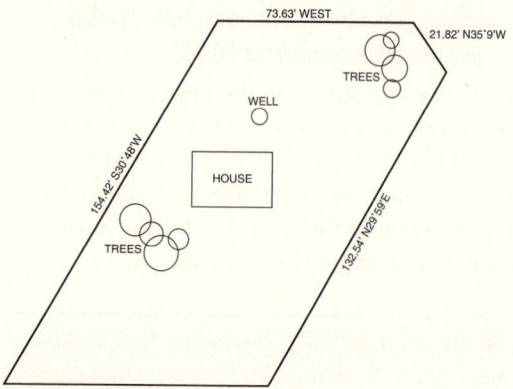

Figure 1.32

Editing Text

Once you have written the text, you can later edit it by using a DDEDIT command located in the Modify menu. You will select Modify, Object and Text from the menu bar.

Pick: **Modify, Object, Text...**

Command: _ddedit

<Select an annotation object>/Undo: **(pick the top line of text)**

After you activate the command, AutoCAD prompts you to select a text object (line of text). The selected text appears in a dialog box like the one shown in Figure 1.33. You can add or change text without erasing the entire line. Use the arrow keys on your keyboard to move within the line of text without deleting. Use the [DELETE] key to delete the letter to the right of the cursor; use the [BACKSPACE] key to delete the letter to the left of the cursor. You can also retype the entire line. Be sure that the text in the dialog box reads: 73.63' WEST. Exit the dialog box.

Pick: **OK**

    <Select an annotation object>/Undo: ⏎ *(to exit the command)*

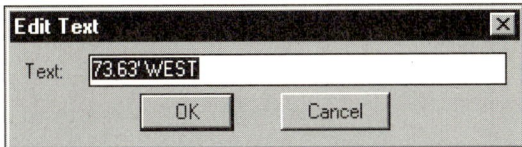

Figure 1.33

Setting the Text Style

The Style command allows you to create a style specifying the font and other characteristics for the shapes of the text in your drawing. A *font* is a character set comprising letters, numbers, punctuation marks, and symbols of a distinctive style and design. AutoCAD supplies several *shape-compiled fonts* that you can use to create text styles. In addition to the AutoCAD shape-compiled fonts, you can also use your own True Type and Type 1 Postscript fonts. Some sample True Type fonts are provided with AutoCAD in the directory *c:\AutoCAD R14\ fonts*. AutoCAD also supports Unicode fonts for many languages that use large numbers of characters.

The Style command allows you to control the appearance of a font so that it is slanted (oblique), backwards, vertical, or upside-down. You can also control the height and proportional width of the letters.

If you don't use the default style, you must create the style you want to use before you add text to your drawing (or later change the properties of the text to that style).

To create a style, you will select Format, Text Style from the menu bar. Once you have created (or selected) a style, that style will remain current for all text until you set a new style as the current one. An AutoCAD shape-compiled font that works well for engineering drawings is the Roman Simplex font.

Pick: **Format, Text Style**

The dialog box shown in Figure 1.34 appears on your screen. It is divided into four different sections: Style Name, Fonts, Effects, and Preview. You can set a new style in the Style Name section, and it will become the default style for any subsequently written text.

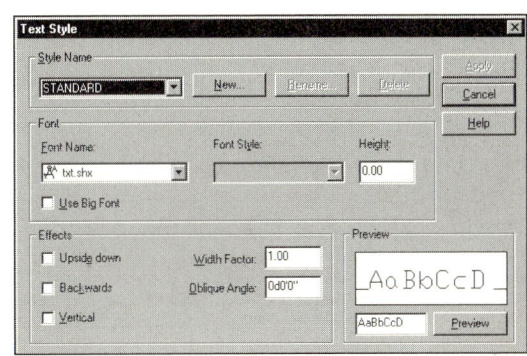

Figure 1.34

You will pick the New button, and give the new style the name Romans.

Pick: **New**

Type: **Romans** *(in the New Text Style dialog box)*

Pick: **OK**

Next you will set the font for the new style, *Romans.shx*. First, scan through the list of fonts available (all AutoCAD shape-compiled fonts and other True Type or Type 1 Postscript fonts on the system should be there) and note the changes in the Preview section of the dialog box. The text in the Preview box will be underlined in red if it is a True Type font, or it will appear as in Figure 1.34 if it is a shape-compiled font. For now, scroll through the list until you see the *Romans.shx* font, preceded by the AutoCAD symbol, and select it. The Font Style option only allows changes to True Type fonts, not to Shape Compiled fonts (*.shx*). You will accept the default text height of 0.00. Setting the text height in the Text Style dialog box causes you not to be prompted for the height of the text when you use the Dtext command.

At this point, you will make no changes to the Effects portion of the dialog box. The dialog box should now appear like that in Figure 1.35. Choose the Apply button to make the new text style current and then Close to exit.

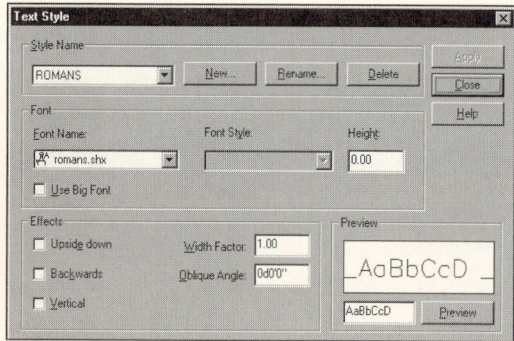

Figure 1.35

Any new text added to your drawing will apply the current text style until you set a different style as current. You will now add a title to the bottom of the drawing.

Type: **Dtext**

Justify/Style/<Start Point>: **J** ⏎

Align/Fit/Center/Middle/Right/TL/TC/TR/ML/MC/MR/BL/BC/BR: **C** ⏎

Center point: *(select a point near the middle bottom of the drawing)*

Height<3.00>: ⏎

Rotation angle <E>: ⏎ *(be sure that your rotation angle is set to E)*

Text: **PLOT PLAN** ⏎

Text: ⏎

Using Mtext

Next you will use the Multiline Text feature to add a block of notes to your drawing. Unlike Dtext, the Mtext command automatically adjusts text within the width you specify. You can create text with your own text editor and then import it with the Mtext command, or you can create text with the Mtext command. As with Dtext, you can easily edit Mtext by using the DDEDIT command after you have already placed the text in the drawing. You can select Mtext by picking the Text button from the Draw toolbar or choose Draw, Text, Multiline Text from the menu.

Pick: **Text icon**

Command: _mtext Current text style: Romans. Text height: 3.00

Specify first corner: *(pick a point to the right of the plot plan drawing where you want to locate the corner of the notes; a window will form)*

Specify opposite corner or [Height\Justify\Rotation \Style\Width]: *(pick the second corner of the window to size the area for the notes)*

If you want to enter only the width of the text you will type in, type *W* at the preceding prompt instead of specifying the corners of the box, and then type the width you want to specify.

The Multiline Text Editor dialog box appears on your screen, as shown in Figure 1.36, containing three index cards: Character, Properties, and Find/Replace. The large blank area of the Multiline Text Editor dialog box is where you will type in the notes for the plot plan.

Figure 1.36

Type: **NOTES:** ⏎ **MINIMUM SETBACK FROM ALL LOT LINES OF 20' REQUIRED.**

You can use the standard Windows Control keys to edit text in the Multiline Text Editor dialog box.

Ctrl-C	Copy selection to the Clipboard
Ctrl-V	Paste Clipboard contents over selection
Ctrl-X	Cut selection to the Clipboard
Ctrl-Z	Undo and Redo
Ctrl-SHIFT-SPACEBAR	Insert a nonbreaking space
⏎	End the current paragraph and start a new line

You can create stacked fractions or stack text with the Mtext command. To do so, separate the text to be stacked with either / or ^. Using / draws a line between the numerator and denominator of fractions. Using ^ stacks the text with no line.

The Character index card allows you to change the current font, style of text (Bold or Italics), or color of text. It also allows you to insert symbols, such as the plus/minus, degree, or diameter symbols.

The Properties card allows you to change the current style, justification, width, and rotation of text. The Find/Replace card is a handy tool for use with existing Multiline text groups. Picking on the Import button allows you to select a text file that you have already created with another text editor.

Pick: **OK *(to exit the Multiline Text Editor dialog box)***

The notes are placed in your drawing, which should now be similar to that in Figure 1.37.

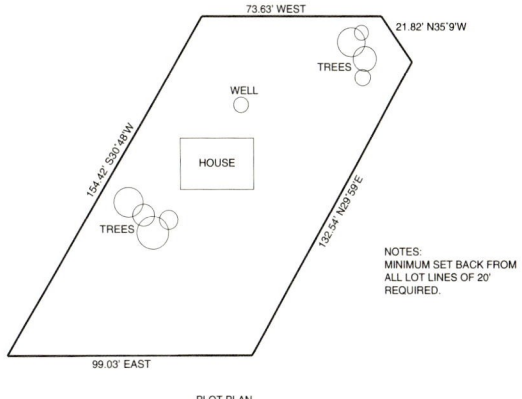

PLOT PLAN

Figure 1.37

Spellchecking Your Drawing

You can use AutoCAD's spell checker to detect spelling errors in your drawing.

Pick: **Tools, Spelling**

Select objects: *(pick on the notes you added)* ⏎

If the text contains spelling errors, the Check Spelling dialog box will appear, as shown in Figure 1.38. If your drawing does not contain any misspellings, the message "Spelling complete" will appear.

Pick: **OK**

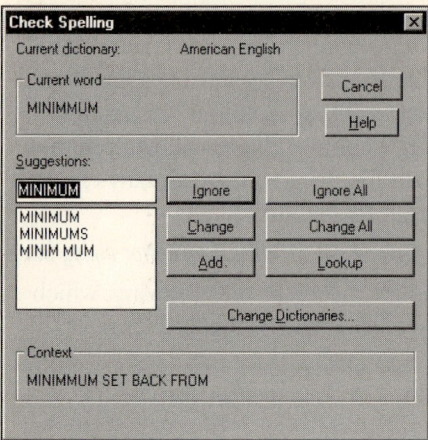

Figure 1.38

Saving Your Drawing

Now, save your drawing.

Pick: **Save icon**

The default file name is always the current drawing name. When you started the new drawing, you named it *plotplan.dwg*. The drawing is saved on the hard drive because you did not specify a drive letter.

Transferring Files

■ *Warning:* To complete this section you will need a blank formatted floppy disk. ■

You can transfer files from one drive to another. To transfer a file, place a blank formatted disk in drive A: (or substitute the appropriate drive letter for your system). To transfer the file named *plotplan.dwg* from the hard drive to a floppy disk in the drive selected, use the Windows Explorer.

Pick: **(the minimize button in the upper right-hand corner of the AutoCAD screen)**

This action minimizes AutoCAD and returns you to the Windows desktop. Click on Start, Programs, and Windows Explorer. The

Windows Explorer dialog box appears on your screen for use in copying your saved AutoCAD files from the hard drive to the specified drive. The various directories and drives on your computer are located on the left side of the dialog box under All Folders. The contents of the directory selected are on the right side of the dialog box, under Contents of. To change the directory, position the arrow cursor over the name of the directory you want to select (in the list on the left) and press the pick button. If you need to step back out of a subdirectory to a directory one level closer to the root, or main, directory, click on the C:\ selection. If you need to select a different drive, click on its name (A:, B:, C:, etc.). Once you have selected the drive and directory where your file is stored, pick on the file you want to copy from the Contents section of the Dialog box. Then, holding the pick button down, drag the file to the drive where you want to save the file. You should see a box showing the file being copied.

Close the Windows Explorer. Then, click on the minimized AutoCAD button on the Windows Taskbar, which returns you to your AutoCAD drawing session.

Exiting AutoCAD

You can use the Exit selection from the File menu to return to Windows, or you can type QUIT at the command prompt. If you have not previously saved your drawing, the File, Exit selection prompts you to save changes, discard changes, or cancel. If you have already saved your drawing, picking on File, Exit causes an immediate exit. If you have not saved your changes and want to, pick on Save Changes in the dialog box. This action will save your drawing to the file name *plotplan.dwg* that you selected when you began the new drawing.

The system returns to the Windows desktop. Now you have completed Tutorial 1.

absolute coordinates	file extensions	relative coordinates
angle brackets	floating	root page
architectural units	flyouts	scientific units
bearing	font	selection set
buttons	fractional units	session
Cartesian coordinate system	graphics cursor (crosshairs)	shape-compiled fonts
case-sensitive	graphics window	Standard toolbar
command aliasing	grid	status bar
command prompt	highlighted	surveyor units
command window	implied Crossing mode	title bar
context-sensitive	implied Windowing mode	toggle
coordinate values	instance	tool tip
default directory	menu bar	toolbars
default units	Object Properties toolbar	transparent command
docked	objects	UCS icon
drag	options	unidirectional text
drawing session	polar coordinates	Windows Control box
engineering units	real-world units	World Coordinate System (WCS)

Circle Center Radius	New Drawing	Save
Dynamic Text	OOPS	Snap
Erase	Multiline Text	Style
Grid	Quit	Undo
Help	Rectangle	Units
Limits	Redo	Zoom All
Line	Redraw All	

EXERCISES

Redraw the following shapes. If dimensions are provided, use them to create your drawing, showing the exact geometry of the part. The letter M after an exercise number means that the dimensions are in millimeters (metric units). If no letter follows the exercise number, the dimensions are in inches. The ∅ symbol indicates that the following dimension is a diameter. Do not include dimensions on the drawings.

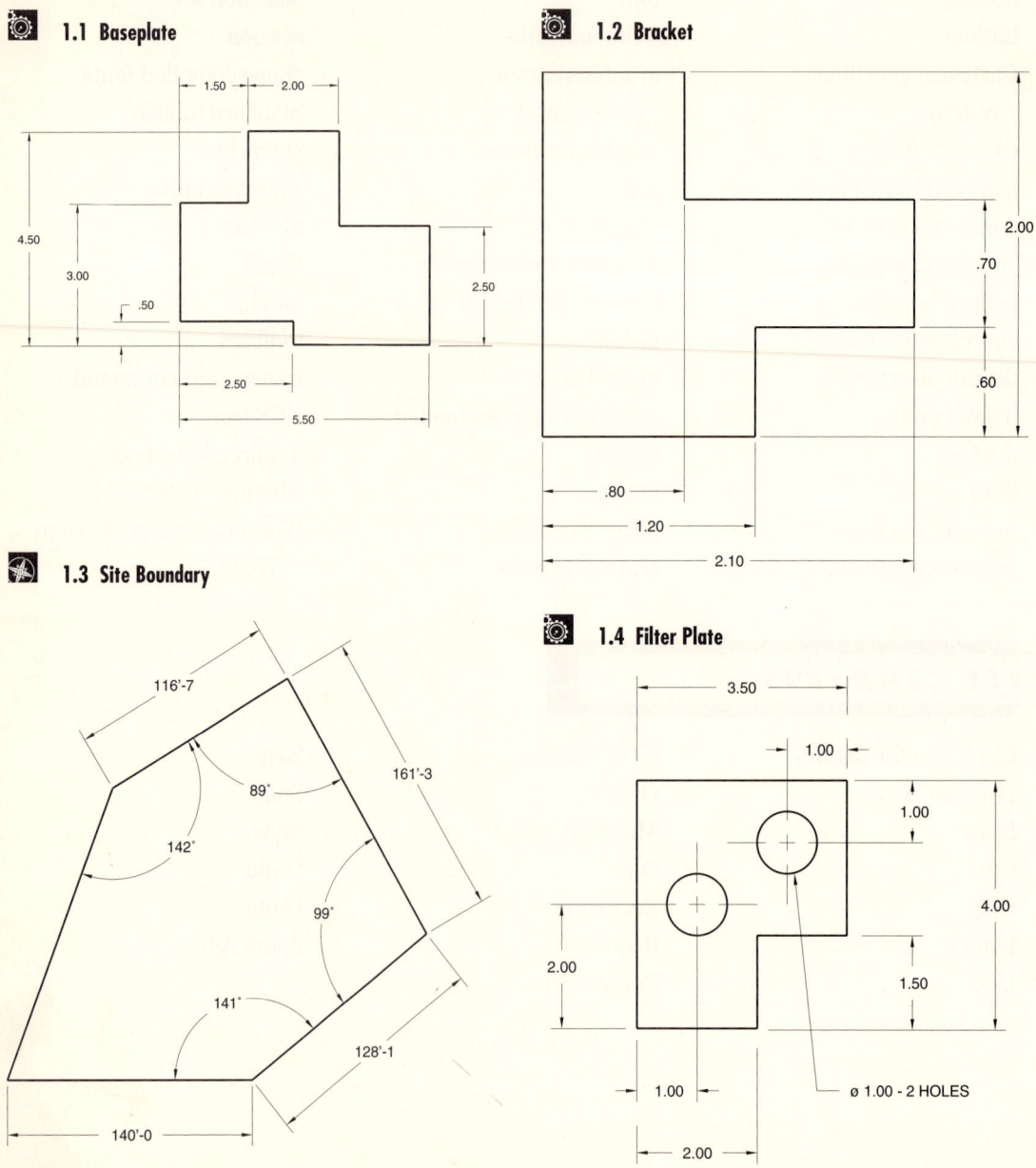

1.1 Baseplate

1.2 Bracket

1.3 Site Boundary

1.4 Filter Plate

1.5M Gasket

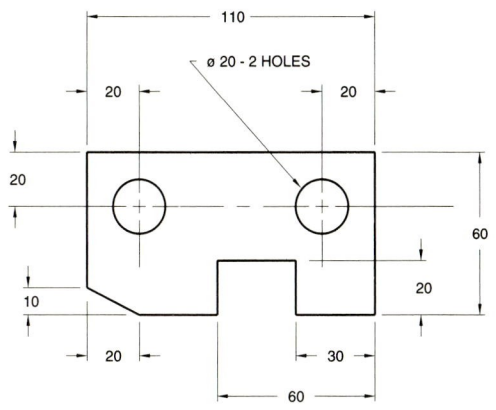

1.6 Spacer

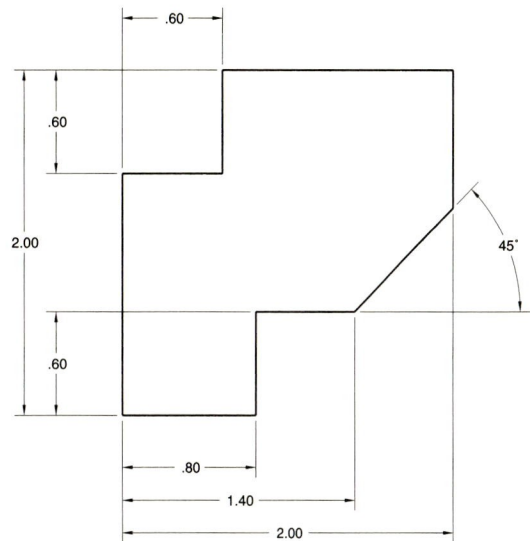

1.7M Guide Plate

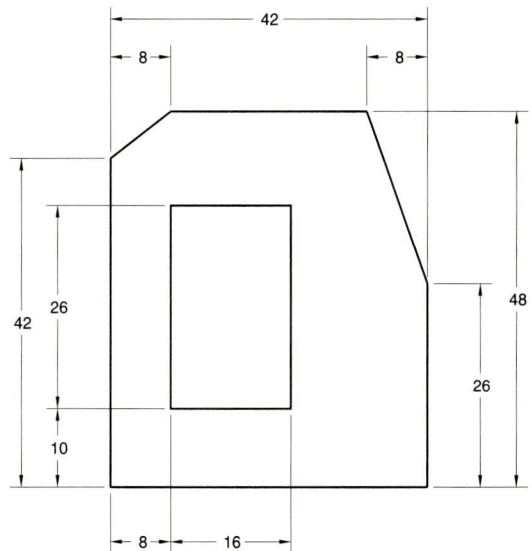

(4) PLATE - 1020 STEEL
2 REQUIRED - FULL SIZE

1.8 Amplifier

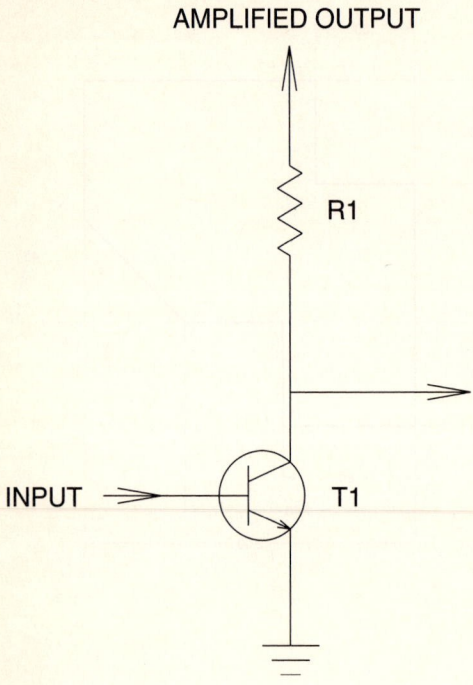

1.9 Dorm Room

Reproduce this drawing using Snap and Grid set to 0.25. The numerical values are for references only. Use Dtext to label the items.

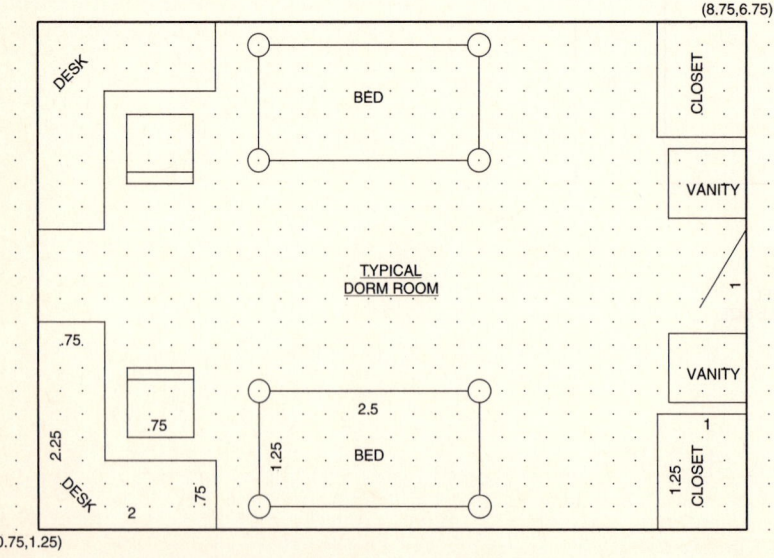

1.10 Template

Draw the shape using one grid square equal to 1/4" or 10mm.

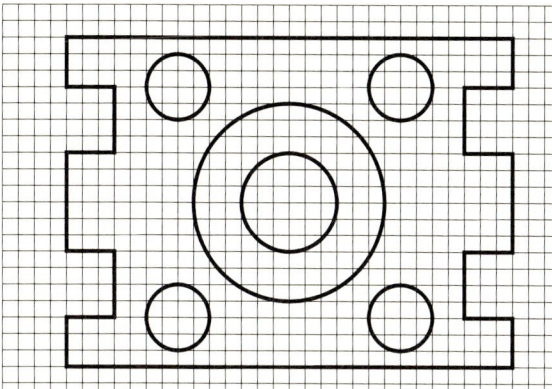

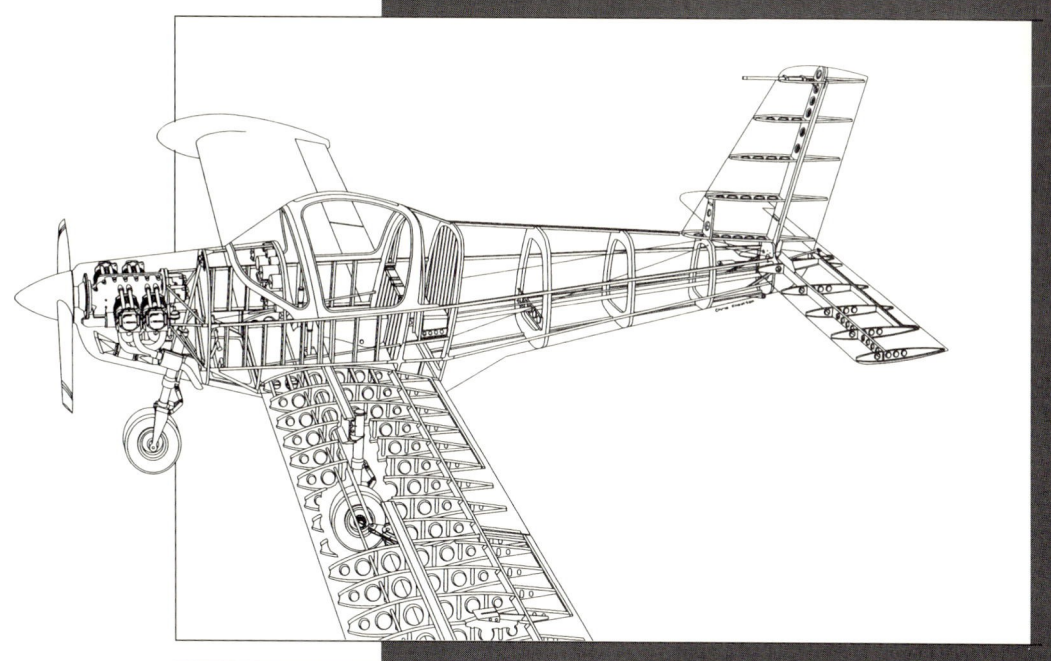

TUTORIAL 2

Basic Construction Techniques

Objectives

When you have completed this tutorial, you will be able to

1. Open existing drawings.

2. Work with existing layers.

3. Draw, using the Arc and Circle commands.

4. Set and use running object snaps.

5. Change the display, using Zoom and Pan.

6. Use the Aerial Viewer.

Introduction

You usually create technical drawings by combining and modifying several different basic shapes called *primitives*, such as lines, circles, and arcs, to create more complex shapes. This tutorial will help you learn how to use AutoCAD to draw some of the most common basic shapes. As you work through the tutorial, keep in mind that one of the advantages of using AutoCAD over drawing on paper is that you are creating an accurate model of the drawing geometry. In Tutorial 3, you will learn to list information from the drawing database. Information extracted from the drawing is accurate only if you created the drawing accurately in the first place.

Starting

Before you begin, launch AutoCAD Release 14. If you need assistance in loading AutoCAD, refer to the Getting Started section in this manual.

■ *Warning:* If the main AutoCAD screen, including the graphics window, Standard toolbar, Object Properties toolbar, status bar, command window, and menu bar, is not on your display screen, check to be sure that you have configured AutoCAD properly for your computer system. If you are still having difficulty, ask your technical support person for help. ■

Opening an Existing Drawing

This tutorial shows you how to add arcs and circles to the subdivision drawing provided with the data files that came with this manual. In Tutorial 3 you will finish the subdivision drawing so that the final drawing will look like Figure 2.1.

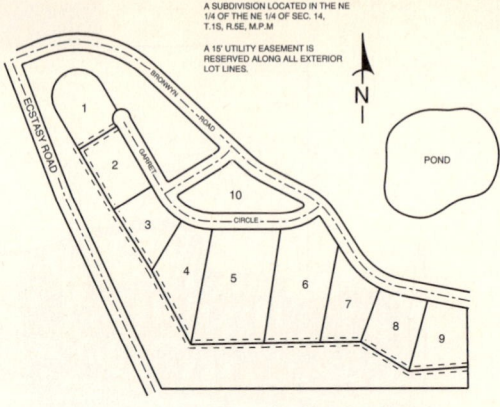

Wannabe Heights Estates

Figure 2.1

To open an existing drawing, use the File, Open selection on the menu bar or pick the icon that looks like an open folder from the Standard toolbar.

Pick: **Open icon**

The Select File dialog box appears on your screen. Use the center portions, which show the default directory and drive, to select the location where your data files have been stored. In the Getting Started section, you should have already created a directory called *c:\datafile*, and copied all the AutoCAD data files into it. If you have not done so, you may want to review Getting Started. If the correct directory is not showing, double-click on *c:* to move to the root directory. Use the scroll bars if necessary to scroll down the list of directories until you come to the appropriate one. The files are shown on the left side of the dialog box. Use the scroll bars to scroll down the list of files until you see the file named *subdivis.dwg*. When you select a file, a preview of the file appears in the box to the right. (AutoCAD Release 14 automatically creates the preview and saves it inside the drawing file when you save the drawing. You can use the Makepreview command to create a preview for drawings created in older releases.) Figure 2.2 shows the Select File dialog box and a preview of drawing *subdivis.dwg*.

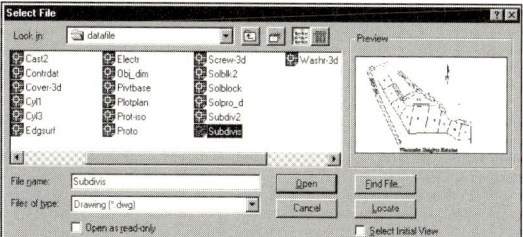

Figure 2.2

If you want to search through existing drawing files, you can use the Find File button near the bottom of the Select File dialog box.

Pick: **Find File**

The Browse/Search dialog box appears on your screen, as shown in Figure 2.3.

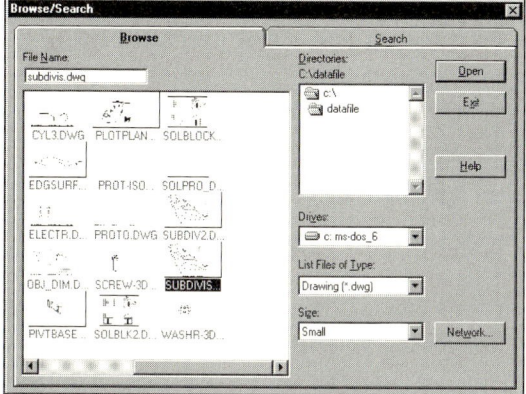

Figure 2.3

Use the directory tree at the right to select the directory where the files you want to view are. Figure 2.3 shows the *datafile* directory. When you have picked on the directory name, all the drawing files in the directory that can be previewed appear in the box to the left. If the images are difficult to decipher, you may change their size to medium or large by select-

ing the Size option located at the bottom right. Double-click on the preview picture of the file you want to open.

Pick: **(the c:\datafile directory from the directory list at the right)**

Double-click: **(on the picture of file subdivis.dwg)**

When you have opened the file, it appears on your screen, as shown in Figure 2.4. Note that it opens with its own defaults for Grid, Snap, and other features. These settings are saved in the drawing file. When you open a drawing, its own settings are used.

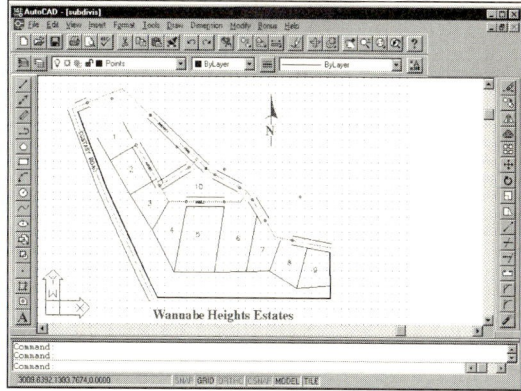

Figure 2.4

Saving as a New File

The Save As command allows you to save your drawing to a new file name. You can select this command from the File selection on the menu bar. Don't use the Save command, because that will save your changes into the original data file.

Pick: **File, Save As**

The Save Drawing As dialog box appears on your screen, similar to Figure 2.5. On your own, use it to select the drive and directory *c:\work* as shown and specify the name for your drawing, *subdivis.dwg*. The new file name is the same as the previous file name, but the directory is different. Thus a new copy of drawing *subdivis.dwg* is saved in the directory *c:\work*.

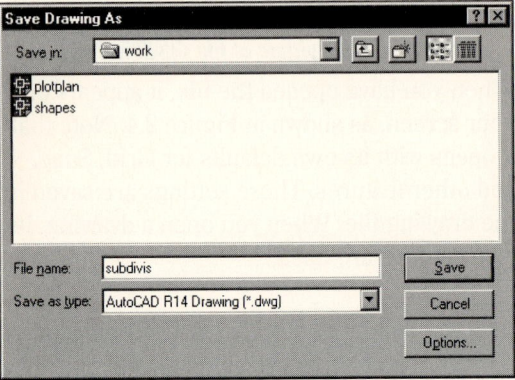

Figure 2.5

The original file *c:\datafile\subdivis.dwg* will remain unchanged on your drive. When you use the Save As command and specify a new file name, AutoCAD sets the newly saved file as current.

Using Layers

You can organize drawing information on different *layers*. Think of a layer as a transparent drawing sheet that you place over the drawing and that you can remove at will. The coordinate system remains the same from one layer to another, so graphical objects on separate layers remain aligned. You can create an unlimited number of layers within the same drawing. The Layer command controls the color and linetype associated with a given layer and allows you to control which layers are visible at any given time. Using layers allows you to overlay a base drawing with several different levels of detail (such as wiring or plumbing schematics over the base plan for a building).

By using layers, you can also control which portions of a drawing are plotted, or remove dimensions or text from a drawing to make it easier to add or change objects. You can also lock layers, making them inaccessible but still visible on the screen; you can't change anything on a locked layer until you unlock it.

Current Layer

The *current layer* is the layer you are working on. Any new objects you draw are added to the current layer. The default current layer is Layer 0. If you do not create and use other layers, your drawing will be created on Layer 0. You used this layer when drawing the plot plan in Tutorial 1. Layer 0 is a special layer provided in AutoCAD. You cannot rename it or delete it from the list of layers. Layer 0 has special properties when used with the Block and Insert commands, which are covered in Tutorial 8.

Layer POINTS is the current layer in *subdivis.dwg*. There can be only one current layer at a time. The name of the current layer appears on the Object Properties toolbar.

Controlling Layers

The Layer Control feature on the Object Properties toolbar is an easy way to control the visibility of existing layers in your drawing. You will learn more about creating and using layers in Tutorial 4. In this tutorial you will use layers that have already been created for you.

To select the Layer Control pull-down feature,

> *Pick: (on the layer name POINTS, which appears on the Object Properties toolbar near the top of the screen)*

The list of available layers pulls down, as shown in Figure 2.6. Notice the special layer 0 displayed near the top of the Layer Control list.

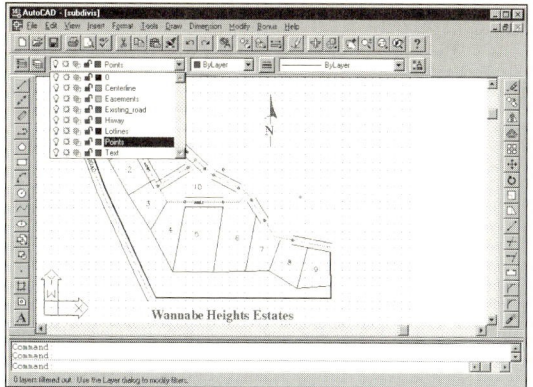

Figure 2.6

Pick: **(on the layer name CENTERLINE from the Layer Control list)**

It becomes the current layer listed on the Object Properties toolbar. Any new objects will be created on this layer until you select a different current layer. Your screen should look like Figure 2.7. Note the layer name CENTERLINE shown at the upper left.

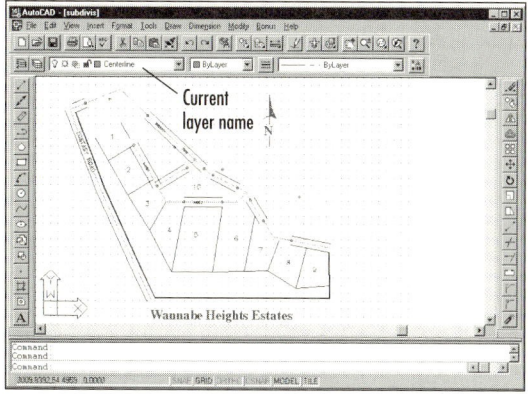

Figure 2.7

Use the Line command you learned in Tutorial 1 to draw a line off to the side of the subdivision drawing. Note that it is green and has a centerline linetype (long dash, short dash, long dash). The line you drew is on Layer CENTERLINE. Now, erase or undo the line on your own; don't add it to your drawing.

Controlling Colors

Each layer can have a color associated with it. Using different colors for different layers helps you visually distinguish different types of lines in the drawing. An object's color also controls which pens are used during plotting. You can use layers of different colors to make setting pen colors and widths easy during plotting.

AutoCAD provides two different ways of selecting the color for objects on your screen. The best way is to set the layer color and draw the objects on the appropriate layer. This method keeps your drawing organized. The other method is to select the Color Control feature on the Object Properties toolbar to the right of the Layer Control. To select the Color Control pull-down feature,

Pick: **(on the word ByLayer, which appears on the Object Properties toolbar)**

The list of available colors pulls down, as shown in Figure 2.8.

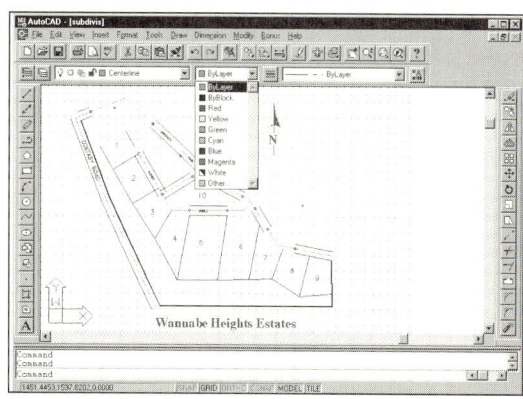

Figure 2.8

Note that the standard colors (yellow, red, green, blue, etc.) are shown. You can also select Other to view the full color palette.

Pick: **Other...**

The Select Color dialog box shown in Figure 2.9 appears on your screen, giving you a full range of colors to choose from.

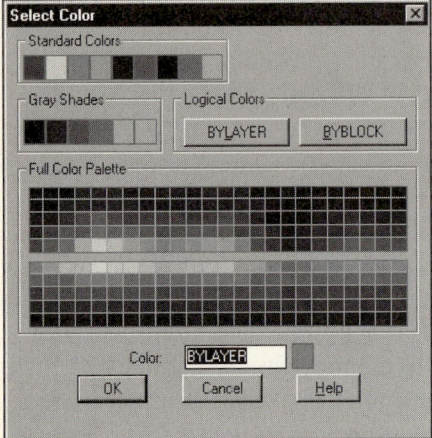

Figure 2.9

The default option for the Color (and also for the Linetype) command is BYLAYER. It's the best selection because, when you draw a line, the color and linetype will be those of the current layer. Otherwise, the color in your drawing can become very confusing. You will pick Cancel to exit the Select Color dialog box without making any changes. The colors for your new objects will continue to be determined by the layer they are created on.

Pick: **Cancel**

Linetype in Layers

Layers can have associated linetypes, as well as colors, as Layer CENTERLINE does. For example, you could create a layer that not only drew all lines in red, but also drew only hidden (short dashed) lines. You will learn how to load the linetypes and create new layers that use them in Tutorial 4.

Layer Visibility

One of the advantages of using layers in the drawing is that you can choose not to display selected layers. That way, if you want to create projection lines or even notes about the drawing, you can draw them on a layer that you will later turn off so that it isn't displayed or plotted. Or you may want to create a complex drawing with many layers, such as a building plan that contains the electrical plan on one layer and the mechanical plan on another, along with separate layers for the walls, windows, and so on. You can store all the information in a single drawing, and then you can plot different combinations of layers to create the electrical layout, first-floor plan, and any other combination you want. Next, use Layer Control to turn *off* Layer POINTS, *freeze* Layer TEXT, and *lock* Layer LOTLINES.

Pick: **(on Layer CENTERLINE, shown on the Object Properties toolbar)**

The list of layers pulls down. Refer to Figure 2.10 as you make the following selections.

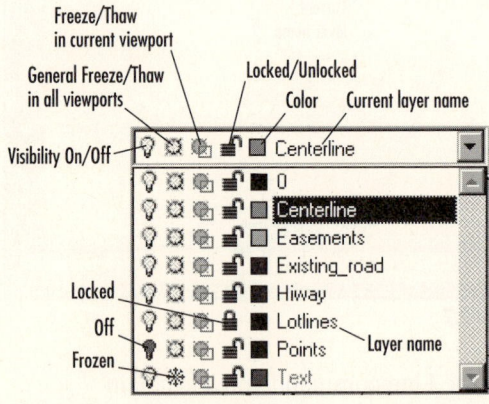

Figure 2.10

Pick: **(on the On/Off icon, which looks like a lightbulb, to the left of Layer POINTS)**

Pick anywhere away from the layer list. Note that the points have been turned off so that they no longer appear. Invisible (off) layers are not printed or plotted, but objects on these layers are still part of the drawing. Next you will freeze Layer TEXT.

Freezing Layers

Freezing a layer is similar to turning it off. You use the Freeze option not only to make the layer disappear from the display, but also to cause it to be skipped when the drawing is regenerated. This feature can noticeably improve the speed with which AutoCAD regenerates a large drawing. You cannot freeze the current layer because that would create a situation where you would be drawing objects that you couldn't see on the screen. The icon for freezing and thawing layers looks like a snowflake when frozen and a shining sun when thawed.

> Pick: *(the General Freeze/Thaw icon to the left of Layer TEXT)*

Pick anywhere in the graphics window for the selection to take effect. Layer TEXT is still on, but it is frozen and therefore invisible. A layer can both be turned off and frozen, but the effect is the same as freezing the layer.

Locking Layers

You can see a locked layer on the screen, and you can add new objects to it. However, you can't make changes to the new or old objects on that layer. This constraint is useful when you need the layer for reference but do not want to change it. For example, you might want to move several items so that they line up with an object on the locked layer but do not want anything on the locked layer to move. You will lock layer LOTLINES so that you cannot accidentally change the lines that are already on the layer.

> Pick: *(the Lock/Unlock icon to the left of Layer LOTLINES)*

Your screen should now be similar to Figure 2.11. Layer CENTERLINE is the current layer. Layer TEXT is frozen and does not appear. Layer POINTS is turned off and does not appear. Layer LOTLINES is locked so that you can see and add to it, but not change it. On your own, try erasing one of the lot lines. A message will appear, saying that the object is on a locked layer. The object won't be erased.

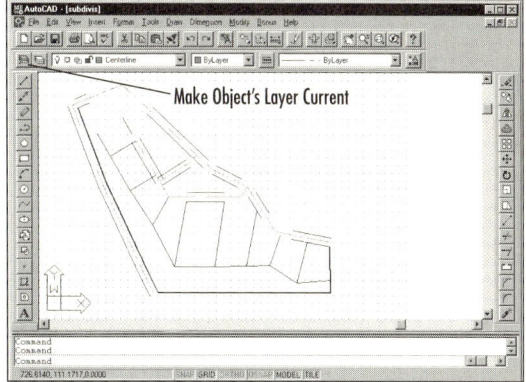

Figure 2.11

Making Object's Layer Current

 A new feature in AutoCAD Release 14 is the Make Object's Layer Current button, labeled in Figure 2.11. You will select a blue road line on your Subdivisions drawing.

> Pick: *(a blue line, representing the edge of the road, on the left-hand side of the Subdivisions drawing)*

Note that the layer name has changed to Existing_Road, the layer of the line you selected. This name is only temporary, for if you draw a new line, the layer name changes back to the current layer (Centerline). This is a powerful feature for you to use when you are unsure about which layer a particular object is on.

Now you will use the Make Object's Layer Current button.

Pick: **(the same blue line previously picked)**

Pick: **(the Making Object's Layer Current button located left of the Layer Control pull-down menu)**

The current layer is changed to Existing_Road, and any changes to the drawing will be done on this layer. Within the command prompt is the line: EXISTING_ROAD is now the current layer.

Now you will change the current layer back to Centerline.

Pick: **(the Layer Control pull-down list on the Object Properties toolbar)**

Pick: **CENTERLINE**

Next you will create the curved sections of the road centerline for the subdivision. The straight-line sections that the curves are tangent to have been drawn to get you started. First you will turn on the object snap for Node.

Using Object Snap

The object snap feature in AutoCAD accurately selects locations based on existing objects in your drawing. When you just pick points from the screen without using object snaps, the resolution of your screen makes it impossible for you to select points with the accuracy with which AutoCAD stores the drawing geometry in the database. You have learned how to pick an accurate point on the screen by snapping to a grid point. Object snap makes it possible for you to pick points accurately on your drawing geometry by snapping to an object's center point, endpoint, midpoint, and so on. Whenever AutoCAD prompts you to select a point, you can use an object snap to help make an accurate selection. Without this command, locating two objects with respect to each other in correct and useful geometric form is virtually impossible. Object snap is one of the most

important tools that AutoCAD provides. The many specialized object snaps are listed on the Object Snap flyout on the Standard toolbar, as shown in Figure 2.12. The tool tip for the default top tool on the Object Snap flyout says *Tracking*.

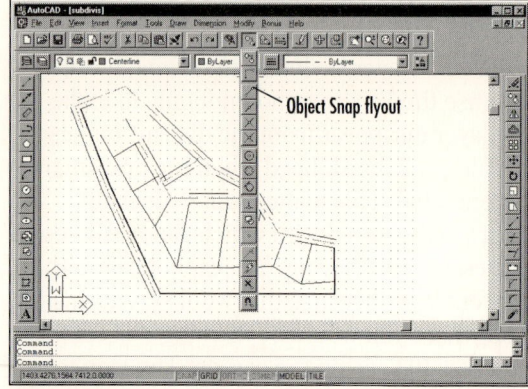

Figure 2.12

Object Snap can operate in two different ways. The first is called override mode. With this method, you select the object snap during a command. The object snap acts as a modifier within the command string to target the next point you select. You activate object snaps from within other commands by picking the appropriate icon from the Standard toolbar. The object locations they select are indicated by red dots on the icons. When you activate an object snap in this manner, it is active for one pick only. Remember, you can use this method only during a command that is prompting you to select points or objects.

■ *TIP* You can also activate an object snap by typing the three-letter name any time you are prompted to enter points or select objects. Refer to the Command Summary for the three-letter codes. ■

AutoCAD R14 has a special feature called AutoSnap™ when Object Snap is active. It will display a marker and description (SnapTip) when the cursor is placed near or on a snap point. This feature helps you to determine what location on the object will be selected.

The second method for using Object Snap is called running mode. With this method, you turn on the object snap and leave it on before you invoke any commands. When a running mode object snap is on, the marker box and SnapTip will appear during any future command when you are prompted for a point location, object selection, or other pick. The SnapTip will tell you which object snap location is being targeted.

 Now you will turn on the running mode object snap. Use the Standard toolbar to select Object Snap Settings from the Object Snap flyout.

Pick: **Object Snap Settings icon**

The Osnap Settings dialog box appears on your screen. It is broken into two index cards, Running Osnap and AutoSnap.

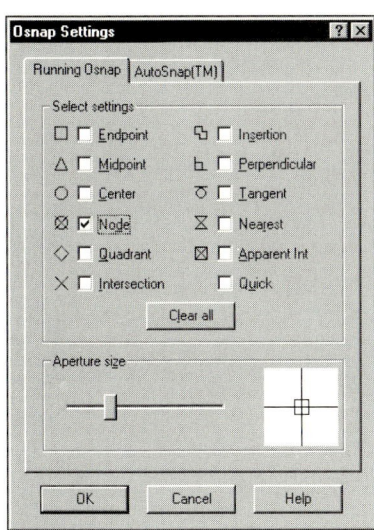

Figure 2.13

Pick: **Node** *(on the Running Osnap card)*

A check appears in the box when it is selected, as shown in Figure 2.13. Node snaps to objects drawn with AutoCAD's Point command. The figure located next to the Node setting represents the AutoSnap marker shape that will appear in the drawing. The Aperture Size feature on the bottom of the dialog box resizes the cursor aperture, which is activated when AutoSnap is turned off.

Next, choose the AutoSnap index card to familiarize yourself with its options. As Figure 2.14 shows, you can change Marker size, color, and select settings. You may want to make the Marker box smaller or larger, depending on your drawing's complexity and size. Try using the slider to make the box smaller and then larger. Next, you may want to change the Marker color to a more noticeable color, such as red. When you have finished, pick OK.

AutoSnap is automatically turned on when you start a new drawing or open an old drawing, and it is recommended that you leave it that way. AutoSnap is a very useful tool and you will use it throughout the rest of this manual.

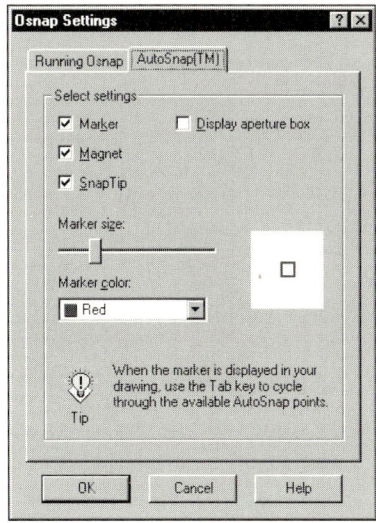

Figure 2.14

Notice that the OSNAP button on the Status bar is now highlighted, meaning it is active. You may double-click the OSNAP button to toggle the use of object snap on or off, similar to the SNAP, GRID, and ORTHO buttons.

Now you are ready to start creating arcs at accurate locations in the drawing. When you are prompted to select, look for the marker box on the crosshairs. When it's there, you know that Snap to Node is being used.

Use the Layer Control button on the Object Properties toolbar to turn Layer POINTS back on. Do so now, before continuing. Be sure that CENTERLINE is the current drawing layer.

Using Arc

The Arc command is on the Draw toolbar, on the Draw menu, or you can type *ARC* at the command line. There are eleven different ways to create arcs; however, only one icon is on the Draw toolbar, the 3 Point Arc. To see the other ten options, choose Draw, Arc from the Menu.

Each Arc command option requires that you input point locations. You can define those point locations by manually typing in the coordinate values or locating the points with the cursor and pressing the pick button on your

pointing device. For the exercises presented in this tutorial, follow the directions carefully so that your drawing will turn out correctly. Keep in mind that if you were designing the subdivision, you might not use all the command options demonstrated in this tutorial. When you are using AutoCAD later for design, select the command options that are appropriate for the geometry in your drawing.

Arc 3 Points

The 3 Points option of the Arc command draws an arc through three points that you specify. The red dots located on the icon represent the three points of definition. This means that three point locations will be necessary for drawing that arc. Remember, to specify locations you can pick them or type in absolute, relative, or polar coordinates. You will draw an arc using the 3 Points option; refer to the points defined on Figure 2.15 as you are selecting. The Snap to Node running object snap and AutoSnap feature will help you pick the points drawn in the data file.

Pick: **Arc icon**

Center/<Start point>: **(select point 1, using the AutoSnap marker as shown in Figure 2.15)**

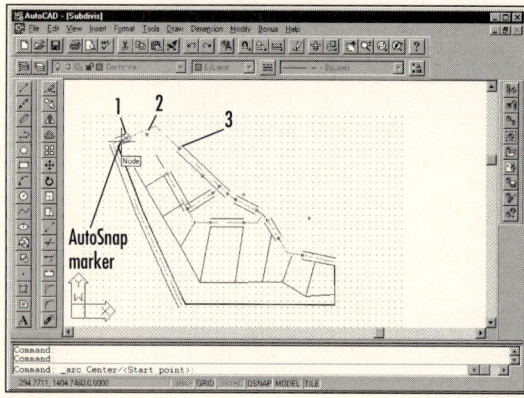

Figure 2.15

As can be seen in Figure 2.15, the AutoSnap marker for Node appears when the cursor is near a node point.

Now you will continue with selecting the points.

Center/End/<Second point>: ***(select point 2)***

AutoCAD enters drag mode, whereby you can see the arc move on the screen as you move the cursor. Many AutoCAD commands permit dynamic specification, or dragging, of the image on the screen.

Move the cursor around the screen to see how it affects the way the arc would be drawn. Recall your use of this feature to draw circles in Tutorial 1.

Endpoint: ***(select point 3)***

The third point defines the endpoint of the arc. The radius of the arc is calculated from the locations of the three points. Your drawing should now show the completed arc, as shown in the upper part of Figure 2.16.

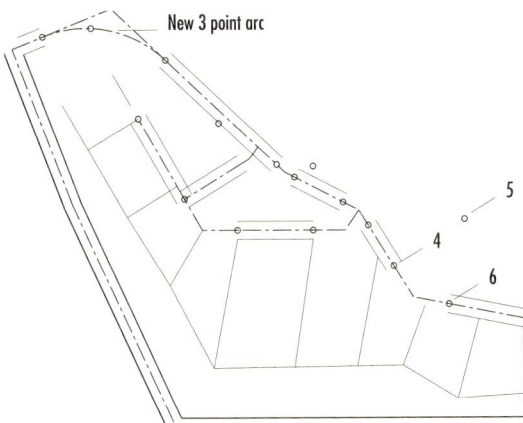

Figure 2.16

Arc Start Center End

Next, you will draw an arc by specifying the start, center, and endpoints.

Pick: **Draw, Arc, and Start Center End**

Use points 4–6 in Figure 2.16 to create this arc. At the prompt

Center/<Start point>: ***(select point 4)***

AutoCAD prompts you for the center point.

Center: ***(select point 5)***

The distance between the start point and the center point will be used as the radius for the arc. You must specify the endpoint location to define the end of the arc. You are in drag mode.

Angle/Length of chord/<Endpoint>: ***(select point 6)***

An arc is always drawn counterclockwise from the start point. Thus you need to define correctly the start point and the center point. Figure 2.16 shows the point locations needed to draw a concave arc. If the start point was located where the endpoint is, a convex arc outside the centerlines would have been drawn. When you have added the arc correctly, your arc should look like that in the lower part of Figure 2.17.

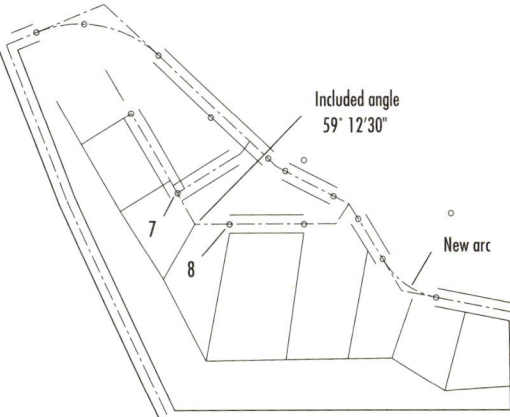

Figure 2.17

On your own, try drawing another arc with the Arc Start Center End option, this time picking point 6 first, then the center, and then point 4. Note that this arc is drawn counterclockwise, resulting in a convex arc. Undo this backward arc by typing *U* ↵ at the command prompt.

■ **Warning:** You can type the option letters at the command prompt to select an arc creation method. Pressing the spacebar, the right mouse button, or ↵ to repeat the previous command will invoke the Arc command. ■

Arc Start End Angle

Arc Start End Angle draws an arc through the selected start and endpoints by using the *included angle* you specify. To draw an arc with the Start End Angle option, refer to Figure 2.17.

> *Pick:* **Draw, Arc, and Start End Angle**
>
> Center/<Start point>: *(select point 7)*

Next, specify the endpoint for the arc.

> Endpoint: *(select point 8)*

The arc is defined by the included angular value (often called the *delta angle* in survey drawings) from the start point to the endpoint. Positive angular values are measured counter-clockwise. Negative angular values are measured clockwise. (Type *d* for ° when typing in surveyor angles. Use the single quote and double quote for minutes and seconds.)

> Included angle: **59d12'30"** ↵

When you have drawn the arc, your screen should look like Figure 2.18.

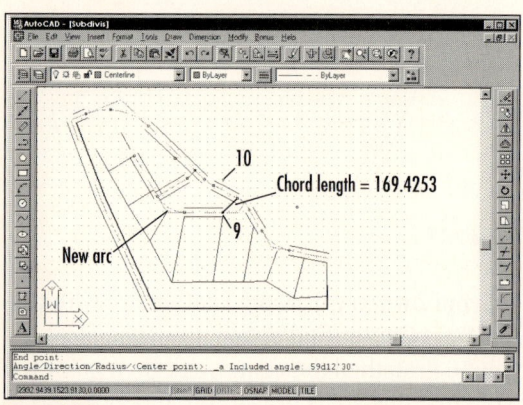

Figure 2.18

Arc Start Center Length

Arc Start Center Length draws an arc specified by the start and center points of the arc and the *chord length*. The chord length is the straight-line distance from the start point of the arc to the endpoint of the arc. You can enter negative values for the chord length to draw an arc in the opposite direction. Figure 2.19 shows what to input to create this type of arc.

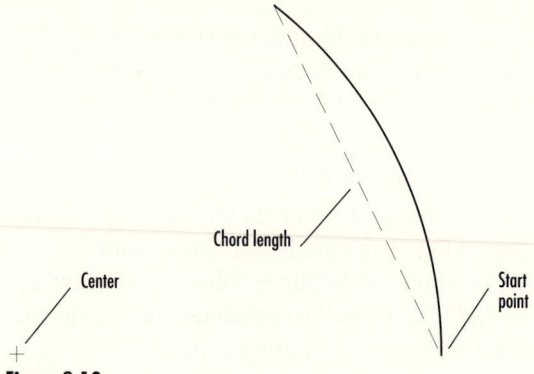

Figure 2.19

You will draw the next arc by using Arc Start Center Length. Refer back to Figure 2.18 for the points to select and the chord length to use.

> *Pick:* **Draw, Arc, and Start Center Length**
>
> Center/<Start point>: *(select point 9)*
>
> Center: *(select point 10)*
>
> Length of chord: **169.4253** ↵

Your new arc should look like the one shown in Figure 2.20.

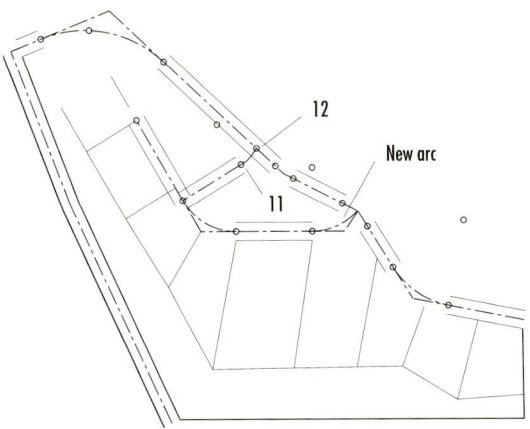

Figure 2.20

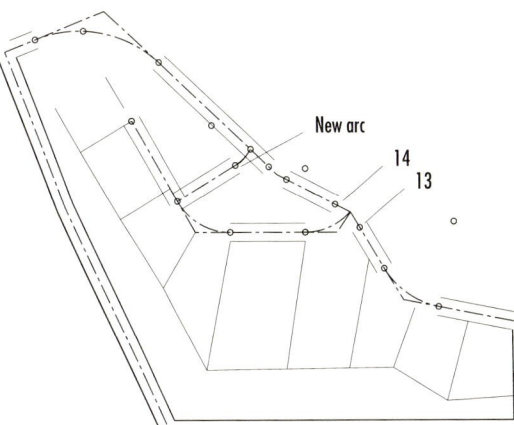

Figure 2.21

Arc Start End Radius

Draw the next arc by using Arc Start End Radius. The locations of the start and endpoints, 11 and 12, are shown in Figure 2.20.

Pick: **Draw, Arc, and Start End Radius**

Center/<Start point>: *(select point 11)*

Endpoint: *(select point 12)*

Radius: **154.87** ⏎

■ *TIP* You can use negative radius values to create a *major arc*. A major arc comprises more than 180 degrees of a circle. ■

Your drawing should be similar to that in Figure 2.21.

Continuing an Arc

Arc Continue allows you to join an arc to a previously drawn arc or line. You will give it a try off to the side of the subdivision drawing and then erase or undo it, as it is not a part of the drawing. To draw an arc that is the continuation of an existing line,

Pick: **Line icon**

and draw a line anywhere on your screen. Then,

Pick: **Draw, Arc, and Continue**

The last point of the line becomes the first point for the arc. AutoCAD is in drag mode, and an arc appears from the end of the line. AutoCAD prompts you for an endpoint.

Endpoint: *(select an endpoint)*

Next you will use Arc Start Center End to draw another arc. Then you will continue an arc from the endpoint of that arc.

Pick: **Draw, Arc, and Start Center End**

Center/<Start point>: *(select a start point)*

Center: *(select a center point)*

Angle/Length of chord/<Endpoint>: *(select an endpoint)*

Pick: **Draw, Arc, and Continue**

Your drawings should look similar to those in Figure 2.22.

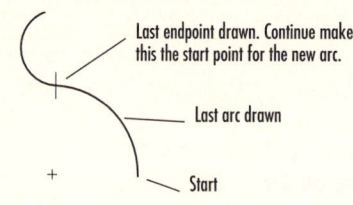

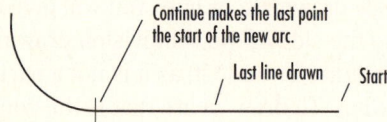

Figure 2.22

On your own, erase or undo the extra arcs and lines that you created.

Quick Selections for Arc

To select the Arc command, you can also type the command alias, *A*, at the command prompt. When you start the Arc command this way, the default for drawing the arc is the 3 Points option you used earlier in this tutorial. If you want to use another option, you can type the command option letter at the prompt. You will start the Arc command by typing its alias. You will then specify the start, end, and angle of the arc by typing the command option letters at the prompt. Refer to Figure 2.21 for the points, 13 and 14, to select.

Command: **A** ⏎

The command Arc is echoed at the command prompt, followed by the prompt,

Center/<Start point>: **(pick point 13)**

Center/End/<Second point>: **E** ⏎

Endpoint: **(pick point 14)**

Angle/Direction/Radius/<Center point>: **A** ⏎

Included angle: **30d54'04"** ⏎

The arc is added to your drawing. However, it may be hard to see because of the small size of the arc, relative to the size of the screen and drawing. AutoCAD provides Zoom commands to change the size of the image on your display.

Using Zoom

The Zoom flyout is located on the Standard toolbar, as shown in Figure 2.23.

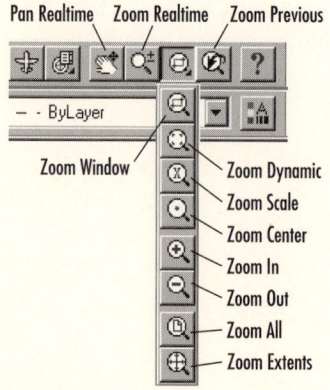

Figure 2.23

To zoom in on an area of the drawing, you can select the Zoom In icon from the Zoom flyout.

Pick: **Zoom In icon**

Your drawing is enlarged to twice its previous size on the screen so that you can see more detail. Your screen should be similar to Figure 2.24.

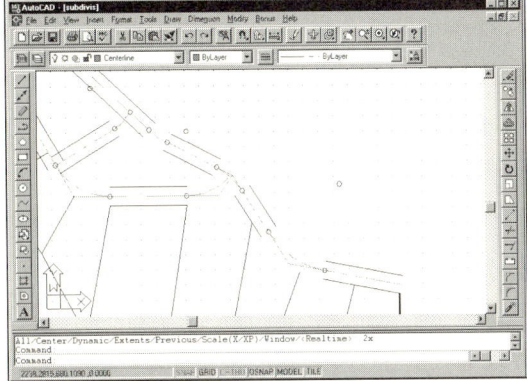

Figure 2.24

 Now you will zoom back out, using the Zoom Out selection from the Zoom flyout.

Pick: **Zoom Out icon**

Your drawing should return to its original size on the screen. The Zoom In and Zoom Out options use the Scale feature of the Zoom command to zoom to a scale of 2X (twice the previous size) and 0.5X (half the previous size).

Zooming Using Scale Factors

 You can also use *scale factors* to zoom when you pick the Scale option of the Zoom command. Scale factor 1.00 shows the drawing limits. Scale

factor 0.5 shows the drawing limits half-size on the screen. Typing *X* after the scale factor makes the zoom scale relative to the previous view. For example, entering *2X* causes the new view to be shown twice as big as the view established previously, as you saw when using Zoom In. A scale factor of 0.5X reduces the view to half the previous size, as you saw when using Zoom Out. Zoom Scale uses the current left corner or (0,0) coordinates as the base location for the zoom. Typing *XP* after the scale factor makes the new zoom scale relative to *paper space*. A scale factor of 0.5XP means that the object will be shown half-size when you are laying out your sheet of paper. You will learn more about paper space in Tutorial 5.

For now, you will select the Zoom command by typing its alias at the command prompt.

Type: **Z** ⏎

All/Center/Dynamic/Extents/Previous/Scale(X/XP)/
    Window/<Realtime>: **2X** ⏎

The objects are enlarged on the screen to twice the size of the previous view. Repeat the command.

Command: ⏎

All/Center/Dynamic/Extents/Previous/Scale(X/XP)/
    Window/<Realtime>: **.5** ⏎

The drawing limits appear on the screen at half their original size. The area shown on the screen is twice as big as the drawing limits. Now, restore the original view.

Command: ⏎

All/Center/Dynamic/Extents/Previous/Scale(X/XP)/
    Window/<Realtime>: **1** ⏎

Zoom Center lets you specify a center point before you enter the zoom scale factor. You will learn more about this option in Tutorial 9.

Zoom Window

You can use Zoom Window to create a window around the area that you want to enlarge to fill the screen. This method lets you quickly enlarge the exact portion of the drawing that you are interested in seeing in detail. You will select this command by using the icon on the Zoom flyout.

Pick: **Zoom Window icon**

The Zoom command is echoed at the command prompt. Note that the Zoom command has an apostrophe in front of it in the command prompt area. Its presence means that the command is transparent and that you can select it during execution of another command. You will zoom in on the area shown in Figure 2.25. To create the window, you will first select a point at the top left corner of the area to be zoomed and then select a point on the diagonal in the lower right corner, as shown in Figure 2.25.

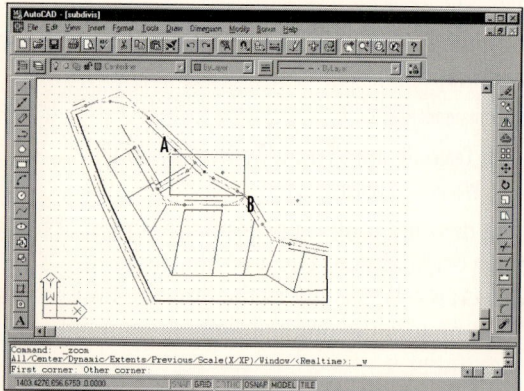

Figure 2.25

First corner: *(select point A)*
Other corner: *(select point B)*

The defined area is enlarged to the full screen size, as shown in Figure 2.26.

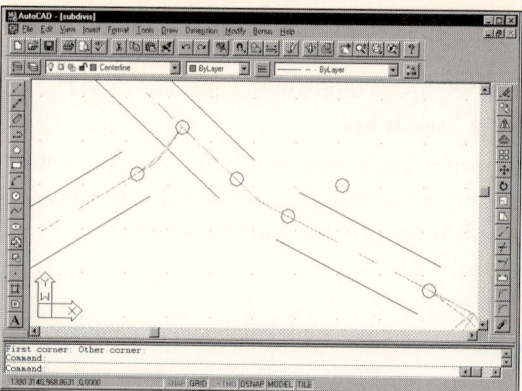

Figure 2.26

Zoom Previous

To return an area to its previous size, you will pick Zoom Previous from the Standard toolbar.

Pick: **Zoom Previous icon**

Your drawing is returned to its original size. Areas can be repeatedly zoomed, that is, you can zoom in on a zoomed area; in fact, you can continue to zoom until the portion shown on the display is ten trillion times the size of the original.

Zoom Realtime

An easy way to zoom your drawing to the desired size is to use the Zoom Realtime feature on the Standard toolbar.

Pick: **Zoom Realtime icon**

Select an arbitrary point in the middle of your drawing and hold down the pick button of your pointing device while dragging the cursor up and down. When you move the cursor upward, AutoCAD zooms in closer to the drawing; when you move the cursor downward, the image zooms out.

Using the Aerial Viewer

 To allow you to zoom in and move around in a large drawing quickly, AutoCAD provides the Aerial Viewer. To select the Aerial Viewer, you will pick the icon that looks like an airplane from the Standard toolbar.

Pick: **Aerial View icon**

The Aerial Viewer appears on your screen, as shown in Figure 2.27. If the Aerial Viewer window is not as large as you would like, you can pick on its corners and drag to increase its size. You can move it around on your screen by picking on its title bar and moving it to a new location while holding down the pick button.

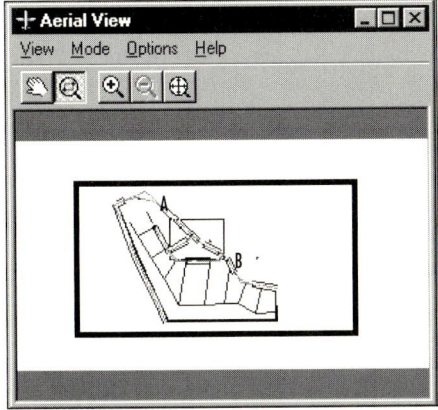

Figure 2.27

■ *Warning:* If Fast Zoom is turned off within the Viewres command, you will get a message that the Aerial Viewer is not available. The Aerial Viewer does not work when you are viewing in Perspective (DView) or shaded. ■

You will use the crosshairs to form a window in the Aerial Viewer around the area of the drawing you would like to enlarge on the screen.

Pick: **(point A in Figure 2.27)**

Pick: **(point B)**

Once you have selected point B, the area enclosed in the window in the Aerial Viewer is enlarged to fill the graphics window, as shown in Figure 2.28. The thick border around the area in the Aerial Viewer shows which portion is the current view. The Aerial Viewer remains on your screen, available for continued use.

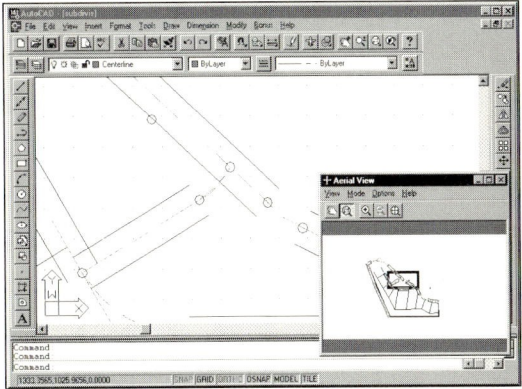

Figure 2.28

 Next, you will select the Pan option from the toolbar in the Aerial Viewer. Be sure to pick the Pan button from the Aerial Viewer toolbar and not the Pan command from the Standard toolbar, as this command is also located there. The two commands have the same effect: moving the view while maintaining the same zoom factor. However, the two versions are implemented differently. Using the Pan icon from the Aerial Viewer toolbar, you can easily position the area of the screen you want to view.

Pick: **Pan icon**

The Pan icon on the Aerial Viewer toolbar is highlighted. Note that now, instead of crosshairs that move around the Aerial Viewer screen to let you select, a box appears. Position the box over the area shown in Figure 2.29 to move the view at the same enlargement to the new area. Then,

Press: (the pick button)

The view on your screen moves to show the area now enclosed in the Aerial Viewer box.

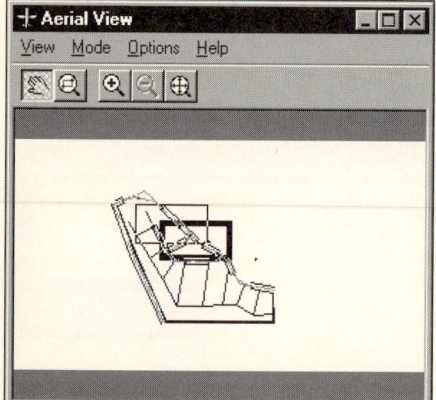

Figure 2.29

Zoom All

 Zoom All returns the drawing to its original size by displaying the drawing limits, or displaying the drawing extents (all of the drawing objects), whichever is larger. Select the Zoom All icon from the Zoom flyout.

Pick: **Zoom All icon**

> ■ *TIP* If you open a drawing that you've saved with the view zoomed in, you can use Zoom All to return to the original drawing limits. ■

The drawing should return to its original size, that is, as it was before you began the Zoom command. Experiment on your own with the other options of the Zoom command and read about them on the Help screen. Zoom Dynamic works much the same way as the Aerial Viewer. Using Zoom Dynamic, you position and size a window on the area of the drawing you want to zoom. When you have the window positioned and sized the way you want it, press the return button or ⏎ to select that area to zoom in on.

The Vmax option (an icon and command prompt option in AutoCAD Release 13) is still available and must be typed at the zoom command prompt. Vmax lets you zoom out as far as possible without causing AutoCAD to *regenerate* the drawing. AutoCAD uses the *virtual screen*, a file from which the screen display is created. If you try to zoom out to areas that are not calculated in the current virtual screen file, AutoCAD will have to regenerate the drawing to calculate the display for the new area. Regenerating can take quite a while on a complex drawing, so being able to zoom out without causing a regeneration is useful.

Using Pan Realtime

 As you saw when you used the Aerial Viewer, the Pan command lets you position the drawing view on the screen without changing the zoom factor. Unlike the Move command, which moves the objects in your drawing to different locations on the coordinate system, the Pan command does not change the location of the objects on the coordinate system. Rather, your view of the coordinate system and the objects changes to a different location on the screen.

Pick: **Pan Realtime icon**

You can now choose a location central to your drawing by holding down the pick button and dragging the drawing around the screen. The drawing should move freely about the drawing area until you let go of the mouse pick button, at which point it will remain stationary. Press [ESC] or ↵ to exit the command.

Using Circle 2 Point, Circle 3 Point, and Circle Tangent Tangent Radius

In Tutorial 1 you learned how to use the Circle command by specifying a center point and a radius value. You can also use the Circle command to draw circles by specifying any two points (Circle 2 Point), any three points (Circle 3 Point), or two tangent references and a radius (Circle Tan Tan Radius).

Before adding the next two circles, you will turn on the Endpoint running mode object snap so that the circles you create will line up exactly with the ends of the existing lot lines.

Pick: **Object Snap Settings icon**

Pick: **(to turn on Endpoint)**

Pick: **OK**

The Endpoint and Node running mode object snaps are turned on. When you see the AutoSnap marker appear on the endpoint of a line, the crosshairs will snap to the marker point. If you see the Node marker, then move the crosshairs until the Endpoint you want is highlighted.

On your own, pick the Layer Control from the Object Properties toolbar. When the list of layers pulls down, pick on the layer name LOTLINES to set it as the current layer before continuing. That ensures that the next circle you create is on layer LOTLINES. It is still locked so you can't change any of the objects already on it, but you can create new objects on this layer. If you need to make corrections, you will have to unlock the layer first.

Circle 2 Point

First you will draw a circle by using Circle 2 Point. Select two points (1 and 2) to be the endpoints of the circle's diameter, as shown in Figure 2.30.

Pick: **Draw, Circle, and 2 Points**

First point on diameter: **(select point 1)**

Because you are in drag mode, you can see the circle being created on your screen as you move the cursor.

Second point on diameter: **(select point 2)**

The circle has been defined by the endpoints you selected for the diameter of the circle, which also are the endpoints of lot lines. Your screen should look like Figure 2.30.

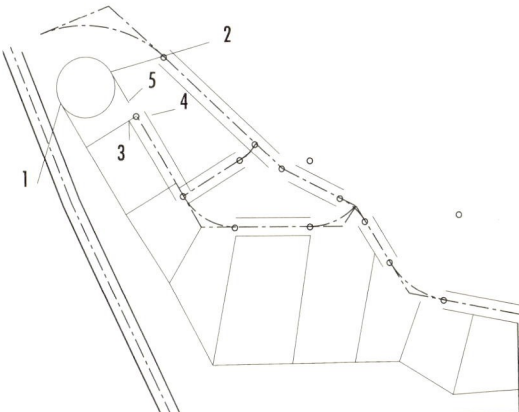

Figure 2.30

Circle 3 Point

To draw a circle using Circle 3 Point, specify any three points on the circle's circumference. Refer to Figure 2.30 for the points (3, 4, and 5) to select.

Pick: **Draw, Circle, and 3 Points**

First point: **(select point 3)**

Second point: **(select point 4)**

Because you are in drag mode again, you can see the circle being created on your screen as you move the cursor.

Third point: (select point 5)

The three points on its circumference have defined the circle. Your screen should look like Figure 2.31.

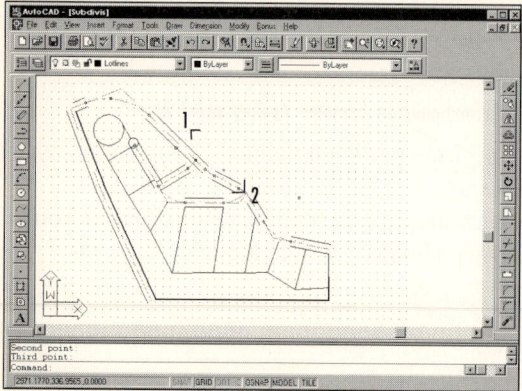

Figure 2.31

You will use Circle Tangent Tangent Radius to draw a circle tangent to two angled centerlines. First, zoom up the area between points 1 and 2, shown in Figure 2.31.

Pick: Zoom Window icon

First corner: (pick point 1)

Other corner: (pick point 2)

The area should be enlarged on your screen, as shown in Figure 2.32.

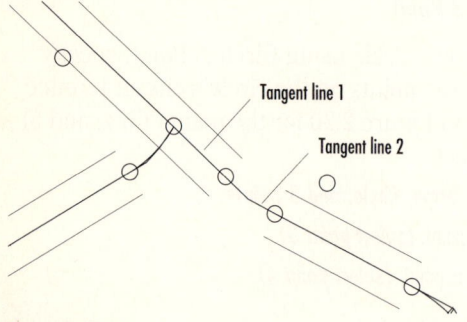

Figure 2.32

Now you will turn off the running object snaps by double-clicking on the OSNAP button on the Status bar. The button should now be grayed out, meaning that all running mode object snaps are now temporarily turned off. You should do so because sometimes object snaps can interfere with the selection of points and the operation of certain commands. As the Circle Tangent Tangent Radius option for drawing circles uses the Tangent object snap, be sure that other object snaps are turned off.

Because the next circle you draw is to be created on Layer CENTERLINE, set that layer current at this time. On your own, use the Object Properties toolbar to select the Layer Control list. Pick on the layer name CENTERLINE to make it current.

Circle Tangent Tangent Radius

Circle Tangent Tangent Radius requires that you specify two objects to which the resulting circle will be tangent and the radius of the resulting circle. This method is frequently used in laying out road centerlines. It involves selecting the two straight sections of the road centerline to which the curve is tangent and then specifying the radius.

Pick: Draw, Circle, and Tan Tan Radius

Note that the AutoSnap marker appears any time the cursor is near a line. Refer to Figure 2.32 for the lines to select. You can select either line first.

Enter Tangent spec: (pick line 1)

Enter second Tangent spec: (pick line 2)

Radius <34.0000>: 267.3098 ⏎

A circle with a radius of 267.3098 is drawn tangent to both original lines, as shown in Figure 2.33.

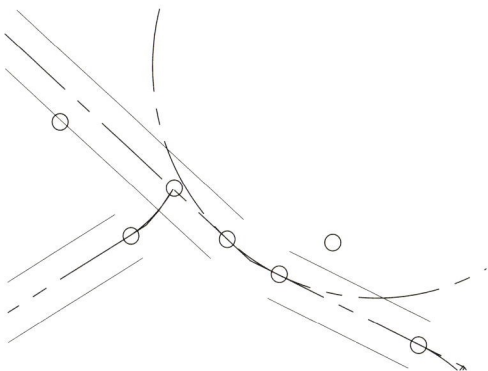

Figure 2.33

You can also use Circle Tangent Tangent Radius to draw a circle tangent to two circles.

Draw two circles off to the side of the subdivision drawing on your own, using Circle 2 Point for the first and Circle 3 Point for the second. Then draw a circle tangent to them.

Pick: **Draw, Circle, and Tan Tan Radius**

Enter Tangent spec: *(select one of the circles)*

Enter second Tangent spec: *(select the other circle)*

Radius<267.3098>: **150** ⏎

A circle with a radius of 150 is drawn tangent to both circles.

> ■ *TIP* If you get the message, *Circle does not exist*, the radius you specified may be too small or too large to be tangent to both lines. ■

Erase the extra circles on your own. Use Zoom All to return to the original view. Save your drawing at this time.

> ■ *TIP* If you have a copy of AutoCAD Release 12 or 13, this latest release allows you to save your drawing to those formats. In the Save Drawing As dialog box, scroll through the Save Drawing Type list for the various options. ■

When you have successfully saved your drawing,

Pick: **File, Exit**

You are now back at the Windows desktop. You will finish the subdivision drawing in Tutorial 3.

KEY TERMS

aperture	layer	regenerate
chord length	lock	running mode
current layer	major arc	scale factors
delta angle	override mode	target area
freeze	paper space	virtual screen
included angle	primitives	

KEY COMMANDS

Aerial Viewer	Circle 3 Point	Snap to Node
Arc 3 Points	Circle Tan Tan Radius	Save As
Arc Continue	Color Control	Zoom In
Arc Start Center End	Layer Control	Zoom Out
Arc Start Center Length	Open Drawing	Zoom Previous
Arc Start End Angle	Pan Realtime	Zoom Realtime
Arc Start End Radius	Running Object Snap	Zoom Scale
Circle 2 Point	Snap to Endpoint	Zoom Window

Redraw the following shapes. If dimensions are given, create your drawing geometry exactly to the specified dimensions. The letter M after an exercise number means that the dimensions are in millimeters (metric units). If no letter follows the exercise number, the dimensions are in inches. Do not include dimensions on the drawings. The ∅ symbol means diameter; R indicates a radius.

2.1M Bracket

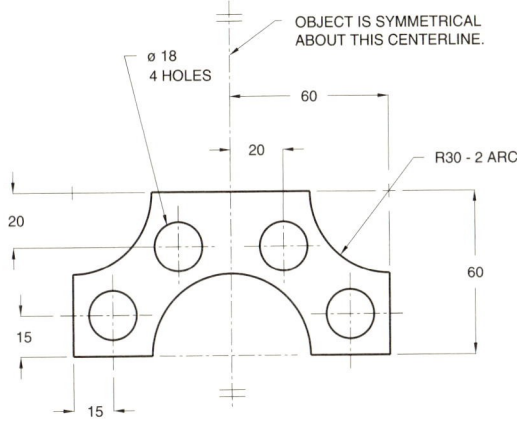

2.2 Gasket

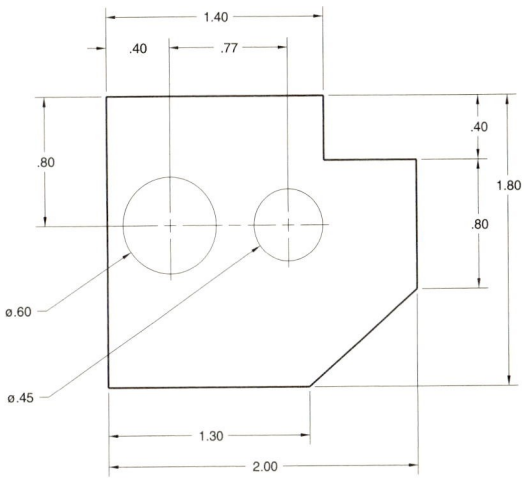

2.3M Flange

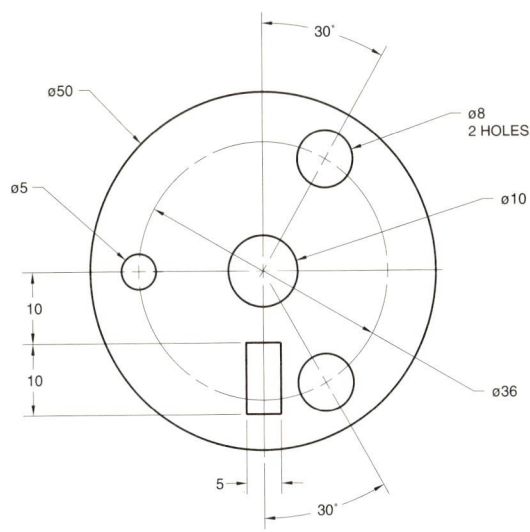

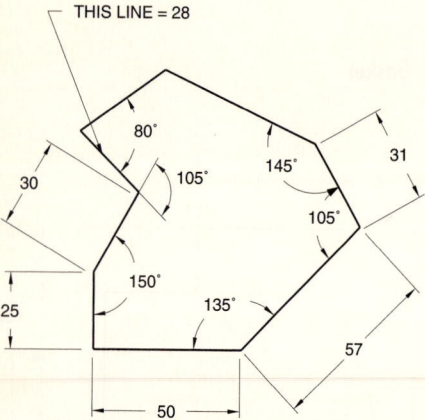

✦ 2.4M Puzzle

Draw the figure shown according to the dimensions provided. From your drawing determine what the missing distances must be (hint: use DIST).

✦ 2.5 Plot Plan

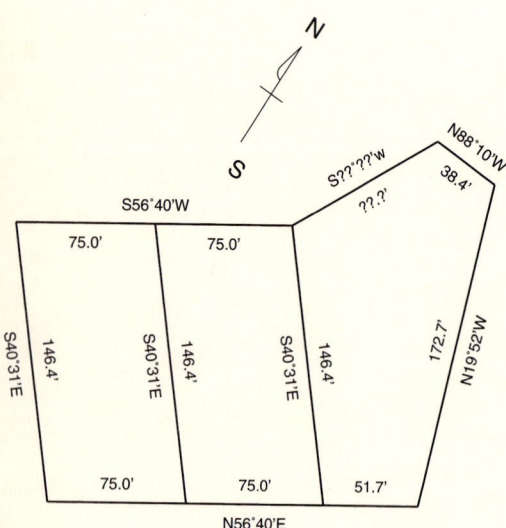

 **2.6 Power Supply**

Draw this circuit, using the techniques you have learned.

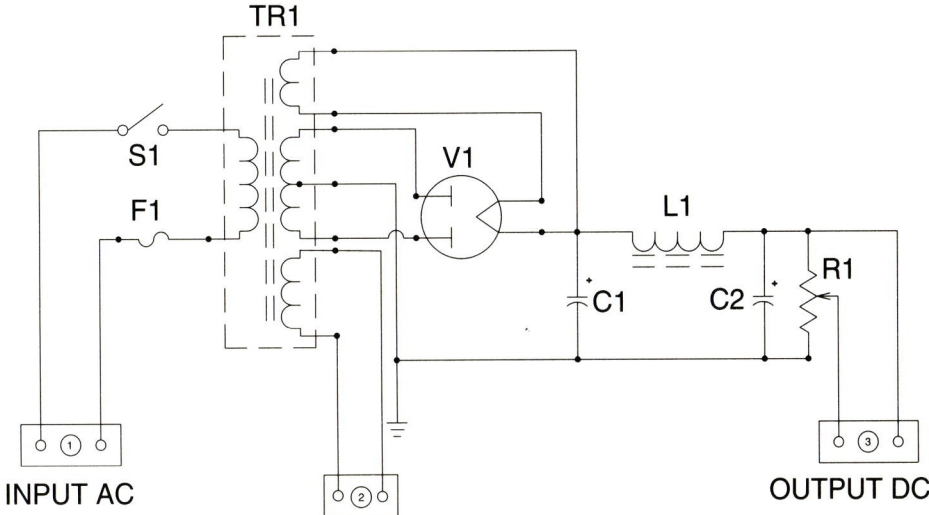

TR1

S1

F1

V1

L1

C1   C2   R1

INPUT AC

OUTPUT DC

POWER SUPPLY

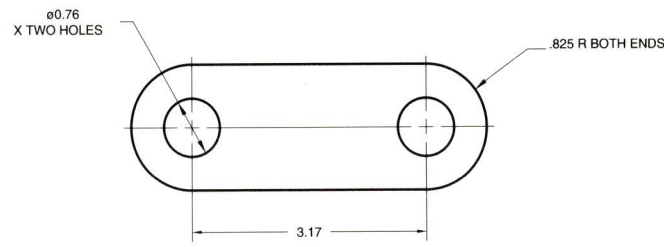 **2.7 Link**

ø0.76
X TWO HOLES

.825 R BOTH ENDS

3.17

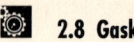 **2.8 Gasket**

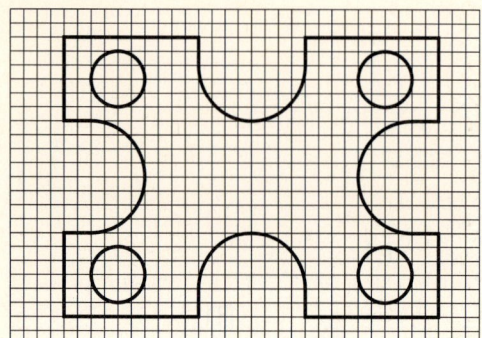

2.9 Shape

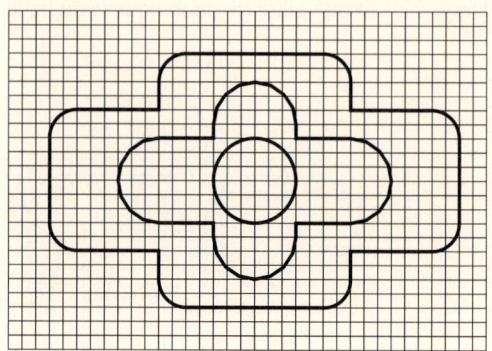

2.10 Lever Crank

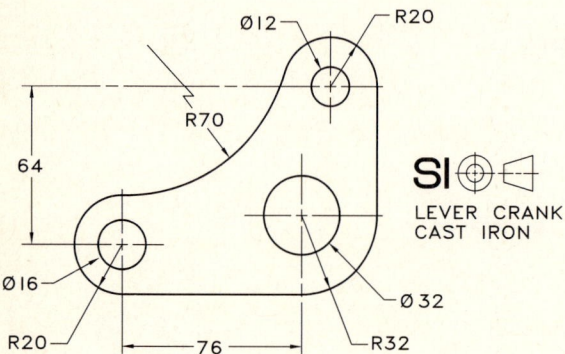

SI ⊕◁
LEVER CRANK
CAST IRON

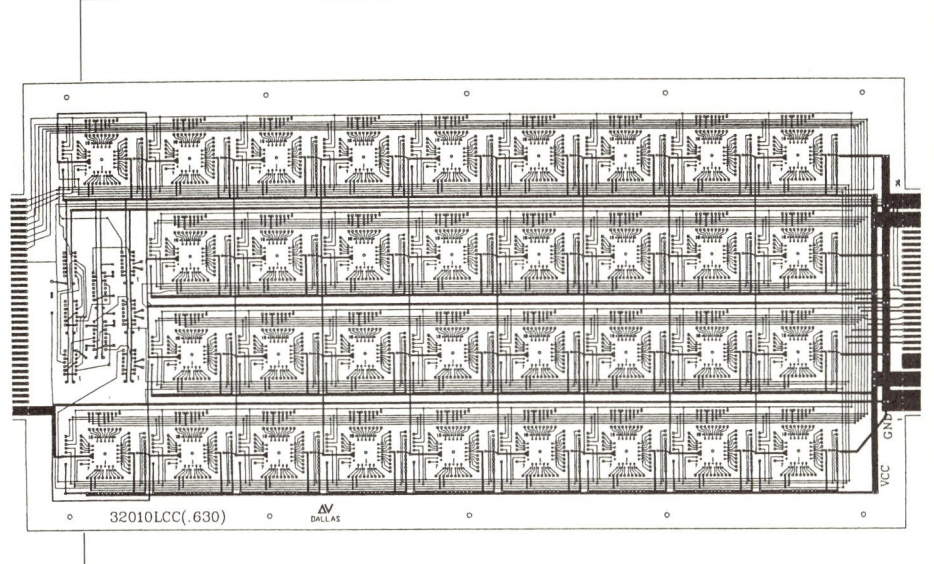

32010LCC(.630)

△V
DALLAS

Basic Editing and Plotting Techniques

Objectives

When you have completed this tutorial, you will be able to

1. Modify your drawing, using the Fillet, Chamfer, Offset, and Trim commands.

2. Create and edit polylines and splines.

3. List graphical objects, locate points, and find areas from your drawing database.

4. Change properties of drawing objects.

5. Create multilines and multiline styles.

6. Print or plot your drawing.

Introduction

In this tutorial you will learn how to modify some of the basic shapes that you've drawn to create a wider variety of shapes required for technical drawings. As you work through the tutorial, keep in mind that one of the advantages of using AutoCAD over drawing on paper is that you are creating an accurate model of the drawing geometry. You will see how to use this accurate drawing database to find areas, lengths, and other information. In later tutorials, you will also learn to use basic two-dimensional (2D) shapes as a basis for three-dimensional (3D) solid models.

In this tutorial you will finish drawing the subdivision that you started in Tutorial 2. A drawing has been provided with the data files that accompany this manual. You will start from it, in case the drawing that you created in Tutorial 2 has some settings that are different from the data file. You will trim lines and arcs, and add fillets, multilines, and text to finish the subdivision drawing. When you've finished the drawing, you will plot your final result.

Starting

Before you begin, launch AutoCAD Release 14 from the Windows Desktop. If you need assistance in loading AutoCAD, refer to the Getting Started section.

Starting from a Template Drawing

Template drawings are used as the basis for a new drawing without changing the original template drawing file. This method is useful when you have standard AutoCAD settings, preferences, and drawn objects that you want to use continuously. Files that can be used as

templates include regular drawing files (*.dwg) and drawing template files (*.dwt). There is really no difference between them except for the file extension name.

In order to continue with the subdivision drawing, you will start a new file, using the drawing file *subdiv2.dwg* as a template.

The Start Up dialog box should be showing at this time. You will select the Use a Template button and More Files... selection from underneath the Select a Template heading.

Pick: **Use a Template**

Select: **More Files...**

Pick: **OK**

The Select template dialog box should be showing, similar to that in Figure 3.1. AutoCAD offers a list of Templates (including standard Title Blocks) to choose from in its Template folder.

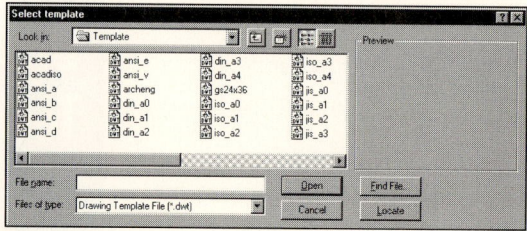

Figure 3.1

For now you will choose the Find File... button, opening the Browse/Search dialog box as you learned in Tutorial 2. You will change the List Files of Type to show *.dwg files, and change the Directories until it displays the drawings located in the *c:\datafile* folder. The dialog box should look similar to Figure 3.2.

Find and select drawing *subdiv2.dwg* from the Browse window.

Pick: **Open**

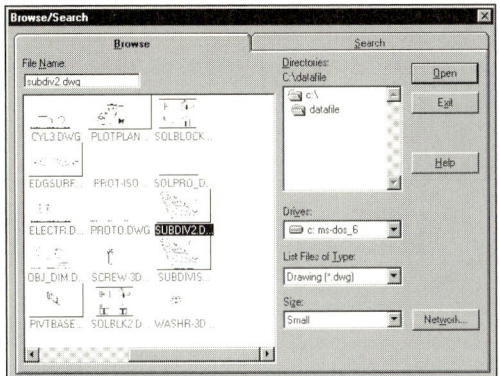

Figure 3.2

You should see the AutoCAD default name "Drawing" appear as the name of your AutoCAD drawing, meaning that the actual *subdiv2.dwg* will not be changed. On your own, save the drawing under a new name in your *c:\work* directory. For the new drawing name,

Type: **MYSUBDIV**

The drawing on your screen should be similar to that shown in Figure 3.3.

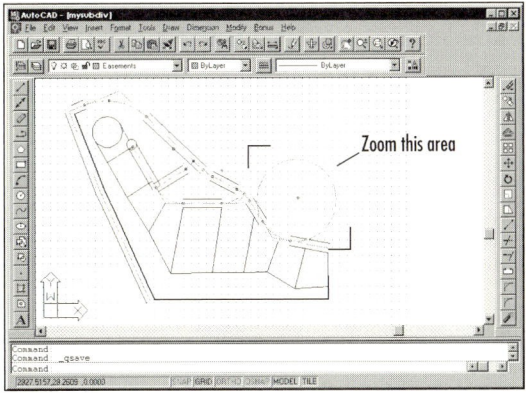

Figure 3.3

On your own, use Zoom Window to enlarge the view of the area indicated in Figure 3.3. When you have finished zooming, your screen will look like Figure 3.4.

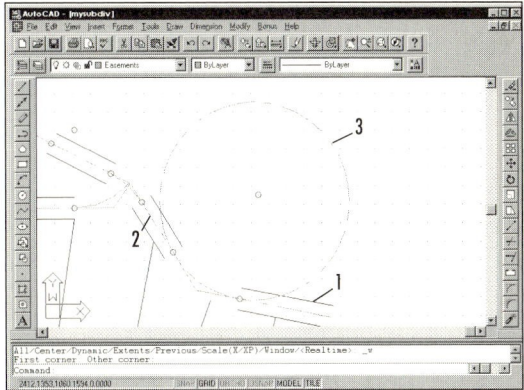

Figure 3.4

Using Trim

The Trim icon is located on the Modify toolbar. Trim removes part of an object in two steps. First, AutoCAD prompts you to select the objects that you will use as cutting edges. The cutting edges are drawing objects that you will use to cut off the portions that you want to trim. The cutting edge objects selected must cross the objects that you want to trim at the point at which you want to trim. Press ↵ to indicate that you have finished selecting cutting edges and want to begin the second step. After you press ↵, AutoCAD prompts you to select the portions of the objects that you want to trim. Pick on the portions you want to remove by using the Trim command.

The Project option of the Trim command gives you three choices for the projection method used by the command.

View   Trims objects where they intersect, as viewed from the current viewing direction

None   Trims objects only where they intersect in 3D space

UCS   Trims objects where they intersect in the current User Coordinate System

The Edge option lets you decide whether to trim objects only where they intersect in 3D space or where they would intersect if the edge were extended. The Undo option lets you undo the last trim without exiting the command; it's similar to the Undo option you used with the Line command. The Project and Edge options are very useful when you are working with 3D models as you will be in Tutorial 9.

You will pick the Trim command from the Modify toolbar and use it to remove the excess portion of the circle. Refer to Figure 3.4.

> *Pick:* **Trim icon**
>
> Select cutting edges: (Projmode = UCS, Edgemode = No extend)
>
> Select objects: *(select lines 1 and 2)*

The cutting edges selected will be highlighted. Note that the command line reported the projection mode set to UCS and the edge mode set to no extension.

> Select objects: ⏎

You have finished selecting the cutting edge. Next, select the portion of the circle to be removed.

> <Select object to trim>/Project/Edge/Undo: *(pick on the circle near point 3)*

You will press ⏎ to end the command,

> <Select object to trim>/Project/Edge/Undo: ⏎

When you are done, your figure should be similar to Figure 3.5.

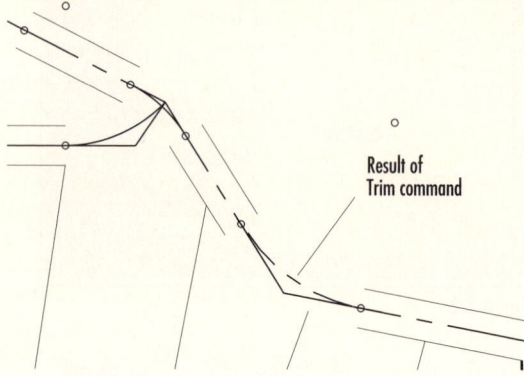

Result of Trim command

Figure 3.5

Refer to Figure 3.6 to understand how the cutting edge works in relation to the object to be trimmed.

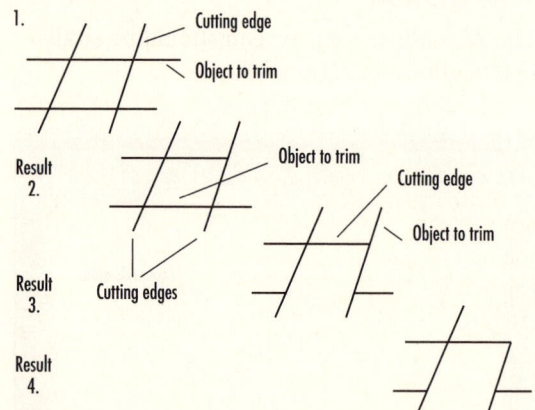

1. Cutting edge Object to trim

Result 2. Object to trim Cutting edge Object to trim

Result 3. Cutting edges

Result 4.

Figure 3.6

When lines come together, as at a corner, you can select one or both of the lines as cutting lines, as shown in Figure 3.6, parts 1 and 3. Again, the cursor location determines which portion of the line is removed.

On your own, restore the previous zoom factor by selecting Zoom Previous from the Standard toolbar. Next, unlock layer LOTLINES so that you can make changes to the objects on that layer. Currently, the layer is locked, so you cannot trim the lines and circles on layer LOT-LINES. Use the Layer Control list, which pulls down from the Object Properties toolbar, to unlock layer LOTLINES on your own.

Note that the current layer showing is EASE-MENTS and that the trims you accomplished affected the lines on the CENTERLINE layer, even though it was not the current layer. Using the Make Object's Layer Current icon that you learned about earlier, change the current layer to CENTERLINE. When you have finished these steps, your screen should look like Figure 3.7.

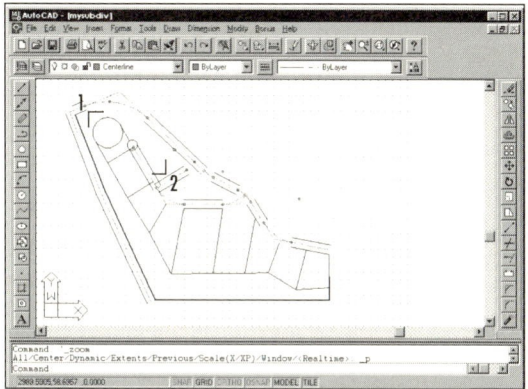

Figure 3.7

Next, use Zoom Window to enlarge the area shown on your own. Refer to Figure 3.7 and pick near point 1 for the first corner of the window and near point 2 for the other corner. When you have finished this step your screen should look like Figure 3.8.

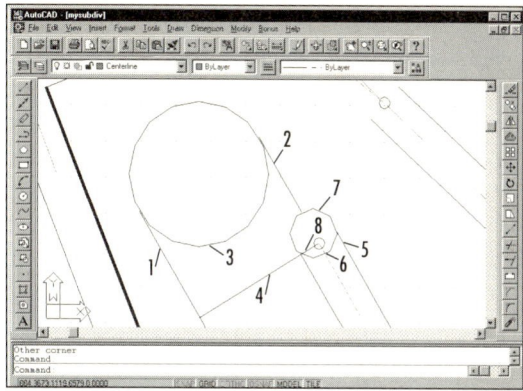

Figure 3.8

> ■ *TIP* When you zoom in on an area, circles may not be shown correctly. On the screen, AutoCAD draws circles by approximating them with a number of straight-line segments. The number of segments used may look fine when the circle first appears on the screen, but when you zoom in on it, it may appear jagged. To eliminate this poor appearance, you can type *REGEN* at the command prompt to cause the display to be recalculated from the drawing database and redisplayed on the screen correctly. ■

Next, you will trim the excess portions of the circles that you created with the 2 Point and 3 Point options. The arcs that remain when you have finished trimming will form the lot line and the cul-de-sac for the road.

Pick: **Trim icon**

Select cutting edges: (Projmode = usc, Edgemode = No extend)

Select objects: *(select lines 1 and 2)*

Select objects: ⏎

<Select object to trim>/Project/Edge/Undo: *(select on the circle near point 3)*

<Select object to trim>/Project/Edge/Undo: ⏎

Command: ⏎ *(to restart the Trim command)*

Select objects: *(select lines 4 and 5)*

Select objects: ⏎

<Select object to trim>/Project/Edge/Undo: *(pick on the circle near point 6)*

<Select object to trim>/Project/Edge/Undo: ⏎

Command: ⏎ *(to restart the Trim command)*

Select objects: *(select the circle near point 7)*

Select objects: ⏎

<Select object to trim>/Project/Edge/Undo: *(pick on the line near point 8)*

<Select object to trim>/Project/Edge/Undo: ⏎

When you have finished trimming the circles, your screen should look like Figure 3.9. If it doesn't, redraw on your own.

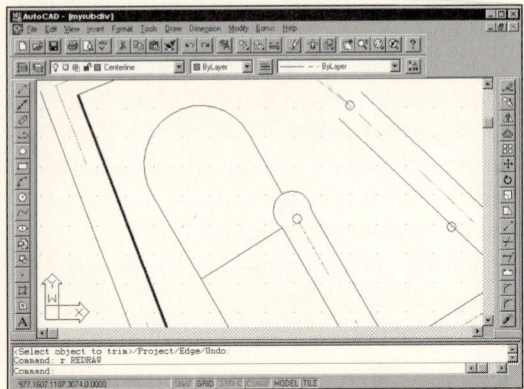

Figure 3.9

On your own, use Zoom Previous to restore the previous zoom factor.

Using Offset

 The Offset command creates a new object parallel to a given object. The Offset command is on the Modify toolbar. You will use Offset to create parallel curves 30 units from either side of the curved centerlines that you drew in Tutorial 2.

Pick: **Offset icon**

To offset an object, you need to determine the *offset distance* (the distance away from the original object) or the *through point* (the point through which the offset object is to be drawn). To specify the offset distance indicated in Figure 3.10,

Offset distance or Through <Through>: **30** ⏎

Select object to offset: *(select curve 1)*

Side to offset? *(pick a point below the curve, like point A)*

Select object to offset: *(select curve 1)*

Side to offset? *(pick a point above the curve, like point B)*

Once you have defined the offset distance, AutoCAD continues to repeat the prompt *Select object to offset:*, allowing you to create additional parallel lines that have the same spacing.

On your own repeat the steps just described to create lines offset 30 units from either side of the remaining curved centerlines so that your drawing looks like that shown in Figure 3.10. If necessary, use the Zoom commands to enlarge the smaller curves so that you can see them better on your screen for making selections. Use Zoom Previous to return to the previous view. Remember that the Zoom commands (with the exception of Zoom All) can be transparent so that you can select them during another command without canceling that command.

If you select the wrong item to offset, either erase the incorrect lines when you have finished or press [Esc] to cancel the command and then start again. When you have finished using the Offset command, press Return or ⏎ to end the command. Note that the new objects created by the Offset command are on the same layer as the object that you selected to offset. Which layer is current doesn't matter; the newly offset object will always be on the same layer as the original object that you selected.

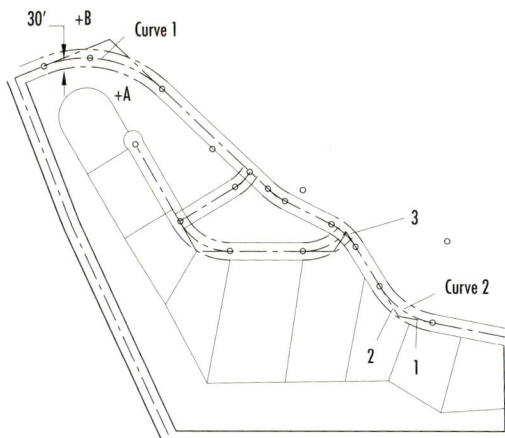

Figure 3.10

Trimming the Remaining Lines

Refer to Figure 3.10 and, on your own, select curve 2 as the cutting edge. Be sure to press ⏎ or the return button when you have finished selecting cutting edges. Trim off the portions of the centerline by picking the points labeled 1 and 2, which extend past the cutting edge. On your own repeat this process and trim all remaining centerlines where they extend beyond the centerline arcs. Note that line 3 doesn't extend past any cutting edges, so it cannot be trimmed. On your own use the Erase command to remove it.

Next, on your own zoom in on the center area of the drawing so that you can trim the excess lines where the roadways join. The center portion of the drawing will be enlarged on your screen.

This time, instead of selecting one cutting edge at a time, you will use implied Crossing to select all the lines in that area as cutting edges. Refer to Figure 3.11 to make your selections.

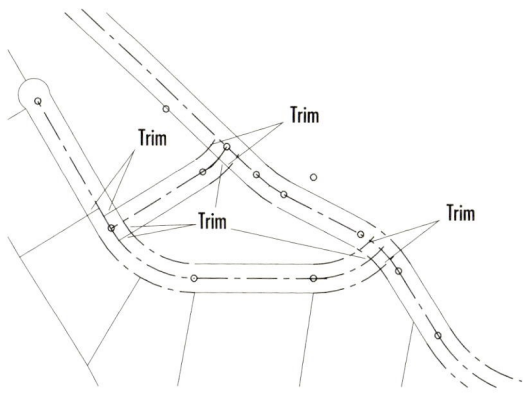

Figure 3.11

Pick: **Trim icon**

Select cutting edges: (Projmode = ucs, Edgemode = No extend)

Select objects: *(pick the first corner of a window near the bottom right of the screen. Be sure not to select any lines or objects.)*

Other corner: *(pick the second corner of the window, at the upper left of the screen)*

All objects that crossed the selection box become highlighted, indicating that you selected them as cutting edges.

Select objects: ⏎

<Select object to trim>/Project/Edge/Undo: *(select the extreme ends of the lines you want to remove and continue doing so until all the excess lines are trimmed)*

<Select object to trim>/Project/Edge/Undo: ⏎

> ■ *TIP* If you pick something to trim by mistake, you can type the option letter *U* (for Undo) and press ⏎ at the prompt. This restores the last object trimmed. ■

When you have finished trimming, use Zoom All to show the entire drawing on the screen. If you have some short line segments left that no longer cross the cutting edges, use Erase to remove them on your own. Your drawing should now look like that shown in Figure 3.12.

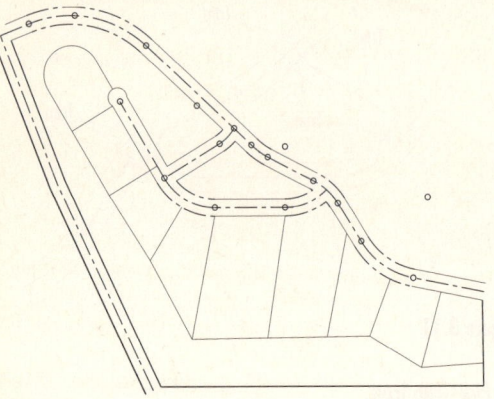

Figure 3.12

Changing Object Properties

Once you've drawn an object in AutoCAD, changing the properties of the object or a group of objects may be useful. Next, you will change the properties of the centerlines that you offset so that they are on the layer LOT-LINES. The new offset lines are the edges of the road and should be on the same layer as the other lot lines.

A quick way to change objects to another layer is to use the Layer Control pull-down list. When you have objects selected and then pick a layer from the Layer Control list, the selected objects will be changed to the picked layer.

Pick: *(Select some, but not all, of the roadway edges that appear as centerlines)*

Now, you will pick on the current layer name listed, CENTERLINE, to pull down the menu of all the layers available.

Pick: **LOTLINES** *(from the list of layers shown)*

You will see the lines you selected change to the color and linetype for layer LOTLINES. Now you will press the escape key a couple of times so that the current layer name will change back to CENTERLINE.

Type: Esc *(twice)*

This same function applies to changing the color and linetype of an object, without having to change its layer. You perform the same steps, only choose a different color or linetype from the Color Control and Linetype Control pull-down menus.

Using the Change Properties Dialog Box

 You can also use the Change Properties dialog box to change an object's properties. The Change Properties dialog box lets you change the linetype, layer, and color of the objects you select, as well as their linetype scale (which you will learn more about in Tutorial 4) and their thickness. You can select the Change Properties dialog box by picking the Properties icon near the right side of the Object Properties toolbar.

Pick: **Properties icon**

Select objects: *(select the roadway edges that have not been changed yet)*

Select objects: ⏎

The Change Properties dialog box appears on your screen, as shown in Figure 3.13. (The Modify Line dialog box will appear if you have selected only one line, otherwise you should see Figure 3.13). You will use it to change the objects you selected from layer CENTERLINE to the layer LOTLINES.

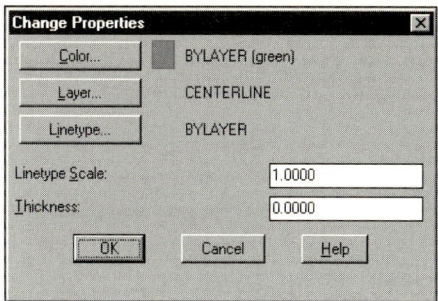

Figure 3.13

Pick: (the button with the word Layer)

A dialog box showing the existing drawing layers appears, as in Figure 3.14. From the list, select layer LOTLINES.

Figure 3.14

Pick: **OK** *(to exit the Select layer dialog box)*

Pick: **OK** *(to exit the Change Properties dialog box)*

You will see the lines you selected change to black (or white, depending on your configuration) and linetype CONTINUOUS, as set by layer LOTLINES's properties.

Next, you will use the Fillet command to add fillets, or small rounded corners, to some of the lots. Before doing so, change your current layer to LOTLINES on your own. Pick on one of the roadway edges that you changed to the layer LOTLINES and note that the current layer name changes to LOTLINES. As you

learned earlier, pick on the Make Object's Layer Current button at the left end of the Standard toolbar. The layer name LOTLINES should now appear on the toolbar as the current layer without any objects selected. Use Zoom Window to enlarge the center area of the drawing, as shown in Figure 3.15.

> ■ *TIP* Be sure to save your drawing whenever you have finished a major step and you are satisfied with it. Take a minute to use the Save icon from the Standard toolbar and save your drawing on your own. ■

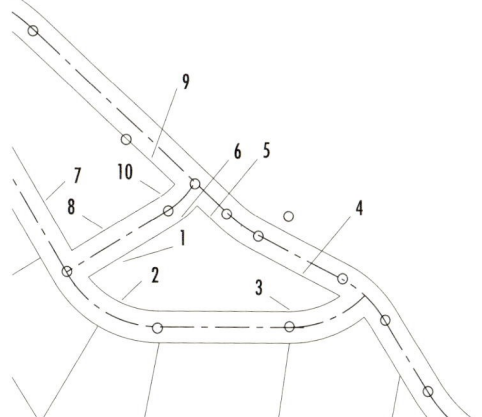

Figure 3.15

Using Fillet

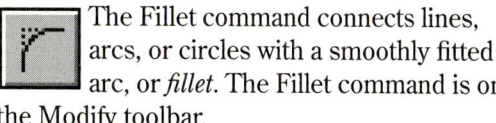 The Fillet command connects lines, arcs, or circles with a smoothly fitted arc, or *fillet*. The Fillet command is on the Modify toolbar.

Pick: **Fillet icon**

Polyline/Radius/Trim/<Select first object>: **R** ⏎

Typing *R* tells AutoCAD that you want to enter a radius. You will enter a radius value of 10.

Enter fillet radius <0.0000>: **10** ⏎

Command: ⏎ *(to restart the Fillet command)*

AutoCAD prompts you to select the two objects. Refer to Figure 3.15 to select the lines.

Command: Fillet Polyline/Radius/Trim/
  <Select first object>: *(select line 1)*

Select second object: *(select line 2)*

A fillet should appear between the two lines, as shown in Figure 3.16.

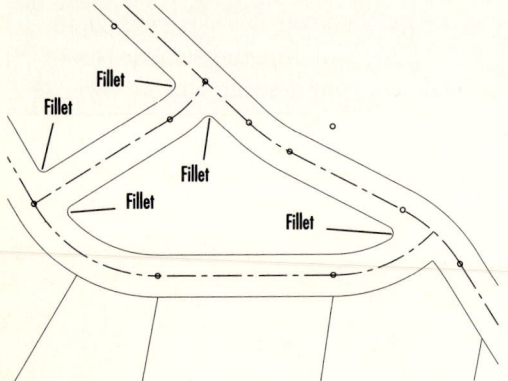

Figure 3.16

Command: ⏎

AutoCAD repeats the previous command when you press ⏎ or the spacebar at the blank command prompt. In this case, it returns to the original Fillet prompt, which allows you to repeat the selection process, drawing additional fillets of the same radius. Create the next fillet between the lines shown in Figure 3.15.

Polyline/Radius/Trim/<Select first object>: *(select line 3)*

Select second object: *(select line 4)*

Continue to fillet lines 5 through 10 on your own.

You can use the Fillet command to fit a smooth arc between any combination of lines, arcs, or circles. Once you have defined the radius value, the direction of the fillet is determined by the cursor location used to identify the two objects. Figure 3.17 shows some examples of

how you can use Fillet to create fillets of different shapes by choosing different point locations.

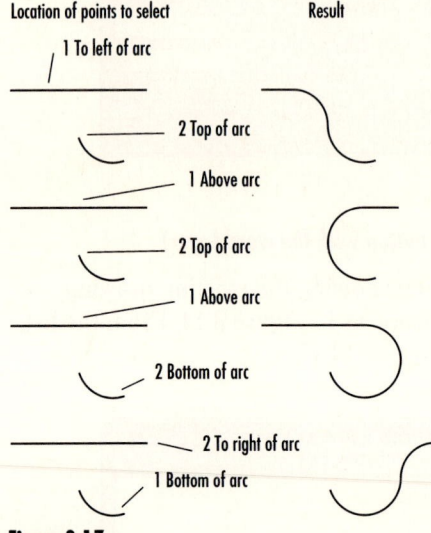

Figure 3.17

Each example starts with a line located directly above an arc, as shown in the column on the left. Picking the objects where indicated yields the results shown in the column on the right.

You can also set the Fillet command so that it does not automatically trim the lines to join neatly with the fillet, which is the default. You do so by selecting the command option Trim by typing the letter *T* at the command prompt and then choosing the option No trim by typing *N*. Experiment with this method on your own.

> ■ **TIP** If you use a radius value of 0 with the Fillet command, you can use the command to make lines intersect that do not meet neatly at a corner. This method also works to make a neat intersection from lines that extend past a corner. You can also use the Chamfer command (which you will learn next) by setting both chamfer distances to 0. ■

Using Chamfer

 The Chamfer command draws a straight-line segment (called a chamfer) between two given lines. *Chamfer* is the name for the machining process of flattening a sharp corner of an object. Use the AutoCAD Help window to read about the options for the Chamfer command on your own. The Chamfer command is on the Modify toolbar.

On your own use the Line command to draw a rectangle, off to the side of the subdivision drawing, as shown in Figure 3.18.

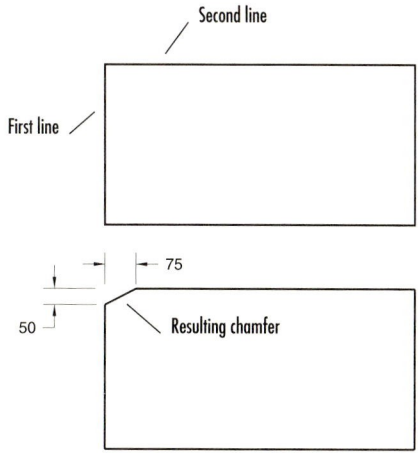

Figure 3.18

Pick: **Chamfer icon**

Polyline/Distances/Angle/Trim/Method/
  <Select first line>: **D** [↵]

Typing *D* tells AutoCAD you want to enter distance values.

Enter first chamfer distance <0.0000>: **50** [↵]
Enter second chamfer distance <0.0000>: **75** [↵]
Command: [↵]

AutoCAD returns to the Chamfer command and prompts you for selection of the first and second lines between which the chamfer is to be added. Refer to the upper part of Figure 3.18 to select the lines.

Polyline/Distances/ Angle/Trim/Method/
  <Select first line>: *(select the first line)*
Select second line: *(select the second line)*

A chamfer should appear on your screen that looks like the one shown in the lower part of Figure 3.18. The distance that each end of the chamfer is from the corner defines the size of the chamfer. As with the Fillet command, once you have entered these distances, you can draw additional chamfers of the same size by pressing [↵] after completing each chamfer to restart the command.

On your own use the commands that you have learned to erase the chamfered rectangle from your drawing. Use Zoom All to restore the original appearance and then save the drawing.

In Tutorial 1 you learned how to draw a Rectangle with the Rectangle icon, with which you had the options of Fillet and Chamfer. The same Fillet and Chamfer principles that you just learned apply to using the Rectangle icon. At this point you may find it useful to practice with the Rectangle icon and changing the Chamfer and Fillet dimensions offered in the command prompt. When you have finished practicing drawing rectangles, erase them before moving on to the next section.

Using Polyline

 The Polyline command draws a series of connected lines or arcs that AutoCAD treats as a single graphic object called a *polyline*. The Polyline command is used on technical drawings to draw irregular curves and lines that have a width. An irregular curve is any curve that does not have a constant radius (a circle and an arc have a constant radius). The Polyline command is on the Draw toolbar and looks like an arc and line connected.

You will use the Polyline command to create the shape for a pond to add to the subdivision drawing. On your own make layer LOTLINES current before continuing.

Pick: **Polyline icon**

From point: *(select any point to the right of the subdivision where you want to locate the pond)*

Current line-width is 0.0000

Arc/Close/Halfwidth/Length/Undo/Width/
    <Endpoint of line>: *(select 9 more points)*

You will use the Close option as you did in Tutorial 1 when drawing closed figures with the Line command.

Arc/Close/Halfwidth/Length/Undo/Width/
    <Endpoint of line>: **C** ⏎

Your screen should show a line made of multiple segments, similar to the one in Figure 3.19.

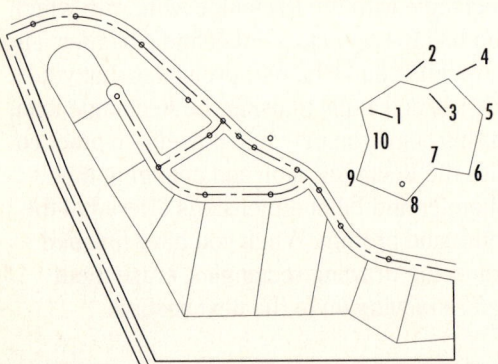

Figure 3.19

A shape drawn by using Polyline is different from a shape created by using the Line command, in that AutoCAD treats the polyline as one line. You cannot erase one of the polyline segments. If you try, the entire polyline is erased. Other Polyline options allow you to draw an arc as a polyline segment (Arc), spec-

ify the starting and ending width (or half-width) of a given segment (Width, Half-width), specify the length of a segment (Length), and remove segments already drawn (Undo).

Edit Polyline

Once you have created a polyline, you can change it and its individual segments with Edit Polyline. The Edit Polyline command is on the Modify menu, under Object, Polyline. You will use Edit Polyline to change the segmented line you just created to a smooth curve.

The Edit Polyline command provides two different methods of fitting curves. The Fit option joins every point you select on the polyline with an arc. The Spline option produces a smoother curve by using either a cubic or a quadratic B-spline approximation (depending on how the system variable Splinetype is set). You can think of the spline operation as working like a string stretched between the first and last points of your polyline. The vertices on your polyline act to pull the string in their direction, but the resulting spline does not necessarily reach those points.

Pick: **Modify, Object, Polyline**

Select polyline: *(select any part of the polyline)*

Open/Join/Width/Edit vertex/Fit/Spline/Decurve/
    Ltype gen/Undo/eXit <X>: **F** ⏎

■ *TIP* You can use Edit Polyline to convert objects into polylines. At the *Select polyline:* prompt, pick the line or arc that you want to convert to a polyline. You will see the message *Entity selected is not a polyline. Do you want to turn it into one? <Y>:.* Press ⏎ to accept the yes response to convert the selected object into a polyline and continue with the prompts for the Edit Polyline command. ■

The straight-line segments change to a continuous curved line, as shown in Figure 3.20.

Figure 3.20

Fit curve connects all the vertices of a 2D polyline by joining each pair of vertices with an arc. You will use the Undo option of the command to return the polyline to its original shape.

Close/Join/Width/Edit vertex/Fit/Spline/Decurve/
Ltype gen/Undo/eXit <X>: **U** ⏎

The original polyline returns on your screen. Now try the Spline option:

Close/Join/Width/Edit vertex/Fit/Spline/Decurve/
Ltype gen/Undo/eXit <X>: **S** ⏎

A somewhat flatter shape, similar to the one shown in Figure 3.21, should replace the original polyline. Among other things, splined polylines are useful for creating contour lines on maps.

> ■ *TIP* If you have a polyline that has already been curved with Fit or Spline, Decurve returns the original polyline. ■

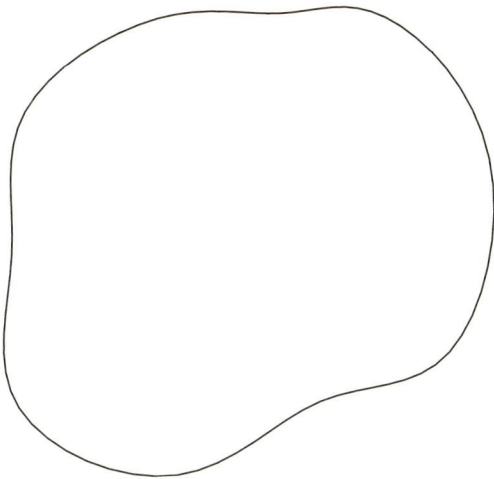

Figure 3.21

Now try the Width option.

Close/Join/Width/Edit vertex/Fit/Spline/Decurve/
Ltype gen/Undo/eXit <X>: **W** ⏎

Enter new width for all segments: **5** ⏎

The splined polyline is replaced with one that is 5 units across. You will accept the default by pressing ⏎ to exit the Edit Polyline command.

Close/Join/Width/Edit vertex/Fit/Spline/Decurve/
Ltype gen/Undo/eXit <X>: ⏎

Your drawing should look something like the one shown in Figure 3.22.

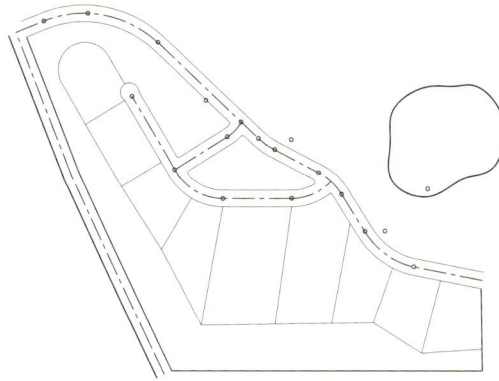

Figure 3.22

Using Spline

The Spline command lets you create quadratic or cubic spline (or NURBS) curves. The acronym NURBS stands for nonuniform rational B-spline, which is the name of the method used to draw the curves. These curves can be more accurate than spline-fitted polylines because you control the tolerance to which the spline curve is fit. Splines also take less file space in your drawing than do spline-fitted polylines.

You can also use the Object option of the Spline command to convert spline-fitted polylines to splines. (The variable Delobj controls whether the original polyline is deleted after it is converted.) The Spline command is on the Draw toolbar. You will draw a spline off to the side of the subdivision and then later erase it.

Pick: **Spline icon**

Object/<Enter first point>: *(pick any point)*

Enter point: *(pick any point)*

Close/Fit Tolerance/<Enter point>: *(pick any point)*

Close/Fit Tolerance/<Enter point>: *(pick any point)*

Close/Fit Tolerance/<Enter point>: *(pick any point)*

Close/Fit Tolerance/<Enter point>: [↵]

Enter start tangent: [↵]

Enter end tangent: [↵]

Pressing [↵] at the tangency prompt causes AutoCAD to calculate the default tangencies. You can make the spline tangent to other objects by picking the appropriate points at these prompts. In addition, the Spline command has a Close option similar to the Line and Polyline commands that you have used.

The spline appears on your screen. Experiment by drawing splines on your own. You can edit Splines by using the Splinedit command on the

Modify menu, under Object. When you have finished experimenting with the Spline command, erase all the splines that you created.

Now, use the commands that you have learned to thaw layer TEXT so that the text returns to your screen. Set TEXT as the current layer and use the Dtext command to add the label POND to the pond. Set the height for the text to 30 units. When you have finished these steps on your own, your screen should look like Figure 3.23.

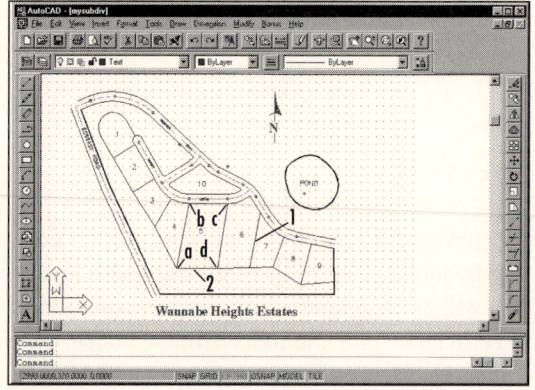

Figure 3.23

Getting Information About Your Drawing

Because your AutoCAD drawing database contains accurate geometry and has been created as a model of real-world objects, you can find information about distances and areas, and locate coordinates. You can also find volumes and mass properties of solid models, which you will learn to do in Tutorial 11. Next, you will use AutoCAD's ability to list objects and make calculations to determine information about the subdivision. You will use the Inquiry flyout on the Object Property toolbar shown in Figure 3.24.

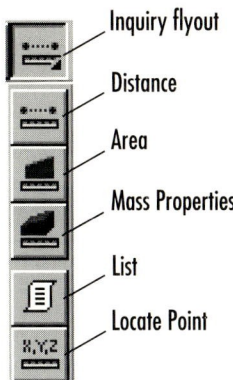

Inquiry flyout

Distance

Area

Mass Properties

List

Locate Point

Figure 3.24

Using List

The List command shows information that AutoCAD holds in the drawing database about the object selected. Different information may be shown, depending on the type of object. The layer, coordinate location for the object, color and linetype if different from the layer, and other information are listed. Refer to Figure 3.23 to make your selections.

Pick: **List icon**

Select objects: *(pick line 1)*

Select objects: [↵]

The AutoCAD text window appears on your screen. If you want, you can resize it by picking on its corners and dragging it to a new size. You can also pick on the title bar and drag it to a new location while you hold the pick button down. Refer to Figure 3.25 for the information listed about the line that you selected. You can use this information to make calculations or to label the lot lines of this subdivision.

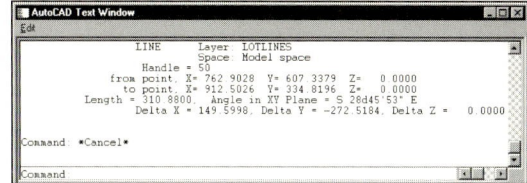

Figure 3.25

Close the text window by clicking on the close button in the upper right corner.

Pick: **List icon**

Select objects: *(pick on the polyline representing the pond)*

Select objects: [↵]

The text window again appears on your screen. It contains all the information about the vertices of the polyline. Keep pressing [↵] until all the information has scrolled by. At the bottom of the information you will see that AutoCAD lists the area and perimeter of the polyline, as shown in Figure 3.26. This information can be very useful for calculating the area of irregular shapes like the pond.

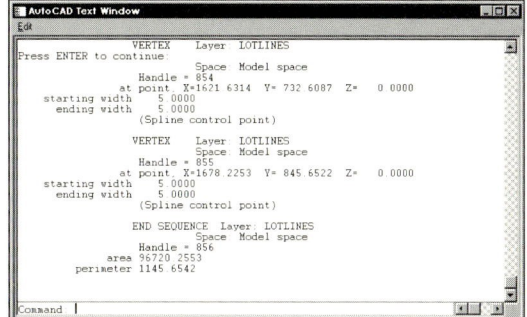

Figure 3.26

Locating Points

 Now turn on the Endpoint running object snap on your own. You will use it with Locate Point to select accurate point locations for the next steps. You can select Locate Point from the Inquiry flyout on the Object Properties toolbar. Locate Point shows the coordinates of the location that you select. When you use this command in conjunction with object snaps, you can find the exact coordinates of an object's endpoint, center, midpoint, and other points.

Pick: **Locate Point icon**

Select point: *(pick on the left end of line 2 in Figure 3.23)*

The command window displays the coordinates of the point, X=912.5026, Y=334.8196, and Z=0.0000

Using Area

 The Area command finds areas that you define by picking points to define a boundary, picking a closed object, or adding and subtracting boundaries. The options for the Area command are First point, Object, Add, and Subtract. To use the default method, First point, you begin by selecting points to define straight-line boundaries for the area that you want to measure. When you press ← to indicate that you have finished selecting, AutoCAD automatically closes the boundary for the area from the last point selected back to the first point. (You can also close the area by picking the first point again if you want.) The area inside the boundary that you specify is reported. By selecting the Object option, you can pick a closed object, such as a polyline, circle, ellipse, spline, region, or a solid object, and its area is reported. You could use this method to find the area of the pond instead of using List as you did previously. The Add

and Subtract options let you define more complex boundaries by adding and subtracting from the first boundary selected. Next, you will use the Area command to find the area of lot 5. Remember that the Endpoint object snap is still active, so AutoCAD finds the nearest endpoint when you select a line. This way the area calculation will be accurate.

Pick: **Area icon**

<First point>/Object/Add/Subtract: *(pick point a on Figure 3.23)*

Next point: *(pick b)*

Next point: *(pick c)*

Next point: *(pick d)*

Next point: ←

The area and perimeter values of lot 5 are listed in the command window.

On your own set the current layer to EASEMENTS before continuing.

Using Multilines

Next, you will use Multiline to add multiple lines to the drawing easily. This feature is especially useful in architectural drawings for creating walls and in civil engineering drawings for highways, among other things, where you can draw the centerline of the road and automatically add the edge of pavement and edges of right-of-way lines a defined distance away. Multiline also allows you to fill the area between two lines with a color. Using the command for Multiline Style, you can define as many as 16 lines, called *elements*, which you will use when drawing multilines. You can save multiline styles with different names so that you can easily recall them. You can also set the multiline justification so that you can create lines by specifying the top, middle, or bottom of the style. You do so by setting the variable Cmljust to 0, 1, or 2, respectively.

Creating a Multiline Style

You can select the Multiline Style command from the Format menu. Multiline Style creates the spacing, color, linetype, and name of the style, as well as the type of endcaps that will be used to finish off the lines.

Pick: **Format, Multiline Style**

The Multiline Styles dialog box appears on your screen, as shown in Figure 3.27. The default style name is STANDARD. Note that it is the current style. To create a new style, you will use the Name portion of the dialog box and type in the new name *EASEMENT*. Pick in the box to the right of Name if the typing cursor is not already shown there. Then you will set the properties of the elements to form this new style.

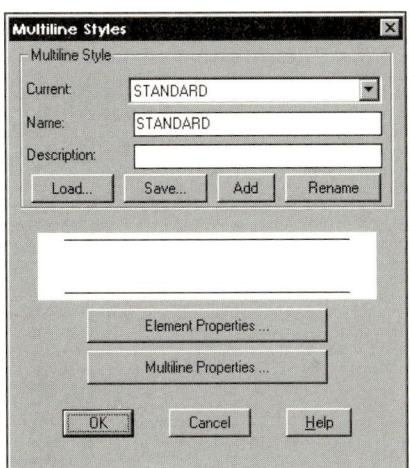

Figure 3.27

Type: **EASEMENT**

The Add button in the Multiline Styles dialog box sets a style in the Name box as current. You will set Style EASEMENT as current.

Pick: **Add**

Pick: **Element Properties**

The Element Properties dialog box appears. You will use it to set the spacing of the elements, as shown in Figure 3.28. To change the spacing for an element, pick the element shown in the list at the top of the dialog box. When it is highlighted, use the box to the right of Offset to set its offset spacing. On your own type *15* for the offset spacing. Now, set the offset spacing for the second element to *-15*. To see the element spacing change for a highlighted item, pick once again on its name in the box at the top of the dialog box. When you have set the element spacing, your dialog box should look like Figure 3.28. You will leave the color and linetype set to BYLAYER, as layer EASEMENTS already has these properties set to cyan and HIDDEN linetype. If you want lines of different colors or different linetypes, you could use the appropriately named portions of the dialog box to set them. When you have finished,

Pick: **OK**

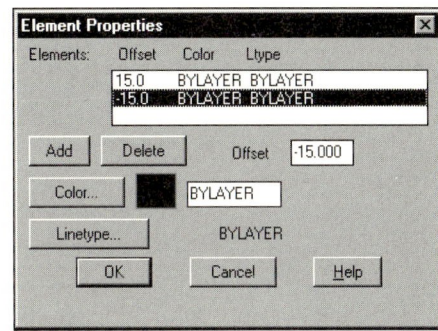

Figure 3.28

Return to the Multiline Styles dialog box. Selecting Multiline Properties will cause the Multiline Properties dialog box to appear on your screen. You can use it to select different types of endcaps, including angled lines or arcs, to finish the multilines you create. You can also use this area to turn on Fill if you want the area between multiline elements to be shaded. Experiment with this area on your own and pick Cancel to return to the Multiline Styles dialog box when you have finished.

The Save button saves a style to the library of multiline styles. When you pick the Save button, the Save Multiline Styles dialog box appears on your screen. AutoCAD saves multiline styles to *c:\AutoCAD R14\support\ acad.mln* by default. You can create your own file for your multiline styles, if you want, by typing in a different name. The Load button loads saved multiline styles from a library of saved styles. Rename lets you rename the various styles you have created. You can't rename the STANDARD style. When you have finished examining the dialog box,

Pick: **Save**

Pick: **OK (to exit the Multiline Styles dialog box)**

Drawing Multilines

 The Multiline command is on the Draw toolbar. You will pick it to add some multilines for easements along the lot lines of the subdivision drawing. You will use the current multiline style, EASEMENT, that you created.

Pick: **Multiline icon**

Now you are almost ready to draw some multilines. Before you do, decide on the justification you will need to use. You can choose to have the multilines you draw align so that the points you select align with the top element, with the zero offset, or with the bottom element. You will set the justification so that the points you pick are the zero offset of the multiline, in this case the middle of the multiline.

Justification/Scale/Style/<From point>: **J** ↵

Top/Zero/Bottom <top>: **Z** ↵

You will see the message *Justification = Zero, Scale = 1.00, Style = EASEMENT.* Now, you are ready to start drawing the easement lines.

Refer to Figure 3.29 for the points to select. Remember that the Endpoint running object snap should still be on.

Justification/Scale/Style/<From point>: **(pick point 1)**

<To point>: **(pick 2)**

Undo/<To point>: **(pick 3)**

Close/Undo/<To point>: **(pick 4)**

Close/Undo/<To point>: **(pick 5)**

Close/Undo/<To point>: **(pick 6)**

Close/Undo/<To point>: **(pick 7)**

Close/Undo/<To point>: **(pick 8)**

Close/Undo/<To point>: ↵

Your drawing should look like Figure 3.29.

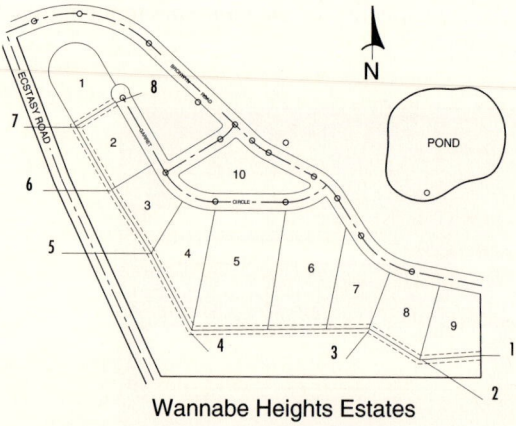

Wannabe Heights Estates

Figure 3.29

> ■ *TIP* If you want to trim a multiline, you must first use the Explode command to change it back into individual objects before you can trim it. Use AutoCAD's Help command for more information about using Explode. ■

On your own, use the Mtext command to add text describing the location of the subdivision and the easements, as shown in Figure 3.30. Save your drawing now.

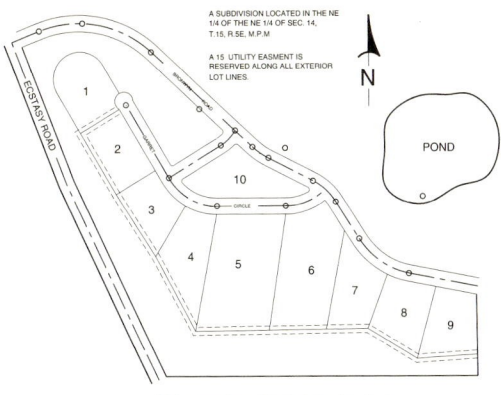

Wannabe Heights Estates

Figure 3.30

■ **Warning:** You should always save your drawing before you plot. If for some reason the plotter is not connected or there is another problem, AutoCAD may not be able to continue. To go on, you would have to reboot the computer and would lose any changes that you had made to your drawing since the last time you saved. ■

The Print/Plot Configuration Dialog Box

Depending on the types of printers or plotters you have configured with your computer system, the Plot command causes your drawing to be printed on your printer or plotted on the device that you configured to use with AutoCAD. The same command is used for printing and for plotting. Selecting Print causes the Print/Plot Configuration dialog box (Figure 3.31) to appear on your screen.

■ **Warning:** The Print/Plot Configuration dialog box will not appear unless the system variable Cmddia is set to 1. If the Print/Plot Configuration dialog box does not appear when you pick the Print icon, type CMDDIA at the

command prompt and be sure that its value is set to 1. ■

Pick: **Print icon**

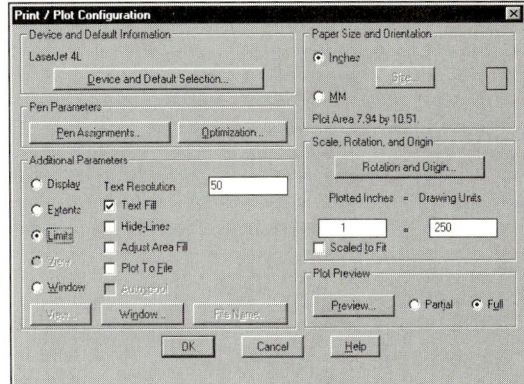

Figure 3.31

Among the things that the Print/Plot Configuration dialog box lets you choose are:

■ The portion of your drawing to plot or print;

■ The plotter or printer where you will send the drawing;

■ The pen or color selections for the plotter or printer;

■ The scale at which the finished drawing will be plotted;

■ Whether to use inches or millimeters as the plotter units;

■ Where your plot starts on the sheet;

■ Whether the plot is rotated or moved on the origin; and

■ The scale relating the number or plotted units to the number of drawing units.

Because there are so many factors, be careful when you start out *not* to change too many things about the plotter at once. That way you can see the effect that each option has on your plot.

Device and Default Selection

Click on the Device and Default Selection box. A dialog box appears that lets you select from the plotters that you have configured. If no printers or plotters are listed, you must use the Printer tab in the Preferences dialog box to add a new printer or plotter configuration. Refer to the Getting Started section for details about configuring your output device. To select a listed device, highlight the name of the device and pick OK. The options displayed for pen selection and other items depend on your particular output device. If an item in the Print/Plot Configuration dialog box appears grayed instead of black, it is not available for selection. You may not be able to choose certain items, depending on the limitations of your printer or plotter.

Additional Parameters

This area of the Print/Plot Configuration dialog box lets you specify what area of the drawing will be printed or plotted. Display selects the area that appears on your display as the area to plot. Extents plots any drawing objects that you have in your drawing. Limits plots the predefined area set up in the drawing with the Limits command. View plots a named view that you have created with the View command. (Note that, if you have not made any views, this area is grayed.) Window lets you go back to the drawing display and create a window around the area of the drawing that you want to plot. The radio button for Window is grayed until you use the Window button at the bottom of the Additional Parameters area to define a window. You can do so by picking from your drawing or by typing in the coordinates.

On your own, select Limits by picking the button to its left because the exact limits of the area you want to plot were already set in the datafile. If you have not set up the size with Limits, then Extents is often useful.

Text Resolution affects the number of dots per inch at which any True Type font text will be plotted. The higher the resolution, the slower the plotting process and vice versa. The Text Fill function affects whether True Type fonts will be filled during plotting. Recall that True Type fonts appear outlined when used in a drawing file and will plot that way unless you change the Text Fill function.

Hide Lines hides the back lines in a 3D drawing. You will not need to use this command for a 2D plot of your subdivision drawing. When Adjust Area Fill is active for a drawing that contains areas solidly filled with color, they are adjusted for the width of the plotter pen so that it does not draw over the boundary edge for the area. Plot To File sends your plot to a file rather than directly to the plotter or printer. If you want this result, check the box to the left of the words. The File Name item will turn black, indicating that you can now use it to display a dialog box in which you can enter the file name. The plot file will be created in the format of whatever output device you have selected in the Device and Default Selection area.

Paper Size and Orientation

Select either inches or millimeters (MM) by picking the appropriate button to the left of the measurement you want to use for your paper. For the drawing you have just set up, pick Inches. A solid circle fills the center of the button to indicate that it is selected. Pick Size to see the dialog box where you can set the paper size. The paper sizes that you can select depend on your printer. Standard 8.5 x 11 paper is called size A; 11 x 17 paper is size B. (Smaller values indicate the image area that your printer can print on the sheet.) Max is the maximum size of paper that your printer will handle. Note the message *Orientation is portrait* (or landscape) near the bottom of the Paper Size dialog box. This message refers to the default orientation for the paper. Portrait orientation is a vertical sheet layout, and landscape is horizontal. (If you have a laser printer and specified that the drawing be rotated when you configured the printer, the paper sizes shown may reflect this instruction.) Certain options may not be available with your model of printer. If so, they will be grayed out.

Scale, Rotation, and Origin

You set the scale for the drawing by entering the number of plotted inches for the number of drawing units in your drawing. If you do not want the drawing plotted to a particular scale, you can fit it to the sheet size by checking the box to the left of Scaled to Fit. Most of the time an engineering drawing should be plotted to a known scale.

If Scaled to Fit is on (a check mark shows in the box), turn it off. On your own type in *1* for Plotted Inches and *250* for Drawing Units if they do not already appear in the boxes. Doing so will give you a scale of 1"=250' for your drawing, because the default units in the drawing represented decimal feet.

Depending on whether you have a printer or plotter, you may need to rotate your drawing to fit the sheet correctly. Most printers require

that a drawing oriented horizontally, such as the subdivision drawing, be rotated 90° to print it correctly. Most plotters will not need to have the drawing rotated. You may have to experiment to find out what will work for your particular printer or plotter.

Use the Rotation and Origin button to place the (0,0) coordinates or bottommost left point in the drawing at the location you specify on the paper. You can move the drawing to the right by using a positive value for the X origin on the paper and move the drawing up on the paper by specifying a positive value for the Y origin. (Keep in mind that choosing to rotate your plot will affect the directions for moving the X and Y origins.) Also, you should be aware that moving the origin for the paper may cause the top and right lines of the drawing not to print if they are outside the printer limits.

Plot Preview

You may choose to have a partial preview, in which you see only the overall border of your graphics window as it will appear on the paper, or a full preview, in which your entire drawing as it will fit on the paper shows on the screen. If you have a very detailed drawing, obtaining a full preview may take some time. The subdivision drawing is not too large to preview in a reasonable amount of time, so pick the radio button next to Full so that you can see approximately how the drawing will appear on the sheet. Click on Preview to show the drawing on the screen. Your screen should look similar to Figure 3.32. Plot rotation is set to 90° in Figure 3.32. You may not need to rotate your drawing for your printer.

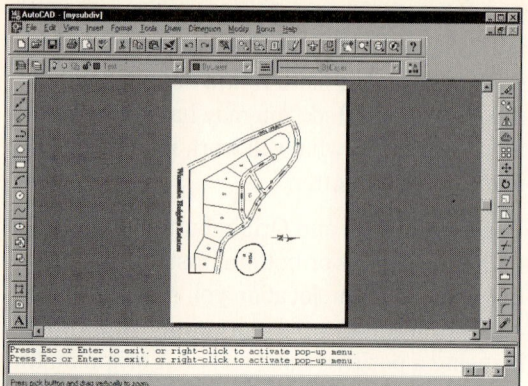

Figure 3.32

The cursor should take on the appearance of a magnifying glass with a plus sign above it and a minus sign below it. It is set on Realtime zoom, allowing you to zoom in or out from the drawing. The following message should be at the command line:

Press Esc or Enter to exit, or right-click to activate the pop-up menu.

If you press the right mouse button, a menu pops up offering different zoom options, pan, and exit. On your own select different com-

mands to familiarize yourself with them. If the drawing appears to fit the sheet correctly, pick Exit from the pop-up menu or press the Escape or Return key. When you have returned to the Print/Plot Configuration dialog box, pick OK to plot your drawing.

You will see a prompt similar to the following, depending on your hardware configuration:

Effective plotting area: 10.50 wide by 8.00 high

Position paper in plotter.

Press RETURN to continue or S to stop for hardware setup.

Press ⏎ if you are ready to print or plot.

If the drawing does not fit the sheet correctly, review this section and determine which setting in the Print/Plot Configuration dialog box you need to change. You should have a print or plot of your drawing that is scaled exactly so that 1" on the paper equals 250' in the drawing.

When you have a successful plot of your drawing, you have completed this tutorial. Exit AutoCAD by choosing

Pick: **File, Exit**

KEY TERMS

chamfer	fillets	through point
cutting edges	offset distance	
elements	polyline	

KEY COMMANDS

Area	List	Plot
Chamfer	Locate Point	Print Polyline
Change Properties	Multiline	Spline
Edit Polyline	Multiline Style	Trim
Fillet	Offset	

Redraw the following shapes. If dimensions are given, create your drawing geometry exactly to the specified dimensions. The letter M after an exercise number means that the given dimensions are in millimeters (metric units). If no letter follows the exercise number, the dimensions are in inches. Don't include dimensions on the drawings.

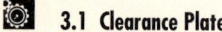

3.1 Clearance Plate

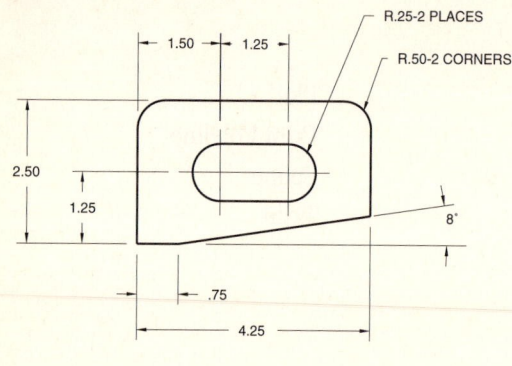

3.2M Bracket

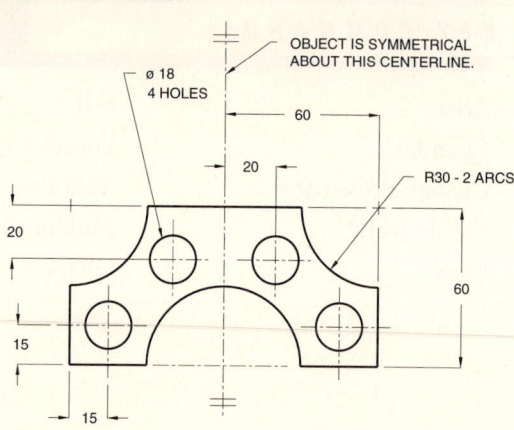

3.3 Roller Arm

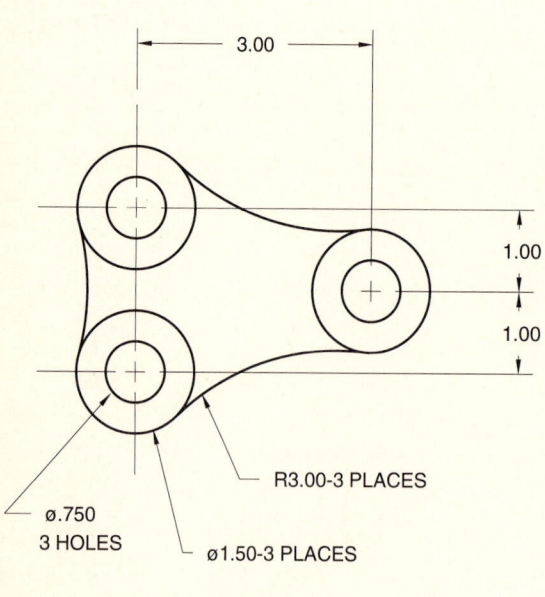

3.4 Roller Support

Hint: Create construction geometry to locate the center of the upper arc of radius 1.16. Turn on the AutoSnap modes for Endpoint and Intersection. Then offset the outer left-hand arc a distance of 1.16 to the outside. Draw a circle with radius 1.16 and its center at the left endpoint of the .56 horizontal line. Where the offset arc and the new 1.16 circle intersect will be the center of the needed arc.

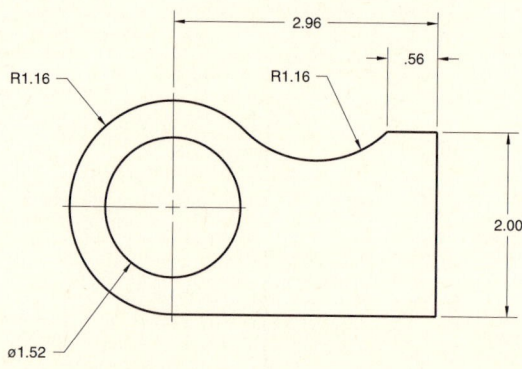

3.5 Slotted Ellipse

Draw the figure shown (note the symmetry). Do not show dimensions.

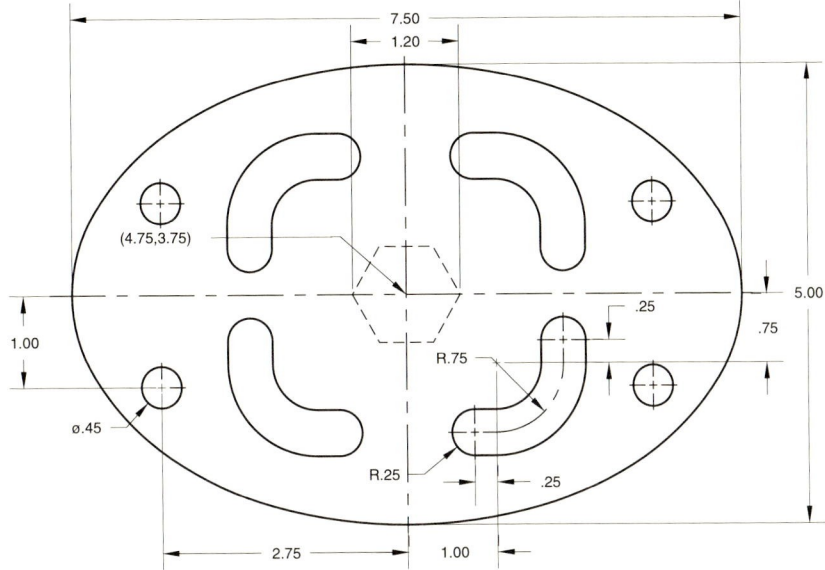

3.6 Roadway

Shown is a centerline for a two-lane road (total width is 20 feet). At each intersection is a specified turning radius for the edge of pavement. Construct the centerline and edge of pavement. Don't show any of the text.

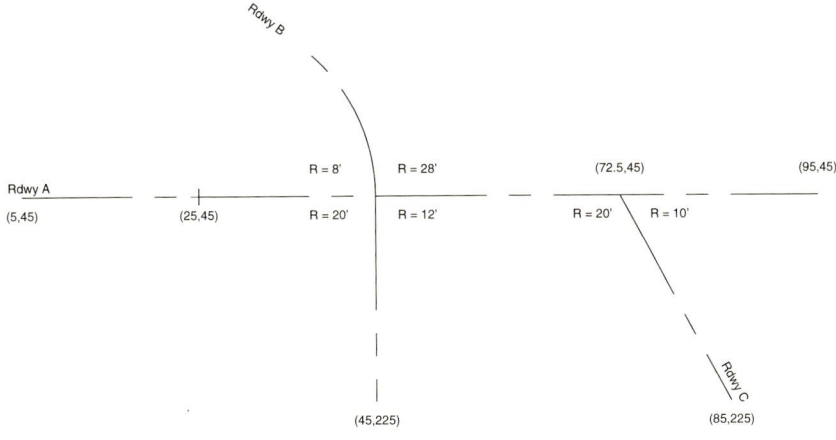

3.7 Circuit Board

Draw the following circuit board, using Polylines with the width option to create wide paths.

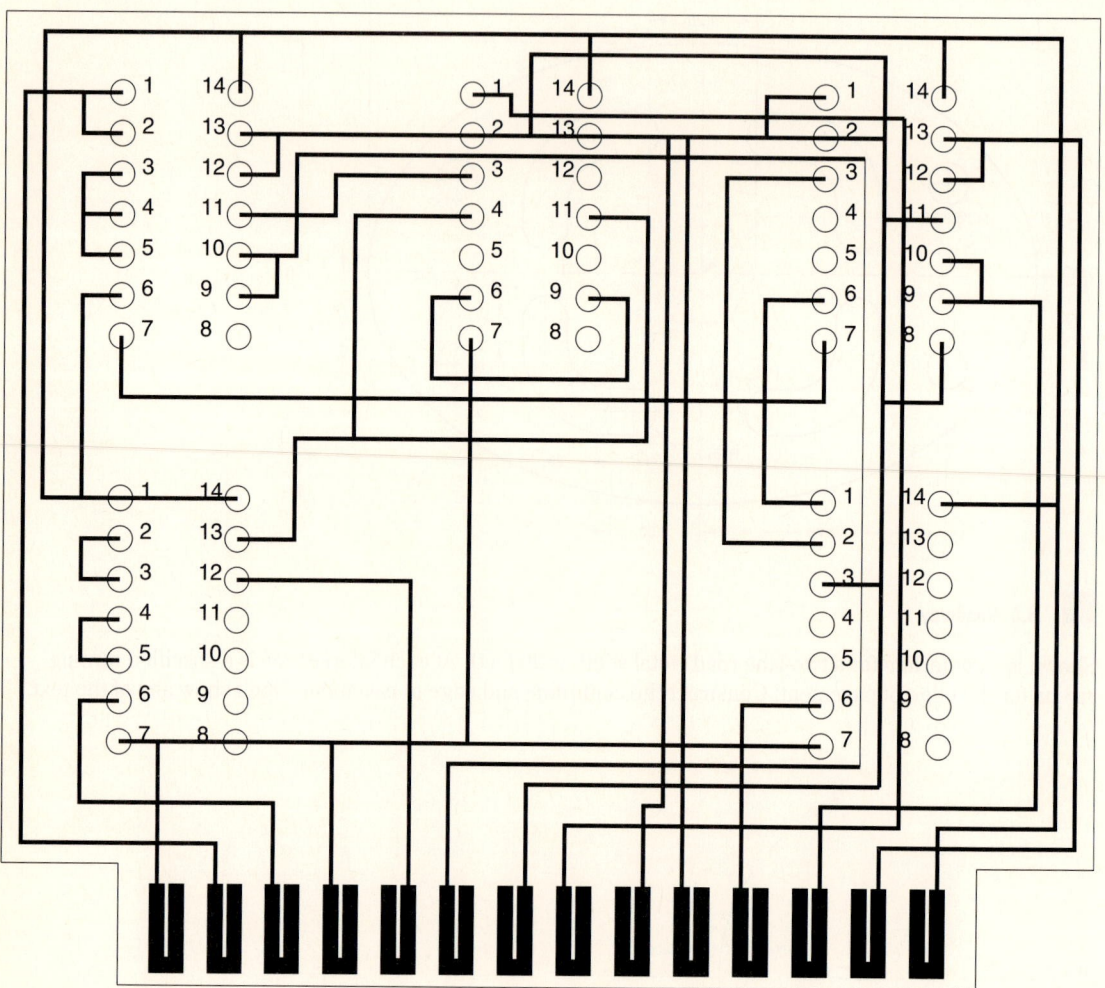

3.8M Shaft Support

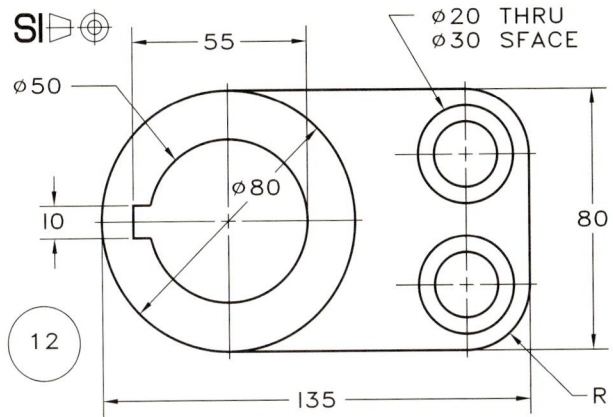

3.9M Puller Base

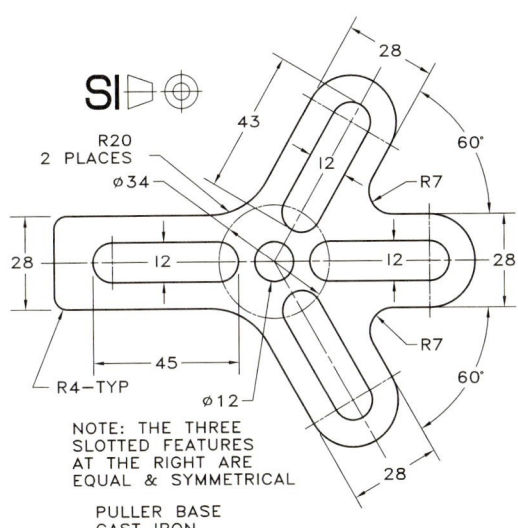

NOTE: THE THREE
SLOTTED FEATURES
AT THE RIGHT ARE
EQUAL & SYMMETRICAL

PULLER BASE
CAST IRON

3.10 Concrete Walkway

Hint: Construct the straight centerlines first. Then use a circle to construct the top R20 arc. Use Circle, TTR to construct the right-hand arc, tangent to the upper arc and a vertical line 45' from the left end. Trim the lines and offset them to finish.

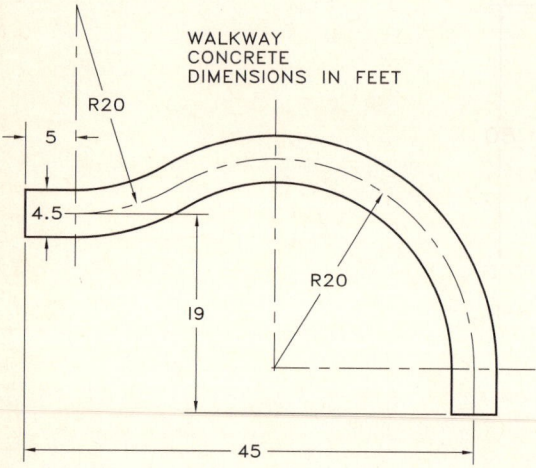

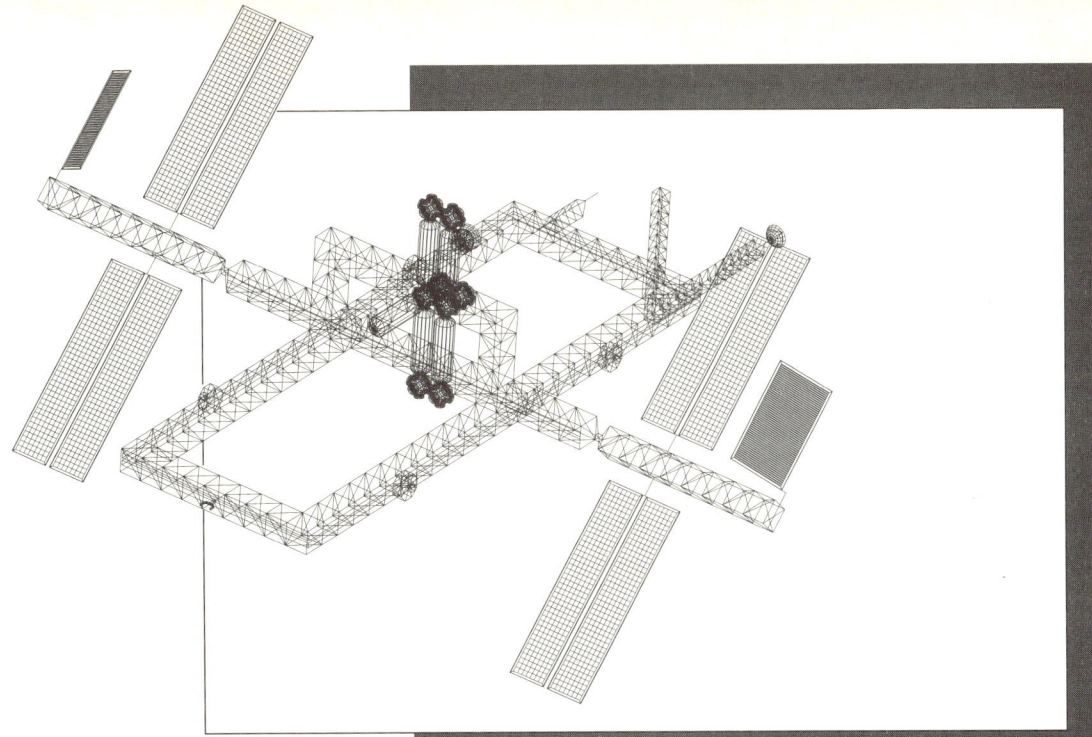

TUTORIAL 4

Geometric Constructions

Objectives

When you have completed this tutorial, you will be able to

1. Draw polygons and rays.

2. Use object snaps to pick geometric locations.

3. Load linetypes and set their scaling factors.

4. Use the Copy, Extend, Rotate, Move, Mirror, Array, and Break editing commands.

5. Build selection sets.

6. Use hot grips to modify your drawing.

Introduction

This tutorial will help you expand your skills in creating and editing drawing geometry by introducing several of the AutoCAD drawing techniques used for geometric constructions. You will learn more about how to use object snaps to select locations, such as intersections, endpoints, and midpoints of lines, based on your existing drawing geometry. This tutorial synthesizes the techniques you have learned so far by showing you how to coordinate the drawing commands to create shapes for technical drawings. In this tutorial you will create three drawings: a wrench, a coupler, and a geneva cam. You will learn how to apply editing commands to create drawing geometry quickly.

Starting

Launch AutoCAD Release 14 from your Windows desktop.

Using the Wizard

Upon launching AutoCAD Release 14, the Start Up dialog box should open, allowing you to Open a Drawing, Start from Scratch, or Use a Wizard. The Use a Template option has already been thoroughly covered in Tutorial 3.

> ■ *TIP* Located at the bottom of the Start Up dialog box is an option for *Show this dialog box at startup* with a check next to it. If you deselect it, then the next time AutoCAD is launched, the Start Up box will not appear. You can also change this setting in the Preferences dialog box, covered in Tutorial 1. ■

Pick: **Use a Wizard**

Your dialog box should appear similar to Figure 4.1, in which you have the option of a Quick Setup or an Advanced Setup. These are described in the Wizard Description box. You will choose the Quick Setup.

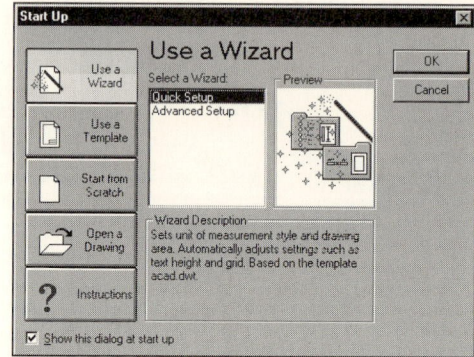

Figure 4.1

Pick: **Quick Setup**

Pick: **OK**

The dialog box should now appear similar to Figure 4.2. In the previous Tutorials, you have set your drawing limits and units, so the options should seem familiar to you.

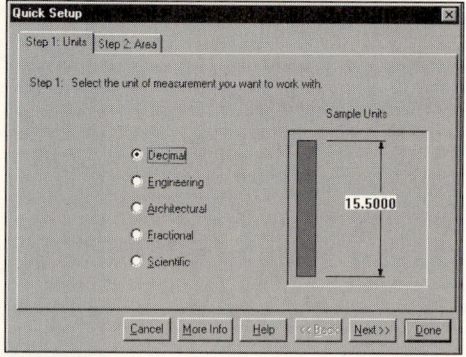

Figure 4.2

The Quick Setup Wizard allows you to set the Units and Limits of your drawing prior to starting a new drawing, rather than after a drawing has been started. There are two index cards for the following options: Units and Area. Pick through both cards to familiarize yourself with them, either by choosing the next button or picking on the card's tab.

The Units card allows you to change the units you want to work in (e.g., Decimal, Engineering, Architectural). The Area card sets the drawing limits.

You will create the wrench drawing using the decimal inches measurements. For example, 4.5 units in your drawing will represent 4.5 inches on the real object. Although AutoCAD's decimal units can stand for any measurement system you want, here you will have them stand for inches. When you create drawing geometry, make the objects in your drawing the actual size they would be in the real world. Do not scale them down as you would when drawing on paper. Keep in mind that one advantage of the accurate AutoCAD drawing database is that you can use it directly to control machine tools to create parts. You would not want your actual part to turn out half-size because you scaled your drawing to half-size. When you plot the final drawing, you specify the ratio of plotted inches or millimeters to drawing units to produce scaled plots.

The limits you will use for the wrench are 12.0000 by 9.0000 units. Check to make sure that the Units card is set for decimal, and the area card is set for 12.0000 x 9.0000. Then,

Pick: **Done**

On your own, pick the save icon and save your drawing as *wrench.dwg*.

■ *Warning:* AutoCAD has three system variables that control the use of dialog boxes. They are Filedia, Cmddia, and Attdia. Filedia controls whether a dialog box is used for file selection. Cmddia controls whether dialog boxes are used with other commands, for example, the Plot command. Attdia controls whether a dialog box is used with the Insert command. Setting these variables to 1 uses the dialog box in each instance. Setting the variables to 0 suppresses the use of the dialog box, so that the command line is used instead. If you have trouble getting the Create New Drawing dialog box to appear, type *FILEDIA* at the command prompt and make sure its value is set to 1. ■

Drawing the Wrench

You will draw the basic shapes required for a wrench. You will do many of the steps on your own, using the commands you have learned in the previous tutorials. Pay careful attention to the directions, being sure that you complete each step before going on.

First, you will draw construction lines in the drawing; you will later change these to centerlines. Refer to Figure 4.3.

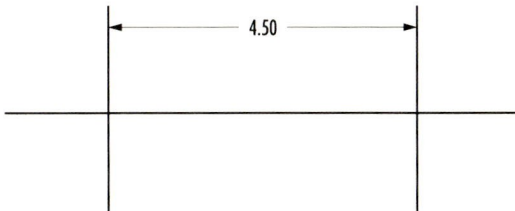

Figure 4.3

On your own, toggle on Grid and Snap (both should be set to 0.5 inches). If necessary, pick Zoom All so that the grid area fills the graphics window. (Depending on the aspect ratio of your screen, the grid may not fill it in all directions.)

Now use Line to draw a horizontal line 7.5 units long near the middle of your graphics window. The line does not have to be at the exact center because you can easily move your drawing objects later if they do not appear centered. You will learn how to move objects in this tutorial.

Next, draw two vertical lines that are 3.00 units long and 4.5 units apart, as shown in Figure 4.3. (Hint: Draw one line and then use the Offset command you learned in Tutorial 3, with an offset distance of 4.5, to create the other line.) The exact location of the vertical lines on the horizontal construction line is not critical, as long as they are 4.5 units apart.

Remember, you cannot select accurately unless you use Snap or an object snap.

 Now use Circle on your own to draw a circle with a 1-inch radius. The center point of the circle should be the intersection of the horizontal line and the right vertical line and should be a point you can use Snap to pick.

Using Copy

 Copy Object copies an object or group of objects within the same drawing. The original objects remain in place, and during the command you can move the copies to a new location. You can pick Copy Object quickly from the Modify toolbar. You will pick Copy Object to create a second circle in your drawing.

Pick: **Copy Object icon**

Select objects: *(select the circle)*

Select objects: ⏎

Pressing ⏎ tells AutoCAD that you have finished selecting objects and want to continue with the command. Be sure that Snap is on again so you can accurately select the center point of the circle, which you created on the snap increment.

<Base point or displacement>/Multiple: *(select the center point of the circle)*

AutoCAD switches to drag mode so that you can see the object move about the screen and then prompts you for the second point of displacement. You can define the new location by typing new absolute or relative coordinate values or by picking a location from the screen. Because you created this object on the snap increment, use the snap and pick the second point of displacement.

The second circle must be centered at the intersection of the horizontal and left vertical lines, which should be a snap location.

Second point of displacement: *(pick the left intersection)*

Your drawing should now look like Figure 4.4.

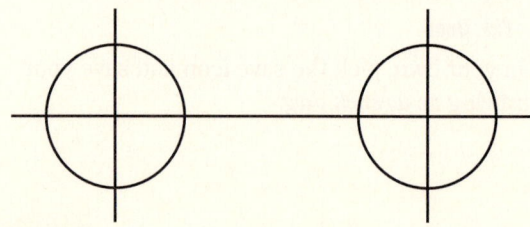

Figure 4.4

Next, use the Offset command on your own to offset the horizontal construction line you created a distance of 0.5 unit to either side of the center construction line to create the body of the wrench. Your resulting drawing should look like Figure 4.5.

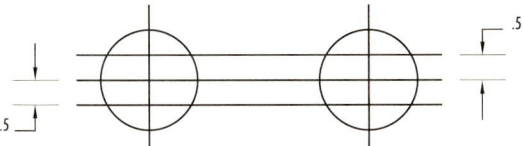

Figure 4.5

On your own use the Trim command to trim the lines that you created with the Offset command so that they intersect neatly with the circles. When you have finished, your drawing should look like Figure 4.6.

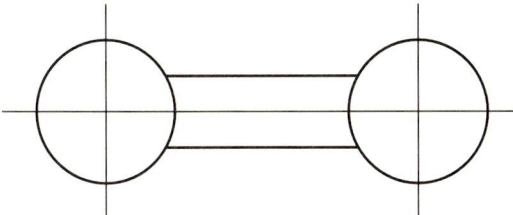

Figure 4.6

Loading Linetypes

Once you have finished removing the excess lines, you will change the middle construction lines in your drawing to centerlines. Before you can use various linetypes in AutoCAD, you must load them into the drawing. You will load the linetype CENTER.

■ *TIP* You can load linetypes when you create and assign linetypes to layers. You will learn to create layers in Tutorial 5. ■

The Linetype icon is on the Object Properties toolbar.

Pick: **Linetype icon**

The Layer and Linetype Properties dialog box appears on your screen, as shown in Figure 4.7. You will use it to load the linetype CENTER in your drawing so that it is available for use. Pick the Load button on the right side of the dialog box.

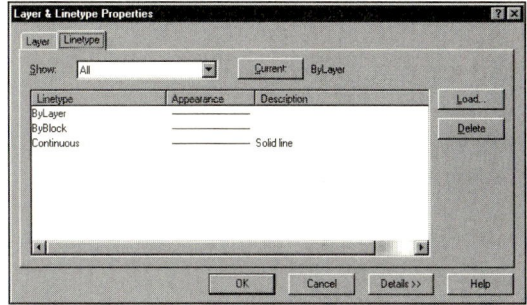

Figure 4.7

Pick: **Load**

The Load or Reload Linetypes dialog box shown in Figure 4.8 appears on your screen. Use the dialog box to highlight the CENTER linetype.

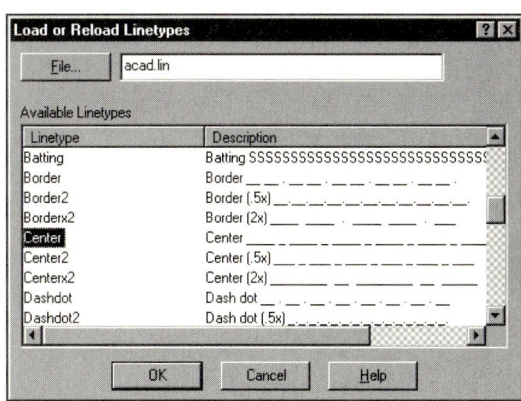

Figure 4.8

Pick: **OK** *(to exit the Load or Reload Linetypes dialog box)*

Pick: **OK** *(to exit the Layer & Linetype Properties dialog box)*

The linetype is loaded into your drawing and ready for use in other commands.

Changing Properties

To change the lines you have drawn to centerlines, you will use one of the methods you learned in Tutorial 3. However, changing objects in this way is not always a good practice. In Tutorial 5 you will learn to create your own layers and set their linetypes, colors, and other properties. Changing the color or linetype of a layer does not affect objects on that layer that have had their color or linetype set to something other than BYLAYER. In general, using layers to set the color and linetype is better.

Pick: **(select the middle horizontal line and the two vertical lines)** ⏎

Small blue squares (called hot grips, which you will learn about later in this tutorial) should be active on the endpoints and midpoints of each line.

Pick: **Linetype Control** *(on the standard toolbar)*

Pick: **Center** ⏎

To remove the hot grips from your screen so that your drawing looks like Figure 4.9, press the ⎋ key two times.

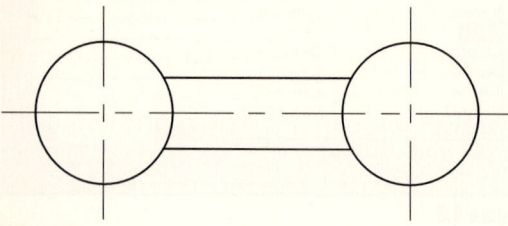

Figure 4.9

Setting the Global Linetype Scaling Factor

The *global linetype scaling factor* is a value that you set. It adjusts the lengths of the lines and dashes used to make up various linetypes. Each linetype pattern is stored in a file with the *.lin* file extension, which is external to your drawing. The lengths of the lines and dashes are set in that external file. But when you are working on large-scale drawings and the distances are in hundreds of units, a linetype with the dash length of 1/8 unit will not appear correctly. To adjust the linetypes, you specify a factor by which the length of the dashes in the *.lin* file should be multiplied. The global linetype scaling factor adjusts the scaling of all the linetypes in your drawing.

Thus when you are working on a large drawing such as the subdivision plat that you created in Tutorials 2 and 3, you must scale the dashed lines of the linetype up in order to make them visible. When you are working on a drawing that you will plot full scale, generally you should set the global linetype scaling factor to 1.00, which is the default. Setting the global linetype scaling factor to a decimal less than one (e.g., 0.75) results in a drawing with smaller dashes comprising the linetypes. In general, the linetype scaling factor you set in your drawing should be the reciprocal of the scale at which you will plot your drawing. When you learn about paper space in Tutorial 5, you will see that you can set the linetype scaling factor both in model space, where you are working now, and in paper space. You will type the command to set the global linetype scale at the command prompt.

Type: **LTSCALE** ⏎

New scale factor <1.0000>: **.75** ⏎

The centerlines should now cross in the center of each circle, as shown in Figure 4.10.

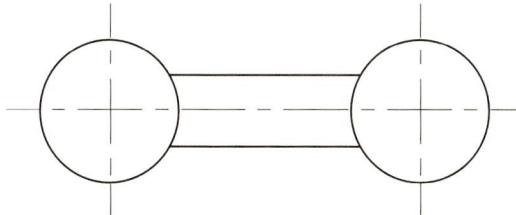
Figure 4.10

Changing an Object's Linetype Scale

You can also use the Linetype Scale area of the Change Properties dialog box to set the line-type scale for any particular object, independent of the rest of the drawing. Next, you will set the linetype scale for the two vertical lines to 1.5, giving them a different linetype scale than the horizontal line. You will use the Object Properties toolbar and pick Properties in order to use the Change Properties dialog box.

Pick: **Properties icon**

Select objects: **(pick the two vertical lines)**

Select objects: ⏎

The Change Properties dialog box appears on your screen. Use the box to the right of Linetype Scale to set the linetype scaling factor for the objects that you previously selected. You will set the linetype scale for the two lines to 1.5. Highlight the text in the box to the right of Linetype Scale and overtype the new value.

Type: **1.5**

Pick: **OK**

Now your drawing should look like Figure 4.11.

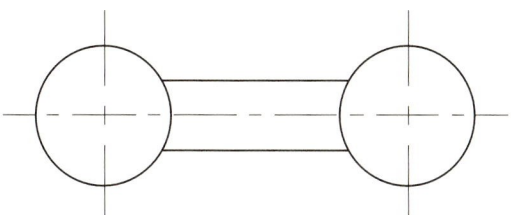
Figure 4.11

■ *TIP* The Dimcenter command creates correctly shown centerlines for circles. You will learn to use it in Tutorial 6. ■

Save your drawing before continuing.

Now you are ready to add the fillets to your drawing, as shown in Figure 4.12. On your own, use the Fillet command, with a radius of 0.5, to add the fillets. (Refer to Tutorial 3 if you need to review Fillet.)

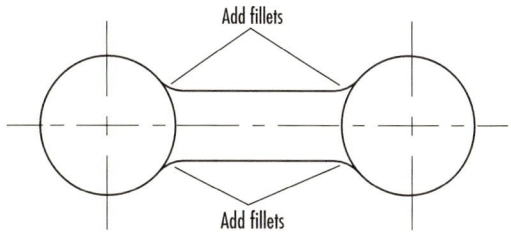
Add fillets

Add fillets

Figure 4.12

Using Polygon

 The Polygon command draws regular *polygons* with 3 to 1024 sides. AutoCAD creates polygons as polyline objects, which you used in Tutorial 3. Polygons act as a single object because they are created as a connected polyline. A regular polygon is one in which the lengths of all sides are equal. The size of a polygon is usually expressed in terms of a related circle. This convention comes from the classic straightedge/compass construction techniques you probably learned in your first geometry course. Polygons are either *inscribed* in or *circumscribed* about a circle. A pentagon is a five-sided polygon. Figure 4.13 shows a pentagon inscribed within a circle and a pentagon circumscribed about a circle.

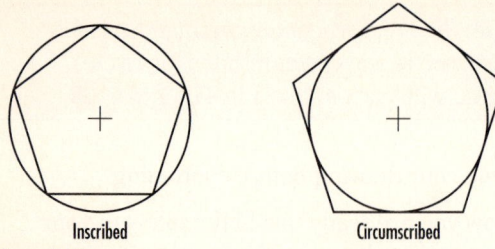

Inscribed   Circumscribed

Figure 4.13

Polygon is on the Draw toolbar. You will add a pentagon to the right side of the wrench. You are going to circumscribe a pentagon about a circle.

> *Pick:* **Polygon icon**
>
> Number of sides <4>: **5** ⏎
>
> Edge/<Center of polygon>: **(select the center point of the right circle)**
>
> Inscribed in circle/Circumscribed about circle (I/C)<I>: **C** ⏎

You are now in drag mode; observe that the pentagon changes size on the screen as you move your cursor. You must specify the radius of the circle, either by picking with the pointing device or by typing in the coordinates. For this object you will specify a radius of one-half inch.

> Radius of circle: **.5** ⏎

A five-sided regular polygon (a pentagon) is drawn on your screen. Your drawing should be similar to Figure 4.14.

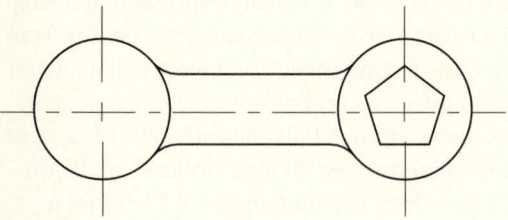

Figure 4.14

Drawing Hexagons

Hexagons are six-sided polygons and are a common shape in technical drawings. The heads of bolts, screws, and nuts often have a hexagonal shape. The size of a hexagon is sometimes referred to by its *distance across the flats*. The reason is that the sizes of screws, bolts, and nuts are defined by the distance across their flat sides. For example, a 16-mm hexhead screw would measure 16 mm across its head's flats and would fit a 16-mm wrench. The distance across the flats of a hexagon is not the same as the length of an edge of the hexagon. Figure 4.15 illustrates the difference between the two distances. If a hexagon is circumscribed about a circle, the diameter of the circle equals the distance across the hexagon's flats. If a hexagon is inscribed in a circle, the diameter of the circle equals the distance across the corners of the hexagon.

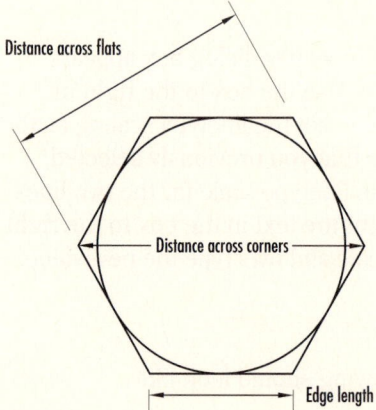

Distance across flats

Distance across corners

Edge length

Figure 4.15

You can use the Polygon command's Edge option to draw regular polygons by specifying the length of the edge. This function is helpful when you are creating side-by-side hexagonal patterns (honeycomb patterns). Note that AutoCAD draws polygons counterclockwise.

If you use the Edge option, the sequence in which you select points will affect the position of the polygon, as shown in Figure 4.16.

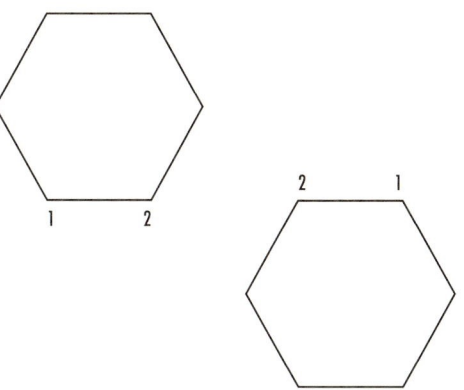

Figure 4.16

Next, you will add the hexagon to the other side of the wrench. You will use the circum-scribed option again.

Command: ⏎ *(or press the return button)*
Number of sides <5>: **6** ⏎
Edge /<Center of polygon>: *(pick the center point of the left circle)* ⏎
Inscribed in circle/Circumscribed about circle (I/C)<C>: ⏎
Radius of circle: **.5** ⏎

A hexagon measuring 1.00 across the flats appears in your wrench. The final drawing should be similar to Figure 4.17.

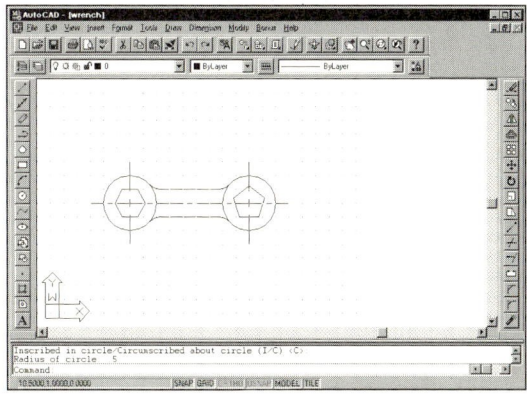

Figure 4.17

You have now completed your wrench. You will save this drawing and begin a new drawing to continue this tutorial.

Pick: **Save icon**

In this next section, you will practice with additional drawing commands and with object snaps. You will start a new drawing and name it *coupler.dwg*.

Pick: **New icon**

Choose Start From Scratch, with English Units.

Pick: **OK**

You are returned to the drawing editor.

Pick: **Save icon**

Type *coupler.dwg* as the name of the new drawing.

Using Object Snaps

As you learned in Tutorial 2, you use object snaps to select locations accurately in relation to other objects in your drawing. Object snaps can operate in two different ways: the override mode and the running mode.

Remember, when an object snap is active, the AutoSnap marker box appears whenever the cursor is near a snap point. When you select a point, it will select the object snap point if the marker box is present, whether or not the cursor is within the marker box.

Object Snap Overrides

You will draw the coupler shown in Figure 4.18, using the object snap overrides. You will use the object snap overrides to position lines and circles relative to this figure. (You may see other ways that you could use editing commands to create parts of this figure, but in this example object snaps will be used as much as possible. When you are working on your own drawings, you will use a combination of the methods you have learned.)

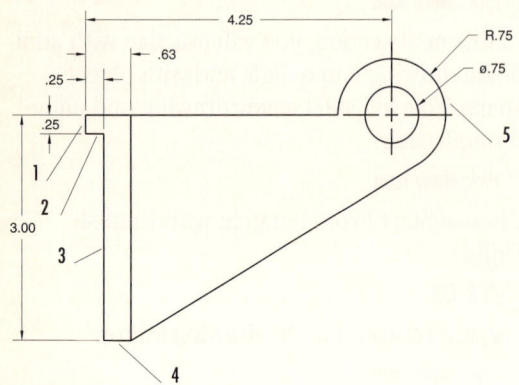

Figure 4.18

You will use the Line and Circle commands to create lines 1–4 and circle 5. On your own set Grid and Snap to 0.25 and be sure that they are turned on. Use Zoom All if necessary so that the grid area fills the graphics window.

Pick: **Line icon**
From point: **3.75,6.5** ⏎
To point: **@.25<270** ⏎
To point: **@.25<0** ⏎
To point: **@2.75<270** ⏎
To point: **@.375<0** ⏎
To point: ⏎
Pick: **Circle icon**
3P/2P/TTR/<Center point>: **8,6.5** ⏎
Diameter/<Radius>: **.75** ⏎

When you have finished drawing these objects, your drawing should look like the one in Figure 4.19.

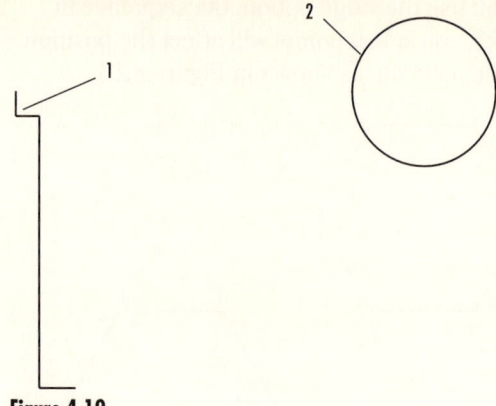

Figure 4.19

Turn Snap off. You will use the object snaps to locate points accurately. Leaving Snap turned on may interfere with selection.

Showing a Floating Toolbar

You will show the floating Object Snap toolbar to make selecting object snaps easy. Recall that the object snaps are also on the Standard toolbar as a flyout selection. You can show their floating toolbar by using the Toolbars selection on the View pull-down menu.

Pick: **View, Toolbars, Object Snap**
Pick: **Close**

The Object Snap toolbar appears on your screen, as shown in Figure 4.20. You can position it anywhere on the screen or dock it to the edge of the graphics window. Refer to Tutorial 1 if you need to review floating windows and how to dock toolbars.

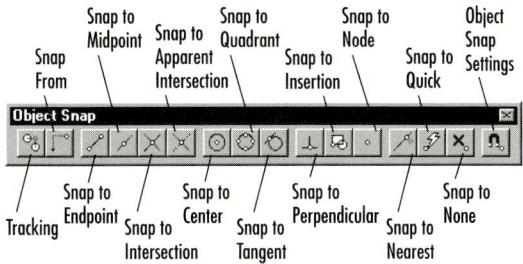

Figure 4.20

Snap to Endpoint

The Endpoint object snap locates the closest endpoint of an arc, line, or polyline vertex. You will draw a line from the endpoint of line 1, shown in Figure 4.19, to touch the circle at the *quadrant point.* The quadrant points are at 0°, 90°, 180°, and 270° of a circle or an arc. You will use the Endpoint and Quadrant object snaps to locate the exact points. You will use the floating Object Snap toolbar to select the object snaps.

> ■ *TIP* You can also type a three-letter code for the object snap at the *Select objects* prompt instead of picking it from the menu. The command-line equivalent of the Endpoint object snap is END. ■

Pick: **Line icon**

From point: **(pick Snap to Endpoint icon)**

Note that AutoSnap shows a colored marker box reflecting the object snap you chose.

Endpoint of: **(place the cursor on the upper end of line 1 and press the pick button, using the AutoSnap Endpoint marker)**

Note that the beginning of the line has jumped to the exact endpoint of the line you chose.

> ■ *TIP* When an object snap running mode is active, you can still use a different object snap during a command. The object snap override takes precedence over the object snap running mode for that pick. ■

Snap to Quadrant

 The Quadrant object snap attaches to the quadrant point on a circle nearest the position of the crosshairs. The command-line equivalent of Snap to Quadrant is QUA. The quadrant points are the four points on the circle that are tangent to a square that encloses it. They are also the four points where the centerlines intersect the circle. Next you will finish the line, using the Quadrant object snap to pick the quadrant point of the circle as the second endpoint for your line. You should see the *To point:* prompt for the Line command.

To point: **(pick Snap to Quadrant icon)**

Quadrant of: **(pick on the circle near point 2 in Figure 4.19)**

To point: ⏎

Your drawing should be similar to Figure 4.21.

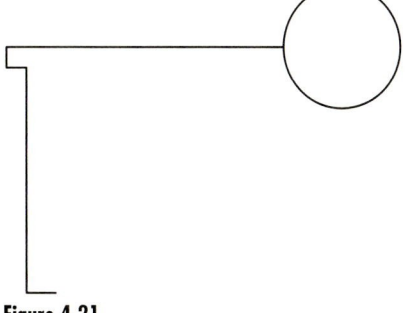

Figure 4.21

Snap to Center

 The Center object snap finds the center of a circle or an arc. You will use it to create concentric circles by selecting the center of the circle that you have drawn as the center for the new circle you will add. The command-line equivalent for Snap to Center is CEN.

Pick: **Circle icon**

3P/2P/TTR/<Center point>: *(pick Snap to Center icon)*

Center of: *(pick on the edge of the circle)*

AutoCAD finds the exact center of the circle and a circle rubberbands from the center to the location of the crosshairs. Note the AutoSnap marker that appeared when your cursor was on the circle. Object snap overrides stay active for only one pick. AutoCAD now prompts you to specify the radius.

Diameter/<Radius> <0.7500>: **.375** ⏎

The circle should be drawn concentric to the original circle in the drawing, as shown in Figure 4.22.

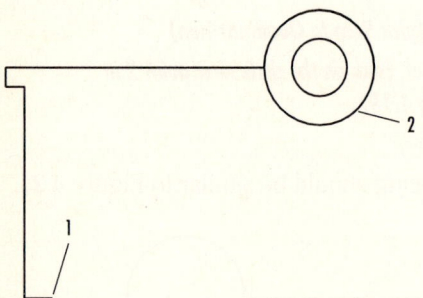

Figure 4.22

To position the next line from the endpoint of line 1 in Figure 4.20 tangent to the outer circle, you will use the Endpoint object snap and then the Tangent object snap.

Pick: **Line icon**

From point: *(pick Snap to Endpoint icon)*

Endpoint of: *(pick near the right endpoint of line 1)*

A line rubberbands from the endpoint of line 1 to the location of the crosshairs. Next, you will use the Tangent object snap to locate the second point of a line tangent to the circle.

Snap to Tangent

 The Tangent object snap attaches to a point on a circle or an arc; a line drawn from the last point to the referenced object is drawn tangent to the referenced object. The command-line equivalent is TAN.

To point: *(pick Snap to Tangent icon)*

Tangent to: *(pick the lower right side of the circle when you see the AutoSnap Tangent marker)*

To point: ⏎

Your screen should be similar to Figure 4.23.

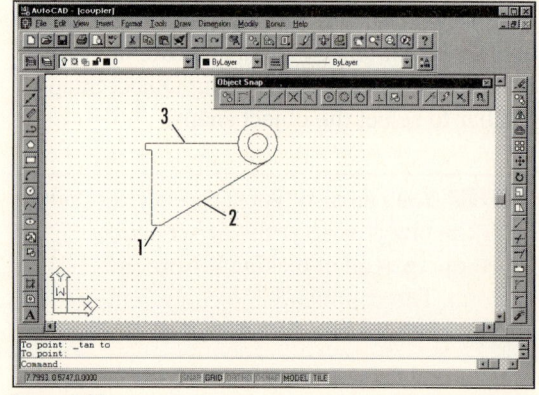

Figure 4.23

> ■ *TIP* To draw a line tangent to two circles, pick Line. At the *From point* prompt, pick Snap to Tangent and pick on one of the circles in the general vicinity of the tangent line you want drawn. You will not see a line rubberband. At the *To point* prompt, pick Snap to Tangent again and pick the other circle. When you select the second object, the tangent is defined and it will be drawn on your screen. ■

Next you will draw a line from the intersection of the short horizontal line and the angled line, perpendicular to the upper line of the object. You will use two more object snaps to draw this line.

Snap to Intersection

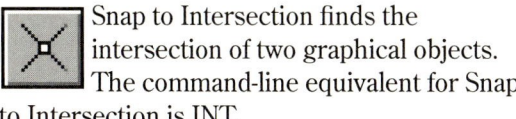Snap to Intersection finds the intersection of two graphical objects. The command-line equivalent for Snap to Intersection is INT.

When you use 3D solid modeling in Tutorial 9, remember that when you are looking for the intersection of two objects with Snap to Intersection, the objects must *actually* intersect in 3D space, not just *apparently* intersect on the screen (one line may be behind the other in the 3D model). To select lines that do not intersect in space, use the Apparent Intersection object snap. It finds the intersection of lines that would intersect if extended and also the apparent intersection of two lines (where they appear to intersect in a view), when one line actually is behind the other in 3D space.

Refer to Figure 4.23 when selecting your points.

Pick: **Line icon**

From point: *(pick Snap to Intersection icon)*

Intersection of: *(pick so that the intersection of lines 1 and 2 is anywhere inside the target area)*

A line rubberbands from the intersection to the current position of the crosshairs. To select the second endpoint of the line, you will use Snap to Perpendicular to draw the line perpendicular to the top horizontal line.

■ *TIP* Note that when object snap Intersection is on and the cursor passes over any object, the AutoSnap marker for intersection (an X) shows up with three dots to its right. These dots mean that if you select the line, AutoCAD will prompt you for a second line or object that intersects with the first. ■

Snap to Perpendicular

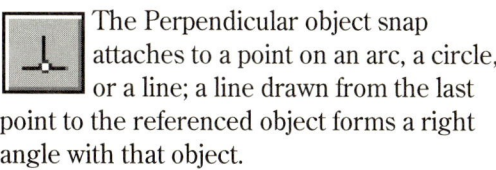The Perpendicular object snap attaches to a point on an arc, a circle, or a line; a line drawn from the last point to the referenced object forms a right angle with that object.

You will invoke the Perpendicular object snap by typing its command-line equivalent, PER.

To point: **PER** ⏎

To: *(pick on line 3)*

To point: ⏎

The line is drawn at a 90° angle, perpendicular to the line, regardless of where on the line you picked. The drawing on your screen should be similar to Figure 4.24. Although this example draws a perpendicular line that touches the target line, it need not touch it. If you choose to draw perpendicular to a line that does not intersect at an angle of 90°, the perpendicular line is drawn to a point that would be perpendicular if the target line were extended.

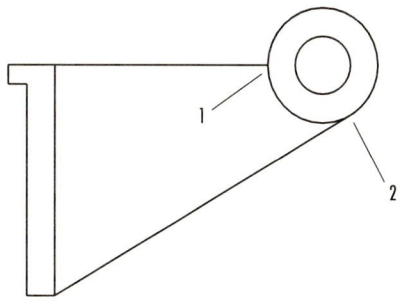

Figure 4.24

■ *TIP* Remember, in order to draw a perpendicular line, you must define two points. When you are defining a perpendicular from a line, no line is drawn until you define the second point. The reason is that infinitely many lines are perpendicular to a given line, but only one line is perpendicular to a line through a given point. Your perpendicular line will be drawn after you have selected the second point. ■

Use the techniques that you have learned to save your drawing now.

Practice on your own with each of the object snaps. Try selecting the rest of the object snaps from the toolbar.

Practicing with Running Mode Object Snaps

You can also use object snap in the running mode, as you did in Tutorial 2 with Snap to Node. When you use object snaps this way, turn the mode on and leave it on. Any time that a command calls for the input of a point or selection, the current object snap is used when you pick. The running mode object snaps are very useful, as they reduce the number of times you must pick from the menu to achieve the desired drawing results. You can use any of the object snaps in either the running mode or the override mode.

■ *Warning:* When using running mode object snaps, be sure to turn them off when you have finished. (You can do so by either deselecting them in the Object Snap Settings dialog box or toggling off the OSNAP button in the Status bar). If you forget, you may have trouble with certain other commands. For example, if you turn on the running object snap for Perpendicular and leave it on and

later try to erase something, you may have trouble selecting objects because AutoCAD will try to find a point perpendicular to every object you pick. ■

Pick: **Object Snap Settings icon**

The Object Snap Settings dialog box appears on the screen, as shown in Figure 4.25.

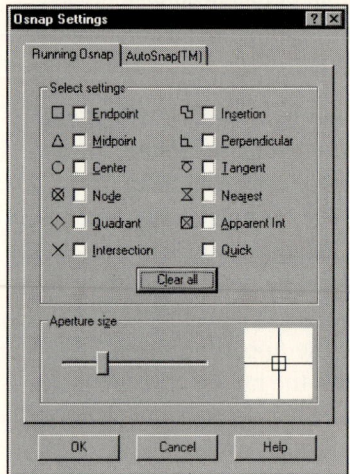

Figure 4.25

Pick the empty box to the left of Intersection on your own. A check appears in the box to indicate that you have selected Intersection. Pick OK to exit the dialog box.

Now Intersection is turned on. Anytime AutoCAD prompts you to select, an AutoSnap marker box will appear when the cursor is near a snap point.

You will use the Break command to break the circle between intersection 1 and intersection 2 in Figure 4.24. You will use the running mode object snap Intersection while selecting points during the Break command. (Note that you could also accomplish the same thing by using Trim.)

Using Break

The Break command erases part of a line (or an arc or circle). The Break icon is on the Modify toolbar. When using the Break command, you can specify a single point at which to break the object or specify two points on the object and the Break command will automatically remove the portion between the points selected. You also can select the object and then specify the two points at which to break it. You will use the 2 Points Select option after selecting the Break icon. Refer to Figure 4.24 to make your selections.

Pick: **Break icon**

Select object: **(pick the large circle)**

Enter second point (or F for first point): **F**

Enter first point: **(pick intersection 1)**

Enter second point: **(pick intersection 2)**

When you select the second break point, the portion of the circle between the two selected points is removed. Your drawing should look like Figure 4.26.

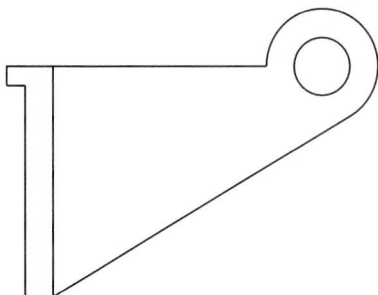

Figure 4.26

The other object snaps are described next. Try them on your own until you are familiar with how they work. When you have finished, return to the Object Snap Settings dialog box and turn off any object snaps you may have left on.

Snap to Apparent Intersection

The Apparent Intersection object snap finds two different types of intersections that the regular Intersection object snap wouldn't find. One type is the point at which two objects would intersect if they were extended. The other type is the point at which two 3D lines appear to intersect on the screen, when they do not in fact intersect in space. The command-line equivalents for Snap to Apparent Intersection are APPINT and APP.

Snap to Nearest

The Nearest object snap attaches to the point object or the point on an arc, circle, or line closest to the middle of the target area of your cursor.

A line drawn with the Nearest function may look like a line that could have been drawn simply using the Line command, but there is a subtle difference. Many of AutoCAD's operations, such as hatching, require an enclosed area: All lines that define the area must intersect (touch). When you draw lines by picking two points on the screen, sometimes the lines don't actually touch. They appear to touch on the screen, but when you zoom in on them sufficiently, you will see that they don't touch. They may only be a hundredth of an inch apart, but they don't touch. When creating a drawing with AutoCAD, you should always strive to create the drawing geometry accurately. The Nearest function ensures that the nearest object is selected. The command-line typing equivalent is NEA.

Snap to Node

The Node object snap finds the exact location of a point object in your drawing. You used it in Tutorial 2 to locate exact points that had already been placed in the drawing with the Point command. The command-line typing equivalent is NOD.

Snap to Quick

You can use the Quick object snap in conjunction with the other object snaps that you have learned, to limit the search through the database and make selection with object snaps faster. For example, if Intersection and Quick are turned on, the search through the drawing database stops as soon as one intersection is found, even if there are several. When you're using Quick, having just one of the type of object you are trying to locate is best. The command-line typing equivalent is QUI.

Snap to Insertion

The Insertion object snap finds the insertion point of text or of a block (you will learn about blocks in Tutorial 8). This method is useful when you want to determine the exact point at which existing text or blocks are located in your drawing. The command-line typing equivalent is INS.

> ■ *TIP* The Locate Point command that you learned in Tutorial 3 lists the coordinates of a selected point. To find the exact insertion point of text or a block, pick Locate Point and then use the Insert object snap and pick on the text or block. Similarly, you can find the coordinates of an endpoint or intersection by combining those object snaps with the Locate Point command. ■

Snap From

The From object snap is a special object snap tool. It establishes a temporary reference point from which you can specify the point to be selected. Usually it is used in combination with other object snap tools. For example, you would use it if you were going to draw a new line that you want to start a certain distance from an existing intersection. To do so, pick the Line command; when AutoCAD prompts you for the starting point of your line, pick the Snap From icon. You will then be prompted for a base point. It is the location of the reference point from which you want to locate your next input. At the base point prompt, pick Snap to Intersection. You will be prompted to select the intersection of two lines. Pick on the intersection. Then you will see the additional prompt *of <Offset>:*, the result of your having picked Snap From. At this prompt, use relative coordinates to enter the distance you want the next point to be from the base point you selected.

Tracking

Tracking is a unique snap tool allowing you to start a line with reference to other locations. For instance, if you have drawn a rectangle and want to draw a line starting at the middle the rectangle, there is no snap for that point.

With Tracking, AutoCAD starts the line with reference to as many points as you select. Tracking will be covered in greater detail in Tutorial 6.

> ■ *TIP* You can turn on more than one object snap at the same time. For instance, you could select both Intersection and Nearest from the Object Snap Settings dialog box. AutoSnap will show different marker boxes for each snap setting chosen. ■

Using Extend

![icon] The Extend command extends the lengths of existing lines and arcs to end at a selected boundary edge. Its function is the opposite of Trim's. Like Trim, Extend has two parts: First you select the object to act as the boundary, and then you select the objects you want to extend. Extend is on the Modify toolbar.

Similar to the Trim command, the command-line options for the Extend command are Project, Edge, and Undo. The default option is just to select the objects to be extended. You can use the Edge option to extend objects to the point at which they would meet the boundary edge if it were longer. This method is useful when the boundary edge is short and the lines to be extended would not intersect it. The Project option allows you to specify the plane of projection to be used for extending. This option is useful when you're working in 3D drawings because it allows you to extend objects to a boundary selected from the current viewing direction or User Coordinate System, even though the 3D objects may not actually intersect the boundary edge but just appear to do so in the view. The Undo option lets you undo the last object extended, while remaining in the Extend command.

Before continuing, change your snap and grid settings to 0.25. Then use the commands that you have learned to erase the lines labeled 1, 2, and 3. Draw a new line 0.5 unit (two snap increments) from the previous leftmost line. Refer to Figure 4.27 when making your selections.

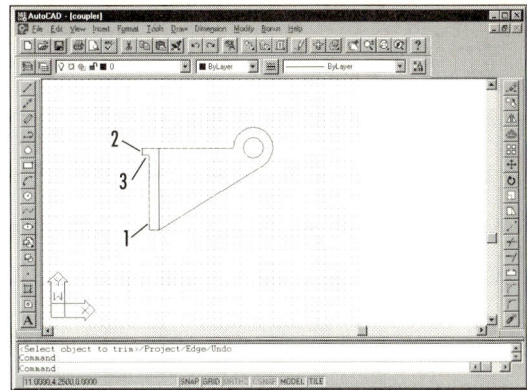

Figure 4.27

When you have completed this step, your drawing should be similar to Figure 4.28.

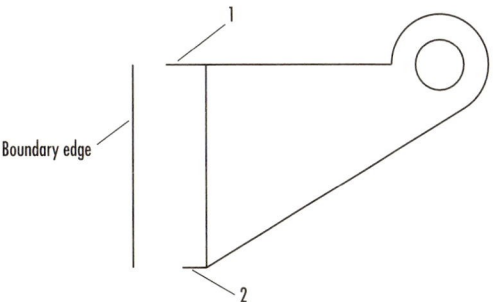

Figure 4.28

Now you will use the Extend command to extend the existing lines to meet the new boundary.

Pick: **Extend icon**

Select objects: *(select boundary edge)*

Select objects: ⏎

You have finished selecting boundary edges. To extend the horizontal lines, select them by picking points on them near the end closer to the boundary edge. (If the points are picked closer to the other end, you will get the message *No edge in that direction,* and the lines won't be extended.)

<Select object to extend>/Project/Edge/Undo: *(select lines 1 and 2)* ⏎

The screen will look like Figure 4.29 once the lines have been extended.

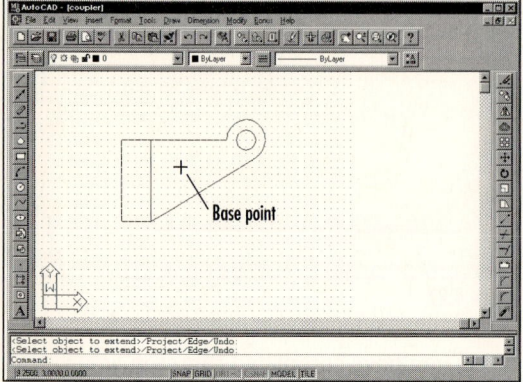

Figure 4.29

Using Rotate

 The Rotate command allows you to rotate a drawing object or group of objects to a new orientation in the drawing. Rotate is on the Modify toolbar. In this section, you will use Rotate to rotate objects in your drawing.

Pick: **Rotate icon**

Select objects: *(select the entire coupler using a Crossing window)*

7 found

Select objects: ⏎

AutoCAD prompts you for the base point. You will select the middle of the object as the base point, as shown in Figure 4.29.

> ■ *TIP* The base point need not be part of the object chosen for rotation. You can use any point on the screen; the object rotates about the point you select. You may often find it useful to select a base point in the center of the object you want to rotate. As the object rotates, it stays in relatively the same position on the screen. ■

Base point: *(pick in the middle of the coupler)*

<Rotation angle>/Reference: **45** ⏎

Your coupler should be rotated 45°, as shown in Figure 4.30. Positive angles are measured counterclockwise. A horizontal line to the right of the base point is defined as 0°. You can also enter negative values.

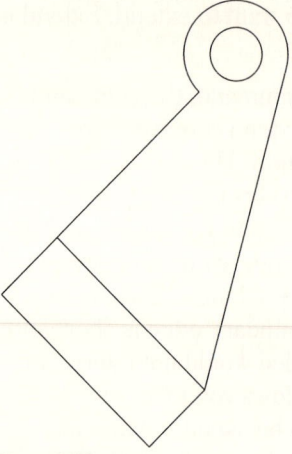

Figure 4.30

On your own, pick the Undo icon from the Standard toolbar to undo the Rotate command and restore the object to its original position before you continue.

Using Move

 The Move command moves existing objects from one location on the coordinate system in the drawing to another location. Do not confuse the Move command and the Pan command. Pan moves your view and leaves the objects where they were. Move actually moves the objects on the coordinate system. Move is on the Modify toolbar.

Pick: **Move icon**

Select objects: *(use Crossing or Window to select the entire coupler)*

Select objects: ↵

Base point or displacement: *(pick the Snap to Endpoint icon)*

Endpoint of: *(select the upper corner of the coupler)*

AutoCAD switches to the drag mode so that you can see the object move about the screen, as shown in Figure 4.31.

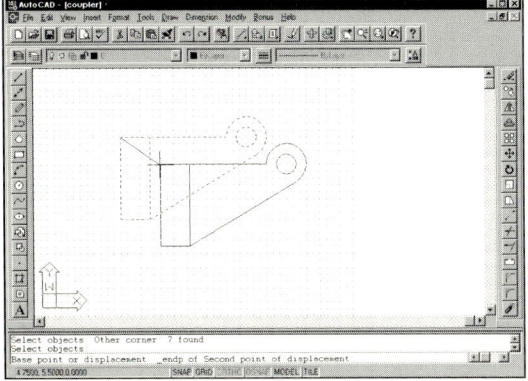

Figure 4.31

Second point of displacement: *(select a point so the coupler is to the lower right of its old location)*

Methods of Selecting Objects

One advantage of AutoCAD is the many ways that you can select graphical objects for use with a command. You have already learned how to use Window and Crossing. You can save lots of time when creating and editing drawings by the clever use of the selection methods. Generally, whenever you are asked to select objects, you can continue selecting until you have all the objects that you want highlighted in one *selection set*. You can combine the various selection modes to select the objects you want. The command then operates on the selection set you have built.

During any command that prompts *Select objects*, you can type the option letters for the method you want to use. AutoCAD allows you to continue selecting, using any of the methods in the following table, until you indicate that you have finished building the selection set by pressing ↵ or the return button. Then the command will take effect on the objects that you have selected.

Name	Option Letter(s)	Method
Picking	none	Picks objects by placing selection cursor and pressing pick button
Select Window	W	Specifies diagonal corners of a box that only selects objects that are entirely enclosed
Select Crossing	C	Specifies diagonal corners of a box that selects all objects that cross or are enclosed in the box
Select Group	G	Selects all objects within a named group, prompts to enter group name
Select Previous	P	Reselects the previous selection set
Select Last	L	Selects the last object created
Select All	ALL	Selects all the objects in your drawing unless they are on a frozen layer
Select Window Polygon	WP	Similar to Window except you draw an irregular polygon instead of a box around the items to select

Name	Option Letter(s)	Method
Select Crossing	WC	Similar to Window Polygon except that all Polygon objects that cross or are enclosed in the polygon are selected
Select Fence	F	Similar to WC except that you draw line segments through all the objects you want to select
Select Add	A	Use after Remove to add more objects to the selection set; can continue with any of the other selection modes once Add has been chosen
Select Remove	R	Selects objects to remove from the current selection set; removal continues until Add or ↵ is used; you can use Window, Crossing, etc., while selecting items to remove
Undo	U	During object selection, deselects the last item or group of items you selected

Selection filters are a special method of selecting objects. You can use them to select types of objects, such as all the arcs in the drawing or objects that you created by setting the color or linetype independently of the layer that they are on.

Type: **FILTER**

The Object Selection Filters dialog box shown in Figure 4.32 appears on your screen. You can use this dialog box to filter the types of objects to select. You can save named filter groups for reuse.

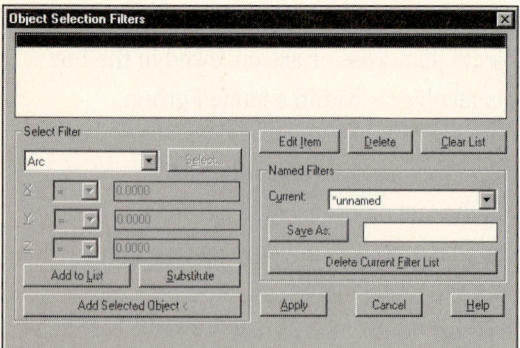

Figure 4.32

Pick: **Arc (below the heading Select Filter)**

The list of drawing object types pops up. You can use these drawing object types to filter the objects in your selection.

Pick: **Arc (so that it is highlighted)**

Pick: **Add to List**

Pick: **Apply**

You return to the command prompt to select the objects in your drawing. Note the message *Applying filter to selection.*

Select objects: *(use Crossing to select all of the objects)*

You should see the message *7 found, 6 were filtered out.* Only the arc object matched the filter list. The lines of the drawing were filtered out.

Select objects: Esc *(to cancel)*

Additional methods of selecting are Auto, Box, Single, and Multiple. Consult AutoCAD's online help about the Select command for definitions of their use.

On your own, create a new drawing called *geneva.dwg.* Discard the changes to *coupler.dwg.*

Creating the Geneva Cam

You will create the geneva cam shown in Figure 4.33, using many of the editing commands you have learned in this tutorial. You will learn to use the Array command to create rectangular and radial patterns.

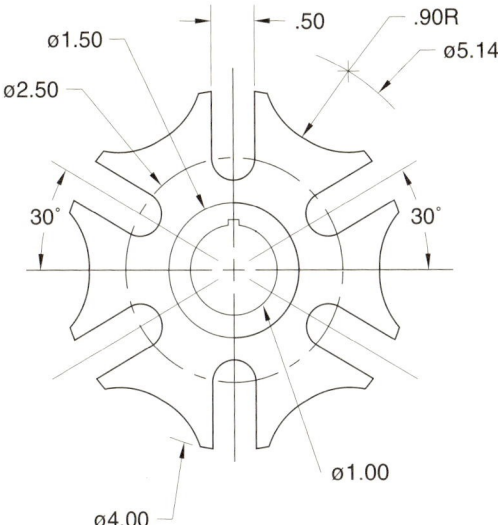

Figure 4.33

On your own, set Snap to 0.25. Be sure that Snap and Grid are turned on by checking to see if the words Snap and Grid are displayed on the status bar. If not, turn them on. Use Zoom All if needed so that the grid area fills the graphics window.

Next you will use the Circle command to create the innermost circle of diameter 1.00. Locate the center of the circle at (5.5, 4.5).

Pick: **Circle icon**

3P/2P/TTR/<Center point>: **5.5,4.5** ⏎

Diameter/<Radius>: **D** ⏎

Diameter: **1** ⏎

On your own, use Offset or the Circle command to draw the 1.5-diameter circle and the outer 4-diameter circle concentric to the circle you drew in the previous step and add the centermarks. (If you use Offset, remember to calculate the offset distance. You will need to subtract the diameter of the inner circle from the diameter of the outer circle and then divide the result by 2 to find the value to use for the offset distance.)

Your drawing should look like Figure 4.34.

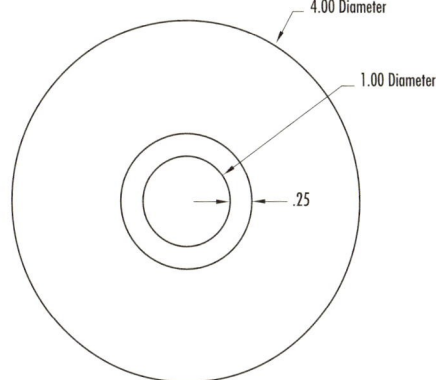

Figure 4.34

On your own, draw vertical and horizontal construction lines through the center of the circles. Next you will use the Offset command to create lines parallel to the vertical centerline.

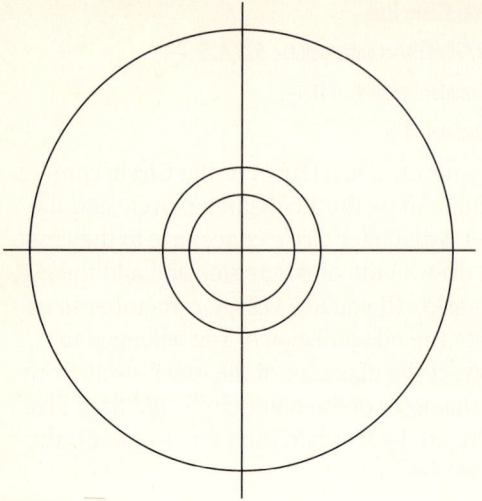

Figure 4.35

Pick: **Offset icon**

Offset distance or Through <Through>: **.25** ⏎

Select object to offset: *(pick the vertical centerline)*

Side to offset? *(pick on the right side of the line)*

Select object to offset: *(pick the same vertical centerline)*

Side to offset? *(pick on the left side of the line)*

Select object to offset: ⏎

Next you will use the Construction Line command to draw a line angled at 60° and infinitely long, as shown in Figure 4.36.

Using Construction Line

 The Construction Line and Ray commands create lines that extend infinitely and are very useful in creating construction geometry. Construction Line, on the Draw toolbar, creates a line (through a point that you select) that extends infinitely in both directions from the first point selected. The Ray command, on the Draw menu, extends infinitely in only one direction.

Pick: **Construction Line icon**

From point: **5.5,4.5** *(or pick the center point of the circles, using Snap)*

Through point: **@3<60** ⏎

Through point: ⏎

The line appears in your drawing, as shown in Figure 4.36. The line will extend as far as you zoom in both directions of extension. If you trim a construction line at one end, it becomes a ray, extending infinitely in only one direction. If you trim it at both ends, it becomes a line. On your own, trim the construction line at the horizontal line so that it only extends upward.

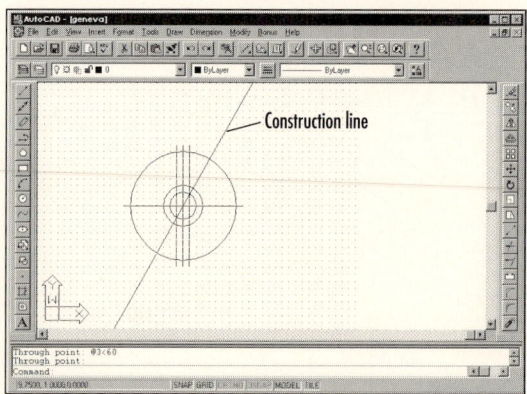

Figure 4.36

Now you will use the Circle command to add the construction circles to your drawing. On your own, make sure that the Running Osnap is on with Center and Intersection selected.

Pick: **Circle icon**

3P/2P/TTR/<Center point>: *(select the center of the circle using the AutoSnap Center marker)*

Diameter/<Radius>: **2.57** ⏎

Command: ⏎ *(or press the return button)*

3P/2P/TTR/<Center point>: *(select the center of the circle using the AutoSnap Center marker to pick the same center point as before)*

Diameter/<Radius> <2.5700>: **1.25** ⏎

Command: ⏎

3P/2P/TTR/<Center point>: *(use AutoSnap Intersection to pick the point where the vertical centerline crosses the 1.25-radius construction circle)*

Diameter/<Radius> <1.2500>: **.25** ⏎

The small circle is added to your drawing between the two lines you offset. Your drawing should look like that in Figure 4.37.

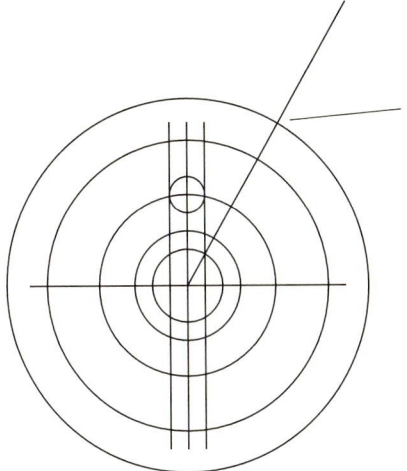

Figure 4.37

Draw the final construction circle by specifying its center where the 60° ray and the outer circle intersect.

Command: ⏎

3P/2P/TTR/<Center point>: *(pick point 1 in Figure 4.37, using the AutoSnap Intersection marker)*

Diameter/<Radius> <.2500>: **.90** ⏎

Your drawing should now look like that in Figure 4.38.

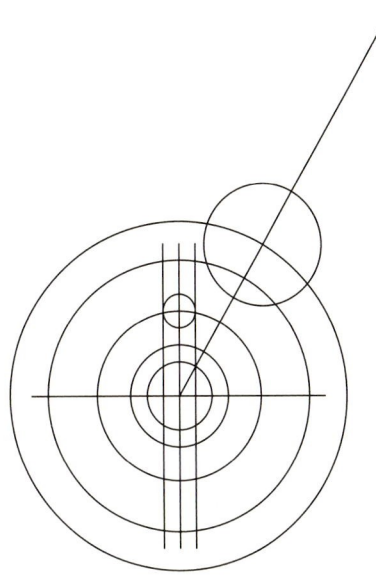

Figure 4.38

The next step is to remove the portions of the lines and circles that you won't need in the Array command.

On your own, use the Trim and Erase commands to remove the unwanted portions of the figure until your drawing looks like the one in Figure 4.39.

> ■ *TIP* If you use the implied Crossing method to select all the objects in the drawing as the cutting edges and then press ⏎, you can quickly trim the object to the shape shown by picking on the extreme ends of the objects you want to remove. Use Erase to remove any of the objects that no longer cross a cutting edge. ■

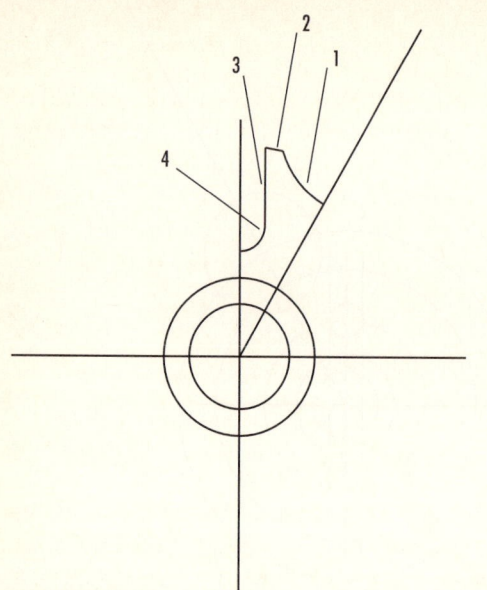

and the distance at which the mirror image is to be drawn. The mirrored image is drawn perpendicular to the mirror line. The mirrored object is the same distance from the mirror line as the original object is, but it's on the other side of the mirror line. The mirror line does not have to exist in your drawing before you use the Mirror command; when asked to specify the mirror line, you can pick two points from the screen that define it.

The vertical centerline will serve as the mirror line as you mirror the lines and arcs of the geneva cam. Refer to Figure 4.39 as you pick points. Be sure that Snap is turned on or use object snaps before picking the mirror line.

Pick: **Mirror icon**

Select objects: *(pick objects 1, 2, 3, and 4)*

Figure 4.39

■ *TIP* If you have a three button mouse, the override object snaps are available with the middle button. A pop-up menu appears on screen with the object snap settings as well as Running Osnap. ■

Using Mirror

 The Mirror command makes a mirror-image copy of the objects you select around a *mirror line* that you specify. The Mirror command is on the Modify toolbar. Its icon is the mirror image of a shape. The Mirror command also allows you to delete the old objects or to keep them.

To mirror an object, you must specify a mirror line, which can be horizontal, vertical, or slanted. A mirror line defines both the angle

■ *TIP* If you select objects that you do not want to mirror, you can use the Remove option of the object selection modes to remove them from the selection set. To do so, type *R* ↵ or pick Remove from the Select Objects flyout on the Standard toolbar while still at the *Select objects:* prompt. Pick the object to be removed from the selection. If you want to add more objects after using Remove, pick Add or type *A* ↵ while still at the *Select objects* prompt. Press ↵ when you have selected the objects you want. Doing so tells AutoCAD that you have finished selecting and want to continue with the command. ■

Select objects: ↵

First point of mirror line: *(select the center of the concentric circles)*

Your cursor is now dragging a mirrored copy of the objects.

> Second point: **(select a point straight down from the center point)**
>
> Delete old objects? **<N>** ⏎

Your screen should look like Figure 4.40.

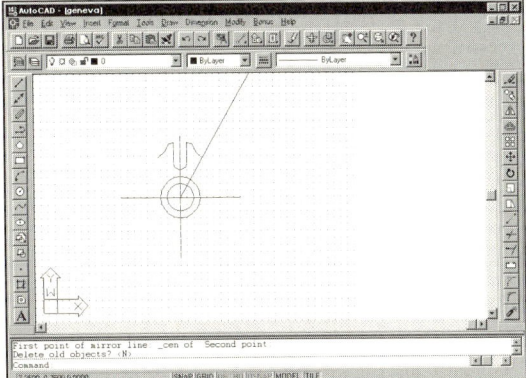

Figure 4.40

> ■ **TIP** Figure 4.41 shows an object mirrored about one of its own edge lines. If you want, you can add mirror lines to the drawing and then erase them after completing the construction; or you can pick any two points from the screen that would define the line you want to specify. ■

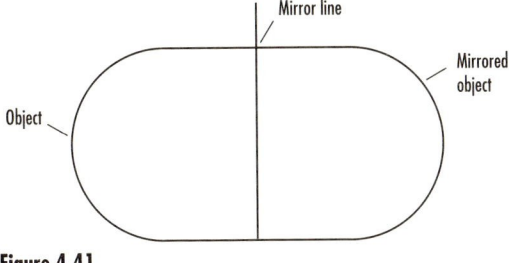

Figure 4.41

Using Array

The Array command copies an object multiple times to form a regularly spaced rectangular or circular pattern. Anytime that you need to create a regularly spaced pattern of objects in your drawing, you can use the Array command to do so quickly. Examples are creating a circular pattern of holes in a circular hub, creating rows of desks in laying out a classroom, or creating the teeth on a gear.

The Array command is on the Modify toolbar. In the Array command, AutoCAD prompts you to select the items you want to array. The array can be either rectangular or polar. A *rectangular array* is composed of a specified number of rows and columns of the items with a specified distance between them. In a *polar array*, the items are copied in a circular pattern around a center point that you specify. The copies can fill 360° of the circle or some portion of 360°. You can choose to have the items rotated as they are copied with the Array command so that they have the proper orientation.

You are now ready to use the Array command to copy the lines and arcs in a radial pattern to finish creating the geneva cam. Refer back to Figure 4.39 for the items to select.

> *Pick:* **Array icon**
>
> Select objects: **(pick the items 1–4 that you mirrored and the mirrored copies of them)**
>
> Select objects: ⏎
>
> Rectangular or Polar array (<R>/P): **P**
>
> Base/<Specify center point of array>: **(pick the center of the concentric circles)**
>
> Number of items: **6** ⏎
>
> Angle to fill (+=ccw, –=cw) <360>: ⏎
>
> Rotate objects as they are copied? <Y>: ⏎

On your own, use Erase to erase the ray from the drawing. Your drawing should look like that in Figure 4.42.

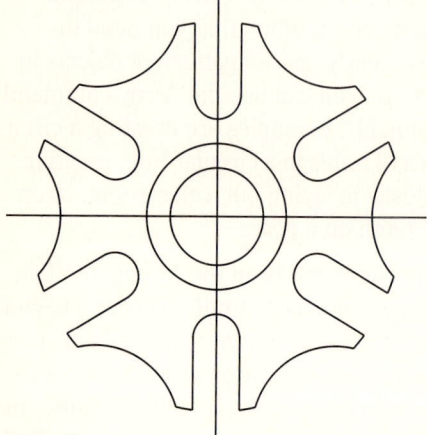

Figure 4.42

Your drawing is now complete; save it on your own.

Making Changes with Hot Grips

A quick way to make changes to your drawing is by using *hot grips*. Hot grips let you grab an object already drawn on your screen and use certain editing commands directly from the mouse or pointing device without having to select the commands from the menu or keyboard. To use the hot grips you must have a two-button mouse as your minimum hardware configuration. You will verify that hot grips are enabled, using the Grips dialog box.

Pick: **Tools, Grips**

The Grips dialog box appears on your screen, as shown in Figure 4.43. You should select the box to the left of Enable Grips if it is not already selected.

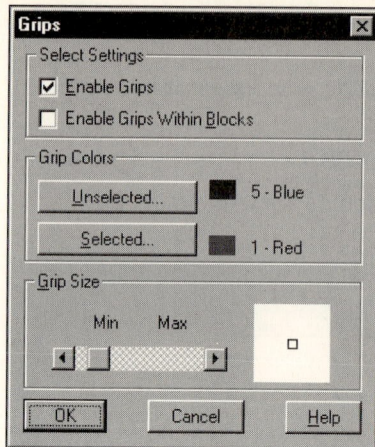

Figure 4.43

AutoCAD allows you to turn grips on or off in your drawing and to change their color and size. To do so on your own, move the slider box or pick on the arrows toward the bottom of the dialog box. Press the pick button on the right arrow several times to enlarge the size of the grips box. When you have finished, restore it to its original size. You can also change the color of the grips and the base or activated grip, and turn grips off completely if you do not want to use them.

Exit the dialog box by picking OK. AutoCAD returns you to the drawing editor.

Activating an Object's Grips

Move the crosshairs over the upper vertical centerline of the geneva cam and press the pick button to select the line. The line becomes dashed and small boxes appear at the endpoints, as shown in Figure 4.44.

Figure 4.44

These boxes are called the hot grips. You can use them to stretch, move, rotate, scale, and mirror the object.

Using Stretch with Hot Grips

Move the crosshairs to the grip at the upper end of the line and press the pick button to select it. It will change to the highlighting color and be filled solid. This *base grip* will act as the base point for the command you will select by pressing the buttons on your pointing device.

In the command line area, you should see the prompt for the Stretch command.

    STRETCH

    <Stretch to point>/Base point/Copy/Undo/eXit:

The upper end of the line, where you picked the base grip, rubberbands from the position of the crosshairs. Move the crosshairs to a point above and to the right of the old location of the corner and press the pick button. The point you selected should move to the new stretched location.

■ *TIP* If you want to cancel using the hot grips, press `Esc` twice to eliminate the hot grips. ■

Using Move with Hot Grips

Now make a crossing box that crosses the entire geneva cam. The hot grips appear as small boxes on all the selected objects. Select the center of the circles as the base grip by moving the crosshairs over the box and picking. You will see the prompt for the Stretch command in the command line area. Press the return button on your pointing device, and a pop-up Hot Grips menu similar to Figure 4.45 will appear. Select Move from the menu.

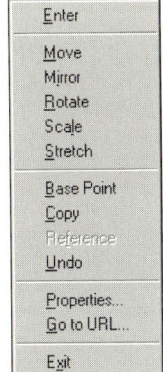

Figure 4.45

■ *TIP* You can also press ← to cycle through the command choices that are available for use with hot grips. ■

The command prompt shows the following options for the Move command:

    MOVE

    <Move to point>/Base point/Copy/Undo/eXit:

Move the crosshairs down and to the left. You will see an outline of the objects attached to the crosshairs. Move them to a new location down and to the left and press the pick button when you have them positioned where you want them.

Using Move with the Copy Option

The hot grips should still be highlighted. Pick the grip at the center of the circles as the base grip. Press the return button on your pointing device to activate the Hot Grips menu and choose Move again.

Now press the return button again and choose Copy, or type in the letter *C* for Copy.

A new prompt, similar to the previous one, appears.

MOVE (multiple)

<Move to point>/Base point/Copy/Undo/eXit:

Move the crosshairs to a location where you would like to make a copy of the object and press the pick button; repeat this procedure to make several copies. When you are finished, press ⏎ or the return button and pick Exit to end the command.

If you were to choose Copy from the menu instead of Move, AutoCAD would copy only the object that shares the Hot Grips box that you selected. To see how this alternative works, perform the same task but select Copy from the menu without selecting Move first.

On your own use the Erase command to erase some of the copies if your screen is too crowded.

Using Rotate with Hot Grips

Now you will use implied Windowing to activate the hot grips for these lines.

> *Pick: (above and to the left of one of the copies of the geneva cam)*

A window box will form as you move the crosshairs away from the point that you selected. Move downward and to the right to enclose all the lines in the drawing in the forming window. When you have the window sized correctly,

> *Pick: (a point below and to the right of the geneva cam)*

The hot grips will appear on the objects in the drawing.

Select the grip at the center of the circles as the base grip by positioning the crosshairs over it and pressing the pick button. You will see it turn the highlighting color and be filled in solid. The Stretch command appears at the command prompt. Press the return button for the Hot Grips menu and select the Rotate command. If Snap is turned on, turn it off so that you can see the effect of the rotation command clearly.

You will see the prompt

ROTATE

<Rotation angle>/Base point/Copy/Undo/Reference/eXit:

You will see the faint object rotating as you move the crosshairs around on the screen. You can press the pick button when the object is at the desired rotation or type in a numeric value for the rotation. Keep in mind that 0° is to the right and positive values are measured counterclockwise.

> *Type:* **45** ⏎

Note that the object rotated so that it is at an angle of 45°.

Using Scale with Hot Grips

The Scale command changes the size of the object in your drawing database. Be sure that you use Scale only when you want to make the actual object larger. Use Zoom Window when you just want it to appear larger on the screen in order to see more detail.

On your own, use hot grips to scale the object. Activate the hot grips, using implied Windowing again. This time, however, use the Crossing option. To use Crossing instead of

Window, start your window box in the lower right of the drawing and select the first point. Move the crosshairs with the window up and to the left. This action specifies a crossing box. Unlike Window, which selects only the objects that are entirely enclosed in the box, Crossing selects anything that crosses the box or is enclosed. When you have done this successfully, the hot grips will appear at the corners and midpoints of the lines.

Pick: **(the grip at the center of the circles as the base grip)**

It becomes solid, filled with the highlighting color. You will see the Stretch command in the command prompt area.

Press: **(the return button)**

Pick: **Scale**

You will see the prompt

SCALE

<Scale factor>/Base point/Copy/Undo/Reference/eXit:

As you move the crosshairs away from the base point, the faint image of the object becomes larger; as you get closer to the base point, it appears smaller. When you are happy with the new size of the object, press the pick button.

> ■ **TIP** To change the scale to known proportions, you can type in a scale factor. A value of 2 makes the object twice as large; a value of 0.5 makes it half its present size. ■

Using Mirror with Hot Grips

On your own, activate the hot grips with implied Windowing again. When the grips appear on the object, select the center grip as the base grip. Press the return button and select Mirror.

The Mirror command uses a mirror line and forms a symmetrical image of the selected objects on the other side of the line. You can think of this line as rubberbanding from the base hot grip to the current location of the crosshairs. Note that, as you move the crosshairs to different positions on the screen, the faint mirror image of the object appears on the other side of the mirror line. You will see the prompt

MIRROR

<Second point>/Base point/Copy/Undo/eXit: B ⏎

The Base point option lets you specify some other point besides the hot grip that you picked as the first point of the mirror line for the object.

Base point: **(pick a point to the left of the figure)**

Move the crosshairs on the screen and note the object being mirrored around the line that would form between the base point and the location of the crosshairs. When you are happy with the location of the mirrored object, press the pick button to select it. The old object disappears from the screen and the new mirrored object remains. End the command by pressing ⏎ or the return button.

Noun/Verb Selection

You can also use hot grips with other commands for *noun/verb selection*. Noun/verb selection is a method of selecting the objects that will be affected by a command first, instead of first selecting the command and then the group of objects. Think of the drawing objects as nouns, or things, and the commands as verbs, or actions. You will use this method with the Erase command to clear your screen.

Pick: **(a point below and to the right of all your drawing objects)**

Other corner: **(pick a point above and to the left of the drawing objects)**

You will see the hot grips for the objects that were crossed by the implied Crossing box appear in your drawing.

Pick: **Erase icon**

Note that you do not have to select the items. They were preselected with the hot grips, and as soon as you picked the Erase icon, the items that you had selected were erased.

Exit AutoCAD. You have completed Tutorial 4.

KEY TERMS

base grip	inscribed	quadrant point
circumscribed	mirror line	rectangular array
distance across the flats	noun/verb selection	selection filters
global linetype scaling factor	polar array	selection set
hot grips	polygon	

KEY COMMANDS

Array	Ray	Snap to Nearest
Break	Rotate	Snap to Perpendicular
Copy	Scale	Snap to Quadrant
Extend	Snap From	Snap to Quick
Grips	Snap to Apparent	Snap to Tangent
Linetype	Intersection	Stretch
Mirror	Snap to Center	
Move	Snap to Insertion	
Polygon	Snap to Intersection	

EXERCISES

The letter M after an exercise number means that the given dimensions are in millimeters (metric units).
If no letter follows the exercise number, the dimensions are in inches. Don't include dimensions on the
drawings.

4.1M Gasket

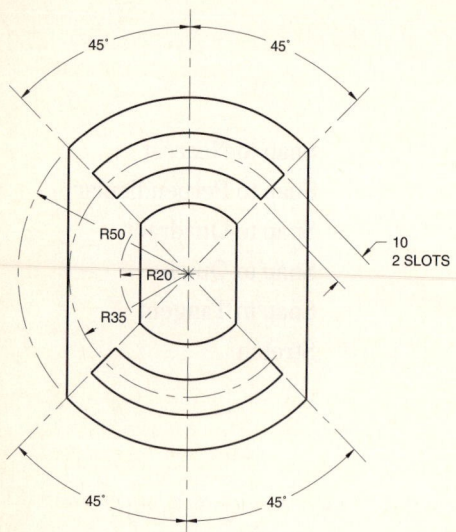

4.2 Starboard Rear Rib

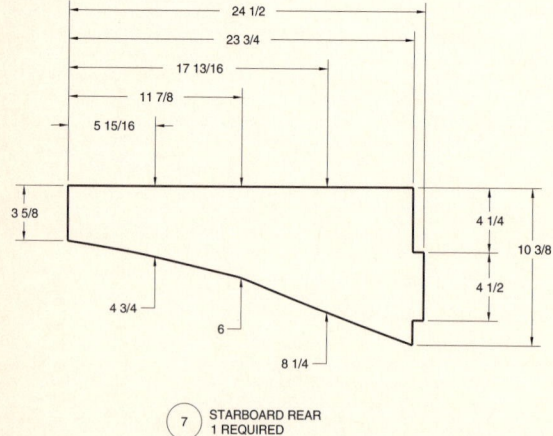

4.3 The Cycle

Starting with the parallelogram shown, use Object snap to draw the lines, arcs, and circles indicated in the drawing. Object snap commands needed are INT, END, CEN, MID, PER, and TAN.

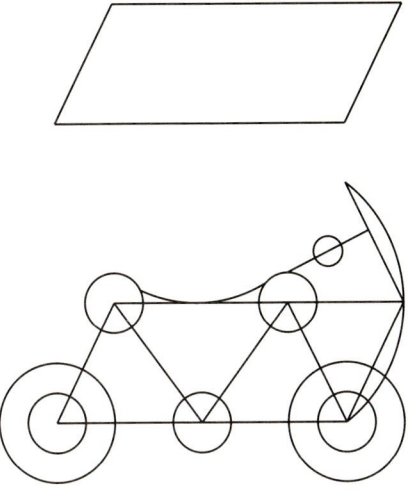

4.4 Park Plan

Create a park plan similar to the one shown here. Use Polyline and Polyline Edit, Spline to create a curving path. Note the symmetry. Add labels with Dtext.

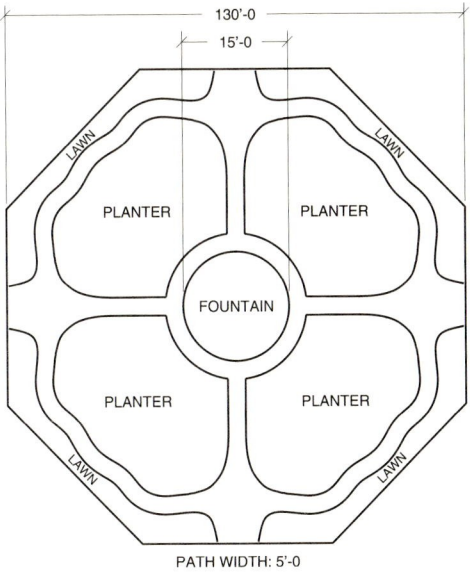

Draw the floor plan according to the dimensions shown. Add text, border, and title block.

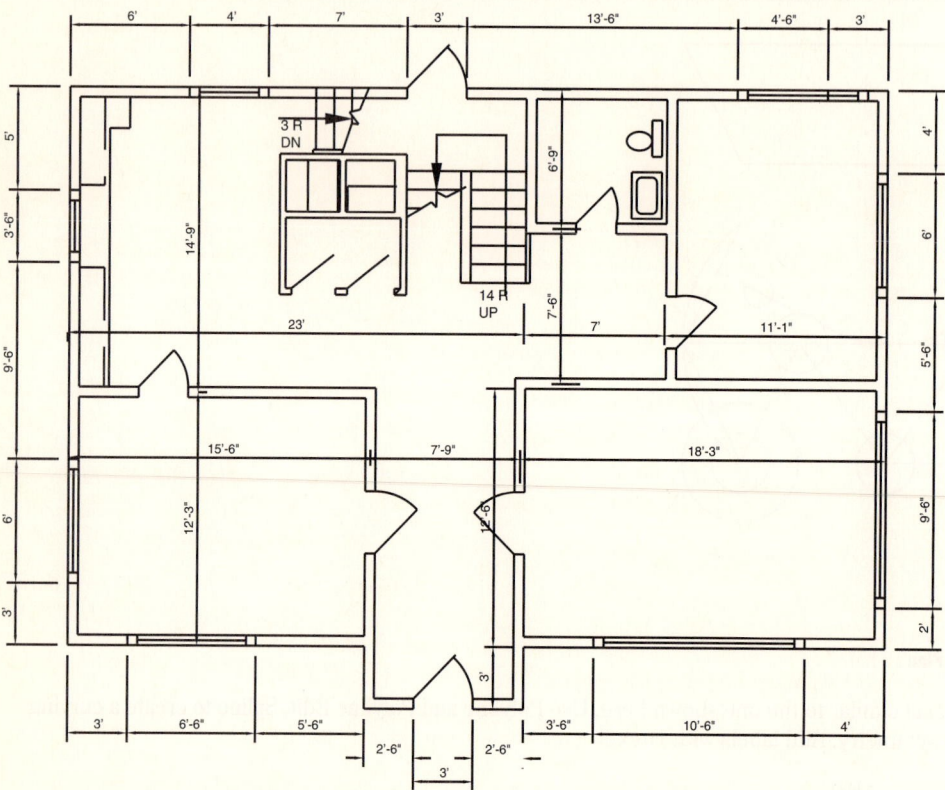

4.6 Interchange

Draw this intersection, then mirror the circular interchange for all lanes.

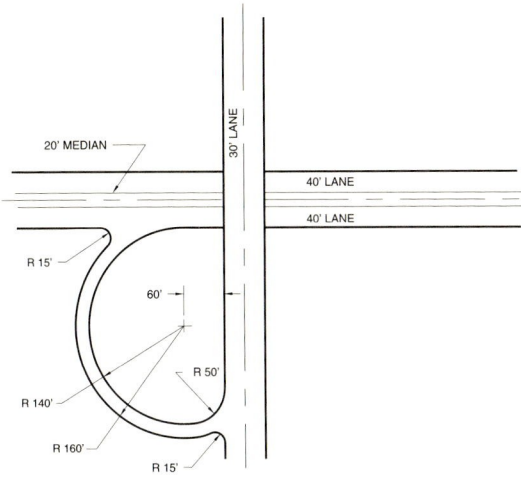

4.7M Plastic Knob

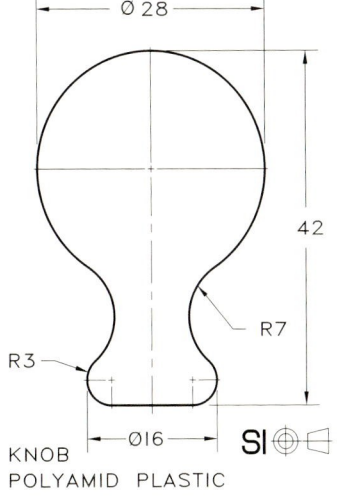

4.8M Foundry Hook

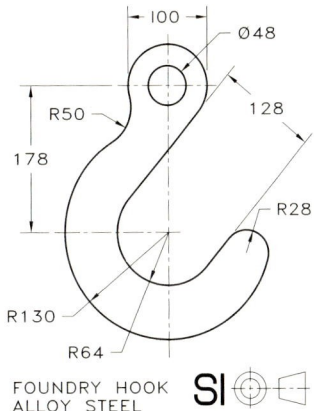

4.9M Support

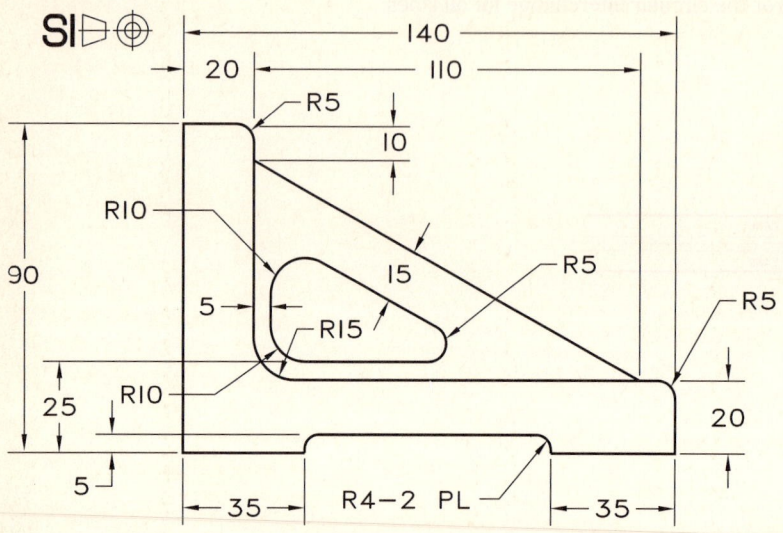

4.10 Hanger

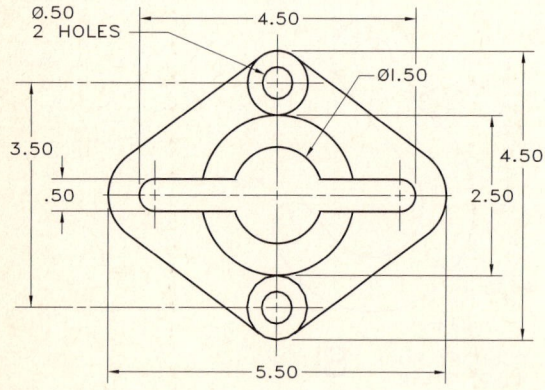

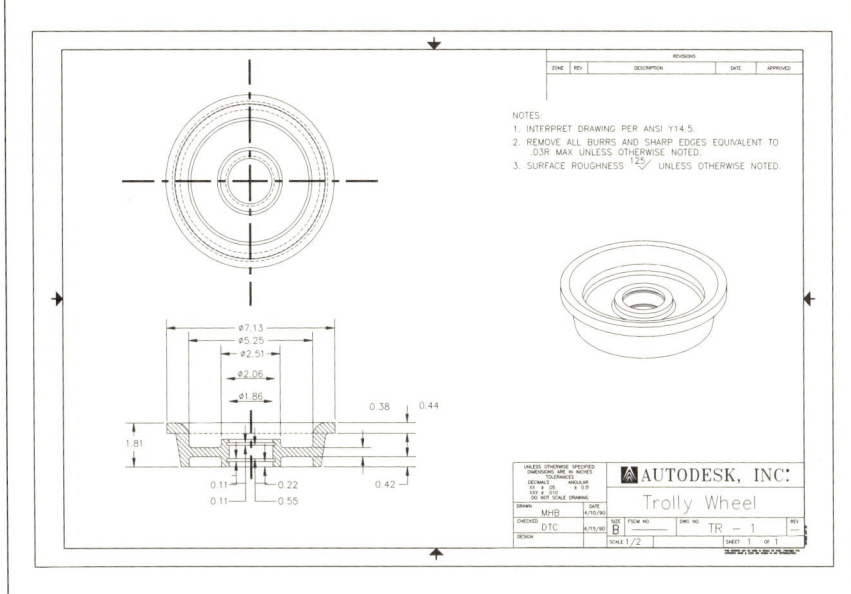

TUTORIAL 5

Template Drawings

Objectives

When you have completed this tutorial, you will be able to

1. Create and save a template drawing for later use.

2. Create a system of basic layers for mechanical drawings.

3. Use the Drawing Aids dialog box.

4. Preset Viewres, Limits, and other defaults in a template drawing.

5. Set up paper space and model space in a template drawing.

6. Insert one drawing into another as a block.

7. Set the style for drawing points.

8. Use the Divide command.

9. Use the Purge command.

10. Plot drawings using paper space.

Introduction

One of the advantages of using AutoCAD is that you can easily rescale, change, copy, and reuse drawings. Up to this point, most of the drawings that you have created began from a drawing called *acad.dwt*, which is part of the AutoCAD program. In *acad.dwt*, many variables are preset to help you begin drawing. A drawing in which specific default settings have been selected and saved for later use is called a *template drawing*. You can use any existing drawing as a template from which to start a new drawing. You may need to establish more than one template drawing to use for different sheet sizes and types of drawings. In this tutorial you will make a drawing containing default settings of your own from which you will start drawings in succeeding tutorials.

Setting up template drawings can eliminate repetitive steps and make your work with AutoCAD more efficient. The amount of time you spend creating one template is roughly the amount of time you will save on each subsequent drawing that you start from the template drawing. Your template drawing will contain your custom defaults for layers, limits, grid, snap, and text size and font for use in future drawings.

You will also use the layer commands to create your own layers for use in your drawings. In Tutorial 2 you used Layer Control to control layers that had already been created in the subdivision drawing. In this tutorial you will learn to create your own layers and set their linetypes and colors. You will also use the Drawing Aids dialog box to set up Grid, Snap, and other drawing tools.

You will enable paper space and set up a viewport in which you can view model space. Using paper space allows you to easily control where your drawing views are placed on the actual sheet of paper when you plot your drawing. You will also insert *c:\work\mysubdiv.dwg* into your current drawing as a block and use the settings you've created there to plot it.

Starting

Before you begin, launch AutoCAD. After choosing Start From Scratch in the Start Up dialog box, you should be in the AutoCAD drawing editor. Be sure that the Draw and Modify toolbars, as well as the Standard and Object Properties toolbars, are turned on.

Pick: **Save icon**

The Save Drawing As dialog box should appear on the screen. Located at the bottom of the Save Drawing As dialog box is a "Save as type:" option allowing you to save the drawing in Release 14, Release 13/LT95, Release 12/L2, and Template format (select the current file type to display a pull-down menu showing these formats). Remember, you may use either a regular AutoCAD drawing (**.dwg*) or an AutoCAD template drawing (**.dwt*) for templates. The only difference between a drawing file (**.dwg*) and a template file (**.dwt*) is the file extension. For this tutorial you will create a template file (**.dwt*). The default choice is to save files in Release 14 (**.dwg*) format, so you must change the selection to Drawing Template File (**.dwt*).

Pick: **Drawing Template File (*.dwt)**

Type: **MYTEMPLATE (in the text box to the right of "File name:")**

Pick: **Save (to exit the dialog box)**

A Template Description dialog box appears in which you may type a description, such as the one shown in Figure 5.1. The measurement pull-down menu at the bottom of the dialog box describes whether the template uses English units or Metric units. For now you will leave it as English and choose OK.

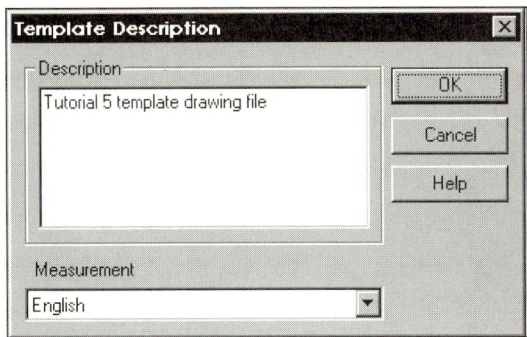

Figure 5.1

Pick: **OK**

The main AutoCAD drawing editor appears on your display screen, with *mytemplate* in the title bar. Remember, you may use either a regular AutoCAD drawing (**.dwg*) or an AutoCAD template drawing (**.dwt*) for templates.

> ■ *TIP* You can use the operating system to rename template files to drawing files and vice versa. ■

Making a Template Drawing

You will set the limits, grid, snap, layers, text size, and text style in this drawing, as well as set up the paper space limits, viewport, and linetype scale. This approach will save time, make plotting and printing easier, and provide a system of layers to help you keep future drawings neatly organized.

Effective Use of Layers

Recall from Tutorial 2 that you can think of layers as clear overlay sheets in your drawing. However, unlike a stack of overlays, the coordinate system always aligns exactly from one layer to the next. Layers are similar to the system of *pin registry* drafting, which is often used

in drafting maps manually. In pin registry drafting, a series of transparent sheets are punched with a special hole pattern along one edge, allowing the sheets to be fitted onto a metal pin bar. The metal pin bar aligns the drawings from one sheet to the next. Each sheet is used to show different information. For example, one sheet may show streets, another may show lot lines, and another may show political boundaries; yet another may show rivers and other geography. When a client wants a map prepared, only the sheets that contain the information needed are attached to the pin bar and printed with a vacuum frame printer. You can use layers for similar purposes, such as to organize the information in your drawing and turn off or freeze information you do not want to show. Unlike pin registry drafting, where the sheets of punched transparent paper sometimes slip, causing misalignment, layers always stay aligned unless *you* change them.

Using layers helps you organize the information in your drawing. To use layers effectively, choose layer names that make sense and separate the objects you draw into logical groups.

Object color in the drawing controls the pen selection when you are printing from AutoCAD. The same control holds for line thickness on printers that are capable of printing different line weights. As discussed in Tutorial 2, the default in AutoCAD is to set color by layer. Using different layers helps standardize the colors of types of objects in the drawing, which in turn helps standardize plotting the drawing. You must use different colors for different types of objects in order to plot more than one color or line thickness effectively.

Using a template drawing helps you maintain a consistent standard for layer names. Using consistent and descriptive layer names allows more than one person to work on the same drawing without puzzling over the purpose of various layers. On networked computer systems, many different people can use or work on a single drawing. A template drawing is an easy way to standardize layer names and other basic settings, such as linetype.

Using Layer

The Layer command controls the color and linetype associated with a given layer. You can also use Layer to control which layers are visible or plotted at any one time and to set the current layer. Remember, only one layer at a time can be current. New objects are created on the current layer. Use the Layers icon on the Object Properties toolbar to create new layers and set their properties (its keyboard alias command is LA).

Pick: **Layers icon**

The Layer and Linetype Properties dialog box appears on the screen, as shown in Figure 5.2.

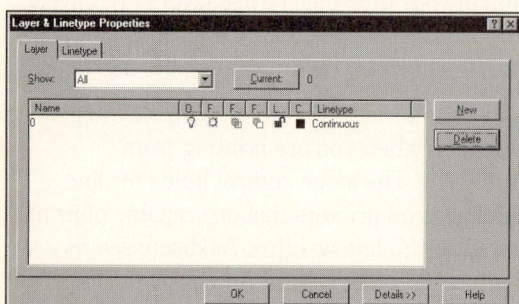

Figure 5.2

One layer name should be listed in the Layer Name column, 0. Layer 0 is a special layer that is provided in AutoCAD. You cannot rename it or delete it from the list of layers. Layer 0 has special properties when used with the Block and Insert commands. You will use Insert later in this tutorial. In Tutorial 8 you will use the Block command and work more with Insert. Layer 0 is the current layer.

You will now create a layer named HIDDEN_LINES for drawing hidden lines and set its color and linetype. Then, you will make it the current layer.

Pick: **New button (located on the right side of the dialog box)**

A new layer appears with the default name Layer1, which should be highlighted. Layer names can be as long as 31 characters. Layer names cannot have spaces in them, nor can they have any illegal DOS characters, such as a period (.), comma (,), or pound sign (#). Letters, numbers, and the characters dollar sign ($), underscore (_), and hyphen (-) are valid.

Type: **HIDDEN_LINES** ⏎

Next, set the color for layer HIDDEN_LINES. To the right of the layer names are the various layer controls mentioned in previous tutorials. Each column title is shortened with the first letter of the function: On (light bulb), three Freeze functions (suns), Lock/Unlock (lock), and Color (small color box). On your own, pick the small box in the color column corresponding to the new layer HIDDEN_LINES.

A dialog box containing color choices pops up on the screen. Figure 5.3 shows the Select Color dialog box.

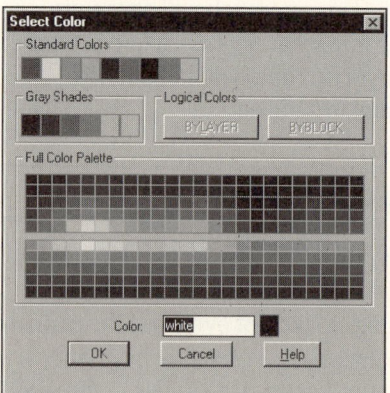

Figure 5.3

Color in Layers

The Select Color dialog box allows you to specify the color for objects drawn on a layer. You will select the color blue for the hidden line layer that you are creating. The color helps you visually distinguish linetypes and layers in drawings. You also use color to select the pen and pen width for your printer or plotter.

You use the Select Color dialog box to select color in other dialog boxes, as well as in the Layer Control dialog box. It has the choices BYLAYER and BYBLOCK on the right side. As you are specifying the color for layer HIDDEN_LINES only, you can't select these choices, so they are shown grayed. Move the arrow cursor into the Standard Colors box, where blue is the fifth color from the left.

Pick: **(the blue Standard Colors box at the top of the dialog box)**

The name of the color that you have selected appears in the box next to the word Color at the bottom of the screen. (If you select one of the standard colors, the name appears in the box; if you select one of the other 255 colors under Full Color Palette, the color number appears.)

Pick: **OK**

Now the color for layer HIDDEN_LINES is set to blue. Check the listing of layer names and colors to verify that blue has replaced white (the default color) in the Color column to the right of the layer name HIDDEN_LINES.

> ■ *TIP* If you selected dark lines on a light background when configuring your video display, white lines will be black on your monitor and color boxes indicating white will be black. ■

Linetype in Layers

The linetype column allows you to set the linetype drawn for the layer. You will select the linetype HIDDEN for your layer named HIDDEN_LINES.

Pick: **(on the word Continuous in the Linetype column and corresponding with the new layer)**

The Select Linetype dialog box shown in Figure 5.4 appears on your screen.

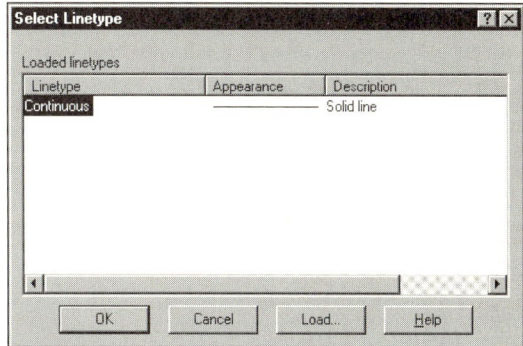

Figure 5.4

Note that only one choice of linetype, CONTINUOUS, is available. Before you can select a linetype in the Layer Control dialog box, you must load it into AutoCAD. You need do so only one time in the drawing, and you do not need to load all the linetypes. To keep your template drawing size small and get a shorter list of linetypes when you set the linetype in the Layer Control dialog box, load only the linetypes that you use frequently in your drawing. You can always load other linetypes as needed during the drawing process. You will select Load, the second button from the right at the bottom of the dialog box, to load the linetypes you want to use.

Pick: **Load**

Loading the Linetypes

The Load or Reload Linetypes dialog box appears, as shown in Figure 5.5. To the right of the selection File is the name of the default file, *acad.lin*, in which the predefined linetypes are stored. You can also create your own linetypes, using the Linetype command. You can store your custom linetypes in the *acad.lin* file or in another file that ends with the extension *.lin*. Below the file name is the list of available linetypes and a picture of each. Use the scroll bar at the right of the list to move down the items until you see the selection HIDDEN. On your own, pick on the name HIDDEN so that it becomes highlighted, and then pick OK to exit the dialog box.

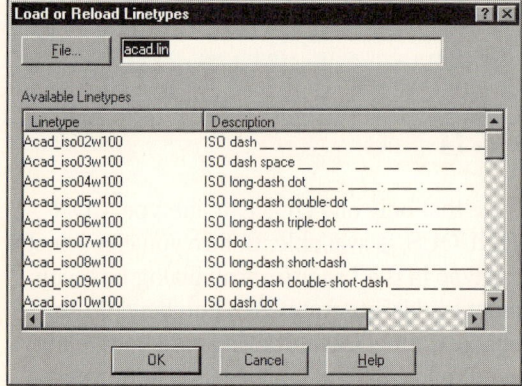

Figure 5.5

■ *TIP* Many of the linetypes have several similar patterns but different scales. Note pattern HIDDEN2 and HIDDENX2 listed near HIDDEN. The pattern HIDDEN2 has dashes half as long as linetype HIDDEN, and HIDDENX2 has dashes that are twice as long as HIDDEN. You can use these different linetypes with the global linetype scaling factor to adjust the scales of families of lines in your drawing. Remember, you can also set the linetype scaling for individual objects with the Linetype or Properties icons on the Object Properties toolbar. ■

The linetype HIDDEN is now listed below the linetype CONTINUOUS in the Select Linetype dialog box. You will use the dialog box to select HIDDEN as the linetype for layer HIDDEN_LINES.

Pick: **HIDDEN**

Pick: **OK**

You return to the Layer Control dialog box. The list of layer names should now show layer HIDDEN_LINES having color blue and linetype HIDDEN. To make HIDDEN_LINES the current layer,

*Pick: (the name **HIDDEN_LINES** in the layer name list; it becomes highlighted)*

While the name is highlighted,

Pick: **Current *(the button above the layer names)***

Pick: **OK**

to exit the dialog box.

■ *TIP* Remember that you can set the current layer quickly by using the Layer Control list, as you did in Tutorial 2. ■

HIDDEN_LINES is now indicated as the current layer on the Object Properties toolbar, and the color blue appears in the square to the left of the layer name. Now, use the Line command to draw a few random lines on the screen on your own. These lines are on the HIDDEN_LINES layer; they will appear blue and have linetype HIDDEN (dashed lines) on your color monitor.

Remember, using layers to control the color and linetype of new objects that you create will work only if BYLAYER is active as the method for establishing object color and object linetype. It is the default, so you should not have to change anything. To check, examine both Color Control and Linetype Control on the Object Properties Toolbar. Both should be set to BYLAYER.

Defining the Layers

Now you will use the Layer Control dialog box to create the remaining layers for your template drawing.

Pick: **Layers icon**

The Layer Control dialog box appears on your screen. From the keyboard, type in the following new layer names, one at a time. Pick the New box before each name.

> ■ *TIP* Many Windows programs identify typing shortcuts by use of an underlined letter. Note the underlined letter *N* in the word New on the button. This underline indicates that you can quickly select this button by typing the key combination Alt-N. Look for the quick key combinations in other dialog boxes. Try it when you create the layers listed below. You can also quickly create several new layers by picking New and then typing their names separated by commas. ■

Pick: **New**
Type: **TEXT** ⏎
Pick: **New**
Type: **VISIBLE** ⏎
Pick: **New**
Type: **THIN** ⏎
Pick: **New**
Type: **CENTERLINE** ⏎
Pick: **New**
Type: **PROJECTION** ⏎
Pick: **New**
Type: **HATCH** ⏎
Pick: **New**
Type: **DIM** ⏎
Pick: **New**
Type: **BORDER** ⏎
Pick: **New**
Type: **VPORT** ⏎

Each time you pick New, a *Layer1* pops up on the list of layer names as the new layer. If the list is long, sometimes you may need to scroll up and down it, using the boxes that are located near the right-hand side of the dialog box.

Setting the Color and Linetype

Now, you will set the colors and linetypes for the layers that you have created. Note that the layers you have created are turned on. The color for all the layers is white (or black depending on your system configuration).

You can select more than one Layer name at the same time by using the Ctrl or SHIFT keys in conjunction with the mouse (as you would to select multiple files in other Windows programs).

Move the arrow cursor to the name THIN and press the pick button to highlight that layer name. While holding down the Ctrl key, select layer HATCH and layer TEXT from the list of layers. You should now have three layers selected.

Pick: (the color box across from one of the selected layers)

The Select Color dialog box, which you used earlier in this tutorial, pops up on the screen. Use the standard colors from the top row. You will pick red as the color for the layers you selected.

Pick: (the red box in the very top row of colored boxes, under the heading Standard Colors)

The word red appears in the Color box, indicating that it is the color choice for the layers that you selected.

Pick: **OK**

Next, you will set the linetype for layer CENTERLINE. First, you will select the layer CENTERLINE from the list of layers, and then you will pick to set the linetype.

Pick: **CENTERLINE** *(to highlight the layer name)*

Pick: (the linetype **CONTINUOUS** *to the right of layer* **CENTERLINE***)*

The Select Linetype dialog box appears on your screen. Do the following steps on your own. Pick the Load button and select the line-type named CENTER to load. Pick OK to return to the Select Linetype dialog box. You will see the linetypes CENTER, CONTINU-OUS, and HIDDEN listed. To select linetype CENTER, pick on the name CENTER. Pick OK to close the Select Linetype dialog box. These actions set the linetype for layer CENTERLINE to linetype CENTER.

Repeat these steps to set the colors and line-types in the following table for the other layers that you have created. When you are setting the colors for the layers listed, keep in mind that cyan is the aqua color on the top row of standard colors, that magenta is a purplish pink color, and that the color white appears black if you have chosen to draw on a light background.

> ■ **TIP** When you are selecting the color and linetype, be sure to highlight only the names of the layers you want to set. ■

Layer	Color	Linetype
0	WHITE	CONTINUOUS
HIDDEN_LINES	BLUE	HIDDEN
TEXT	RED	CONTINUOUS
VISIBLE	WHITE	CONTINUOUS
THIN	RED	CONTINUOUS
CENTERLINE	GREEN	CENTER
PROJECTION	MAGENTA	CONTINUOUS
HATCH	RED	CONTINUOUS
DIM	CYAN	CONTINUOUS
BORDER	WHITE	CONTINUOUS
VPORT	MAGENTA	CONTINUOUS

Before you exit the Layer Control dialog box, set the current layer to VPORT. Highlight layer name VPORT on your own and

Pick: **Current**

When you have finished creating the layers and setting the colors and linetypes, the dialog box on your screen should be similar to Figure 5.6.

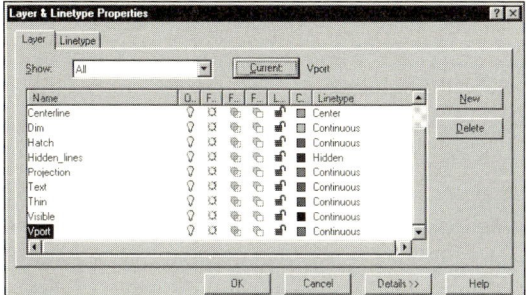

Figure 5.6

Pick: **OK**

You have now created a basic set of layers that you can use in future drawings. When you return to the Layer Control dialog box in the future, you will note that the list of layer names has become alphabetized. (The variable Maxsort controls the number of layers that will be sorted in the Layer Control dialog box. It is set at 200 as the default. If you do not want your layer list to be sorted, you can type *MAX-SORT* at the command prompt and then follow the prompts to set its value to 0.)

Save your drawing before continuing.

Setting Drawing Aids

Next, you will select the Drawing Aids dialog box from the Tools menu and use it to set Snap and Grid.

Pick: **Tools, Drawing Aids**

The Drawing Aids dialog box shown in Figure 5.7 appears on your screen.

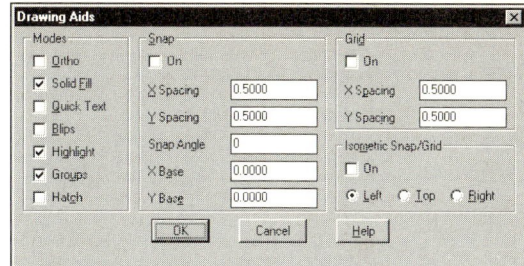

Figure 5.7

You will set the X spacing for the snap to 0.25.

Pick: **(highlight the text in the input box next to X Spacing under Snap by clicking and dragging)**

Type: **.25** ⏎

The Y Spacing box automatically changes to match the X Spacing box when you press ⏎. If you want unequal spacing for X and Y, you can change the Y spacing also.

At the left of the dialog box are seven boxes for Ortho, Solid Fill, Quick Text, Blips, Highlight, Groups, and Hatch. A check appearing in the box to the left of any name indicates that the setting is active.

> ■ *TIP* Blips are little crosses that appear on the screen when you select a point. Sometimes they can be useful. The AutoCAD R14 default setting is off. However, when you open a drawing from the datafile directory, Blips are turned on. If you find them annoying, you can turn them off here. Otherwise, Redraw All removes them from the screen. ■

Move to the On buttons below Snap and Grid and turn them on. Change the values for Grid (previously set to 0.5) to match Snap (0.25).

Pick: **OK (to exit from the dialog box)**

Using Zoom All

You should see a grid of dots on your screen. You will use Zoom All if the grid does not fill the entire screen. Do not be concerned if the grid does not fill entirely to the right edge of the screen. Your drawing limits are currently set to 12 x 9; if your graphics window on the screen does not have the same *aspect ratio* (the ratio of the height to the width), only one dimension will be filled completely. If you want, you can figure out the aspect ratio of your graphics window and always set your drawing limits so that the grid fills the entire screen, but that really is unnecessary. When you start a drawing from your template, you will set the drawing limits at that time to a value large enough for the particular part you will draw. Then you can use the Zoom command to view the new drawing limits. For now,

> *Pick:* **Zoom All icon**

Selecting the Default Text Font

AutoCAD offers various fonts for different uses. The best font for lettering engineering drawings is called romans, for Roman Simplex. Select text font *romans* as your default font.

> *Pick:* **Format, Text Style**

The Text Style dialog box, which you used in Tutorial 1, appears, showing AutoCAD's choices of Style name, Fonts, and Effects.

> *Pick:* **New**
>
> *Type:* **MYTEXT**

Pick OK or press ⏎. Next, you will choose the font name *romans.shx* in the *Font name* pull down menu. You will accept the defaults in all other areas of the dialog box. Remember, accepting the defaults now means you will be prompted for these values when you create the text; this method offers you a lot of flexibility when drawing. Do not become confused between style names and font names. A font is a set of characters with a particular shape. When you create a *style*, you can assign it any name you want, but you must specify the name of a font that already exists for the style to use.

Setting the Viewres Default

The Viewres command controls the number of line segments used to draw a circle on your monitor. It does not affect the way that circles are plotted—just how they appear on the screen. Have you noticed that when you use Zoom Window to enlarge a portion of the drawing, circles may appear as octagons? The reason is that Viewres is set to a low number. The default setting is low to save time when you draw circles on the screen. With faster processors and high-resolution graphics, you can use a larger value. You will type *VIEWRES* at the command prompt.

> Command: **VIEWRES** ⏎
>
> Do you want fast zooms? <Y>: ⏎
>
> Enter circle zoom percent (1-20000) <100>: **5000** ⏎

> ■ *TIP* If you are using a slower computer system, you may notice that performance on your computer slows down. It may be because of this setting. If you need to, you can reset Viewres to a lower number. Type *REGEN* at the command prompt to regenerate circles that do not appear round. ■

Save your drawing now. Before you continue creating the template drawing and make a paper space viewport to be used for plotting, you may want to complete the next section to determine the limits of your output device. In order to center your drawing exactly on the sheet of paper when plotting it, you must know the limits of your printer or plotter. If you already know the limits of your output devices, skip to the topic "Creating a Paper Space Viewport."

Determining the Limits of Your Output Device

Output devices, such as printers and plotters, cannot plot or print all the way to the edge of a sheet of paper. Here is a simple test you can perform to determine the limitations of your output device.

Be sure that you have saved your drawing *mytemplate.dwt*. Begin a new drawing and call it *test.dwg*. You will set the drawing limits to 11 x 8.5 by typing the Limits command.

> Command: **LIMITS** ⏎
>
> ON/OFF/ <Lower left corner><0.0000,0.0000>: **0,0** ⏎
>
> Upper right corner <12.0000,9.0000>: **11,8.5** ⏎

Next, you will draw a horizontal line from point (0,0) to point (11,0) and then a vertical line from (0,0) to (0,8.5). These two lines show the width and height of an 8.5" x 11" sheet of paper.

> *Pick:* **Line icon**
>
> From point: **0,0** ⏎
>
> To point: **11,0** ⏎
>
> To point: ⏎
>
> Command: ⏎ *(to restart the Line command)*
>
> From point: **0,0** ⏎
>
> To point: **0,8.5** ⏎
>
> To point: ⏎

Use the command Plot and plot or print the drawing limits on 8.5" x 11" paper (size A). In the Print/Plot Configuration dialog box, you will see the defaults you selected when you configured your output device. Select Limits, under Additional Parameters, as the graphics window to plot. Be sure that the drawing origin is set to (0,0) and that the scale of plotted inches to drawing units is 1=1, not Scaled to Fit. If either is not set correctly, make corrections in the dialog box as you learned to do in Tutorial 3. When the settings are correct, pick OK to plot your drawing.

Your output shows as much of the two lines you have drawn as will fit on the paper at full scale. Measure the actual length of the lines that were plotted to determine the limits of your output device. Where the two lines intersect at point (0,0) in your drawing is the origin for the paper; in other words, it is the spot closest to the lower left corner of the paper that the printer can reach. Knowing this location will help you figure out how to center drawings correctly on the sheet of paper for your printer. (If you have difficulty determining the limits and origin of your output device, ask your technical support person for help.)

If you rotate your plot, your printer driver may locate the drawing origin by the upper right corner, not the lower left. If one or both lines fail to appear on your plot, draw more lines and symbols to help you determine where the driver is locating the origin. You will use this information later in the tutorial.

> ■ *TIP* Setting your system printer to a default landscape orientation from the Windows Control Panel eliminates the need to rotate your plot and will allow both lines to plot correctly at the paper's origin. ■

For the purposes of the tutorials, we will use the values 10.00" x 7.75" as the limits of the output device. You should substitute the correct limits for your output device. If your output device uses more than one paper size, determine the limits for each paper size.

■ *Warning:* If you created *test.dwg*, be sure to open your saved drawing, *mytemplate.dwt*, before continuing or you will have difficulty during the remainder of this tutorial. ■

Creating a Paper Space Viewport

AutoCAD Release 14 allows you to set up your drawings by using a method similar to hand drafting to create plots to any exact scale on your paper. Up to this point you have been working in *model space* to create your drawing geometry. When you are ready to plot, you can use *paper space* to lay out the views of your drawing on the "sheet of paper." Paper space is basically two-dimensional; model space is three-dimensional. Your drawing geometry is created in model space, whereas paper space is used for things like borders, title blocks, and viewports.

The following is an example of the differences between model space and paper space. Imagine that you are watching the Rose Bowl on television. The actual Rose Bowl game is being played in Pasadena, California, at the stadium. On your television set is a picture of the game being taken by a cameraperson, who is actually at the game. If the cameraperson zooms in, the objects in the view of the Rose Bowl become larger on your TV screen. In your AutoCAD drawing, model space is like being at the game; paper space is like the picture on your television set. The actual game is three-dimensional; the picture on your TV screen is two-dimensional. If you have one of the picture-within-a-picture TV sets, you can even show more than one picture in separate "windows" on your TV. These windows on the TV are like "viewports" in paper space. You can create more than one viewport, called *floating viewports,* in paper space if you want. These viewports can be overlapping or separated from each other, as you want. You can create any number of these viewports to help you lay out your drawing on the paper sheet the way you want. Each viewport contains a view of the model space drawing at the zoom factor and line of sight that you specify. Paper space view-

ports are very useful for plotting the drawing, adding drawing details, showing an enlarged view of an object, or showing multiple views of an object.

Keep the television example in mind as you do the next steps.

In 3D model space, you can also create multiple viewports, called *tiled viewport*s, on your screen. The number you can create depends on your hardware configuration. These viewports can't overlap each other; they must meet exactly at the edges. They are called tiled viewports because they are like floor tiles that you would lay edge to edge. Tiled viewports are created when you highlight the Tile button or set Tilemode to 1. You will not be using tiled viewports during these tutorials. You will use paper space viewports.

Enabling Paper Space

Before you can use paper space, you must turn Tilemode off. Tilemode is a toggle. Double-clicking the Tile button on the status bar turns it off if it is on, and on if it is off. The default is on, as signified by the highlighted word TILE on the status bar. (If you have trouble using the status bar, type *TILEMODE* at the command prompt and set its value to 0 for off.)

> Double-click: **TILE button** *(so that it becomes grayed out)*

Now that Tilemode is turned off, leave it off. You will use the PAPER and MODEL buttons on the status bar or the aliases, PS and MS, to switch between paper space and model space from now on. Turning Tilemode on disables paper space.

Don't be alarmed when your grid disappears from the screen. This is normal. Entering paper space is analogous to changing from being at the Rose Bowl game to looking at the blank TV screen. In order to see model space (or the game in the Rose Bowl example), you must create a viewport (i.e., turn on the TV).

Note the change in the UCS icon. It now shows the paper space icon, which looks like a triangle. Your screen should be similar to Figure 5.8; note that PAPER appears where MODEL used to be on the status bar.

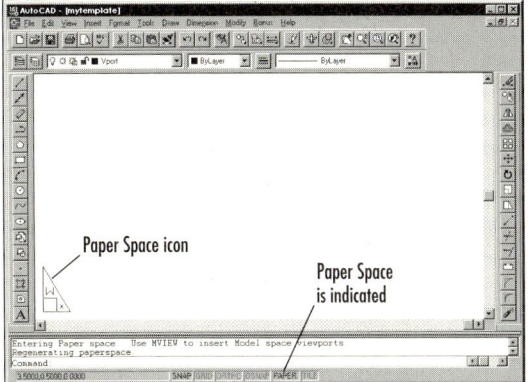

Figure 5.8

Next, you will create a viewport so that you can see model space.

Sizing the Viewport

You will size the viewport so that it will fit inside the limits of your output device when paper space is used full size for plotting. The values used in this tutorial are general and may work for your printer or plotter. However, you can substitute the limit values that you have determined for your specific output device.

Figure 5.9 shows an example of how to determine the drawing coordinates for the viewport that you will create.

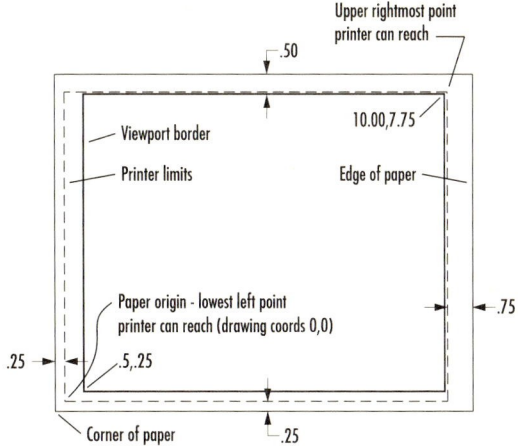

Figure 5.9

The outer line in Figure 5.9 represents the edge of the paper. The distance from the edge that a printer can reach is different for each printer. The distance from the left edge isn't necessarily the same as it is from the right edge or the top and bottom edges. The limits of the printer may also be different for each sheet size.

The dashed line represents the limits of the printer. (When you preview a plot in the Print/Plot Configuration dialog box, the outer line in the preview shows the printer limits.) The thick line represents the viewport border that you will create. In this example, the printer places the lowest corner of the drawing at a location 0.25" above the bottom of the paper and 0.25" in from the left edge of the paper when you choose to plot the origin at (0,0) (in the Print/Plot Configuration dialog box). The printer in this example can only reach to 0.5" from the top border and 0.75" from the right edge of the sheet. In order to create a viewport that is centered on the paper, you would specify (0.5,.25) for the lower left corner of the viewport (which will move the left edge of the viewport an additional 0.5" so that it is 0.75" from the left edge of the paper and move the bottom of the viewport an additional 0.25 so that it is 0.5" from the bottom). Specify (10.00, 7.75) for the upper right corner of the viewport (that is as far as the printer will reach and is 0.75" from the right edge and 0.5" from the top of the paper). This will produce a viewport that is centered on your sheet when you print the drawing from paper space, and the area from (0,0) to (10.00,7.75) will be full size on a sheet of 8.5" x 11" paper.

When you are doing the steps listed below, use your printer limits in place of the suggested values. Decide what values you will need to use in order to center your drawing on the paper. Remember, each style of printer will have its own limits. Take the time to set up your template drawing correctly so that, when you print your drawings, they will look their best.

■ **TIP** If you have access to several different printers or work with several different sheet sizes, having several different template drawings is very useful. ■

Your current layer should now be VPORT (see the toolbar). If it is not, set the layer to VPORT on your own before continuing.

Pick: **View, Floating Viewports, 1 Viewport**

You will create a viewport to correspond to the output device limits. The values (0.5,.25) and (10.00,7.75) may work for you, but if you have determined the exact size that your output device can print, substitute those values when specifying the viewport boundary.

ON/OFF/Hideplot/Fit/2/3/4/Restore/
    <First Point>: **.5,.25** ⏎
Other corner: **10.00,7.75** ⏎

You will see the magenta lines of the viewport boundary on the screen with the grid contained inside them, similar to Figure 5.10. The grid inside the viewport is in model space; you

will be seeing it through your paper space viewport, as though it were the Rose Bowl on your television set!

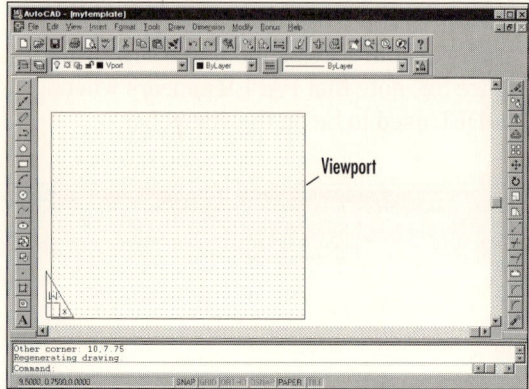

Figure 5.10

Using Limits

The Limits command lets you predefine a boundary in your drawing. When the limits are turned on, you cannot draw outside the specified area. However, you can turn the limits off at any time if you want to draw outside this area. The Limits command can be useful in producing plots located exactly where you want them on the sheet. You can define separate limits in the drawing when you are in paper space and when you are in model space. Next, you will set the limits for paper space. You will set them to the size of your printer's limits because you can't print outside that area. Returning to the Rose Bowl analogy, the paper space limits are like the size of the screen on your TV. Whatever size it is, the only way to make it larger is to buy a larger TV. Set the Limits command in paper space to reflect the limits of your printer.

Setting the Paper Space Limits

The Limits command is under Format on the menu bar. Use limits in your template drawing to represent the edge of the printer limits in paper space. (The viewport represents the area inside the printer limits in which you want to center your drawing.) Setting the upper right corner of the limits slightly beyond the viewport is useful because then, when you use Zoom All in paper space, the viewport border will be slightly in from the edge of the AutoCAD drawing screen, which can make it easier to select.

> *Pick:* **Format, Drawing Limits**
>
> Reset Paper Space limits:
>
> ON/OFF/<Lower left corner> <0.0000,0.0000>: **0,0** ⏎
>
> Upper right corner <12.0000,9.0000>: **10.5,8** ⏎

You will use ⏎ to restart the Limits command and turn the limits on. Then, you will use Zoom All to show the entire limits area on the screen.

> Command: ⏎
>
> ON/OFF/<Lower left corner> <0.0000,0.0000>: **ON** ⏎
>
> *Pick:* **Zoom All icon**

Next, you will switch to model space by picking the PAPER button from the status bar. (If you have trouble with the status bar, type the alias, MS ⏎ at the command prompt to switch to model space.)

> *Double-click:* **PAPER button**

to return to model space where you can create your drawing geometry. Note that the button now says MODEL and that the UCS icon has returned to your screen. When you see the UCS icon, you are in model space. In the Rose Bowl analogy, you are now at the Rose Bowl. Changes that you make in model space are in the "real" 3D world. Changes that you make in

paper space are on the 2D TV screen. Both model space and paper space can have settings for Grid, Snap, and Limits. The settings can be different for model space and for paper space.

> ■ *TIP* When the grid is turned on in both model space and paper space, two patterns of grid dots, which don't necessarily align, are produced, which can be confusing. Generally, you will want to turn off the paper space grid before you return to model space. ■

Now you have a viewport border in your template drawing. It is on a separate layer so that you can freeze it when you do not want it to print. Your template drawing should now look like Figure 5.11.

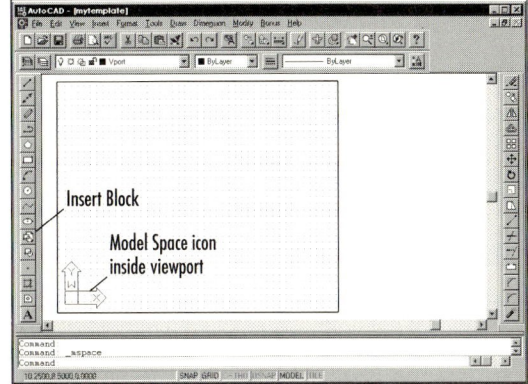

Figure 5.11

On your own, save drawing *mytemplate.dwt* before continuing.

Inserting an Existing Drawing

 You can insert any drawing into any other AutoCAD drawing. You will use Insert Block to insert the subdivision drawing that you finished in Tutorial 3 into the current template drawing. This way you will be able to see the effects of using paper space viewports in the next steps. The Insert Block button on the Draw toolbar is shown in Figure 5.11.

Pick: **Insert Block icon**

The Insert dialog box appears on your screen, as shown in Figure 5.12. You will pick the File button to insert an existing drawing into the current drawing. Only the portion of the drawing created in model space will be inserted.

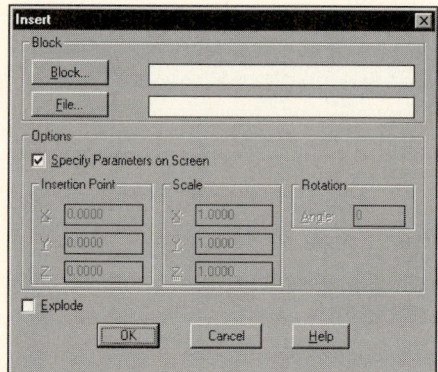

Figure 5.12

Pick: **File**

Use the Select Drawing File dialog box that appears on your screen to select the file *mysubdiv.dwg* that you created in Tutorial 3, and pick Open to return to the Insert dialog box. You may need to change to the directory where it is located.

Note the check in the Options box to the left of Specify Parameters on Screen. This selection returns you to your drawing to select the inser-

tion point, scale, and rotation for the drawing you are inserting. If it is not selected, the grayed-out areas of the dialog box for insertion point, scale, and rotation are selectable. You could use them to type in the values you want to use. If there is no check in the Specify Parameters on Screen box, pick the box now.

Pick: **OK (to return to the drawing)**

Insertion point: **0,0** ⏎
X scale factor <1>/Corner/XYZ: ⏎
Y scale factor <default=X>: ⏎
Rotation angle <0>: ⏎

The image of the subdivision does not show in the viewport because the size of the subdivision is much too large for the viewport at the current zoom factor. In the Rose Bowl analogy, this situation is the same as if the cameraperson zoomed into an area that did not show anything—for example, a blank wall. In order to see the Rose Bowl (or the subdivision) again you must zoom out. To make the subdivision fit inside the window,

Pick: **Zoom All icon**

Now the entire subdivision drawing should appear inside the viewport, as shown in Figure 5.13.

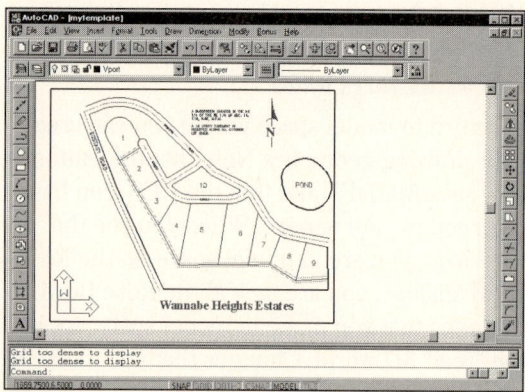

Figure 5.13

However, if you were to plot the drawing as shown in Figure 5.13 you couldn't be sure of the scale at which it would be plotted. In order to have the drawing at a particular scale, you must establish a relationship between the number of units in model space and the number of units in paper space. You do so in a manner similar to the way you'd determine the plot scale to include in the title block of a drawing. In the Rose Bowl analogy, you have to go to Pasadena (model space) and tell the camera-person exactly how much to zoom in or out. This way when you were back in your living room watching TV, you could hold up a ruler on the TV screen and there would be an exact scale between the real size of the stadium and the size shown on your set.

Using Zoom XP

The next step will be to establish a relationship between the number of units in model space and the number of units in paper space. You will set up the drawing so that one unit in paper space equals 250 units in the subdivision in model space (a scale of 1" =250'). To establish this relationship, you will use the Zoom command to specify the XP (times paper space) scale factor. The XP scale factor is a ratio. It is the number of units from the object in paper space divided by the number of units in model space. For example, if you want the model space object to appear twice its size on the paper, specify 2XP as the scale factor (2 paper space units/1 model space unit). You can also think of the zoom XP factor as the value or size of one model space unit when shown in paper space. To specify a scale of 1"=250' for the subdivision, you will set the XP scaling factor to

0.004 (1 unit in paper space/250 units in model space). You will do so by typing the alias for the Zoom command.

Command: **Z** ⏎

All/Center/Dynamic/Extents/Left/Previous/VMax/Window/ Scale (X/XP): **.004XP** ⏎

Your screen should be similar to Figure 5.14. Note the image is now smaller.

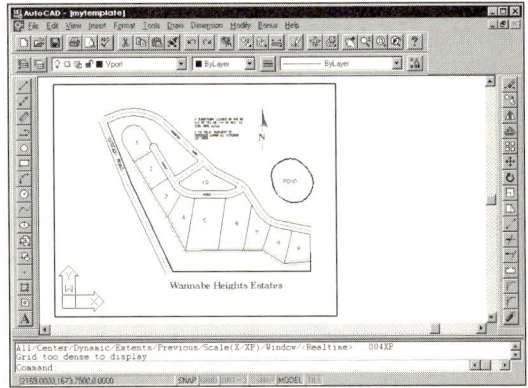

Figure 5.14

You have set up the Zoom XP scale factor so that if you plot paper space at a scale of 1=1, the model space object is shown at a scale of 1"=250' on the paper. Remember, in the original subdivision drawing, each unit represented one foot. In order to preserve the relationship between model space and paper space, be sure that if you use Zoom Window to enlarge your drawing, you use Zoom Previous, not Zoom All, to return to this XP size before you plot. Additionally, you can always use Zoom and later specify a different XP scale factor if you want to.

MVSetup to calculate the Zoom XP scale factor. To do so, type *MVSETUP* at the command prompt. You will see the options Align, Create, Scale viewports, Options, Title block, and Undo. Type *S* ⏎ to scale the viewports. AutoCAD will prompt you to select the viewports to scale. Pick on the viewport border (the magenta line in your drawing) and press ⏎. AutoCAD prompts you to enter the ratio of paper space units to model space units. Start by entering the number of paper space units and pressing ⏎, then enter the number of model space units. The MVSETUP program converts these units to a ratio and determines the correct value, which it will then use in the Zoom XP command. ■

Using Pan Realtime

You can use the Pan command (on the Standard toolbar) to drag the model space drawing around in the paper space viewport without changing the scale or the location of the drawing on the model space coordinate system.

> *Pick:* **Pan Realtime icon**
>
> Displacement: *(pick a point near the center of the subdivision and drag the image around until it is centered in the viewport)*

Exit the Pan command by pressing the return button on your pointing device and selecting Exit from the pop-up menu. (You can also press ESC or ⏎ to exit.)

Creating a Second Floating Viewport

You can use floating viewports to create enlarged details, location drawings, or additional views of the same object. (You will learn more about this method when you create solid models in Tutorial 9.) You will type the Mview command to create another viewport.

> *Type:* **Mview**

The Mview command, similar to choosing *1 Viewport* from the menu, also has options to turn viewports on and off. You can't see any objects in viewports that are turned off, regardless of whether the objects' layers are frozen or thawed. The option Hideplot allows you to select viewports in which you want hidden lines removed from the view when plotting in paper space. The Fit options fit a viewport to the available paper space graphics window. The options for 2, 3, and 4 viewports create patterns of multiple viewports in the drawing. Restore converts tiled viewports to floating viewports of similar configurations.

At the prompt for first point, you will pick a location for one corner of the viewport. Refer to Figure 5.15 for the placement of the viewport you will create.

> Switching to paper space
>
> ON/OFF/Hideplot/Fit/2/3/4/Restore/<First point>:
>   *(pick near point 1 for the lower left corner of the viewport)*
>
> Other corner: *(pick near point 2 for the upper right corner to locate the viewport)*

A second viewport is added to the drawing. It shows the entire subdivision drawing, zoomed so that all of the drawing limits fit inside the viewport. Note also that the second viewport you created has its own set of crosshairs and its own UCS icon. Only one viewport can be active at a time. To make a viewport active, move the arrow cursor into that viewport and press the pick button. When a viewport is active, its border becomes highlighted and the crosshairs appear completely inside it.

Next, you will use the Zoom Window command to enlarge an area inside of the smaller viewport you created.

Pick: *(inside the smaller viewport to make it active or press ⒸⓉⓇⓁ-R)*

Pick: **Zoom Window icon**

First corner: *(pick the first corner so that you enlarge the rounded end of lot 1)*

Other corner: *(pick to create a window around lot 1)*

The area inside the window you selected will be enlarged inside the viewport, as shown in Figure 5.15.

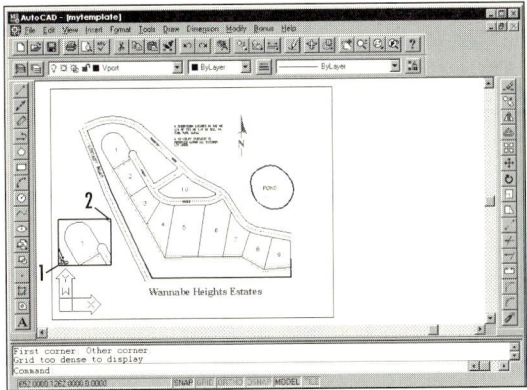

Figure 5.15

The viewport border is on layer VPORT. If you do not want the boundary of the viewport plotted in your drawing, you can freeze this layer.

Adding a Title Block and Text to Paper Space

You can easily add a title block to your drawing using lines and text in paper space. The advantage of adding the title block, notes, and border to paper space is that the measurements there are the same as you would make them on a regular sheet of paper. For instance, the standard size for text on 8.5" x 11" drawings is 1/8". To add text of this height to paper space, you would set the text height to 1/8" and create the text. However, if you wanted to add text to model space so that it would appear 1/8" high on the final plot, you would have to recognize that model space is going to be plotted at 1/250th of its actual size. So you would need to create the text 31.25 units high in model space in order for it to be 1/8" high on the final plot. Also, in model space you can draw in only one viewport or another after you have enabled them. Using paper space, you can draw across the viewports.

Return to paper space on your own by double-clicking the MODEL button on the status bar or by typing the alias PS and pressing ⏎ before adding the lines and text to make up the title block.

You will see the paper space icon return to your screen and the button that you picked now says PAPER.

Next, use Layer Control on the Object Properties toolbar and set BORDER as the current layer on your own. Once you have selected BORDER as the current layer, turn layer VPORT off by picking the selection across from VPORT that looks like a lightbulb, so that it darkens. The viewport borders no longer appear because they are on a layer that has been turned off. Before continuing, check the toolbar to verify that BORDER is the current layer. If you need help to do so, refer to Tutorial 2.

> ■ *TIP* You cannot edit a viewport when its layer is turned off. You can still edit the contents of the viewport, that is, the objects that are in model space. If at some time you are trying to change or erase a viewport while in paper space and you cannot get it to work, check to see if the layer where the viewport border was created is turned off or frozen. ■

You will use the commands you've learned to add borders and a title block on your own. First, draw the lines for the border of the drawing shown in Figure 5.16.

Next, set Snap to 0.125, which will be useful for positioning lettering and the lines for the title block. Use the Offset command to offset a line across the bottom of the viewport 0.375 unit (three snap increments) up from the bottom line of the viewport. Then use the Circle command to draw a circle around the area from the

second viewport, which is going to serve as a detail of lot 1. Be sure to add the border lines to paper space. Lines added to model space will show up in every viewport (unless you control the visibility with viewport layer visibility control). Keep in mind that paper space represents the sheet of paper on which you are laying out the drawing. Things like borders belong on the sheet of paper and not in model space.

When you have finished, your drawing should look like that in Figure 5.16.

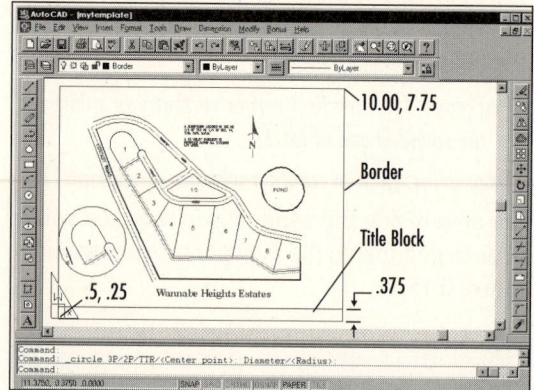

Figure 5.16

Using Divide

The Divide command places points along the object that you select, dividing it into the number of segments you specify. You can also choose to have a block of grouped objects placed in the drawing instead of points. You will use the Divide command to place points along the line that you just drew, dividing it into three equal segments. Refer to Figure 5.16.

Type: **DIVIDE**

Select object to divide: *(pick the line you offset to form a title strip)*

<Number of segments>/Block: **3** ⏎

Because the Point Style is set at just a dot, you probably won't be able to see the points that mark the equal segment lengths. You will use the Point Style dialog box to change the display of points in the drawing to a larger style so you can see them easily.

Pick: **Format, Point Style**

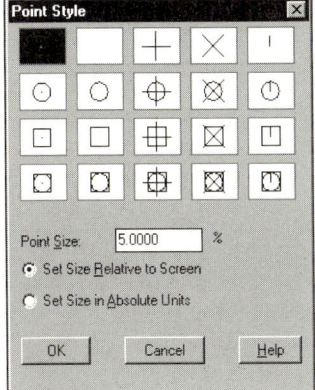

Figure 5.17

The Point Style dialog box shown in Figure 5.17 now appears on your screen. On your own, select one of the point styles that has a circle or target around the point so that it is easily seen. To exit the dialog box,

Pick: **OK**

You will need to tell AutoCAD to recalculate the display file for your drawing in order to see this change. You do so with the Regen command. You will type the command at the prompt.

Command: **REGEN** ⏎

The points should appear larger on the screen now. Your drawing should look similar to that in Figure 5.18.

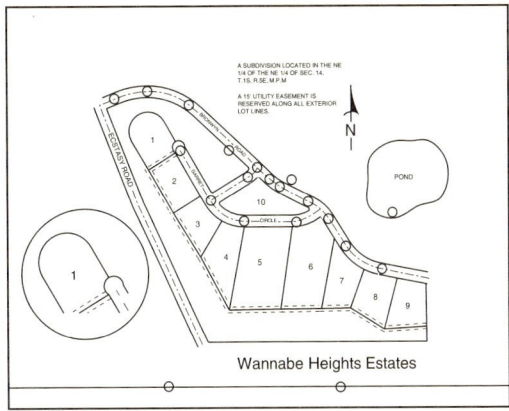

Wannabe Heights Estates

Figure 5.18

You can use the Node object snap to find point objects in your drawing. On your own, select Node and Perpendicular from the running object snap dialog box (by double-clicking on OSNAP) to draw lines dividing the title area exactly into thirds.

Pick: **Line icon**

From point: *(target one of the points with AutoSnap)*

Draw a line straight down from the point by using the Perpendicular object snap.

To point: *(pick the bottom line of the viewport, the AutoSnap Perpendicular marker should appear)*

Now repeat this process to draw another line at the other point. On your own use the Point Style dialog box to change the point style back and then use the Regen command to regenerate the points on the screen.

> ■ *TIP* The Measure command is similar to the Divide command. However, instead of specifying the number of segments into which you want to have an object divided, with Measure you specify the length of the segment you would like to have. Like Divide, Measure puts points or groups of objects called blocks (which you will learn about in Tutorial 8) at specified distances along the line. ■

On your own, use the Object Properties toolbar to set TEXT as the current layer. You will use the Dtext command to add titles to the drawing.

Type: **Dtext**

Justify/Style/<Start point>: **C** ⏎

Center point: **5.25, .375** ⏎

Height <0.2000>: **.125** ⏎

Rotation angle <0>: ⏎

Text: **SUBDIVISION** ⏎

Text: ⏎

The word *SUBDIVISION* appears, centered on the point you selected. The centering is horizontal only; otherwise the letters would appear above the point selected for the center. If you want both horizontal and vertical centering, pick the Justify option and then Middle.

Now, repeat this process using the Justify, Left option of the Dtext command to position the words *DRAWN BY: YOUR NAME* in the left part of the title block. The default Start point option prompts you for the bottom left starting point for the text that you will enter. Use the Justify, Right option to right justify the words *SCALE: 1"=250"* in the area to the right. The Justify R option prompts you for the lowest, rightmost point for the text that you will enter.

On your own, add text identifying *DETAIL A*, as shown in Figure 5.18.

Now you have completed a simple title block for your drawing and are ready to plot. Your drawing should look like that in Figure 5.19. Save it before you go on.

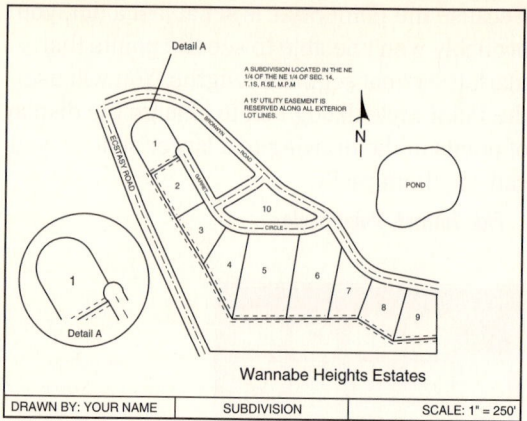

Figure 5.19

Plotting from Paper Space

You have correctly set up your drawing in paper space and used the Zoom command with the XP scaling factor to establish the number of units in model space that you want to equal one unit in paper space; you are now ready to plot. Be sure that you are still in paper space for plotting your drawing. If you are not, your plotted drawing may not fit on the sheet correctly and the title block won't be shown. Because you used paper space, you will type 1=1 for the scaling in the Plot command.

Pick: **Print icon**

Use the Print/Plot Configuration dialog box that appears on your screen to select the limits of your drawing as what to plot. Select size A paper and plot the drawing at a scale of 1=1 on the paper.

> ■ *TIP* AutoCAD offers a command for Batch plotting, named BATCHPLT, allowing for plotting a series of drawings from a system prompt rather than from AutoCAD. Examine the help topics within AutoCAD for more information. ■

Now that you understand paper space, you are ready to erase the subdivision and the detail viewport from the drawing and save the drawing to use as a template for starting future drawings.

If you have the grid on in paper space, turn it off on your own by pressing F7 or by double-clicking the word GRID on the status bar while in paper space. Turn layer VPORT on and use the Erase command to pick on the edge of the viewport border containing Detail A. Viewports are much like any other drawing object when you are in paper space; you can scale, stretch, move, and erase them as desired. The viewport and its contents are erased. Also erase the text referring to Detail A.

Then continue on your own and return to model space by typing *MS* ↵ at the command prompt or by double-clicking on the PAPER button on the status bar. Now use the Erase command on your own to erase the entire sub-division, which should act as one object when you select it.

Using Purge

The Purge command eliminates unused layers, styles, blocks, and other named objects. A named object is just that: any type of AutoCAD object that can have a name, such as layers, linetypes, views, blocks, and others. Note that, when you inserted the subdivision into your drawing, it automatically brought along all of its layers that contained drawing objects and other settings. To eliminate these unwanted layers from your template drawing, you will use the Purge command. Purge is also very useful for keeping your drawing database as small as possible. A small database is very important when you are working with 3D solids, which can result in very large files. Be careful not to purge any of the layers that you created for the template.

Command: **PURGE** ↵

Purge unused

Blocks/Dimstyles/LAyers/LTypes/SHapes/STyles
/Mlinestyles/All: **A** ↵

At the Purge layer prompt, respond Y to each of the subdivision layers and text styles on your own. Press ↵ to accept No as the default for the template drawing layers and styles. (Refer to pages 140 and 142 for the layers and styles that you created.)

Set the current layer to VISIBLE. You will save this drawing with VISIBLE set as the current layer. Then, when you begin a new drawing from this template, you will be ready to start drawing on the layer for VISIBLE lines. Now save your drawing to the file name *mytemplate.dwt*.

Pick: **Save icon**

Now you have completed *mytemplate.dwt*. Be sure to keep a copy of the drawing on your own floppy disk. You also should keep a second copy of your drawings on a separate floppy as a backup disk, in case the first disk becomes damaged.

You can easily use the Edit Text command to make changes to the standard information you provide in the title block. Refer to Tutorial 1 if you need to review the Edit Text command.

Beginning a New Drawing from a Template Drawing

You can use any AutoCAD drawing as a starting point for a new drawing. The settings that you have made in drawing *mytemplate.dwt* will be used to start future drawings. An identical template drawing, called *proto.dwt*, is in your data files. If you want to use the template you just created, substitute *mytemplate.dwt* whenever you are asked to use *proto.dwt*. Next, you will start a new drawing from the template provided with the data files.

Pick: **New**

Pick: **Use a Template**

Pick: **More Files...**

Use the dialog box shown in Figure 5.20 to select the correct drive and directory and pick *proto.dwt*.

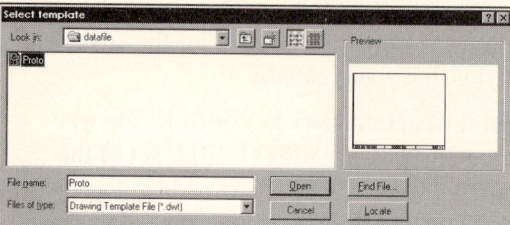

Figure 5.20

> ■ *TIP* You can also use any standard drawing file (**.dwg*) as a template file. To do this, pick in the Files of Type box and select Drawing (**.dwg*) as the type of file. Then make your selection from the list of files. ■

When you have finished making this selection, pick Open to return to the drawing editor. Now when you choose Save, AutoCAD prompts for a new drawing name, leaving the original template file untouched.

Pick: **Save icon**

Type: **TRY1** *(next to File Name)*

Pick: **OK**

Now you can work in the new drawing (called *try1.dwg*) from a copy of the drawing called *proto.dwt*. The drawing *proto.dwt* remains unchanged so that you can use it to start other drawings. Your current drawing name is now *try1.dwg*. For future drawings in this book, use *proto.dwt* or your drawing *mytemplate.dwt* as a template drawing unless you are directed to do otherwise.

Changing the Title Block Text

You created the border, text, and lines of the title block in paper space. To make a change to any of them, you must first return to paper space. Then, you will use the Edit Text selection (Ddedit command) to change the title block text so that it is correct for the new drawing that you are starting.

Command: **PS** ⏎

Type: **DDEDIT** ⏎

<Select an annotation object>/Undo: *(select the text Drawing Title)*

The Edit Text box appears on your screen with the text you selected. Use the ⬅, BACKSPACE, and/or DELETE keys to remove the word DRAWING TITLE. Change the entry to TRY1. When you have finished editing the text,

Pick: **OK** *(to exit the dialog box)*

<Select a text or ATTDEF object>/Undo: ⏎ *(to end the command)*

> ■ *TIP* You can change the style of text that has already been added to your drawing by picking the Properties icon from the Object Properties toolbar. At the prompt, select a text object and press ⏎. The Modify Text dialog box will appear on your screen. Use the Style area near the lower right of the dialog box to pull down the list of available styles and make a new selection. Remember that you must create styles with the Style command before you can use them. You can also use Modify Text to change the height, width factor, rotation, obliquing angle, and location of the text entry. ■

Remember to return to model space before continuing to draw when using your template.

Exit AutoCAD and discard the changes to drawing *try1.dwg*.

KEY TERMS

aspect ratio

floating viewports

model space

paper space

pin registry

style

template drawing

tiled viewports

KEY COMMANDS

Divide

Insert Block

Layer

Measure

Mview

Purge

Regen

Tile

Viewres

Draw the following objects. The letter M after an exercise number means that the given dimensions are in millimeters (metric units).

 5.1 Amplifier Circuit

Draw the amplifier circuit. Use the grid at the top to determine the sizes of the components. Each square = 0.0625.

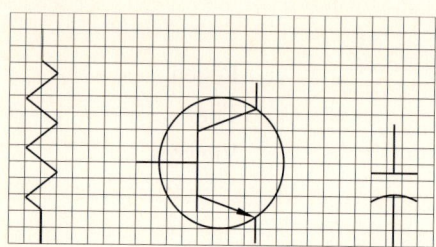

RESISTOR   NPN TRANSISTOR   CAPACITOR

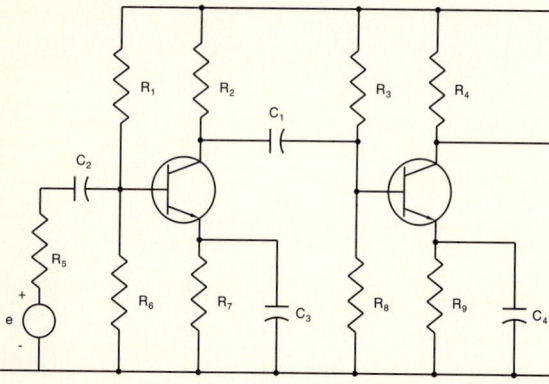

 5.2 Support

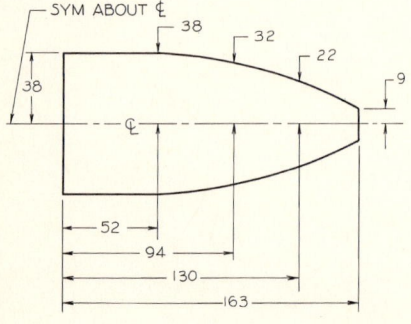

5.3 Vee Block

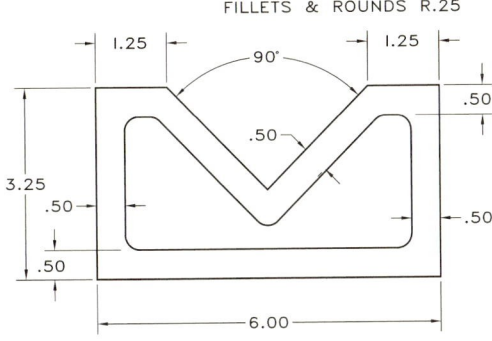

FILLETS & ROUNDS R.25

5.4M Tee Handle

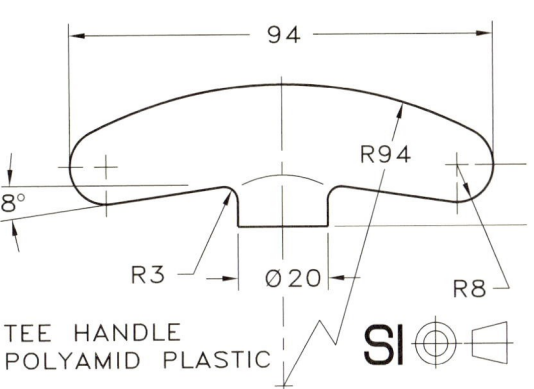

TEE HANDLE
POLYAMID PLASTIC

5.5M Grab Link

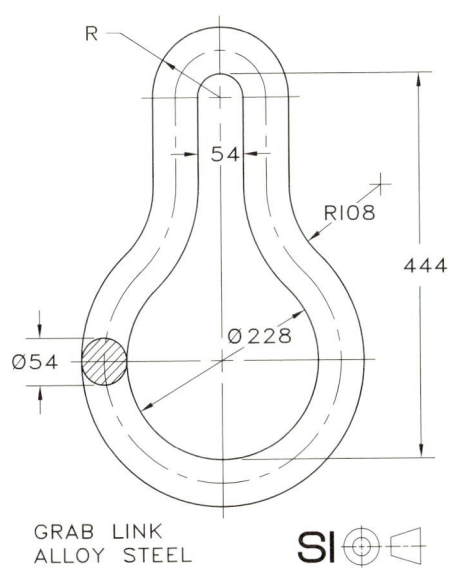

GRAB LINK
ALLOY STEEL

5.6 Idler Pulley Bracket

Create the front view for the object shown.

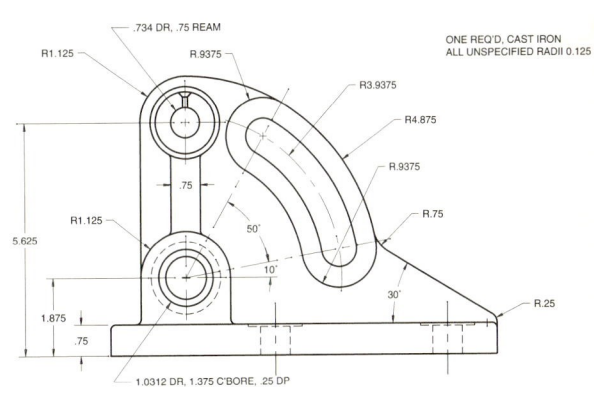

5.7 Hanger

Draw the object shown. Set Limits to (–4,–4) and (8, 5). Set Snap to 0.25 and Grid to 0.5. The origin (0, 0) is to be the center of the left circle. Do not include the dimensions.

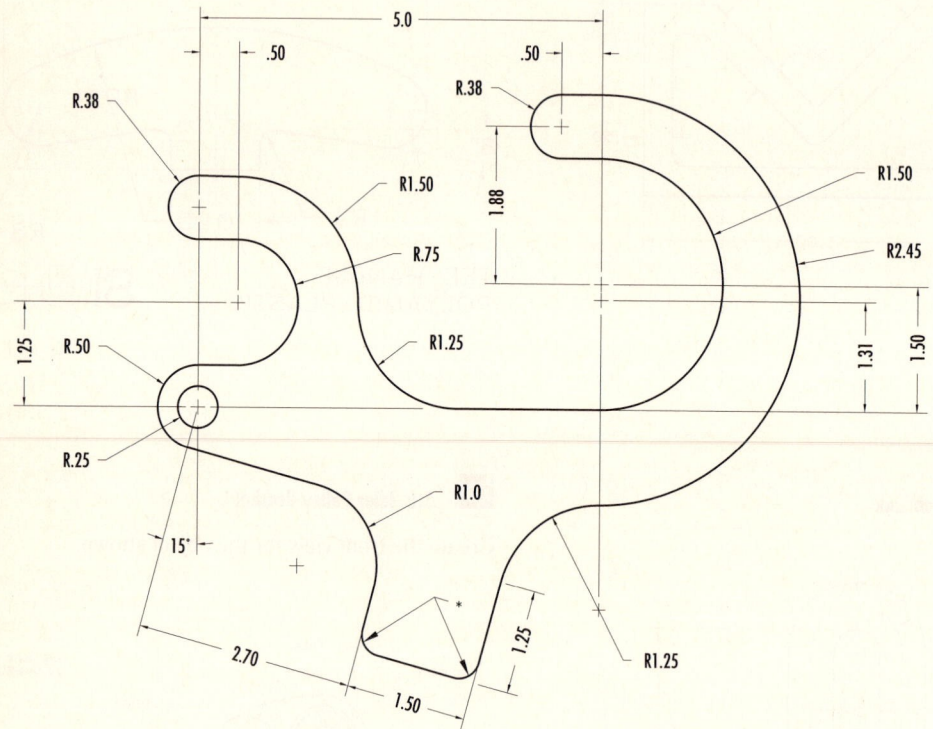

*Corners have R.25

5.8 Window Schedule

Redraw the window schedule shown below. Experiment with different fonts.

WINDOW SCHEDULE

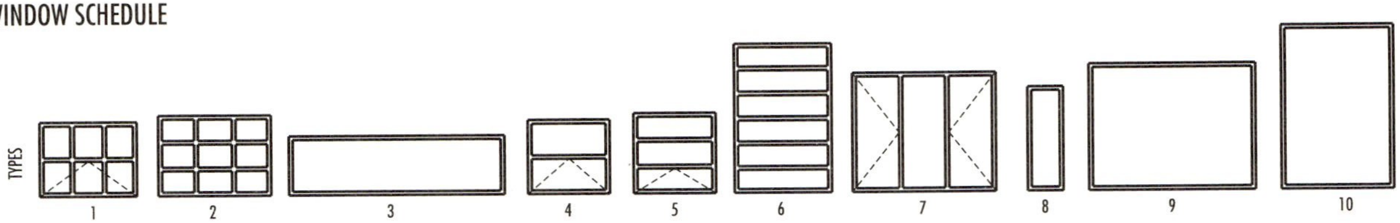

SYM	TYPE	SIZE H	SIZE W	MATL FIN	FRAME	FIN SCRN AREA	AREA		VENT	GLAZING MAT'L	DETAILS H	J	S	REMARKS
A	1	5'-0"	6'-4"	MTL	ST	MTL	ST	Y	32.5 SF	65 SF				
B	2	5'-5"	7'-4"	MTL	ST	MTL	ST	N		40.5 SF				FIXED GLAZING
C	3	4'-0"	14'-8"	MTL	ST	MTL	ST	N		58.8 SF				FIXED GLAZING
D	4	7'-0"	5'-0"	MTL	ST	MTL	ST	N		70 SF				FIXED GLAZING
E	5	5'-5"	4'-6"	MTL	ST	MTL	ST	N		48.6 SF				FIXED GLAZING
F	6	5'-0"	5'-4"	MTL	ST	MTL	ST	Y	13.3 SF	26.5 SF				
G	7	5'-5"	5'-4"	MTL	ST	MTL	ST	Y	9.9 SF	29.7 SF				
H	8	8'-0"	6'-4"	MTL	ST	MTL	ST	N		100.8 SF				FIXED GLAZING
I	9	5'-6"	6'-4"	MTL	ST	MTL	ST	N		126.1 SF				FIXED GLAZING
J	10	8'-0"	9'-4"	MTL	ST	MTL	ST	Y	24.8 SF	74.4 SF				
K	11	5'-6"	4'-8"	MTL	ST	MTL	ST	N		25.3 SF				FIXED GLAZING
L	12	6'-0"	3'-0"	WD	ST	WD	ST	N		36 SF				FIXED GLAZING
M	13	7'-6"	11'-6"	WD	ST	WD	ST	N		86.3 SF				FIXED GLAZING WITH CURVED GLASS
N	14	10'-0"	8'-0"	MTL	ST	MTL	ST	N		160 SF				FIXED GLAZING
O	15	10'-0"	14'-0'	MTL	ST	MTL	ST	N		140 SF				FIXED GLAZING
P	16	6'-0"	6'-6"	WD	ST	WD	ST	N		32.5 SF				FIXED GLAZING

5.9 Saw Blade

Draw the saw blade shown below. Use Array and Polyline. Do not include dimensions. Use the Arc option of the Polyline command to create the thick arc. Add the arrow to the arc by making a Polyline with a beginning width of 0 and a thicker ending width that shows the blade's rotation.

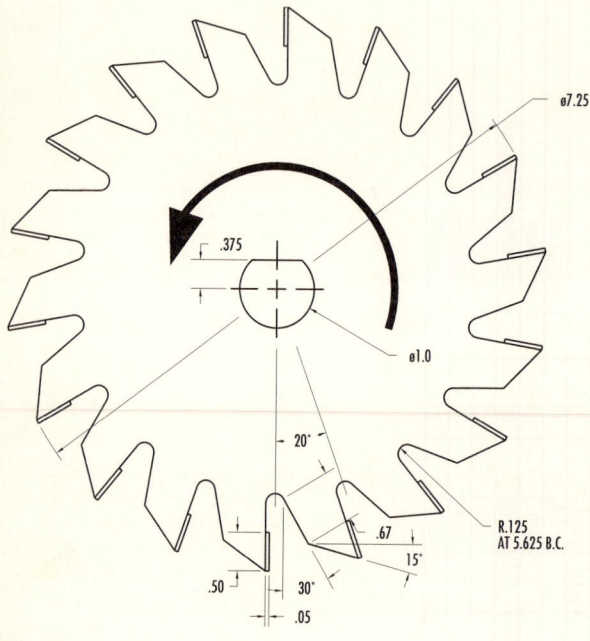

5.10M Five Lobe Knob

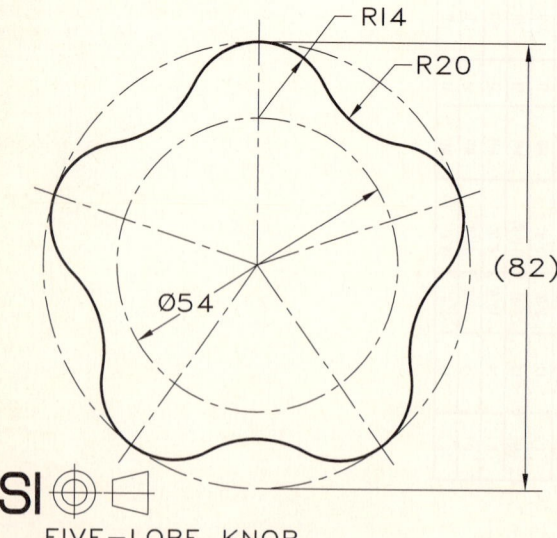

FIVE−LOBE KNOB

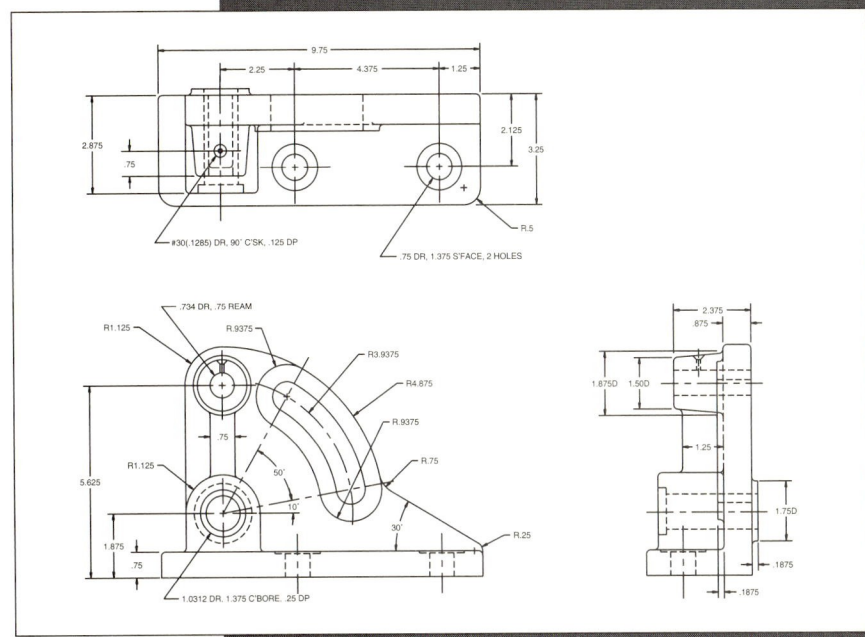

TUTORIAL 6

2D Orthographic Drawings

Objectives

When you have completed this tutorial, you will be able to

1. Use AutoCAD to create 2D orthographic views.

2. Use Ortho for horizontal and vertical lines.

3. Create construction lines.

4. Draw hidden, projection, center, and miter lines.

5. Set the global linetype scaling factor.

6. Draw ellipses.

7. Add correctly drawn centermarks to circular shapes.

Introduction

In this tutorial you will apply many of the commands that you have learned in the preceding tutorials to create orthographic views. *Orthographic views* are two-dimensional (2D) drawings that you use to depict accurately the shape of three-dimensional (3D) objects. You will learn to look at a 3D object and draw a set of 2D drawings that define it. In Tutorial 9 you will learn to create a 3D solid model of an object.

The Front, Top, and Right-Side Orthographic Views

Technical drawings usually require front, top, and right-side orthographic views to define completely the shape of an object. Some objects require fewer views and others require more. All the objects in this tutorial require three views. Each orthographic view is a 2D drawing that shows only two of the three dimensions (height, width, and depth). Thus no individual view contains sufficient information to define completely the shape of the object. You must look at all three views together to comprehend the object's shape. For this reason it is important that the views be shown in the correct relationship to each other.

Figure 6.1 shows a part and Figure 6.2 shows the front, top, and right-side orthographic views of the part.

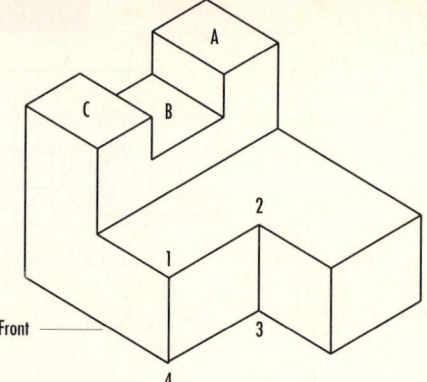

Figure 6.1

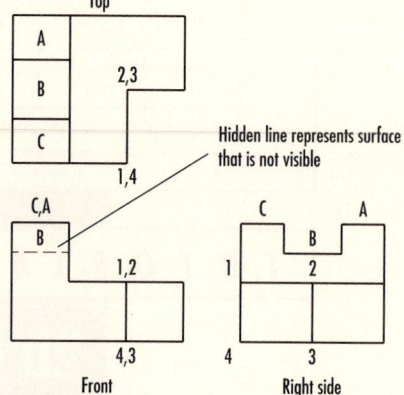

Figure 6.2

Three rectangular surfaces, labeled A, B, and C, are located on the left side of the top view. Which surface is the highest? The top view alone is not sufficient to answer this question. The three surfaces must be located on the other views for you to comprehend the relationships among them. The right-side view (often just referred to as the side or profile view) shows that surfaces A and C are the same height and that surface B is lower.

The side view shows the relative locations of surfaces A, B, and C, but the surfaces appear as straight lines; therefore you need the top view to see the overall shape of the surfaces.

You need both the top view and the side view to define the size, shape, and location of the surfaces.

Look at surface 1-2-3-4 in the right-side view of Figure 6.2. Its shape is shown in the right-side view, but it appears as a straight line in the front and top views because it's perpendicular to the views—like a sheet of paper viewed edge on. A plane surface appears as a straight line when viewed from a direction where the surface is perpendicular to the viewing plane. Surfaces that are perpendicular to two of the three principal orthographic views are called *normal surfaces* (normal meaning 90°). Normal surfaces show the true size of the surface in one of the principal views. All surfaces in the object shown in Figure 6.1 are normal surfaces.

All surfaces must be drawn in all views unless clearly labeled as partial views. Surface B is shown in the front view of Figure 6.2 with a *hidden line*. A *hidden line* represents a surface not directly visible, that is, hidden from view by some other surface on the object.

View Location

The locations of the front, top, and side views on a drawing are critical. The top view must be located directly above the front view. The side view must be located directly to the right of the front view. An alternative position for the side view is to rotate it 90° and align it with the top view. By aligning the views precisely with each other, you can interpret them together to understand the 3D object they represent. Because views are shown in alignment, you can *project* information from one view to another. Refer to Figures 6.3 and 6.4.

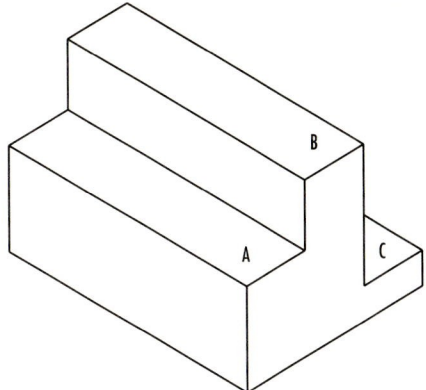

Figure 6.3

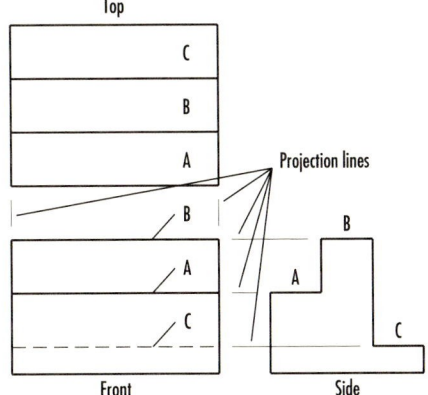

Figure 6.4

Surfaces A, B, and C in Figure 6.3 appear as three straight lines in the front view. Which line represents which surface? Because the front and side views are aligned, you can draw horizontal lines from the vertices of the surfaces in the front view to locate them in the side view. These lines are called *projection lines* and each surface is located between its projection lines in both views. You can locate surfaces between the front and top views by using vertical projection lines. Without exact view alignment, you couldn't relate the lines and surfaces of one view to those of another view, making comprehension of the drawing difficult or impossible.

Starting

Before you begin, launch AutoCAD. You will begin this tutorial by drawing an orthographic view for the adapter in Figure 6.5. The file *proto.dwt* has been provided with the data files so that you can be sure that you are using the same settings as are used in the tutorial.

To start a drawing from a template as you did at the end of Tutorial 5,

> *Pick:* **Start From Template (from the Start Up dialog box)**
>
> *Pick:* **More Files**

Use the dialog box that appears on your screen to select the file *proto.dwt* from the data files that accompany this text (be sure to change the file type to **.dwt* in order to find *proto.dwt*). When you have finished selecting it, pick Open to return to the drawing editor and then pick the Save icon.

> *Pick:* **Save icon**
>
> *Type:* **ADAPTORT**
>
> *Pick:* **Save**

The main AutoCAD drawing editor should reappear on your display screen with *Adaptort* in the title bar. On your own, be sure that Grid and Snap are set to 0.25 spacing and turned on; the grid should be on your screen and the Snap button should be darkened on the status bar.

Figure 6.5 shows the adapter that you will create. All dimensions are in decimal inches.

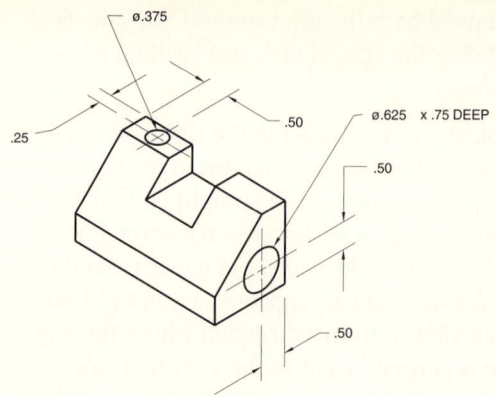

Figure 6.5

Deciding the Model Space Limits

Recall from Tutorial 5 that you can use the Limits command in paper space to set the limits of the paper area that you want to draw on. You can also use the Limits command in model space to set the limits of the area that you will use to create your real-world model of the object. The limits set in paper space and in model space do not have to be the same. Think back to the Rose Bowl analogy in Tutorial 5: your TV screen may be 27", but the actual football stadium is much larger than that. In the template drawing you set the limits for paper space to the printer limits. Set the limits in model space to the value necessary to fit the views of the object that you are creating. You can change the limits at any time during a drawing session by using the Limits command.

Examine the size of the part and the amount of space the views will require. Then set the model space limits to allow a big enough area to create the views. The adapter is 3 inches wide, 1.5 inches high, and 1.5 inches deep. The slot in the top of the adapter is 0.5 inch deep. The default limits for model space are 12 x 9. The adapter that you will draw does not require much space; therefore you will leave the limits set to 12 x 9.

Viewing the Model Space Limits

Use the command Zoom All in model space to show the drawing limits or extents (of the drawing objects), whichever is larger, in the viewport.

Pick: **Zoom All icon**

Move the cursor around inside the viewport. You should see that the coordinates and grid match the limits of the drawing.

Using Ortho

You can use the Ortho command to restrict Line and other commands to operate only horizontally and vertically. This feature is very handy when you are drawing orthographic views and projecting information between the views. You toggle the Ortho command on and off by double-clicking on the ORTHO button on the status bar or by pressing the F8 function key so you can easily activate it when you are in a different command.

Double-click: **Ortho button**

The name on the Ortho button becomes dark to indicate that it is active. You can also use the Drawing Aids dialog box to toggle the Ortho command.

Pick: **Tools, Drawing Aids**

Note that a check now appears in the box to the left of Ortho. If it is not selected, pick it now.

Pick: **OK (to return to the drawing editor)**

The Ortho button should now be darkened on the status bar of your screen.

Next, you will draw the horizontal and vertical construction lines, as shown in Figure 6.6. These lines will represent the leftmost and bottommost margins of your orthogonal views. The coordinates are given for the lines you will create to ensure that your results will be the same as in the tutorial. In general, you can create the construction lines at any location and then move the views if necessary. Don't worry if views are not perfectly centered when you begin your drawing. You will use AutoCAD to center the views after you've drawn them and you see how much room is needed.

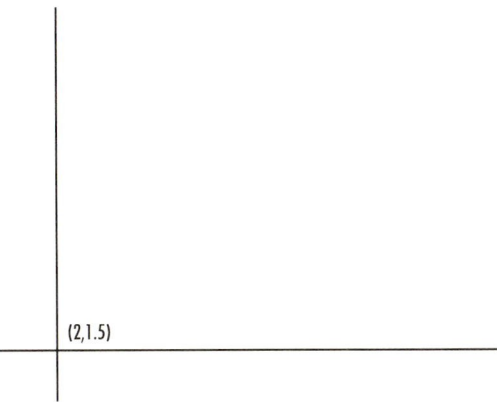

(2,1.5)

Figure 6.6

Drawing Construction Lines

 Construction lines extend infinitely. The default method for drawing a construction line is to specify two points through which it passes. When you use the Horizontal option, the line appears parallel to the X-axis through the point you select. The Vertical option is the same, but creates a line parallel to the Y-axis. By selecting the Angle option, you can specify the construction line by entering the angle and a point through which the line passes. The Bisect option lets you define an angle by three points and create a construction line that bisects it. Finally, the Offset option allows you to specify the offset distance or through point, as when you use the Offset command, to create an infinite construction line.

Pick: **Construction Line icon**

Hor/Ver/Ang/Bisect/Offset/<From point>: **2,1.5** ⏎

Through point: *(pick to the right to define a horizontal line)*

Through point: *(pick above to define a vertical line)*

Through point: ⏎

Two infinite construction lines appear in your drawing: One is vertical through point (2,1.5), and the other is horizontal through the same point.

Next, you will use the Offset command to create a series of parallel horizontal and vertical lines to define the overall dimensions of each view, as shown in Figure 6.7. Then, you will trim the lines to remove the excess portions. If you need to, review the Offset command in Tutorial 3.

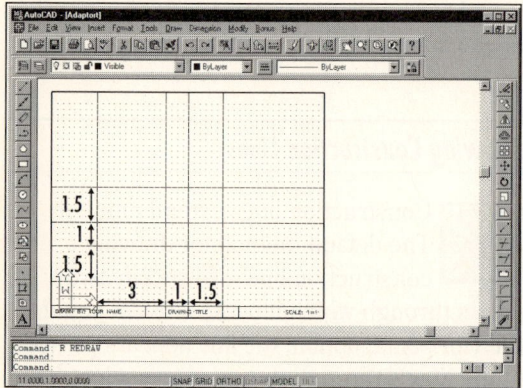

Figure 6.7

For the first horizontal line,

Pick: **Offset icon**

Offset distance or Through<Through>: **1.5** ⏎

Select object to offset: *(pick the horizontal line)*

Side to offset? *(pick any point above the horizontal line)*

A new line is created, parallel to the bottom line and exactly 1.5 units away. You will end the command with the ⏎ key because the next line will be a different distance away.

Select object to offset: ⏎

You will restart the Offset command by pressing ⏎ so that you are prompted again for the offset distance.

Command: ⏎

Offset distance or Through<1.5000>: **1** ⏎

Select object to offset: *(pick the newly created line)*

Side to offset? *(pick any point above the line)*

A line appears 1.00 unit away from the line you selected.

Select object to offset: ⏎

Now repeat this process on your own until you have created all the horizontal and vertical construction lines according to the dimensions shown in Figure 6.7. The lines are parallel to the horizontal line at distances of 1.5 (the given height of the object), 2.5 (the 1.5-inch height and an arbitrary 1-inch spacing between the front and top views), and 4 (the 1.5-inch height plus 1 plus the 1.5-inch depth of the object). Your screen should look like Figure 6.7.

You will define the areas for the front, top, and side views by using the Trim command to remove excess lines. You will pick Trim from the Modify toolbar. You will use implied crossing with the Trim command to select all the lines as cutting edges. Recall that, when construction lines have one end trimmed, they become rays. When trimmed again, they become lines. Use Figure 6.8 to determine which construction lines to trim.

Pick: **Trim icon**

Select cutting edge(s)...

Select objects: *(start your selection at corner A, shown in Figure 6.8, then pick corner B)*

Select objects: ⏎

<Select object to trim>/Undo: *(select segments 1–24 in the order in which they are numbered in Figure 6.8)* ⏎

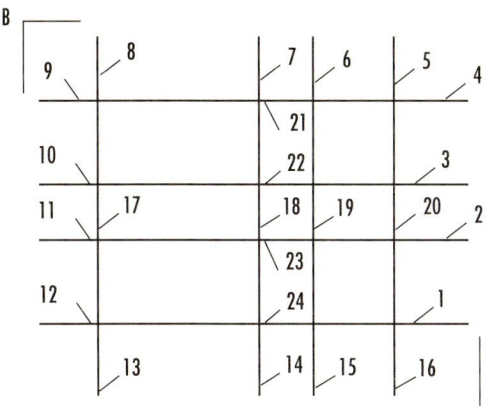

Figure 6.8

Figure 6.10

The overall dimensions of the views are established and aligned correctly. Before continuing, redraw the screen on your own.

Next, draw the slot in the front view. Use the following points or create the lines on your own by looking at the dimensions specified on the object in Figure 6.5.

Pick: **Line icon**
From point: **3,3** ⏎
To point: **3,2.5** ⏎
To point: **4,2.5** ⏎
To point: **4,3** ⏎
To point: ⏎

On your own use the Trim command from the Modify toolbar and remove the center portion of the top horizontal line.

Now, your drawing should look like Figure 6.11.

Next, use the Erase command to remove the unwanted lines, as shown in Figure 6.9.

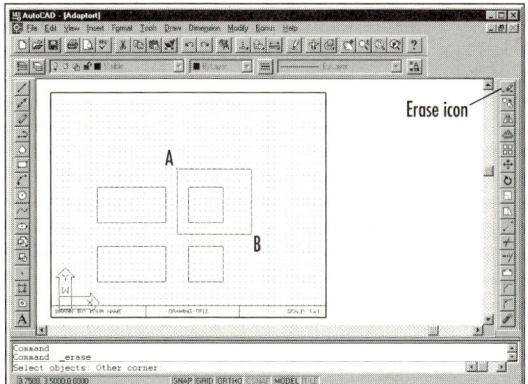

Figure 6.9

Pick: **Erase icon**
Select objects: *(pick corner A, as shown in Figure 6.9)*
Other corner: *(pick corner B)*
Select objects: ⏎
Your drawing should be similar to Figure 6.10.

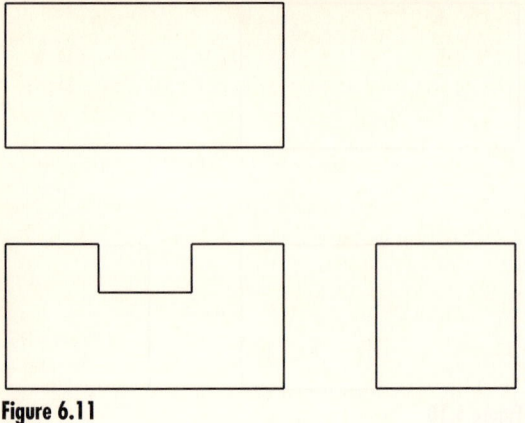

Figure 6.11

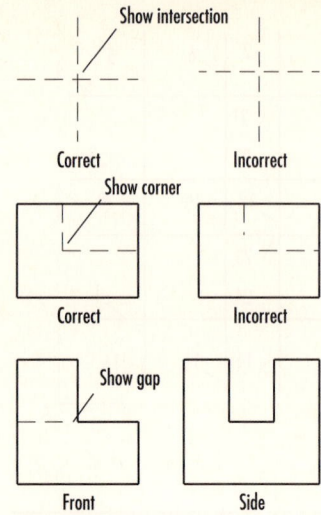

Figure 6.12

Hidden Lines

You will use hidden lines to represent the lines that are not visible in the side view. Remember, each view is a view of the entire object drawn from that line of sight. All surfaces are shown in every view. A hidden line in the drawing represents one of three things:

1. An *intersection* of two surfaces that is behind another surface and therefore not visible;

2. The *edge view* of a hidden surface; or

3. The outer edge of a curved surface that is hidden, which is also called the *limiting element* of a contour.

You should follow at least three general practices when drawing hidden lines to help prevent confusion and to make the drawing easier to read.

■ Clearly show intersections, using intersecting line segments.

■ Clearly show corners, using intersecting line segments.

■ Leave a noticeable gap (about 1/16") between aligned continuous lines and hidden lines.

See Figure 6.12.

These hidden line practices are sometimes difficult to implement. If you change one hidden line using the global linetype scale command, so that it looks better, all the other hidden lines take on the same characteristics and may be adversely affected. In general, hidden line practices are not followed as strictly as they once were, partly because with CAD drawings plotted on a good-quality plotter, the thick visible lines can easily be distinguished from the thinner hidden lines. The results of a reasonable attempt to conform to the standard are considered acceptable in most drawing practices.

Hidden lines are usually drawn in a different color than the one used for the continuous object lines on the AutoCAD drawing screen. This practice helps distinguish the different types of lines and makes them easier to interpret. Also, you control printers and plotters by using different colors in the drawing to represent different thicknesses of lines on the plot. You can use any color, but be consistent. Make all hidden lines the same color. The best approach is to set the color and linetype by layer and draw the hidden lines on that separate layer with the correct properties. That's why BYLAYER is the default choice for color and linetype in AutoCAD. A separate layer for hidden lines already exists in the template drawing from which you started *adaptort.dwg*.

Drawing Hidden Lines

Now you will create a hidden line in the side view of your drawing to represent the bottom surface of the slot.

On your own, set layer HIDDEN_LINES as the current layer.

You will draw a horizontal line from the bottom edge of the slot in the front view into the side view. This line will be used to project the depth of the slot to the side view. Verify that the Ortho button on the status bar is highlighted.

For the following lines you will draw, using the running object snap mode will be quite helpful. On your own select the Intersection snap in the Object Snap Settings dialog box (either double-click on OSNAP or select the icon within the object snap flyout on the Standard toolbar).

Pick: **Line icon**

From point: **(pick the lower right corner of the slot in the front view; an AutoSnap marker should appear at that point)**

To point: **(pick any point to the right of the side view)**

To point: ⏎

Your screen should look like Figure 6.13.

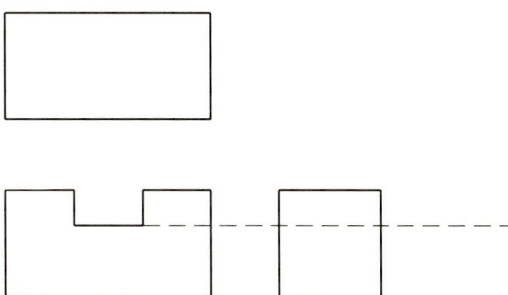

Figure 6.13

On your own, trim the projection line so that only the portion within the side view remains.

Your drawing should look like that shown in Figure 6.14.

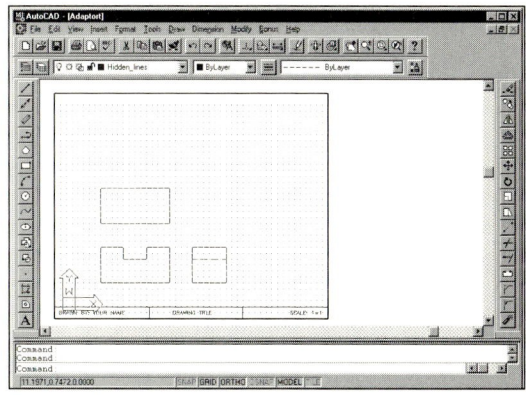

Figure 6.14

Next, you will use the Line command to project the width of the slot from the front view into the top view using vertical lines.

Do the following steps on your own. First, set the current layer to VISIBLE for the lines you will create. To help make the projection lines straight, be sure that Ortho is on. Turn on the running object snap Perpendicular.

Now you are ready to draw the lines for the slot on layer VISIBLE.

> *Pick:* **Line icon**
>
> From point: *(target the upper left corner of the slot in the front view)*
>
> To point: *(pick a point on the upper line of the top view, using the AutoSnap Perpendicular marker)*
>
> To point: ⏎
>
> Command: ⏎
>
> From point: *(target the upper right corner of the slot in the front view)*
>
> To point: *(pick a point on the upper line of the top view, using the AutoSnap Perpendicular marker)*
>
> To point: ⏎

Use the Trim command to remove the excess lines. Your drawing should be similar to Figure 6.15.

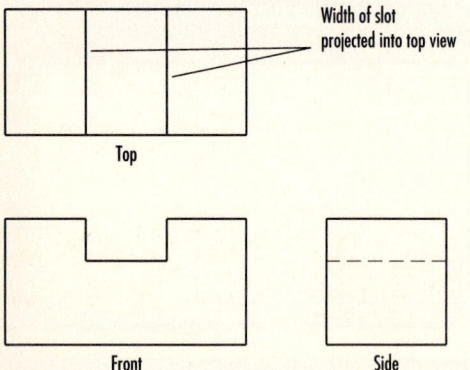

Figure 6.15

Save *adaptort.dwg* before you continue.

Line Precedence

Different types of lines often align with each other within the same view, as illustrated in Figure 6.16.

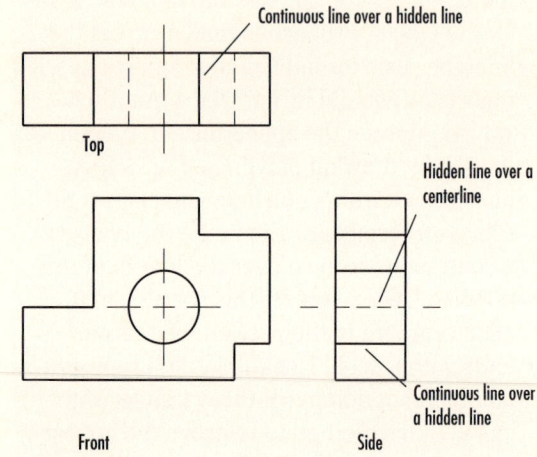

Figure 6.16

The question arises, Which type of line takes *precedence?* That is, Which type do you draw? The rule is that continuous lines take precedence over hidden lines, and hidden lines take precedence over centerlines. Note that in the side view of Figure 6.16, the short end segments of the covered-up centerline show beyond the edge of the object. This practice is sometimes used to show the centerline underlying the hidden line. If you show the short end segments where a centerline would extend, be sure to leave a gap so that the centerline does not touch the other line, as that makes interpreting the lines difficult.

AutoCAD doesn't determine line precedence for you. You must decide which lines to show in your 2D orthographic views. If you draw a line over the top of another line in AutoCAD, both lines will be in your drawing. If you are using a plotter, both will plot, making a darker or thicker line than should be shown. On your screen you may not notice that there are two lines because one line will be exactly over the other.

Slanted Surfaces

Orthographic views can only distinguish *inclined surfaces* (surfaces that are tipped away from one of the viewing planes) from normal surfaces (surfaces that are parallel to the viewing plane) if the surfaces are shown in profile. Inclined surfaces are perpendicular to one of the principal viewing planes and tipped away, or foreshortened, in the other views. As illustrated in Figure 6.17, you can't tell by looking at the top and side views which surfaces are inclined and which are normal.

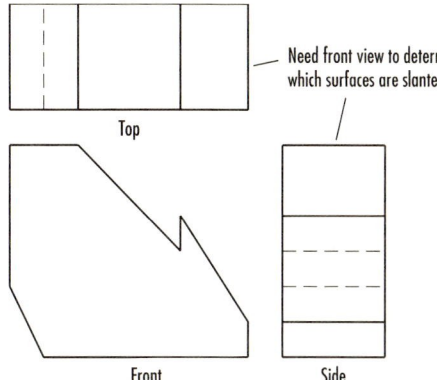

Need front view to determine which surfaces are slanted

Top

Front   Side

Figure 6.17

The front view is required, along with the other two views, to define the object's size and shape completely. In the next steps you will add a slanted surface to the adapter so that the object looks like that shown in Figure 6.18.

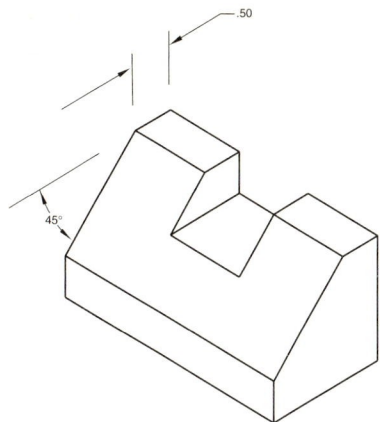

.50

45°

Figure 6.18

You will now use the Line command to add the slanted surface to the side view using relative coordinates. You will locate a point 0.5 inch to the left of the top right corner of the right-side view and draw a slanted line from this point for the 45° surface.

Pick: **Line icon**

From point: **7,3** ⏎

To point: **@3<−135** ⏎

To point: ⏎

The distance 3 was chosen because the exact distance is not known and 3 is obviously longer than needed, as the entire object is only 1.5 inches high. Your screen should look like Figure 6.19.

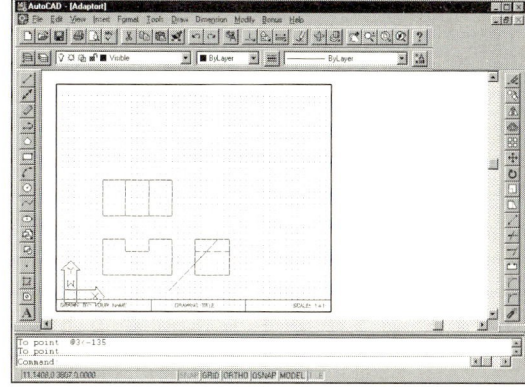

Figure 6.19

Before continuing, use the Trim command on your own to trim this line and any lines above the slanted surface and then set the current drawing layer to layer PROJECTION. When you have finished, your drawing should look like that in Figure 6.20.

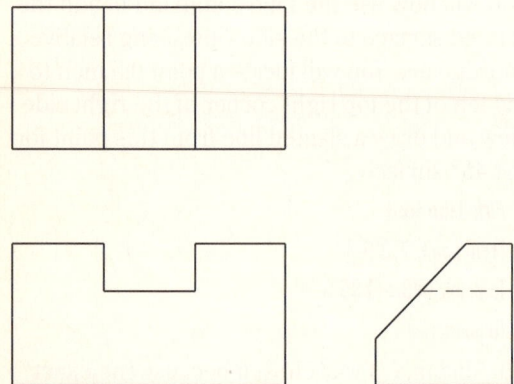

Figure 6.20

Top-View to Side-View Projection

You can project information from the top view to the side view and vice versa by using a 45° *miter line*. You can draw the miter line anywhere above the side view and to the right of the top view, but it is often drawn from the top right corner of the front view, as shown in Figure 6.21.

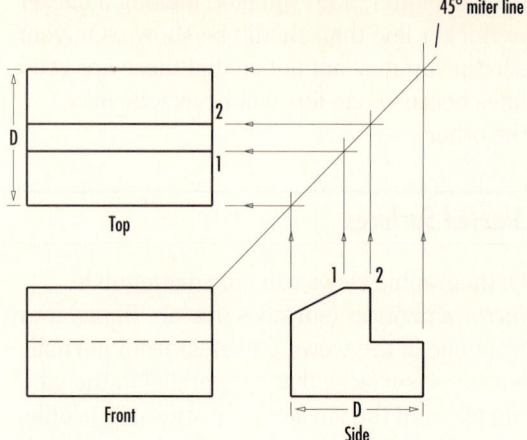

Figure 6.21

To project information from the side view to the top view, you would draw vertical projection lines from the points in the side view so that they intersect the miter line. In Figure 6.21, points 1 and 2 are projected. Then you would project horizontal lines from the intersection of the vertical lines and the miter line across the top view.

Drawing the Miter Line

You will draw a 45° ray, starting where the front edge of the top view and the front edge of the side view would intersect. Remember, the Ray command is located in the Draw menu. On your own, make sure that Ortho is off.

Using Tracking

 Tracking constrains your selections to line up with the X-, Y-, or Z-coordinates of the selected tracking entities as appropriate. You can use the Tracking object snap to start a line from any point on the screen, based on reference points. This method helps you find an intersection where two objects would meet if extended. You will use it to find the point where the bottom edge of the top view and the left edge of the side view would intersect. Refer to Figure 6.22. The OSNAP should still be on, set to Intersection.

Pick: **Draw, Ray**

From point: *(pick Tracking icon, located in the Object Snap flyout on the Standard toolbar)*

First tracking point: *(pick on intersection 1, using the AutoSnap marker to aid selection)*

Next point (Press ENTER to end tracking): *(pick on intersection 2)*

Next point (Press ENTER to end tracking): *(press ⏎ or the return button to end tracking)*

Through point: **@3.5<45** ⏎

Through point: ⏎

The ray that you will use for the miter line is added to the drawing, as shown in Figure 6.22.

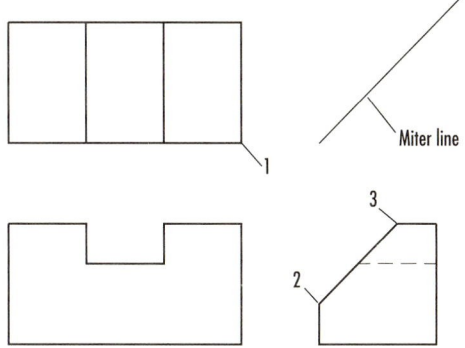

Figure 6.22

Next, you will project the corner point (point 3 in Figure 6.22) created by the slanted surface in the side view to the miter line by drawing a vertical ray from the intersection. Turn Ortho on before you pick points.

Pick: **Draw, Ray**

From point: *(pick point 3 in Figure 6.22)*

Through point: *(pick anywhere above point 3)*

Through point: ⏎

The ray is added to your drawing, extending vertically from point 3. Next, you will project another ray from where the vertical ray intersects the miter line.

Pick: **Draw, Ray**

From point: *(pick where the vertical ray meets the miter line)*

Through point: *(pick a point to the left of the miter line)*

Through point: ⏎

Your drawing should look like Figure 6.23.

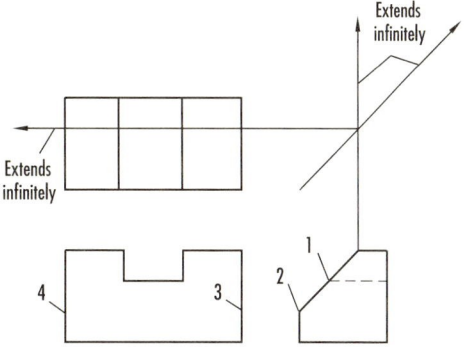

Figure 6.23

On your own, project point 1 to the top view.

Now you will use Tracking to project point 2 into the front view. Make sure that OSNAP is on with Perpendicular selected. To aid in Tracking, make sure that SNAP and ORTHO are toggled off.

Pick: **Line icon**

From point: *(pick the Tracking icon from the Object Snap flyout)*

First tracking point: *(pick point 2 using the AutoSnap Intersection marker)*

Next point (Press ENTER to end tracking): *(pick line 3 using the AutoSnap Perpendicular marker)*

Next point (Press ENTER to end tracking): ⏎

To point: *(pick line 4 using the AutoSnap Perpendicular marker)*

To point: ⏎

The front view should now have a solid line representing the lower part of the block's slope. Now you only have to trim the Rays used in projecting the surface depths to the top view. Remember, when rays are trimmed they become regular lines. On your own use the Trim command and implied Crossing and trim the lines so that your screen looks like Figure 6.24.

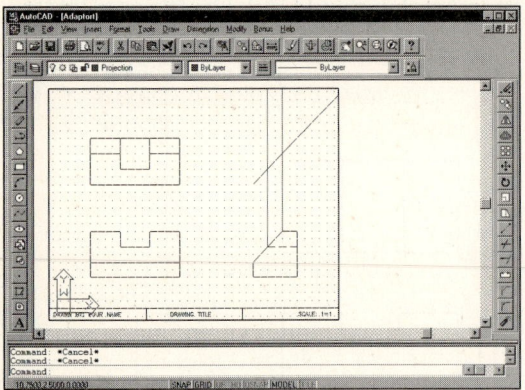

Figure 6.24

Now, you will change the new lines in the top and front views to the layer VISIBLE.

> *Pick: (new lines in top and front views)*
>
> *Select: (layer VISIBLE from the Layer Control pull-down menu on the Object Properties toolbar)*

When you have completed the visible lines, use the techniques you have learned to set layer VISIBLE as the current layer and freeze layer PROJECTION. Leaving the projection lines frozen in the drawing is useful because, if you need to change something, you can just thaw the layer instead of having to re-create the projection lines.

Sizing and Positioning the Drawing

Use Zoom XP, as you learned in Tutorial 5, to establish a relationship between the model space drawing and the size it is to be on paper. When you use Zoom All, the entire drawing or the limits area is fit into the viewport, which does not give you any particular scale for the end drawing. You will use Zoom XP with a value of 1 to set 1 unit in paper space equal to 1 unit in model space. You will type the alias Z for the Zoom command.

> Command: **Z** ↵
>
> All/Center/Dynamic/Left/Previous/Scale (X/XP)/Window/
<Realtime>: **1XP** ↵

The drawing is zoomed so that 1 unit in paper space is equal to 1 unit in model space.

Next, use the Pan Realtime command on your own to position the views you drew inside the viewport. Position the views so that they appear centered in the viewport. When you have finished, your screen should look like Figure 6.25.

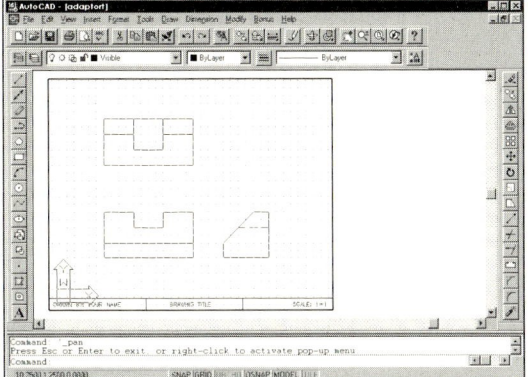

Figure 6.25

Now toggle the running mode object snap off by double-clicking OSNAP.

Double-click: **OSNAP** *(on the status bar)*

■ **TIP** It is good practice to frequently save your drawings on your disk. If your system experiences a power failure and you lose the drawing in memory, you can always retrieve a recent version from your disk. ■

Drawing Holes

Figure 6.26 shows an object with two holes and the ways in which they are represented in front and top views.

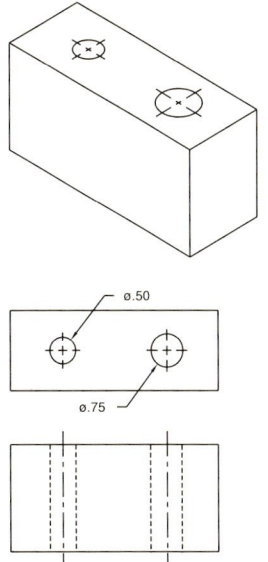

Figure 6.26

The *diameter symbol* ∅ indicates a diameter value. If no depth is specified for a hole, the assumption is that the hole goes completely through the object. As no depth is specified for these holes, the hidden lines in the front view go from the top surface to the bottom surface.

You will add two holes to the orthographic views that you have drawn to represent the adapter, as shown in Figure 6.27.

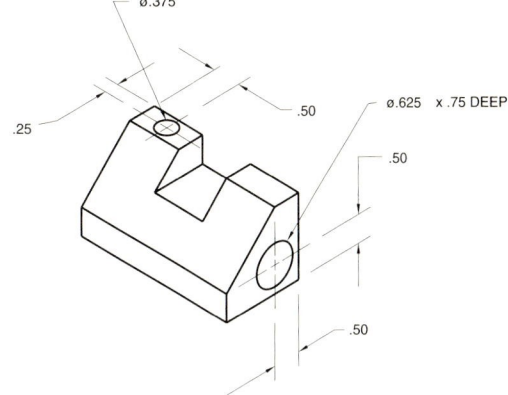

Figure 6.27

In Figure 6.27, the 0.625-diameter hole has a depth specification of 0.75. That is, the hole is 0.75 deep, drilled from the surface of the object. A 120° conical point is added to the bottom of any hole that does not go completely through an object. The reason is that twist drills, the type used most often to drill holes, have a conical point. The 120° drill point is not included as part of the depth of the hole.

Centerlines for holes must be included in all views. A centermark and four lines extending beyond the four quadrant points are used to define the center point of a hole in its *circular view* (the view in which the hole appears as a circle). A single centerline, parallel to the two hidden lines, is used in the other views, called *rectangular views* because the drill hole appears as a rectangle. Centerlines should extend beyond the edge of the symmetrical feature by a distance of at least 3/8" on the plotted drawing.

You will start with the top view and add the circular view of the 0.375 diameter hole, as specified in Figure 6.27. As indicated in the figure, the hole's center point is located 0.5 from the left surface of the view and 0.25 from the back surface.

Use the Drawing Aids dialog box to change the Snap spacing to 0.25; be sure that it is on.

Pick: **Circle icon**

3P/2P/TTR<Center point>: **2.5,5.25** ⏎

Diameter/<Radius>: **D** ⏎

Diameter: **.375** ⏎

Drawing Centerlines and Centermarks

Next, you will draw the centerlines for the circle. Engineering drawings involve the use of two different thicknesses of lines: thick lines are used for visible lines, cutting plane lines,

and short break lines; thin lines are used for hidden lines, centerlines, dimension lines, section lines, long break lines, and phantom lines. In AutoCAD, you use color to tell the plotter or printer which thickness to use when printing lines.

Because centerlines in the circular view should be thin, you need to draw them on a new layer. Make layer THIN current. You could also use layer CENTERLINE, but it has a CENTER linetype. Because the Center Mark command automatically creates the dashes at the center of the circle by drawing short lines, centermarks will usually look better if they are not drawn with a linetype that already contains a dash (e.g., the centerline linetype). This is why you will use layer THIN, which has a continuous linetype.

> ■ *TIP* You can also use the Properties icon to change the linetype for your centermarks to Continuous. ■

The Dimcen variable controls the size and appearance of the centermark that you add to the circle. The *absolute value* of the Dimcen variable determines the size of the centermark. The *sign* (positive or negative) determines the style of the centermark. Figure 6.28 shows the different styles of centermarks that you can create by setting Dimcen to a positive or negative value.

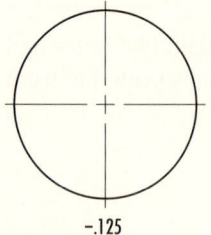

−.125   +.125

Figure 6.28

Setting Dimcen to zero will cause no center-mark to be drawn. (This lack of a centermark is useful when using dimensioning commands such as Radial, which automatically add a centermark. You will learn more about dimensioning and adding centermarks in Tutorial 7.) Usually the cross at the center should be 1/8" wide, but as this hole is relatively small, you will need a smaller centermark; you will use the value –0.05.

Command: **DIMCEN** ⏎

New value for DIMCEN <0.0900>: **–.05** ⏎

Next, show the Dimension toolbar so that you can select the Center Mark command.

Pick: **View, Toolbars, Dimension**

The Dimension toolbar appears on your screen, as shown in Figure 6.29. Position it so that it is convenient for you to use when selecting commands.

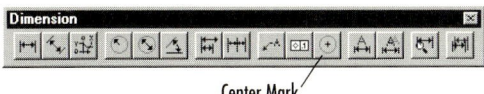

Center Mark

Figure 6.29

Next, you will use Center Mark to draw the circle's centerlines on layer THIN. Use the Center Mark icon from the Dimension toolbar whenever you are drawing a centerline in the circular view (where the hole appears round).

Pick: **Center Mark icon**

Select arc or circle: (*pick on the outer edge of the circle*)

■ *TIP* Use Zoom Window and turn Snap off if you need to make targeting the objects and intersections easier. ■

The circular view centerlines should appear in the drawing. Next, you will use hidden lines and the Intersection object snap to project the

width of the hole into the front view. On your own, set the current layer to HIDDEN_LINES. Now turn the running mode Intersection object snap on by double-clicking on OSNAP. Because you are going to project straight lines, be sure that Ortho is active.

Pick: **Line icon**

From point: (*pick the intersection of the circle's horizontal centerline with the left edge of the circle*)

To point: (*pick Snap to Perpendicular icon*)

per to (*pick any point on the lowest horizontal line in the front view*)

To point: ⏎

Repeat this procedure to draw the projection line for the right side of the hole and for the centerline. (Use Zoom Previous or Zoom All, if necessary, to return your screen to full size.)

Now that you have finished this step, the figure on your screen should be similar to Figure 6.30.

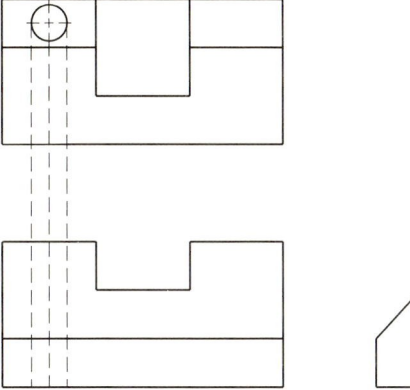

Figure 6.30

Now use the Properties icon from the Object Properties toolbar to change the layer for the middle of the three lines you just drew to layer CENTERLINE and save your drawing. Next, you will break the line so that it is the right length for a centerline in the front view.

Breaking the Centerline

You will break the middle vertical line so that you can use it as a centerline by selecting the Break icon on the Modify toolbar. Centerlines should extend at least 0.375" past the edge of the cylindrical feature when the drawing is plotted. You will extend the centerline 0.5" past the bottom of the front view to meet this criterion.

Pick: **Break icon**

Select object: (select the middle vertical line)

The first point of the break is the point that you want to use as the top of the centerline. The second point is past the end, where the line extends into the top view, so that all of the centerline extending into the top view is broken off. Refer to Figure 6.31. On your own, be sure that Snap is on and that OSNAP is off to make selecting easier.

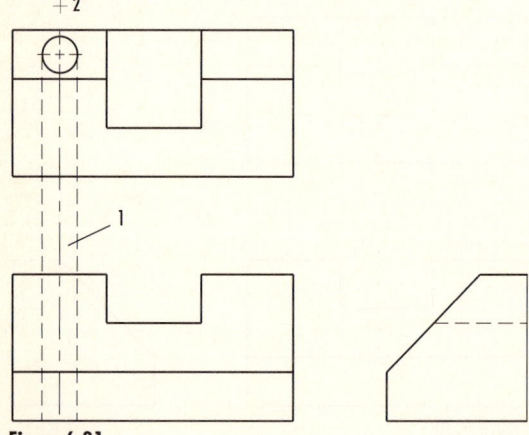

Figure 6.31

Enter second point (or F for first point): **F**

Enter first point: *(select break point 1 on the line, .5 units away from the top of the object in the front view)*

Enter second point: *(select point 2, past the top end of the line)*

■ *TIP* If you pick on the end of the line, you may not get a location past the end of the line for the second point of the break; instead, short segments of the line may remain and look like random points in your drawing. ■

Now use hot grips with the Stretch command to stretch the centerline down 0.5" past the bottom of the front view on your own. Refer to Tutorial 4 if you need to review using hot grips and the Stretch command.

Your drawing should look like Figure 6.32.

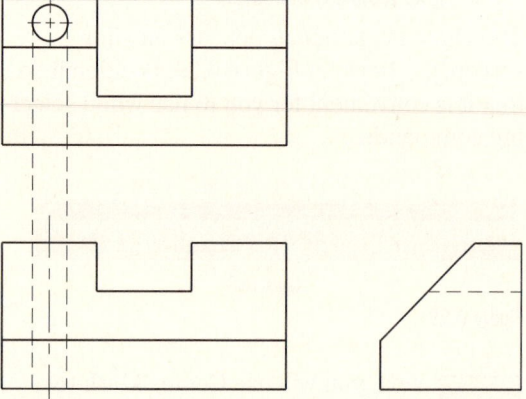

Figure 6.32

Before continuing, trim the two hidden lines where they extend past the front view of the object on your own.

To complete the hole in the side view, you will project the location of the hole into the side view. Then you will use hot grips with the Move, Copy option to copy the lines from the front view to the side view.

On your own, thaw layer PROJECTION, set it as the current layer, and draw a projection line from the top view where the vertical centerline intersects the circle, and project it out past the miter line. Use the OSNAP (Intersection) mode and Ortho to help you. Be sure that Snap is off. Refer to Figure 6.33. Then project the intersection of the horizontal projection line with the miter line into the side view, creating a vertical projection line.

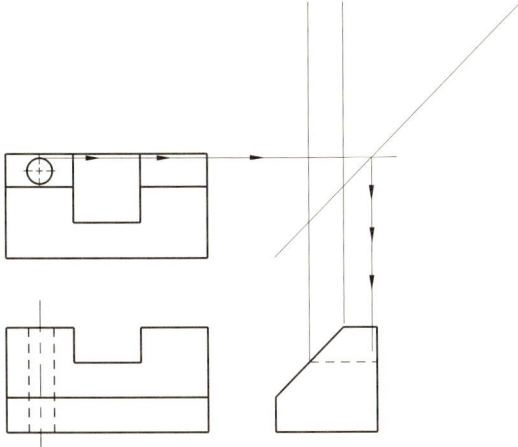

Figure 6.33

Because holes are symmetrical, the side view of the hole appears the same as the front view, except for the location. You will use hot grips with the Move, Copy option to copy the two hidden lines and the centerline from the front view to the side view.

Activate the hot grips for the two hidden lines and the centerline by picking on them on your own.

Pick: *(the grip on the top endpoint of the right-hand hidden line as the base grip)*

** STRETCH **

<Stretch to point>/Base point/Copy/Undo/eXit: *(pick the right mouse button)*

Pick: **Move**

** MOVE **

<Move to point>/Base point/Copy/Undo/eXit: *(pick the right mouse button again)*

Pick: **Copy**

** MOVE (multiple) **

<Move to point>/Base point/Copy/Undo/eXit:

The lines you selected appear faintly, attached to the crosshairs. (OSNAP should still be active, with intersection selected). Target the intersection between the top line in the side view and the projection line locating the side of the hole, as shown in Figure 6.34. You may want to use Zoom Window to help you choose the correct intersection. The three lines are copied to this location. Press the return button to exit the Copy command.

Pick: **Exit**

If you press ↵ or the space bar, you will exit automatically.

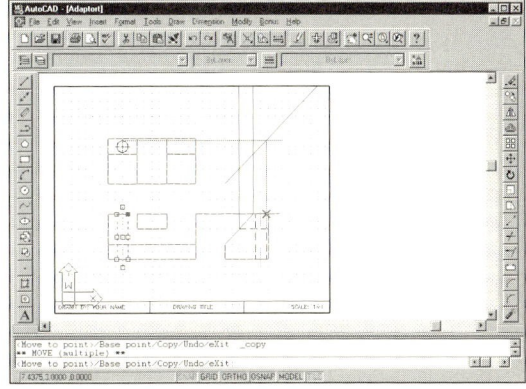

Figure 6.34

Leave the projection lines and the miter line; you will need them to project the other hole from the side view. Your drawing will look like that in Figure 6.35. Redraw your screen before you continue.

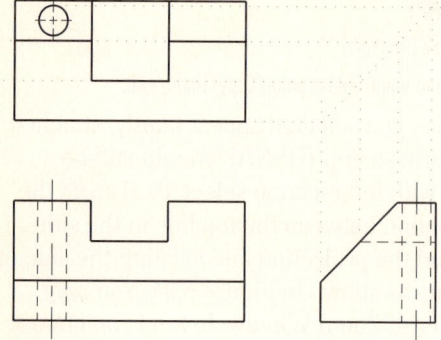

Figure 6.35

Next you will add the circular view of the 0.625 hole to the side view and then project the width and centerline of the hole into the front view, as shown in Figure 6.36. Before you continue, set layer VISIBLE as the current layer.

The center point of the hole is (6.75,2) and the radius is 0.3125. Draw the circle on your own.

Now, make layer THIN current and use the Center Mark command to add centerlines. Return to the layer HIDDEN_LINES and use the Intersection object snap to project the hole's edge lines into the front view on your own.

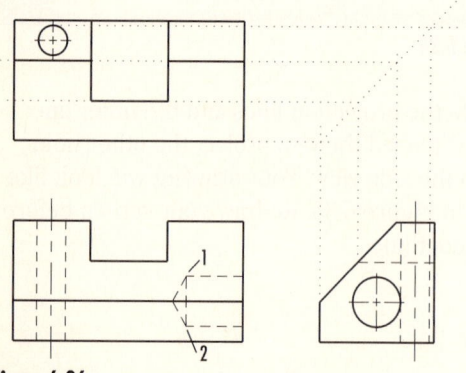

Figure 6.36

Add the 0.75 depth of the hole by using Offset to create a vertical line parallel to the right surface of the front view on your own. If you need help using Offset, refer to Tutorial 3. Then use the Trim command to remove the excess lines. Refer to Figure 6.36. Next, change the line you offset to layer HIDDEN_LINES on your own.

Next you will add the 120° conical point to the front view. First you will enlarge the area you are working on so that it fills your screen.

Pick: **Zoom Window icon**

Zoom the front view so that you can add the conical point. Be sure that Ortho is off.

Pick: **Line icon**

From point: *(target intersection 1 in Figure 6.36)*

To point: **@.75<240** ⏎

To point: ⏎

Command: ⏎

Line from point: *(target intersection 2)*

To point: **@.75<120** ⏎

To point: ⏎

Use the Trim command to remove any excess lines on your own, so that your drawing looks like Figure 6.36.

Pick: **Zoom Previous icon**

to return to your original display screen area.

Use the Make Object's Layer Current icon to change the current layer to CENTERLINE. Project a line from the side view, where the centerline crosses the edge of the circle, to the front view, at least 0.5 inch beyond the drill point.

Note that the horizontal centerline in the front view coincides with an edge line. Refer to the section at the beginning of this tutorial and Figure 6.16 on line precedence. When you are printing, the object's edge line takes precedence over the centerline, but on the screen AutoCAD displays the last line drawn on top of the others.

On your own, use Break to remove the excess centerline, as you learned to do earlier, so that only a 0.5 inch tail remains at each end of the hole. Showing this short tail of centerline is optional. If you feel that it makes the drawing difficult to interpret, you may leave it off entirely. If you show it, you must leave a gap of about 1/16" on the plotted drawing between the visible (or hidden) line and the short centerline tail.

To get rid of the excess blip marks,

Pick: **Redraw All icon** *(on the Standard toolbar)*

Now, project the center of the hole from the side view to the top view so that you can copy the hole from the front view. Be sure that layer CENTERLINE is current and that Snap to Intersection running mode and Ortho are on.

Pick: **Line icon**

From point: *(select the point where the centerline touches the top of the hole in the side view)*

To point: *(select a point slightly past the miter line)*

To point: ⏎

Command: ⏎

Line from point: *(select the point where the projected line intersects with the miter line)*

To point: *(select a point left of the top view)*

To point: ⏎

Setting the Global Linetype Scaling Factor

You can change the linetype scale factor for all linetypes at once by setting the global linetype scaling factor. All standard AutoCAD linetypes are defined in the file *acad.lin*. Recall from Tutorial 4 that each linetype is defined by the distance that each dash, gap, and dot is to be drawn. Because linetypes are defined by specific distances, you may need to adjust the lengths of the dashes and gaps for use in your drawing. The Ltscale command lets you adjust all the linetype lengths by the scaling factor

you specify. A setting of 2 for Ltscale makes the dashes and gaps twice as long as the original pattern; a setting of 0.5 makes them half as long. When plotting your drawing to a particular scale, you usually set the Ltscale factor to the reciprocal of the plot scale. For instance, if you are going to plot the drawing at 1"=10", you set Ltscale to 10. Ltscale affects all linetypes in your drawing at the same time (although you can adjust it for paper space and model space).

■ *Warning:* When you are working in metric units, the linetypes may not always appear correctly. A line may appear to be the correct color but not the correct pattern. The reason is that the lengths defined in the linetype file are defined in terms of inches. Incorrectly shown metric lines may have changed pattern, but the spacing is so small that you can't see it. Use Ltscale to adjust the spacing. For metric units, try a value of 25.4. ■

You may need to adjust your lines with Ltscale to make the CENTER linetype visible and the HIDDEN linetype have shorter dashes.

Command: **LTSCALE** ⏎

New scale factor <1.0000>: **.65** ⏎

Your drawing should look like Figure 6.37.

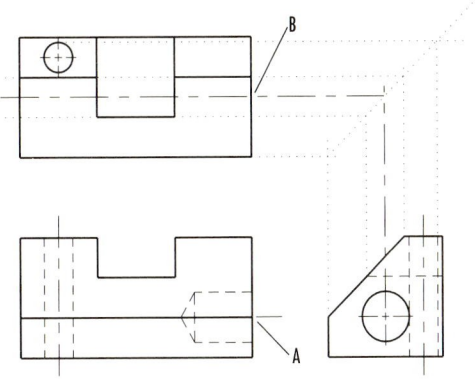

Figure 6.37

Now you are ready to use hot grips and Move, Copy to copy the hidden lines from the front view into the top view. You will use a procedure similar to last time, except that you will change the base point. When prompted for the base point, you will select the intersection of the centerline with the right edge of the object in the front view. Refer to Figure 6.37.

Pick: **(on the hidden lines forming the hole and drill point in the front view to activate the hot grips, but not the centerline)**

Pick: **(any grip as the base grip)**

Press: **(the right mouse button)**

Pick: **Base Point**

Base point: **(target point A)**

** STRETCH **

<Stretch to point>/Base point/Copy/Undo/eXit: **(press the right mouse button)**

Pick: **Move**

** MOVE **

<Move to point>/Base point/Copy/Undo/eXit: **(press the right mouse button)**

Pick: **Copy**

A faint copy of the lines that you have selected will be attached to the crosshairs. Move the crosshairs to the point in the top view where the projection line for the center of the hole intersects with the right side of the top view. You will target this point and press the pick button to select it as the location for the copy.

** MOVE (multiple) **

<Move to point>/Base point/Copy/Undo/eXit: **(select point B)**

An AutoSnap marker should have appeared at point B representing the intersection. You will see the copy appear in the top view, as shown in Figure 6.38. Press ⏎ to end the command.

On your own, use the Break command again to shorten the projected centerline in the top view so that it extends about 0.5" past the edge of the hole. Erase the centerline that is projected from the side view. Then freeze layer PROJEC-TION and redraw. When you redraw, your screen should be similar to Figure 6.38.

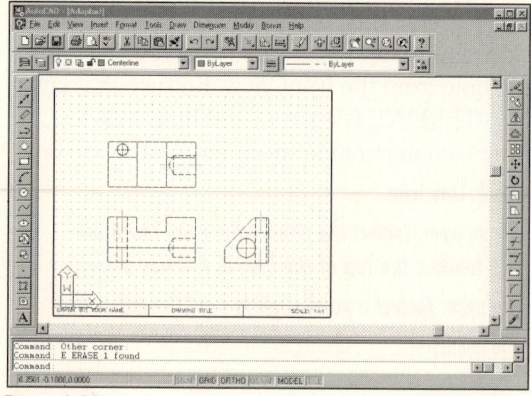

Figure 6.38

You have completed this orthographic drawing. Save the drawing on your disk as *adaptort.dwg*.

Projecting Slanted Surfaces on Cylinders

Next, you will work with projecting slanted surfaces on cylinders. You will begin by opening the existing drawing *cyl1.dwg* from the data files for this manual.

Begin a new drawing called *cyl1orth.dwg* from the template drawing *cyl1.dwg*. Your screen should be similar to Figure 6.39.

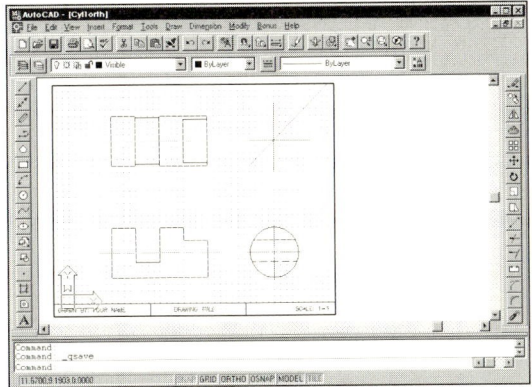

Figure 6.39

On your own, add a slanted surface to the front view, as shown in Figure 6.40. The top of the slanted surface is located 1.5 inches from the right end of the cylinder, and the bottom of the surface is at the horizontal centerline. The top point of the slanted surface, the intersection of the vertical centerline and the edge of the cylinder in the side view, is labeled 1 in Figure 6.40. The bottom edge of the slanted surface is labeled 2, 3 and is located directly on the horizontal centerline in the side view.

Next, remove all excess lines left from the shallow surface in the front view. Remove the same surface from the top and side views; you will be creating a different surface in its place. Your drawing should look like Figure 6.40.

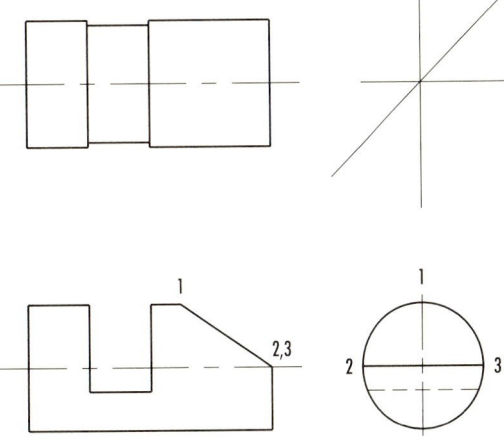

Figure 6.40

What is the shape of the slanted surface in the top view? The front view of the surface appears as a straight slanted line. The side view of the surface is a semicircle, but in the top view the surface is an ellipse, as shown in Figure 6.41. A circular shape seen from an angle other than straight on is an ellipse. Since the locations of points 1, 2, and 3 on the ellipse are known, you can use the Ellipse command to draw the shape in the top view.

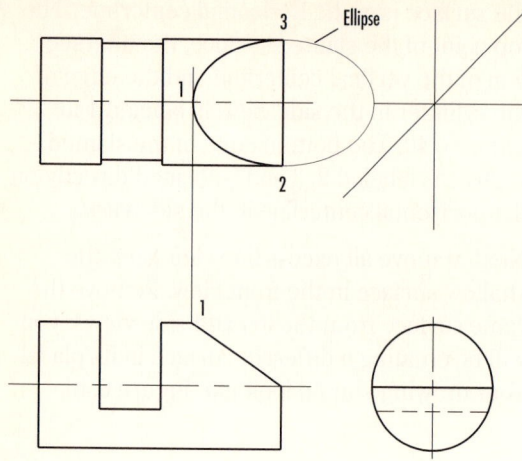

Figure 6.41

One way to describe an ellipse is to create a circle and then tip the circle away from your viewing direction by a rotation angle. This method requires you to specify the angle of rotation instead of the endpoint of the second axis.

You will draw an ellipse using two endpoints and an angle of rotation. The rotation angle must be between 0° and 89.4°.

See Figure 6.42 to determine the location of the points.

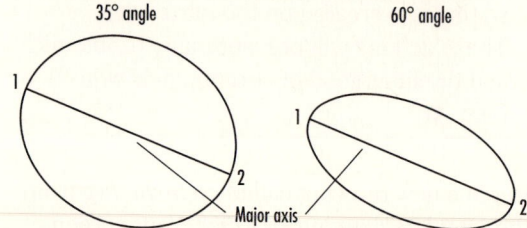

Figure 6.42

Using Ellipse

The Ellipse command draws ellipses. AutoCAD provides three different ways to specify an ellipse. The various methods can be found under Ellipse on the Draw menu. The Ellipse icon is also on the Draw toolbar.

Next, you will practice using the Ellipse command by drawing some ellipses off to the side of the drawing. You will erase them when you are through. To draw an ellipse by specifying three points,

Pick: **Ellipse icon**

Arc/Center/<Axis endpoint 1>: *(select a point)*

Axis endpoint 2: *(select a point)*

<Other axis distance>/Rotation: *(select a point)*

An ellipse is created on your screen, using the three points that you selected. The ellipse has a *major axis*, the longest distance between two points on the ellipse, and a *minor axis*, the shorter distance across the ellipse. AutoCAD determines which axis is major and which is minor by examining the distance between the first pair of endpoints and comparing it to the distance specified by the third point.

Pick: **Ellipse icon**

Arc/Center/<Axis endpoint 1>: *(select point 1)*

Axis endpoint 2: *(select point 2)*

This construction method defines the distance between points 1 and 2 as the major axis (diameter) of the ellipse. This circle will be rotated into the third dimension by the specified rotation angle.

<Other axis distance>/Rotation: R ⏎

Rotation around major axis: **35** ⏎

On your own, erase from your screen the ellipses that you created.

The other two construction methods for ellipses are analogous to the two methods already described, but they use radius values rather than diameter values. That is, the center point of the ellipse is known and is used as a starting point instead of an endpoint. Next, you will draw an ellipse by specifying a center point and two axis points; refer to Figure 6.43 for the locations of the points.

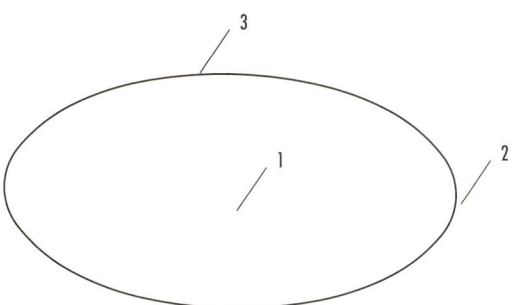

Figure 6.43

Command: ↵ *(to restart the command)*

Arc/Center/<Axis endpoint 1>: **C** ↵

The center of the ellipse is the intersection of the major and minor axes. You can enter a coordinate or use the cursor to select a point on the screen.

Center of ellipse: *(select point 1)*

Next, you must provide the endpoint of the axis. The angle of the ellipse is determined by the angle from the center point to this endpoint.

Axis endpoint: *(select point 2)*

Now you must specify the distance measured from the center of the ellipse to the endpoint of the second axis (measured perpendicular to the first axis) or a rotation angle.

<Other axis distance>/Rotation: *(select point 3)*

AutoCAD uses the point that you selected to decide whether the first axis is a major or minor axis. The ellipse is drawn on your screen.

Next, you will draw an ellipse by specifying a center point, one axis point, and an angle of rotation. Refer to Figure 6.44.

Pick: **Ellipse icon**

Arc/Center/<Axis endpoint 1>: **C** ↵

Center of ellipse: *(pick the point marked 1 in Figure 6.44)*

Axis endpoint: *(pick point 2)*

<Other axis distance>/Rotation: **R** ↵

Rotation around major axis: **30** ↵

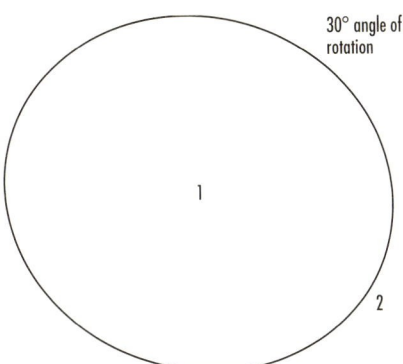

Figure 6.44

Now that you have practiced with the Ellipse command, you are ready to add the ellipse to your drawing. Before you go on, erase the ellipses you drew when practicing.

On your own, project point 1 into the top view, as shown in Figure 6.45. To do so, change to layer PROJECTION. The Intersection running mode object snap should still be on. Use it and the Line command to draw a vertical line from point 1 in the front view extending into the top view past the center. When you have finished, change the current layer back to layer VISIBLE.

Pick: **Ellipse icon**

Arc/Center/<Axis endpoint 1>: **C** ↵

Center of ellipse: *(pick the point marked C in Figure 6.45)*

Axis endpoint: *(pick point 1 in the top view, where your projection line crosses the centerline)*

<Other axis distance>/Rotation: *(pick point 2 in the top view)*

Your ellipse should look like the one shown in Figure 6.45.

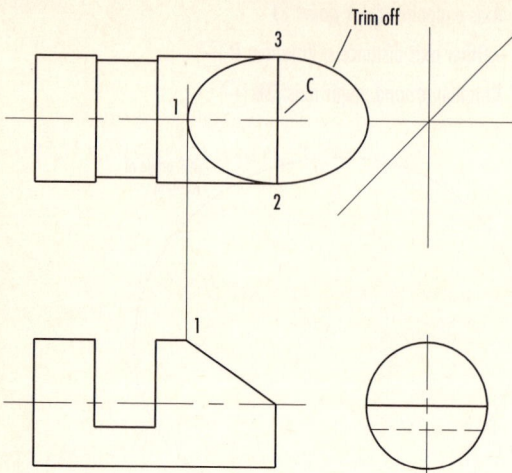

Figure 6.45

Use the Trim command to remove the unnecessary portion of the ellipse. The remaining curve represents the top view of the slanted surface. Remove the projection line from the screen by freezing layer PROJECTION. Your screen should be similar to Figure 6.46.

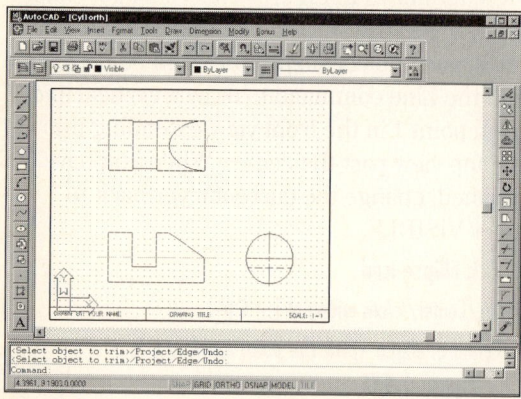

Figure 6.46

Circular shapes appear as ellipses when tipped away from the direction of sight. Not all curved surfaces are uniform shapes like circles and ellipses; some are irregular. You can create irregularly curved surfaces by using the Polyline command and the Spline option that you learned in Tutorial 3. To project an irregularly curved surface to the adjacent view, identify a number of points along the curve and project each point. Use the Polyline command to connect the points. Then use the Pedit, Spline option to create a smooth curve through the points.

Save the drawing as *cylorth.dwg* and print the drawing from paper space. You have now completed Tutorial 6. Turn off any extra toolbars that you may have open, leaving only the Draw and Modify toolbars on the screen. Exit AutoCAD.

KEY TERMS

absolute value
circular view
diameter symbol
edge view
hidden line
inclined surface

intersection
limiting element
major axis
minor axis
miter line
normal surface

orthographic views
precedence
project
projection lines
rectangular view
sign

KEY TERMS

Center Mark
Construction Line

Dimcen
Ellipse

Ortho
Tracking

Draw front, top, and right-side orthographic views of the following objects. The letter M after an exercise number means that the problem's dimensional values are in millimeters (metric units). If no letter follows the exercise number, the dimensions are in inches.

6.1 Base Block

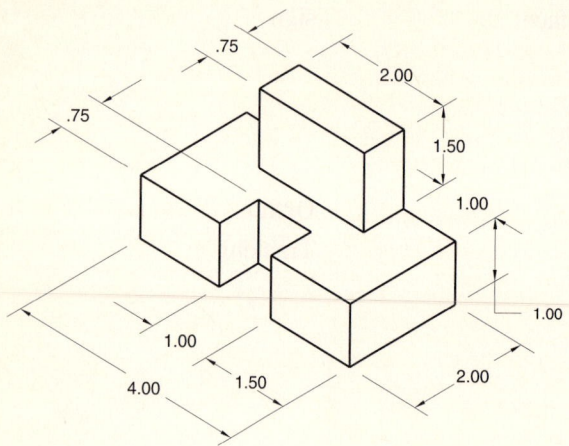

6.2M Shaft Guide

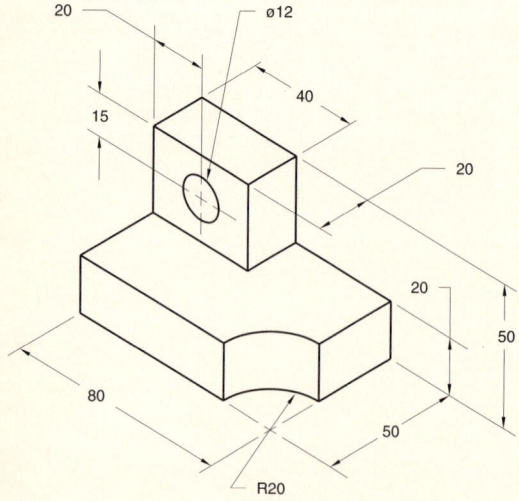

6.3 Piston Guide

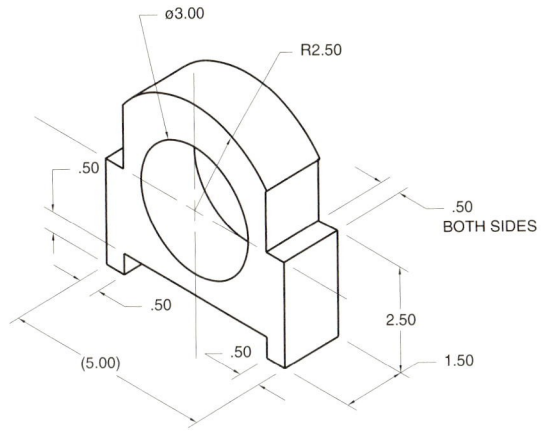

- ø3.00
- R2.50
- .50
- .50 BOTH SIDES
- .50
- 2.50
- .50
- (5.00)
- .50
- 1.50

6.4M Lock Catch

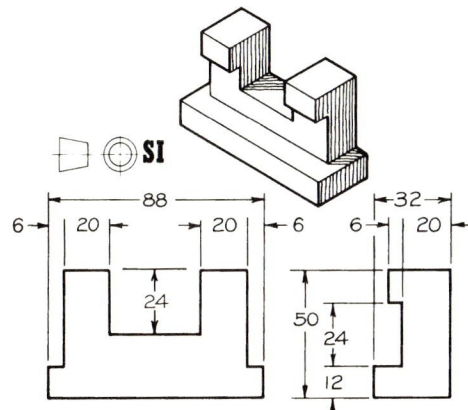

SI

- 88
- 6
- 20
- 20
- 6
- 6
- 32
- 20
- 24
- 50
- 24
- 12

6.5 Bearing Box

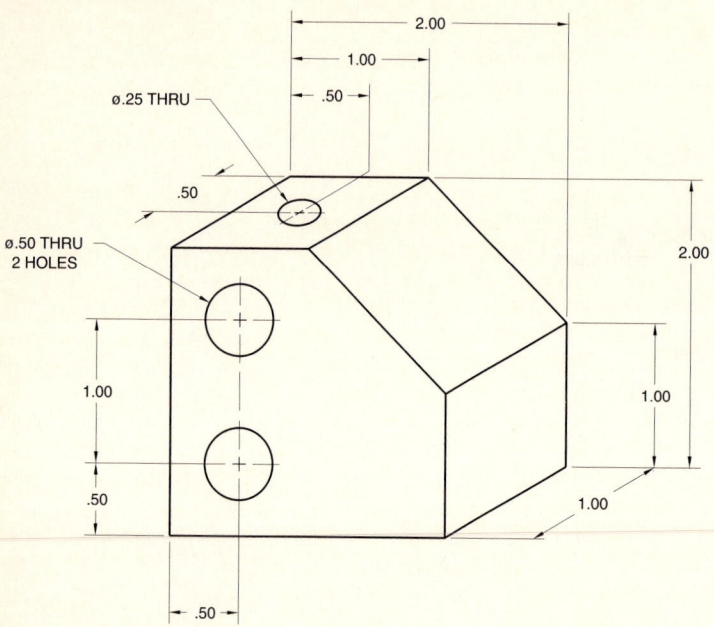

- 2.00
- 1.00
- .50
- ø.25 THRU
- .50
- ø.50 THRU 2 HOLES
- 2.00
- 1.00
- 1.00
- 1.00
- .50
- .50

6.6M Bushing Holder

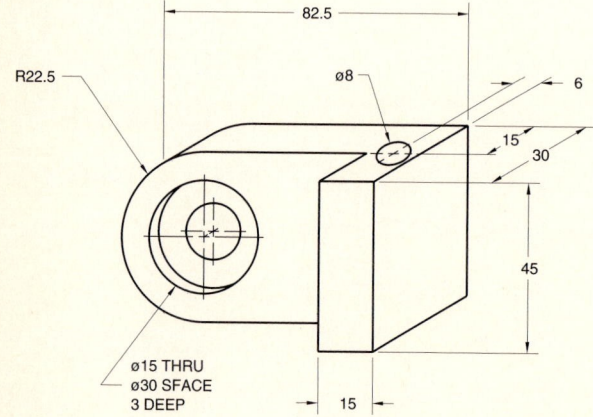

- 82.5
- R22.5
- ø8
- 6
- 15
- 30
- 45
- ø15 THRU
- ø30 SFACE
- 3 DEEP
- 15

6.7 Shaft Support

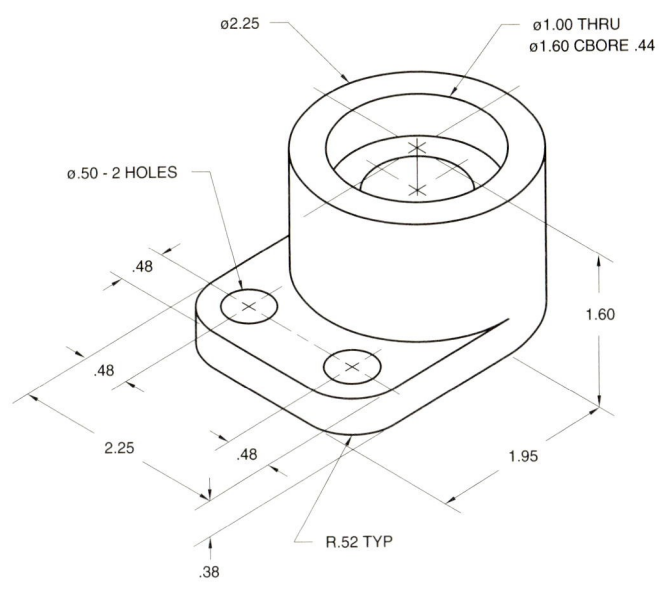

ø2.25

ø1.00 THRU
ø1.60 CBORE .44

ø.50 - 2 HOLES

.48

.48

1.60

2.25

.48

1.95

R.52 TYP

.38

6.8M Stop Plate

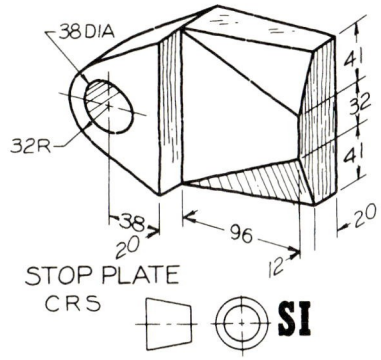

38 DIA

32R

41

32

41

38

20

96

20

12

STOP PLATE
C R S

SI

6.9 Lift Guide

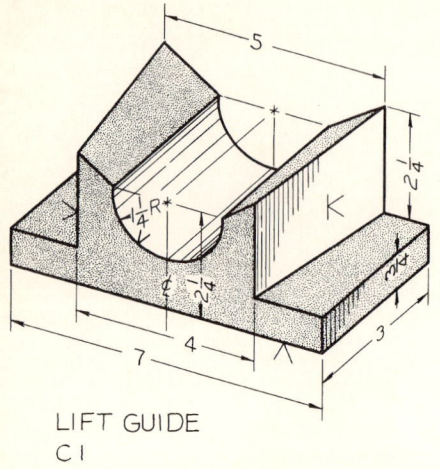

LIFT GUIDE
C I

6.10M Clamp

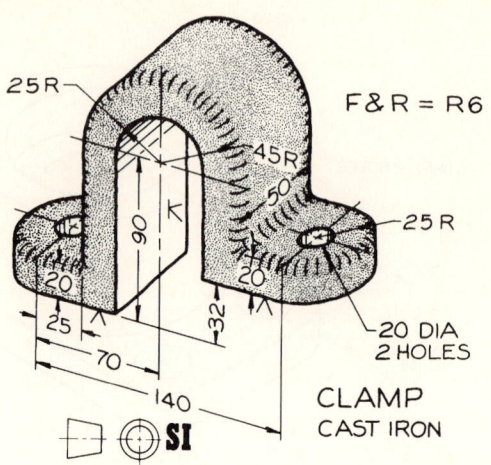

F & R = R6

CLAMP
CAST IRON

20 DIA
2 HOLES

SI

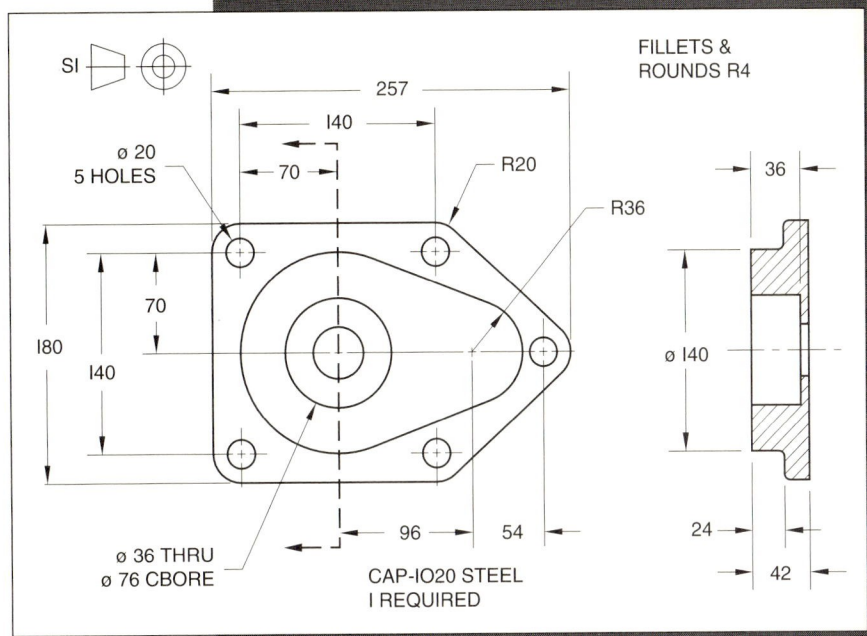

SI

FILLETS &
ROUNDS R4

257

140

ø 20
5 HOLES

70

R20

R36

36

70

180

140

ø 140

ø 36 THRU
ø 76 CBORE

96

54

CAP-IO20 STEEL
I REQUIRED

24

42

TUTORIAL 7

Basic Dimensioning

Objectives

When you have completed this tutorial, you will be able to

1. Understand dimensioning nomenclature and conventions.

2. Control the appearance of dimensions.

3. Set the dimension scaling factor.

4. Locate dimensions on drawings.

5. Set the precision for dimension values.

6. Dimension a shape.

7. Use baseline and continued dimensions.

8. Save a dimension style and add it to the template drawing.

9. Use associative dimensioning to create dimensions that can update.

Introduction

In the preceding tutorials you have learned how to use AutoCAD to define the *shape* of an object. Dimensioning is used to show the *size* of the object in your drawing. The *dimensions* you specify will be used in the *manufacture* and *inspection* of the object. Figure 7.1 shows a dimensioned drawing.

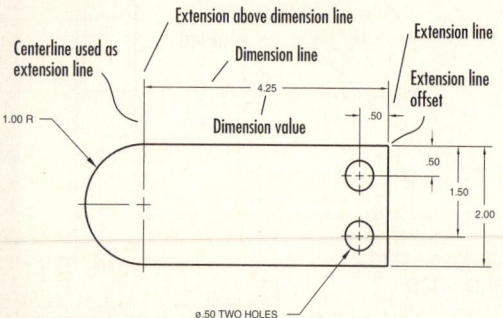

Figure 7.1

For the purpose of inspecting the object, a *tolerance* must be stated to define to what extent the actual part may vary from the given dimensions and still be acceptable. In this tutorial you will use a general tolerance note to give the allowable variation for all dimensions. Later, in Tutorial 14, you will learn how to specify tolerances for specific dimensions and geometric tolerances.

Nomenclature and Conventions

Dimensions are used to describe accurately the details of a part or object so that it can be manufactured. In engineering drawings, dimensions are always placed outside the object outline, unless placing the dimension on the object would result in a drawing that is easier to interpret. *Extension lines* relate the dimension to the feature on the part. There should always be a gap of 1/16″ on the plotted

drawing between the feature and the beginning of the extension line, called the *extension line offset*, as shown in Figure 7.1. Centerlines can also be extended across the object outline and used as extension lines without leaving a gap where they cross object lines.

Dimension lines are drawn between extension lines and have arrowheads at each end to indicate how the dimension relates to the feature on the object. Dimensions should be grouped around a view and be evenly spaced to give the drawing a neat appearance. The dimension values should never touch the outline of the object. *Dimension values* are usually placed near the midpoint of the dimension line, except when it is necessary to stagger the numbers from one dimension line to the next so that all the values do not line up in a row. Staggering the numbers makes the drawing easier to read.

Because dimension lines should not cross extension lines or other dimension lines, begin by placing the shortest dimensions closest to the object outline. Place the longest dimensions farthest out. This way you avoid dimension lines that cross extension and other dimension lines. It is perfectly acceptable for extension lines to cross other extension lines.

When you are selecting and placing dimensions, think about the operations used to manufacture the part. When possible, provide *overall dimensions* that show the largest measurements for each dimension of the object. Doing so tells the manufacturer the starting size of the material to be used to make the part. The manufacturer should never have to add shorter dimensions or make calculations to arrive at the sizes needed for anything in the drawing. All necessary dimensions should be specified in your drawing. However, no dimensions should be duplicated, as this may lead to confusion, especially when an inspector is determining whether a part meets the specified tolerance.

AutoCAD's Semiautomatic Dimensioning

AutoCAD's semiautomatic dimensioning feature does much of the work in creating dimensions for you. You can create the extension lines, arrowheads, dimension lines, and dimension values automatically. To get the most benefit from the dimensioning feature, use associative dimensions. Associative dimensions are linked to their locations in the drawing by information stored on the layer DEFPOINTS that AutoCAD creates. As a result, associative dimensions automatically update when you modify the drawing. Although you can also use nonassociative dimensions, that isn't good practice because these dimensions are created from separate line, polyline, and text objects and do not contain information allowing them to update. *Associative dimensioning* is the default in AutoCAD.

You should create drawing dimensions in model space. You can add dimensions to paper space, but dimensions placed in paper space are associated to paper space locations, whereas the drawing should be in model space. If the drawing in model space is updated or moved, dimensions placed in paper space will not move with it, making updating your drawing difficult.

Starting

Start by launching AutoCAD Release 14. In the Start Up dialog box, choose Start From Template and use the file *proto.dwt* located in the data files as a template. In the drawing editor,

Pick: **Save icon**

Name your drawing *obj-dim.dwg* and pick Save. You return to the drawing editor with the border and settings you created in the template drawing on your screen.

Dimensioning a Shape

Review the object in Figure 7.1. You will draw this shape and then dimension it.

Layer VISIBLE should be the current layer in the drawing. If it is not, set layer VISIBLE as the current layer. Set Grid for 0.25 spacing and Snap for 0.25 spacing. Be sure that both are turned on.

On you own, use the commands that you learned in the preceding tutorials to draw the object shown in Figure 7.1 according to the specified dimensions. You don't have to draw the centerlines now. Create the drawing geometry *exactly*, in order to derive the most benefit from AutoCAD's semiautomatic dimensioning capabilities. Draw the object now. When you have finished, your screen should look like Figure 7.2.

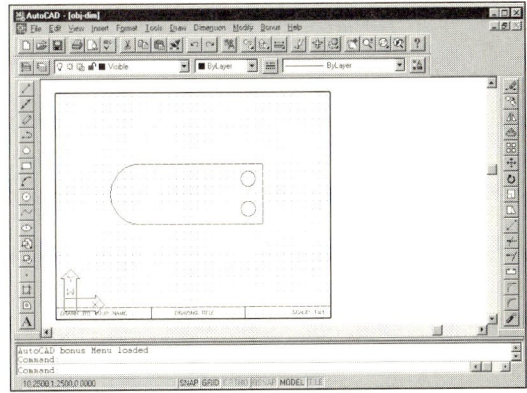

Figure 7.2

Using the DIM Layer

In order to produce a clear drawing of a subject, you draw the outline and visible lines of the object with a thick-width pen when printing or plotting. The dimension, hidden, center, and hatch lines are plotted or printed with a thin-width pen. Thus your eye is drawn first to the bold shape of the object and then to the details of its size and other features.

As you recall from previous tutorials, in AutoCAD the plotted or printed pen width is set according to the color of the object on your drawing screen. For you to be able to select a thin pen for the dimension lines, they must be a different color than you used for object lines (to which you will assign a thick pen when plotting). Also, you will frequently want to turn off all of the dimensions in the drawing. Having dimensions on a separate layer makes color assignment and turning off dimensions easy tasks.

On your own, use the Object Properties toolbar to set DIM as the current layer.

Dimension Standards

There are standards and rules of good practice that specify how dimension lines, extension lines, arrowheads, text size, and various aspects of dimensioning should appear in the finished drawing. Mechanical, electrical, civil, architectural, and weldment drawings, among others, each have their own standards. Professional societies and standards organizations publish drawing standards for various disciplines. The American National Standards Institute (ANSI) publishes a widely used standard for mechanical drawings to help you create clear drawings that others can easily interpret.

For example, there should be a 1/16" space between the extension lines and the feature from which they are extended. The extension line should extend 1/8" past the last dimension line. Arrowheads and text should be approximately 1/8" tall on 8.5" x 11" drawings. The dimension line closest to the object outline should be at least 3/8" from the object outline on your plotted drawing. Each succeeding dimension line should be at least 1/4" from the previous dimension line.

Drawing Scale and Dimensions

When adding dimensions to your drawing, it is important to have already determined the scale at which you will plot your drawing on the sheet of paper. That way you will know how to set up the dimension features to produce the correct sizes on the final plot. If you are going to plot from paper space, you should set the model space Zoom XP scaling factor *before* you dimension your drawing. It is already set to 1XP in *proto.dwt*, from which you started the current drawing. You should check it by repeating the Zoom XP scale command option or by typing MVSETUP and choosing Scale viewports from the command line options.

On your own, switch to model space (if you are not already there) and verify that your drawing is zoomed to 1XP before continuing.

> **■ TIP** Dimensions take a lot of drawing space, especially between the views of a multiview drawing. Often you may need to zoom your drawing to half-size (.5XP) or some other smaller scale to have room for dimensions. Keep this possibility in mind for future drawings. Do not change your XP scale factor for this tutorial. **■**

Turning on the Dimension Toolbar

Dimensioning commands can be quickly selected from one of two places, the Dimension pull-down menu or the Dimension toolbar. You can quickly select the dimensioning commands from the Dimension toolbar, which you will use for this Tutorial.

Pick: **View, Toolbars, Dimension (then close the Toolbars dialog box)**

The Dimension toolbar floats on your screen, as shown in Figure 7.3.

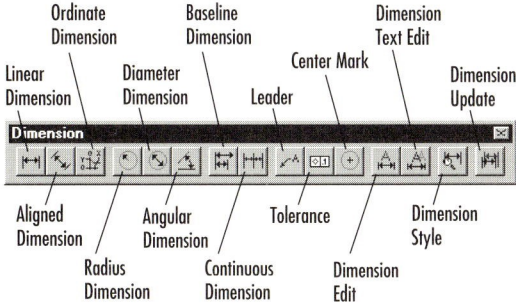

Figure 7.3

Using Dimension Styles

 Many features controlling the appearance of the dimensions are set by *dimension variables (dim vars)*, which you can set using the Dimension Styles dialog box. You can call this dialog box from the Format menu, by typing Ddim at the command prompt, or by picking the Dimension Style icon from the Dimension toolbar. Dimension variables all have names, and you can set each one by typing its name at the command prompt. However, the dialog box allows you to set many dimension variables at the same time, so you will select it from the Dimension toolbar.

Pick: **Dimension Style icon**

The Dimension Styles dialog box appears on your screen, as shown in Figure 7.4.

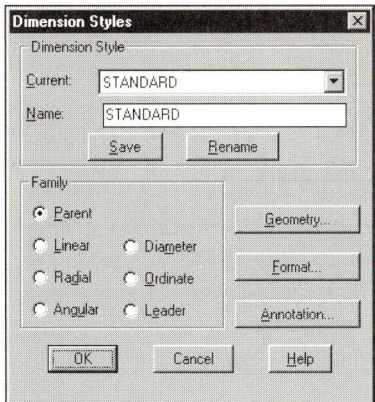

Figure 7.4

Creating a Named Dimension Style

The Dimension Styles dialog box allows you to change the dimension variables that control the appearance of the dimensions. Note that the current style name is STANDARD. This basic set of features is provided as the default.

You can create your own *dimension style* with a name that you specify. This way you can save different sets of dimension features that will be useful for different types of dimensioning standards—such as mechanical or architectural. You will create a dimension style named MECHANICAL by typing it in the box to the right of Name in place of the text STANDARD. Highlight STANDARD and to replace it,

Type: **MECHANICAL**

Pick: **Save (from the buttons below where you typed in MECHANICAL)**

Note that MECHANICAL now appears as the current dimension style at the top of the dialog box instead of STANDARD. Near the bottom of the dialog box you will see the message *Created MECHANICAL from STANDARD.* MECHANICAL is now the current style and will be affected by changes that you make to the dimension variable settings.

Dimension styles can work three ways: as parent styles, child styles, or overrides. A parent style is the default selection for creating a new style. Setting the features of a parent style sets the characteristics for all its children, until you make a different selection for a child.

After you create a parent style and set its characteristics, you can pick on a child type, such as linear, radial, angular, diameter, ordinate, or leader (listed in the Family section of the Dimension Styles dialog box) and then change the features for that child type. Doing so allows you to vary the style for that particular type of dimension. For example, if you set a child style for radial dimensions, it can be different from the parent style. With that parent style current, whenever you create a radial dimension, the child style for radial is automatically applied to vary the appearance of that radial dimension from the parent style.

When you change a style and pick the Save button, the changes are applied to all existing dimensions in the drawing that you created with that style. If you make changes to a style and do not pick the Save button, the changes are used as overrides that take effect on newly created dimensions, but not on existing dimensions having that style.

You will continue creating your parent style for MECHANICAL. (Be sure that the Parent radio button is still active.) To set the dimension variables, you will use the selections for Geometry, Format, and Annotation on the right side of the dialog box. The Geometry button allows you to control how the dimension lines, extension lines, arrowheads, and centermarks are drawn.

Pick: **Geometry**

The Geometry dialog box appears on your screen on top of the previous Dimension Styles dialog box. It should look like Figure 7.5.

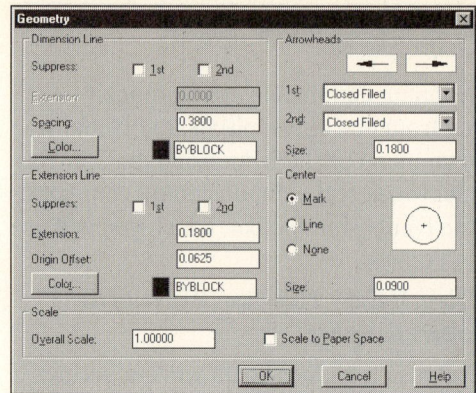

Figure 7.5

Setting the Features

In this dialog box you can change many features of dimensions. An important dimensioning variable is the scaling factor that is near the bottom of the dialog box. You should set this variable first before going on to set other size features, as it controls the sizes of many features at the same time. If you later determine that a different setting would be better, you can change the scaling factor at any time.

The Dimension Scaling Factor

There are essentially two ways to scale the dimension features: relative to model space or relative to paper space. Think about a drawing that is to be zoomed to a scale of 1 plotted inch in paper space to 2 model units. If you want your arrowheads and text to be 1/8" on the plotted sheet, then at this scale the arrows need to be 1/4" high in the model space drawing so that when reduced by half they will be 1/8". You could double all the settings that control sizes of the dimension elements, such as arrowhead size, but as there are many of them, so this method would be time-consuming. Instead, you can automatically scale all the size features by setting the overall scale.

Overall scale sets a scaling factor to use for all of the dimension variables that control the sizes of dimension elements. Its value appears in the box to the right of Overall Scale. Setting the overall scale to 2 doubles the size of all the dimension elements in the drawing so that, when plotted half-size, your arrowheads, text, and other elements will maintain the correct proportions on the final plot. (You can also type Dimscale at the command prompt and enter the value by which the dimension elements are multiplied.)

Picking Scale to Paper Space sets the scaling for dimension elements so that the values you enter in the dialog box are the sizes on the drawing when paper space is plotted at a scale of 1=1. The effect is to multiply all the dimension size features by the reciprocal of the zoom XP factor.

■ *TIP* Using paper space scaling for dimensions can be tricky, but effective. When you are using Scale to Paper Space, AutoCAD uses the zoom scaling factor to determine how to scale the dimension elements. If you zoom in on the view in model space and add a dimension, the dimension elements will be smaller than the dimensions you add when you are not zoomed in. When you zoom back out, the dimensions appear smaller. You can force the dimension to update, but controlling the placement of dimensions when they update in this way is difficult. A better method is to switch to paper space, then zoom in on your view. Before adding the dimension, switch back to model space so the dimension will be in model space. To return to the previous zoom factor, switch back to paper space and select Zoom Previous. ■

For this tutorial, you will select a default overall scale factor of 1 for the dimensions, and you will not scale to paper space. Because you will plot your drawing fullsize, zoomed to 1XP, the dimension features will be shown at the size that you set them in the dialog box. When you use Overall Scale to size the dimension features, zooming does not affect the size of dimension elements.

Next, you will continue to use the Geometry dialog box to set the standard sizes for dimension features used in engineering drawings. The settings that you will make are based on an 8.5" x 11" sheet size. Larger drawing sheets require the use of larger sizes.

Dimension Line Visibility and Size

In the Dimension Line area at the upper left of the dialog box, you can control the appearance of the dimension lines. You can choose to suppress the first or second dimension line when the dimension value divides the dimension line into two parts. You can use the box to the right of Extension to set the value for a distance that the dimension line should extend beyond the extension line. This area is grayed and cannot be selected unless you are using *oblique stroke arrows* as the type of arrowhead. The value entered in the box to the right of Spacing controls the distance between successive dimensions added using baseline dimensioning. The default value, 0.38, will work for now. (The ANSI standard states that successive dimensions should be at least 1/4" or 0.25" decimal apart, so 0.38 will meet this criterion.) Next, you will use the Color selection to set the color by layer.

Pick: **Color**

The standard Select Color dialog box appears on your screen.

Pick: **BYLAYER**

Pick: **OK**

You are returned to the Geometry dialog box.

Arrow Size and Style

To set the arrow size, use the upper right area of the dialog box. On your own, change the value in the box to the right of the word Size. Use the text box to replace the default value, 0.1800, with 0.125, the decimal equivalent of 1/8". Refer to Figure 7.6.

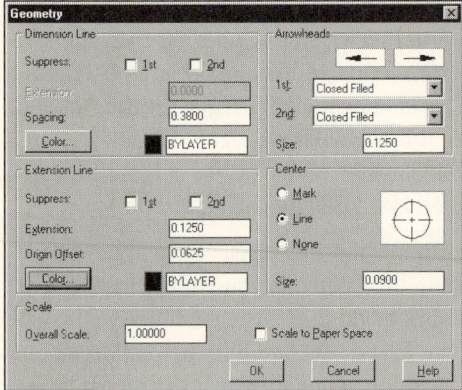

Figure 7.6

You can set the style of the arrow two different ways. One way is to pick on the selection Closed Filled or on the downward-pointing triangle to the right of the Closed Filled selection. A list of the available arrow styles from which you can choose appears on your screen. You can also pick on the *image tile*, or active picture of the arrow, near the top of the Arrowheads area and scroll through the available arrow styles.

> Pick: **(on the image tile of either arrow)**

The tile changes to show a Closed Blank for the arrow on that side. Note that Closed Blank now appears in the box to the right of whichever arrow you picked. (If you picked the first arrow, both changed). On your own, continue picking on the picture of the arrow until it returns to the style Closed Filled.

Continue to refer to Figure 7.6 for the remaining selections.

Extension Line Visibility and Size

In the Extension Line area of the dialog box, you can suppress the first, second, or both of the extension lines while you are dimensioning. To do so, you would pick in the box to the left of 1st to suppress the first extension line, for example. Leave this setting unchanged. The value in the box to the right of the word Extension sets the distance that extension lines extend past the dimension line. The usual setting for this on 8.5" x 11" paper is 0.125". Use the same method you used when setting the arrow size to change the value to 0.125. The box to the right of Origin Offset controls the distance from the edge of the selected feature at which the extension line starts when you're dimensioning. As it is commonly 1/16", leave this value, 0.0625 (the decimal equivalent of 1/16"), unchanged. Next, you will set the color for extension lines so that they are drawn the color of the layer on which they appear.

> Pick: **Color**

Use the Select Color dialog box to pick the BYLAYER button on your own. Pick OK to return to the Geometry dialog box. When you have returned to the Geometry dialog box, you will notice that the color shown in the box to the right of Color is now cyan and that the box to its right says BYLAYER.

Centermark Size and Style

When you create radius or diameter dimensions, you can use the Center area of the Geometry dialog box to control the size and style of the centermark or to suppress it so that no centermark is created. If the value for the centermark size is too large, AutoCAD will not draw full centermarks; instead, it will draw just the tick mark in the center of the circle. Picking the radio button to the left of Line will cause full centermarks to be drawn, and picking the radio button to the left of None will cause no centermarks to be drawn. Picking Mark will cause a tick to be drawn at the center.

Pick: **Line**

The button will be filled in, and the image tile will show a full set of centermarks, as shown in Figure 7.6. You can also set the centermark style by using the image tile on the right of this area.

Pick: **(on the image tile showing the full centermarks)**

The centermarks disappear from the circle and the radio button for None becomes highlighted. You will pick on the picture again to cycle back so that just a centermark or tick will be drawn.

Pick: **(on the image tile where there are now no centermarks)**

A tick or mark appears at the center of the circle.

Pick: **(on the image tile again so that full centermarks are shown)**

Leave the setting to show the full centermarks. You can set the size of the centermark by using the box to the right of the word Size. The value for Size sets the distance from the center to the end of one side of the center cross of the centermark. To make a smaller center cross, set it to a smaller value; for a larger center cross, set the size larger. Leave the value for your centermarks set to the default value, 0.09, for now. You can also control centermarks by typing Dimcen at the command prompt, as you did in Tutorial 6. When you have finished making the selections as shown in Figure 7.6, you are ready to return to the previous dialog box.

Pick: **OK**

You return to the Dimension Styles dialog box. Next, you will set up the format for the appearance of text and dimensions added to the drawing.

Pick: **Format**

The Format dialog box shown in Figure 7.7 appears on your screen. It contains selections that allow you to control the placement of the text, arrows, and leader lines relative to the dimension and extension lines.

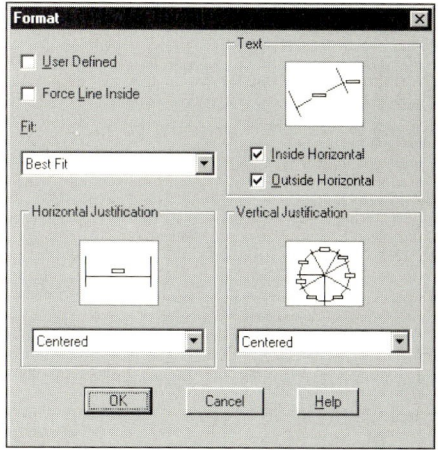

Figure 7.7

Fit

The upper left area of the dialog box contains selections for user-defined text placement and for forcing a line to be drawn between the extension lines, even when text and arrows are placed outside the extension lines. If you select User Defined, when you pick or type the location for the dimension line, the dimension value will appear at that position. The default option for Fit, Best Fit, places *both* the text and arrows inside the extension lines, if possible. If that is not possible, *either* the text or the arrows will be placed inside the extension lines. If neither will fit, both the arrows and text are placed outside the extension lines.

Generally, these are all accepted practices, as long as the dimension can be correctly interpreted. Leave this selection set to Best Fit.

Text Orientation

The Text area of the Format dialog box controls whether text always reads horizontally (i.e., is unidirectional text) or whether it will be aligned with the dimension line. You can control this individually for dimension values placed inside the extension lines or outside the extension lines. Use the image tile for this area to scroll through these options to see the effect each would have. When you have finished experimenting, return both settings to a horizontal orientation.

Horizontal Justification

Using the Horizontal Justification area of the dialog box, you can choose to center text, place it near the first or second extension line, or place it over the first or second extension line. Usually, the default choice, Centered, is used for mechanical drawings. Pick on the image tile in the Horizontal Justification area to see the effects of these selections. When you are finished, return the setting to the default choice, Centered.

Vertical Justification

The final area of the Format dialog box controls the vertical justification of text relative to the dimension line. The options for vertical justification are Centered, Above, Outside, and JIS (Japanese Industrial Standard). For ANSI standard mechanical drawings, select Centered to create dimension values that are centered vertically, breaking the dimension line. The ISO (International Standards Organization) standard usually uses the options for text above or outside the dimension line. The JIS option orients the text parallel to the dimension line and to its left. (To see a picture of this con-

vention, pick JIS from the menu and pick the Help button at the bottom right of the dialog box. Help for Format appears on your screen. Pick on the highlighted option Vertical Justification to see its help screen. Remember to exit Help when you are done.) When you are finished examining this portion of the dialog box, return its setting to Centered on your own.

Pick: **OK**

You return to the Dimension Styles dialog box. Next, you will set up the appearance of annotation, or text, for the dimensions.

Pick: **Annotation**

The Annotation dialog box shown in Figure 7.8 appears on your screen.

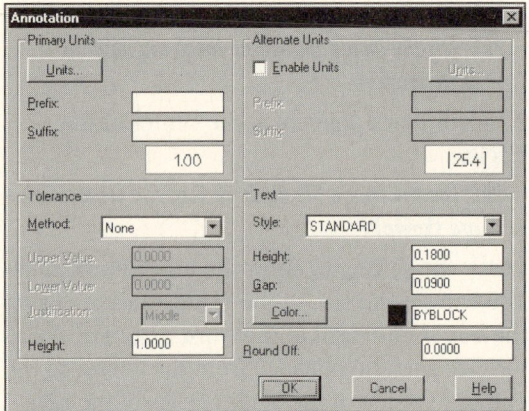

Figure 7.8

Primary Units

You use the Primary Units area of the dialog box to set up the number of decimal places, prefix, and suffix for dimensions. When dimensioning a drawing, you must consider the precision of the values to be used in the dimensions. Specifying a dimension to four decimal places, which is AutoCAD's default, implies that accuracies of 1/10,000th of an inch are appropriate tolerances for this part. Standard practice is to specify decimal inch dimensions to two decimal places (an accuracy of 1/100th of an inch) unless the function of the part makes a tighter tolerance desirable.

Next, you will set up the display of units in dimension values.

Pick: **Units**

The Primary Units dialog box shown in Figure 7.9 appears on your screen. You can use it to select the type of units and set the number of decimal places for dimension values. It also contains selections to suppress leading or trailing zeros in the dimension value. You can also suppress dimensions of 0 feet or 0 inches. In decimal dimensioning, common practice is to show leading zeros in metric dimensions and to suppress them in inch dimensions when dimensions are less than 1 unit. For this tutorial you will leave the type of units set to Decimal.

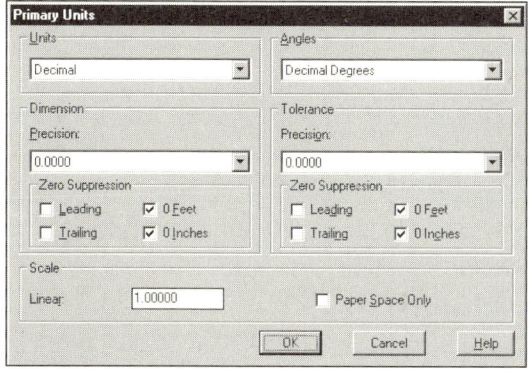

Figure 7.9

Setting Dimension Precision

To set the precision, you will pull down a list of decimal places, as shown in Figure 7.10.

Pick: **0.0000 (below Precision in the Dimension area)**

The current selection shows four places after the decimal, which is AutoCAD's default but is not good engineering practice in dimensioning. Select 0.00 to set the number of decimal places to two, which is common practice when greater

dimension precision is not required. You will learn about using tolerances in Tutorial 14. For now, leave the tolerance precision set as it is.

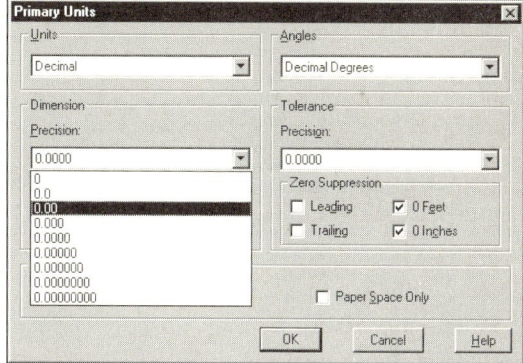

Figure 7.10

Angles

The type of units used to dimension angles can be set using the Angles area of the Primary Units dialog box. To select, pull down the list and select from the types of units shown. (You can further control the appearance of angular dimensions by selecting the Angular child style in the Dimension Styles dialog box.)

Zero Suppression

Zero Suppression is available for both dimensions and tolerances, in four categories: leading 0s, trailing 0s, 0 feet, and 0 inches. Selecting the box to the left of any of these words suppresses those zeros; a check appears in the box when you select it. Pick the boxes to the left of Leading to suppress leading 0s for both dimensions and tolerances.

Scale

The Scale area of the Primary Units dialog box is used to specify a scaling factor by which all the actual dimension values are multiplied. This factor is different than the overall scaling factor because it is the number shown as the dimension affected, not the sizes of the arrowheads and other elements. Leave the value to the right of Linear in the Scale area set to 1.0000.

Pick: **OK** *(to exit the Primary Units dialog box)*

Alternate Units

Alternate units are useful when you want to dimension a drawing with more than one system of measurement. For example, dimensioning drawings with both metric and inch values is fairly common. When you use alternate units, AutoCAD automatically converts your drawing units to the alternate units, using the scaling factor you provide. You can enable alternate units by picking the box to the left of Enable Units in this portion of the Annotation dialog box. Until you select Enable Units, the Units button is grayed out. To set the scale for the alternate units, pick Enable Units, pick Units, and then use the linear scale selection in the Alternate Units dialog box that appears. You can also type a prefix and suffix for the alternate units in the appropriate boxes. The prefix you enter will appear in front of the dimension value; the suffix, after the dimension value.

You can control the appearance of the alternate units by picking on the image tile where you see the value [25.4]. The default places the alternate units inside square brackets following the normal units. You will not enable alternate units now.

Tolerance Method

You use the Tolerance portion of the Annotation dialog box to specify limit or variance tolerances with the dimension value. You can set the tolerance method to Symmetrical, Deviation, Limits, Basic, or None. The values shown in the boxes identified in Figure 7.8 are image tiles that display the units and tolerance method selected. Pick on the image tile just

above the Tolerance box (in the Primary Units area) to scroll through the selections for tolerance so that you can see the effect of these settings. For now leave the setting for the tolerance method set to None.

Text

You can control the text style, height, color, and the gap between the end of the dimension line and start of the text with the Text area of the Annotation dialog box. You will select style MYTEXT.

Pick: **(on STANDARD)**

On your own, use the list of text styles to select MYTEXT as the style for dimension text. Remember, in order for styles to be listed here, you must have previously created them in the drawing with the Style command. Style MYTEXT was previously created in the template drawing you started from. Next you will set the text height to 0.125, the standard height for 8.5" x 11" drawings.

On your own, highlight the value to the right of Height in the Text area of the Annotation dialog box.

Type: **.125**

Now you will set the color for the dimension text to BYLAYER.

Pick: **Color**

On your own use the dialog box that appears to select BYLAYER to set the color for dimension text and return to the Annotation dialog box.

Pick: **OK** *(to exit the Annotation dialog box)*

Now you have set up your basic style for the dimensions.

Saving the Parent Style

In order to save your changes to dimension style MECHANICAL, you must be sure to pick the Save button near the top of the Dimension Styles dialog box, just under the name of the style. If you do not pick Save to save the changes to the dimension style and then you pick OK to close the dialog box, the changes will be applied as overrides. When a change is used as an override, any new dimensions that are created will use those changes, but dimensions that have already been created with that style name will not be affected.

Pick: **Save**

Note the message *Saved to MECHANICAL* at the bottom of the dialog box. The selections you have made apply to dimension style MECHANICAL only. As you dimension your drawing, you can return to this dialog box and create a new style if you need to create dimensions with a different appearance.

Pick: **OK (to exit the Dimension Styles dialog box)**

Using Dimstyle

As you have seen, you can change many settings to affect the appearance of the dimensions you add to your drawings. You can set these variables either through the Dimension Styles dialog box or by setting each individual dimension variable at the command line. You can use the Dimstyle command to list and set dimension styles and variables at the command line. If any dimension variable overrides are set, they will be listed after the name of the dimension style. There should not be any overrides set for style MECHANICAL because you have saved your settings to the parent style.

Command: **DIMSTYLE** ⟵

dimension style: **MECHANICAL**

Dimension Style Edit: (Save/Restore/STatus/Variables/
  Apply/? <Restore>: **ST** ⟵

The dimension variables and their current settings are listed in the text window on your screen. On your own, use the scroll bars to scroll up entries and press ENTER to see more entries.

Associative Dimensioning

The Dimaso variable controls whether associative dimensioning is turned on or off. Associative dimensioning inserts each dimension as a block or group of drawing objects relative to the points selected in the drawing. If the drawing is scaled or stretched, the dimension values automatically update. This feature is very useful, and Dimaso generally should be on. Also, dimensions created with Dimaso on are automatically updated if you change their dimension styles. When Dimaso is turned on (which is the default), the entire dimension acts as one object in the drawing. Dimensions created with Dimaso turned off cannot be updated, but their individual parts, such as arrowheads or extension lines, can be erased or moved. Make sure that Dimaso is on when you view your dimension style variables.

When you are finished, exit the text window by clicking on the close box in the upper right-hand corner.

Now you are ready to start adding dimensions to your drawing. Look at the status bar. You will see that the coordinates still display the cursor position with four decimal places of accuracy, even though you selected to display only two decimal places in your dimensions. AutoCAD can keep track of your drawing and the settings you have made in the drawing database to a precision of at least fourteen decimal places. However, the dimensions, which you will add to your drawing in the next steps, will be shown only to two decimal places, as set in the current style. The display of decimal places on the status bar is set independently of the dimension precision (using the Units command you learned in Tutorial 1). This feature is useful because you can still create and display an accurate drawing database while working on the design, yet dimension according to the often lower precision required to manufacture acceptable parts.

Adding Linear Dimensions

The Linear Dimension command measures and annotates a feature with a horizontal or vertical dimension line. The value inserted into the dimension line is the perpendicular distance between the extension lines. If you use the Linear Dimension command and dimension a line drawn at an angle on the screen, the value that AutoCAD returns is just the X- or Y-axis component of the length. The dimension is adjusted to horizontal or vertical, depending on the points selected.

Be sure that Snap is on to help you locate the dimensions 0.5 unit away from the object outline (thus meeting the criterion that a dimension must be at least 3/8", or 0.375, from the object outline). On your own, turn the Intersection and Quadrant running mode object snaps on. You will use them to select the exact intersections in the drawing for the extension lines so that the dimensions will be drawn accurately.

You will dimension the horizontal distance from the end of the block to the center of the upper hole. Refer to Figure 7.11 for your selections.

Pick: **Linear Dimension icon**

First extension line origin or press ENTER to select: *(pick point 1 at the top right-hand corner)*

Second extension line origin: *(pick on the upper small hole, using AutoSnap Quadrant to locate point 2)*

Dimension line location (Text/Angle/Horizontal/ Vertical/Rotated): *(pick a point two snap units above the top line of the object)*

Dimension text = 0.50

The dimension should appear in your drawing, as shown in Figure 7.11.

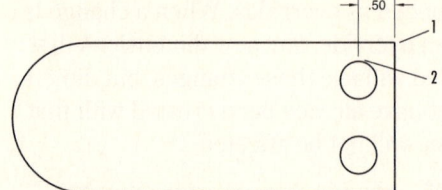

Figure 7.11

Next, you will create the diameter and radius dimensions in your drawing. Usually you will not want the text and arrows inside the dimension when creating diameter and radius dimensions. Using the child settings makes it easy to make certain types of dimensions vary somewhat from the parent style. You will activate the Dimension Styles dialog box by using the Dimension toolbar.

Pick: **Dimension Style icon**

The Dimension Styles dialog box appears on your screen. You will use it to set the child styles for radial and diameter dimensions to be different from the parent style MECHANICAL. The style MECHANICAL should be listed as the current style at the top of the Dimension Styles dialog box.

Setting a Child Style

To create a child style for the diameter that is different than the parent,

Pick: **Diameter**

as shown in Figure 7.12.

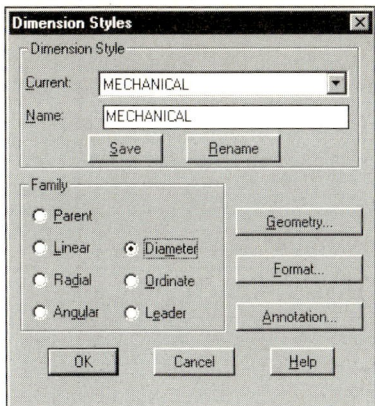

Figure 7.12

You will want most of the settings to remain the same as the parent style; you will change only the items that should differ for radial dimensions and other dimensions. You can think of parent styles and child styles as similar to inherited genetic characteristics. Children inherit appearance from their parents, but each child's appearance may be different from that of the parents. That is why you set the general appearance first using the Parent selection. Then you set the specific differences for each child's appearance after the parent style has been created. *Once a child style has been set, changing the parent style will no longer change the child style.*

Pick: Format

The Format dialog box will appear. Use the methods that you have learned to complete the dialog box as shown in Figure 7.13. Pick User Defined for the placement of text and Leader for the fit. These selections will allow you to place a leader to locate the dimension value off the object when you add a diameter dimension.

When you have finished this step on your own, pick OK to return to the Dimension Styles dialog box.

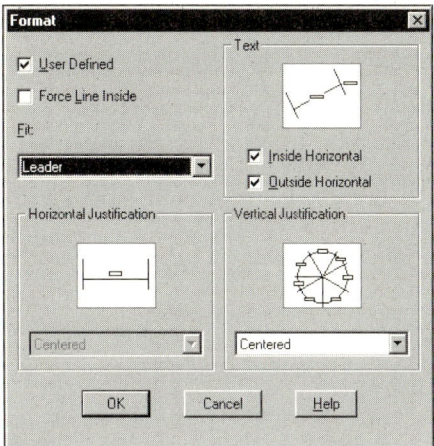

Figure 7.13

Adding Special Text-Formatting Characters

When you are using AutoCAD's shape fonts (e.g., Roman Simplex) to create text, you can type in special characters during any text command and when entering the dimension text.

When using AutoCAD's shape fonts, you can add special text characters to the dimensioning text by inserting the code %% (double percent signs) before the text. The most common special characters have been given letters to make them easy to remember. Otherwise, you can use any special character in a text font by typing its ASCII number. (Most word processor documentation includes a list of ASCII values for symbols in an appendix.)

The most common special characters and their codes are listed in the following table.

Code	Character	Symbol
%%C	Diameter symbol	∅
%%D	Degree symbol	°
%%P	Plus/minus sign	±
%%O	Toggles on and off the overscore mode	Examplē
%%U	Toggles on and off the underscore mode	Example
%%%	Draws a single percent sign	%
%%N	Draws special character number *n*	*n*

You will use the special character %%C to draw the diameter symbol ahead of the diameter dimensions.

Next, you will type in the special character %%C as a prefix in the Annotation dialog box. This way, whenever you create a diameter dimension, the ∅ symbol will automatically precede the value. Adding the ∅ symbol ahead of the dimension value for diameter dimensions (and R for radial dimensions) is common in metric dimensioning and is becoming common in decimal inch dimensioning. Inch dimensions often have the abbreviations DIA (for diameter) and R (for radius) following the value.

Pick: **Annotation**

In the Primary Units area of the dialog box, use the text entry box for Prefix and type in the special character code.

Type: **%%C (in the text box to the right of Prefix)**
Pick: **OK (to return to the previous dialog box)**

Saving the Changes

You must save the changes to the style in order to use that modified style. You cannot use child styles as overrides. To save the changes,

Pick: **Save**

Next, you will set up the radial child type similarly.

Pick: **(the radio button to the left of Radial)**
Pick: **Format**

On your own set the dialog box so that User Defined is used for the placement of text and Leader is used for fit. When you have finished this step, pick OK to return to the Dimension Styles dialog box.

Pick: **Annotation**

In the Primary Units area of the dialog box, for the prefix,

Type: **R**
Pick: **OK (to exit the Annotation dialog box)**
Pick: **Save**
Pick: **OK (to exit the Dimension Styles dialog box)**

Now you are ready to add the diameter dimensions for the two small holes.

Creating a Diameter Dimension

 The Diameter Dimension selection adds the centermarks and draws the *leader line* from the point you select on the circle to the location you select for the dimension value. The leader line produced is a radial line, which, if extended, it would pass through the center of the circle. Keep in mind that this happens only when you set the dimension variables so that you control the placement of the text to use a leader and be user defined, as you did when setting up the child styles for the diameter and radial dimensions.

You will pick the Diameter Dimension icon from the Dimension toolbar. After you have entered the dimension, you will use the Text option of the Diameter Dimension command to add an additional line of text below the dimension value. Refer to Figure 7.15 for the placement of the dimension.

Pick: **Diameter Dimension icon**

Select arc or circle: ***(pick on the lower of the two small circles)***

> ■ *TIP* Be sure that Snap and Ortho are off to make selecting the circles and drawing angled leader lines easier. ■

Dimension line location (Mtext/Text/Angle): **M** ⏎

The Edit MText dialog box shown in Figure 7.14 appears on your screen.

Figure 7.14

You will use the dialog box to add the text *x 2 HOLES* below the <> brackets indicating the default text. When you add the dimension to your drawing, the diameter value for the hole contained in your drawing database will replace the <> brackets.

On your own, position the cursor after the <> brackets and press ⏎ to create a new line for text.

Type: **x 2 HOLES**
Pick: **OK**

> ■ *TIP* You can replace the angle brackets with text or with a value for the dimension. If you do so, however, your dimension values won't automatically update when you change your drawing. A better practice is to create your drawing geometry accurately and accept the default value that AutoCAD provides, represented by the angle brackets. ■

Continue creating the diameter dimension. Next, AutoCAD will prompt you for the dimension line location.

Dimension line location (Mtext/Text/Angle): ***(pick a point below and to the left of the circle at about 7 o'clock, .5 outside the object outline)***

Use the grid as a visual reference to help you position the dimension text 0.5 unit away from the object outline.

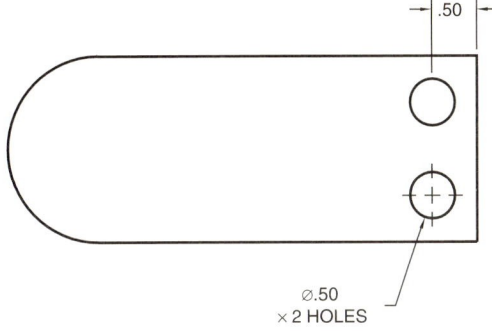

Figure 7.15

Next, you will add a centermark for the upper of the two small holes.

Pick: **Center Mark icon**

Select arc or circle: *(pick on the edge of the top circle)*

The centermark is added to the drawing, as shown in Figure 7.16.

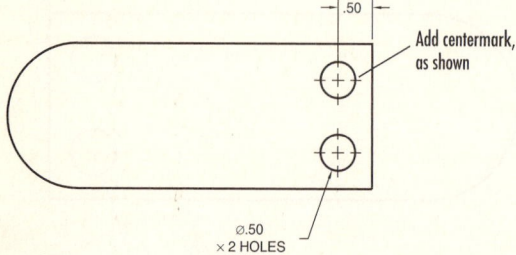

Figure 7.16

Now you will use the Radius Dimension command to create the radius dimension and centermarks for the rounded end. You will use the Radius Dimension icon on the Dimension toolbar to select the command.

Pick: **Radius Dimension icon**

Select circle or arc: *(pick on the rounded end)*

Dimension line location (Mtext/Text/Angle): *(pick a point above and to the left of the circle at about 10 o'clock, .5 outside the object outline)*

The radial dimension and centermark for the arc are added. Your drawing should look like that in Figure 7.17.

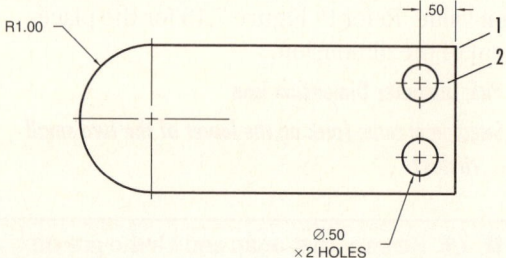

Figure 7.17

Next, you will add a linear dimension for the vertical distance from the upper edge of the part to the center of the top hole. Refer to Figure 7.17 as you make selections.

Pick: **Linear Dimension icon**

First extension line origin or RETURN to select: *(pick point 1 at the top right-hand corner)*

Second extension line origin: *(target point 2)*

Dimension line location (Mtext/Text/Angle/Horizontal/Vertical/Rotated): *(pick a point three snap units to the right of the object)*

The dimension added to your drawing should look like that in Figure 7.18.

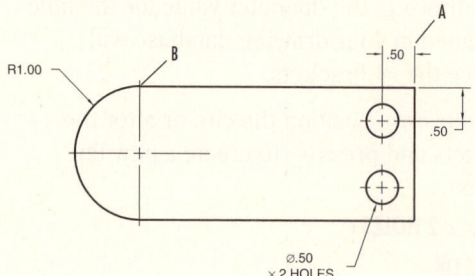

Figure 7.18

Next, you will use baseline dimensioning to add the horizontal dimension between the edge of the part and the rounded end.

Baseline Dimensioning

Baseline and chained dimensioning are two different methods of relating one dimension to the next. In *baseline dimensioning*, as the name suggests, each succeeding dimension is measured from one extension line or baseline. In *chained* or *continued dimensioning*, each succeeding dimension is measured from the last extension line of the previous dimension. Baseline dimensioning can be the more accurate method because the tolerance allowance is not added to the tolerance allowance of the preceding dimension, as it is in chained dimensioning. However, chained dimensioning often may be preferred because the greater the tolerances allowed, the cheaper the part should be to manufacture. The more difficult the tolerance is to achieve, the more parts will not pass inspection. Figure 7.19 depicts the two different dimensioning methods. Note that, if a tolerance of +/– 0.01 is allowed, the baseline dimensioned part can be as large as 4.26 or as small as 4.24. However, with chained dimensions, an acceptable part could be as large as 4.27 or as small as 4.23.

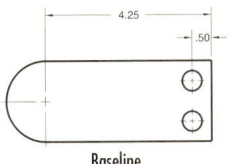
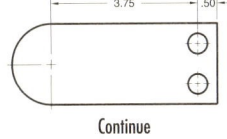

| Baseline | Continue |

Figure 7.19

AutoCAD provides commands to make baseline and chained dimensioning easy.

 Use the Baseline Dimension command to create the next dimension. Adding dimensions with Baseline or Continue is preferable because adding a second dimension by using only the Linear Dimension command will draw the extension line a second

time. That will give a poor appearance to your drawing when it is plotted. Refer to Figure 7.18 as you make selections.

You will use the *Select* option to tell AutoCAD which dimension you want to use as the base dimension. As you have added several dimensions since determining the horizontal location for the upper small hole, you will need to specify that it is the base dimension before you can create the baseline dimension. If you were continuing from the previous dimension, you would not need to do this.

Pick: **Baseline Dimension icon**

Specify a second extension line origin or (Undo/<Select>): ↵

Select base dimension: *(pick A for the base dimension)*

Specify a second extension line origin or (Undo/<Select>):
　　(pick intersection B)

Specify a second extension line origin or (Undo/<Select>): ↵

Select base dimension: ↵

The new dimension should appear, as shown in Figure 7.20. Note that AutoCAD automatically based the location for the dimension on the settings of your dimension style variables.

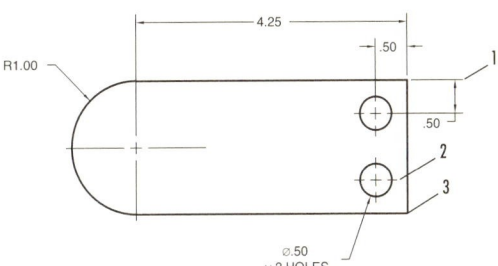

Figure 7.20

Next you will add baseline dimensions for the vertical location of the lower hole and for the overall height. Refer to Figure 7.20 for the points to select.

Pick: **Baseline Dimension icon**

Specify a second extension line origin or (Undo/<Select>): ↵

Select base dimension: *(pick 1 for the base dimension)*

Specify a second extension line origin or (Undo/<Select>): *(pick intersection 2)*

Specify a second extension line origin or (Undo/<Select>): *(pick intersection 3)*

Specify a second extension line origin or (Undo/<Select>): ⏎

Select base dimension: ⏎

Your drawing should now look like that in Figure 7.21.

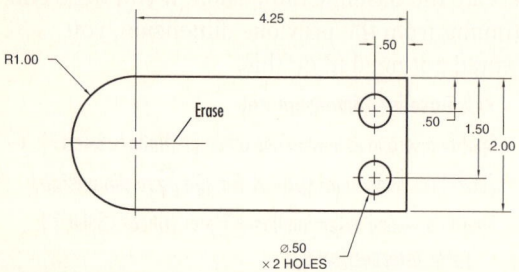

Figure 7.21

Using Xplode

The Xplode command lets you change dimensions, blocks, polylines, and other grouped objects back to their individual components and at the same time to control the color, layer, and linetype of the components. You will use Xplode to erase just the single right line of the centermark for the rounded end. You will select the Layer option of the Xplode command and the default layer, DIM, for the exploded object's layer.

AutoCAD also has a command named Explode, which doesn't have any options. It simply changes the grouped objects back to individual objects, regardless of the X-, Y-, and Z-scaling factors. (In previous versions of AutoCAD the X-, Y-, and Z-scaling factors of a block had to be the same or it could not be exploded.) Refer to online help to investigate the differences between these two commands further on your own.

Because the rounded end is an arc, not a full circle, its centermark should extend only for the half-circle shown. The right portion of the centermark should be erased. Because associative dimensions are created as blocks, they are all one object; no part can be erased singly. To eliminate the extra portion of the centermark on the rounded end, you will type the Xplode command.

Command: **XPLODE** ⏎

Select objects to XPlode.

Select objects: *(pick on the dimension for the rounded end)*

Select objects: ⏎

All/Color/LAyer/LType/Inherit from parent block/<Explode>: **LA** ⏎

XPlode onto what layer? <DIM>: ⏎

You will see the message *Object exploded onto layer DIM* at the command prompt. The appearance of the drawing doesn't change, but now you are able to erase the individual line at the right of the centermark for the rounded end.

On your own, erase the extra line from the centermark for the rounded end and save your drawing at this time.

Switch to paper space on your own and add the note *ALL TOLERANCES ± .01 UNLESS OTHERWISE NOTED*. Use the %%P special character to make the ± sign. Below the tolerance note add notes indicating the material from which the part is to be made, such as *MATERIAL: SAE 1020*, and a note stating *ALL MEASUREMENT IN INCHES*. Use the Edit Text command to change the text in your title block, noting the name of the part, your name, and the scale, to complete the drawing. Save your drawing on your own at this time.

Saving As a Template

Any drawing that you have created can be used as a template from which to start new drawings. You will use Save As and save your *obj-dim.dwg* with a new name and file type so that you can use it as a template to start other drawings.

On your own, erase the object and all dimensions and set layer VISIBLE as the current layer. Use Save As to save this drawing as *dimtemplate.dwt* (by changing the file type to a *Drawing Template File *.dwt*). Drawings started from *dimtemplate.dwt* will have the dimension styles that you created earlier in this tutorial available for use. In the Template Description, describe this template as a dimensioning template and leave measurement with English units.

Dimensioning the Adapter

You will continue to work with the orthographic views of the adapter from Tutorial 6 by adding dimensions to the drawing. Pick the New icon from the Standard toolbar to start a new drawing. In the dialog box, pick on Use a Template and select the file *adapter1.dwg* from the data files as the template drawing. Then pick the Save icon to name the drawing *adapt-dm.dwg*.

Your screen should look like Figure 7.22.

This drawing is similar to the one you did in Tutorial 6. Here, the holes have not been added and the views have been moved apart to make room for dimensions.

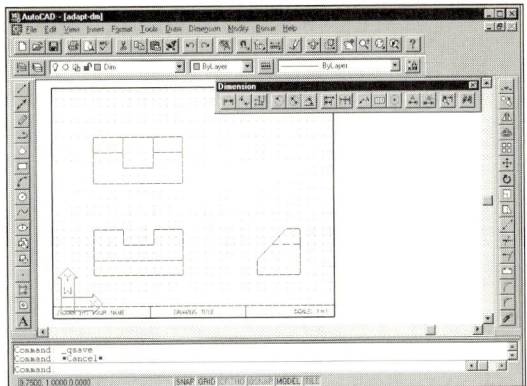

Figure 7.22

Dimensions are usually placed between views when possible, except for the overall dimensions, which are often placed around the outside.

Pick: **Dimension Style icon**

Use the dialog box to select dimension style MECHANICAL. To select it, pick on the name STANDARD, as shown in Figure 7.23 and then select MECHANICAL from the pull-down list of dimension styles. Its name should then appear as the current style. If it is not the current style, pick it before selecting OK.

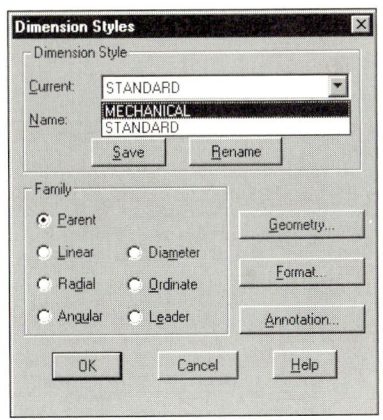

Figure 7.23

Pick: **OK *(to exit the dialog box)***

You will verify that associative dimensioning is turned on. Later in the tutorial you will use the feature to automatically update dimensions, and you can't update dimensions unless you created them with Dimaso turned on. You will type Dimaso at the command prompt.

Command: **DIMASO** ⏎

New value for DIMASO <On>: *(if the default value is Off, then type On ⏎; otherwise, accept the default)* ⏎

Turn on the Intersection running object snap. Check the Object Properties toolbar to be sure that you are on layer DIM. If not, make it the current layer on your own.

Add the horizontal dimension that shows the width of the left portion of the block. The shape of this feature shows clearly in the front view, so add the dimension to the front view between the views. You will use the *ENTER to select* option, which allows you to pick an object from the screen instead of specifying the two extension line locations. AutoCAD will automatically locate the extension lines at the extreme ends of the object you select.

Pick: **Linear Dimension icon**

First extension line origin or ENTER to select: ⏎

Select object to dimension: *(pick the top left-hand line in the front view)*

Dimension line location (Mtext/Text/Angle/Horizontal/Vertical/Rotated): *(pick 0.5 unit above the object outline)*

Your drawing should be similar to Figure 7.24.

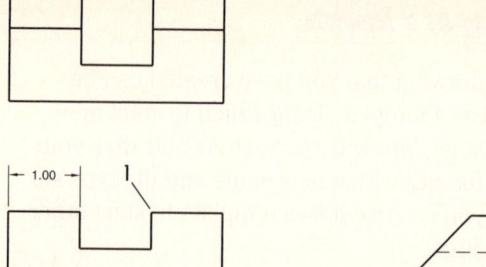

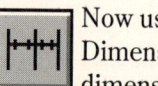

Figure 7.24

Using Continue Dimension

Now use AutoCAD's Continue Dimension command to add a chained dimension for the size of the slot. You will use the *Select* option again to choose which dimension is to be continued. The Baseline Dimension command has a similar option so that, if you have added other dimensions, you can go back and select a different dimension to act as the base dimension. When using this feature, pick near the extension line that you want to have continued (or be the baseline). Refer to Figure 7.24.

Pick: **Continue Dimension icon**

Specify a second extension line origin or (Undo/<Select>): ⏎

Select continued dimension: *(pick the right extension line of the existing dimension)*

Specify a second extension line origin or (Undo/<Select>): *(pick point 1 using AutoSnap intersection)*

Specify a second extension line origin or (Undo/<Select>): ⏎

Select continued dimension: ⏎

The chained dimension should appear in your drawing, as shown in Figure 7.25.

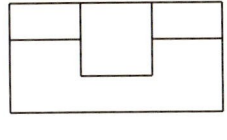

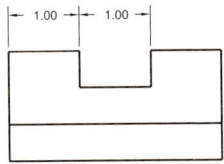

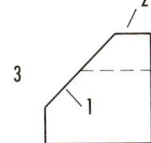

Figure 7.25

Adding the Angular Dimension

Before you add the angular dimension, you will use the Dimension Styles dialog box to change the style for the Angular child dimension.

Pick: **Dimension Style icon**

Pick: **(the radio button next to Angular)**

Pick: **Annotation**

Pick: **Units (in the Primary Units area)**

The precision for the angle should be set to 0, rather than 0.00. On your own, make the change from 0.00 to 0 by picking the box that shows 0.00, if necessary. Otherwise, pick OK to exit the Units Dialog box, then pick OK again to exit the Annotation dialog box. In the Dimension Styles dialog box, be sure to pick Save before you exit the dialog box so changes are saved to the MECHANICAL dimension style.

 Add the angular dimension for the angled surface in the side view, referring to Figure 7.25.

Pick: **Angular Dimension icon**

Select arc, circle, line or press ENTER: *(pick line 1)*

Second line: *(pick line 2)*

Dimension arc line location (Mtext/Text/Angle): *(pick near point 3)*

The angular dimension is added to the side view, as shown in Figure 7.26. You also have the option during the command of pressing ⏎ instead of selecting the first line. If you do so, AutoCAD will prompt for three points to define the angle. You do not have to use the %%D special character to make the degree sign; AutoCAD inserts it automatically unless you override the default text.

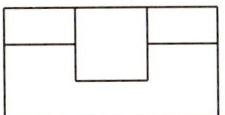

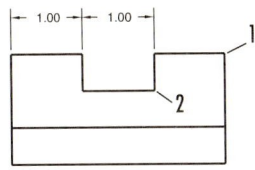

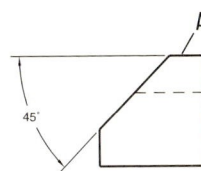

Figure 7.26

Add the dimension for the depth of the top surface in the side view.

Pick: **Linear Dimension icon**

First extension line origin or ENTER to select: ⏎

Select object to dimension: *(pick the short horizontal line labeled A in Figure 7.26)*

Dimension line location (Mtext/Text/Angle/Horizontal/Vertical/Rotated): *(pick 0.5 unit above the top line of the side view)*

Now add the dimension for the height of the slot in the front view. Refer to Figure 7.26.

Pick: **Linear Dimension icon**

First extension line origin or ENTER to select: *(pick the intersection labeled 1)*

Second extension line origin: *(pick the right-hand bottom corner of the slot, labeled 2)*

Dimension line location (Mtext/Text/Angle/Horizontal/Vertical/Rotated): *(pick 0.5 unit away from the right side of the front view)*

The dimension for the slot should appear in your drawing, as shown in Figure 7.27.

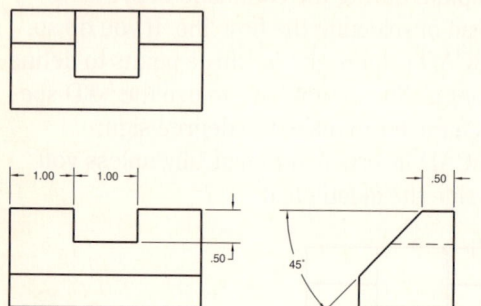

Figure 7.27

On your own, add the overall dimensions to the outsides of the views so that your drawing looks like Figure 7.28.

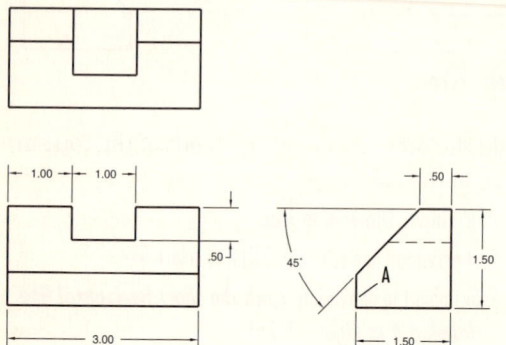

Figure 7.28

Adding the Tolerance Note

Use the Pan Realtime command on your own to move the views of the object up in the viewport to make room near the bottom for drawing notes. Next, switch to paper space. Use the Object Properties toolbar to make layer TEXT current.

Now use the text command to add 0.125" text stating *ALL TOLERANCES ARE ± .01 UNLESS OTHERWISE NOTED*. Use the %%P special text character to create the symbol. Add a second note below the first stating *MATERIAL: SAE 1020*. Add a third note that says *ALL MEASUREMENTS ARE IN INCHES*. Edit the drawing title.

If you want, plot your drawing while in paper space with the drawing limits at the scale of 1=1.

On your own, switch back to model space.

Save your drawing *adapt-dm.dwg* before you continue.

Your finished drawing should be similar to Figure 7.29.

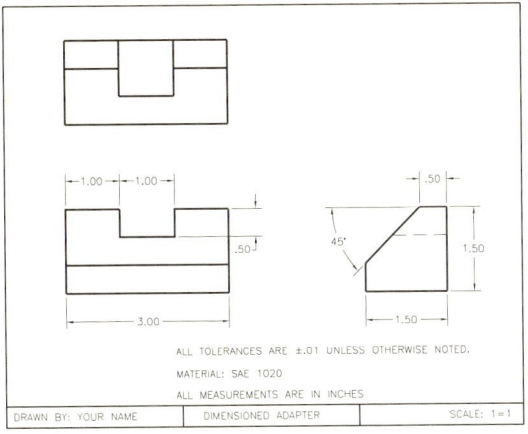

Figure 7.29

> ■ **TIP** You can trim and extend linear and ordinate dimensions with the Trim and Extend commands. When trimmed or extended, the dimension acts as though there were a line between the endpoints of the extension line origins. Trimming and extending takes effect on the dimension at that location. ■

Updating Associative Dimensions

Dimensions created with associative dimensioning (Dimaso) turned on are an AutoCAD block object. Blocks are a group of objects that behave as a single object. If you try to erase a dimension that you created with associative dimensioning on, the entire dimension is erased, including the extension lines and arrowheads. Because these dimensions have the special properties of a block, you can also cause them to be updated automatically.

Using Stretch

The Stretch command is used to stretch and relocate objects. It is similar to stretching with hot grips, which you did in Tutorial 4. You *must* use implied Crossing, Crossing, or Crossing Polygon methods to select the objects for use with the Stretch command. As you select, objects entirely enclosed in the crossing window specified will be moved to the new location rather than stretched. Keep this in mind and draw the crossing box only around the portion of the object you want to stretch, leaving the other portion unselected to act as an anchor.

You will use the Stretch command to make the adapter wider. When you do so, the appropriate dimensions automatically update to the new size. Note that the overall width of the front view is currently 3.00. Refer to Figure 7.30.

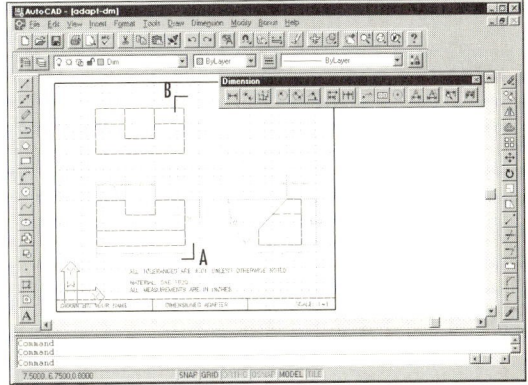

Figure 7.30

Pick: **Stretch icon**

Select objects to stretch by crossing-window or -polygon...

Select objects: *(use Crossing and select point A)*

Other corner: *(select point B)*

Select objects: ⏎

Base point or displacement: *(target the bottom right-hand corner of the front view)*

Second point of displacement: *(move the cursor over .5 units to the right and pick)*

> ■ *TIP* If you have trouble stretching, double-check to be sure that you are not still in paper space. Also, the Stretch command works only when you use the Crossing option. If you type the command, you must be sure to select Crossing or type *C* and press ← afterward, or use implied Crossing (by drawing the window box from right to left), as you did in the previous step. ■

Note that the overall dimension now reads 3.50 and that it has updated automatically. The result is shown in Figure 7.31.

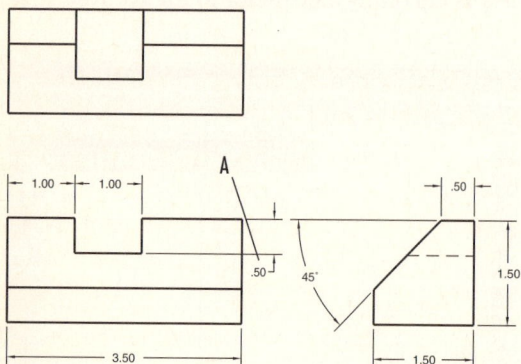

Figure 7.31

Modifying Dimensions

You can modify individual dimensions and override their dimension styles by using the Properties icon on the Object Properties toolbar. Refer to Figure 7.31.

Pick: **Properties icon**

Select objects: *(pick on the vertical dimension labeled A)* ←

The Modify Dimension dialog box shown in Figure 7.32 appears on your screen. As with other objects, you can modify the layer, color, linetype, thickness, and linetype scale of a dimension. Near the bottom center of the dialog box is a selection for style that contains the name MECHANICAL. You can pick on the name to pull down the list of styles that you have already created in the drawing. If you want to change the dimension so that it uses a different style, you can select the new style by using this area.

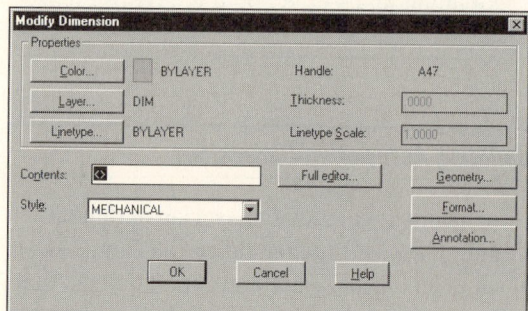

Figure 7.32

Near the center of the dialog box is a button labeled Full editor. Picking the button opens the Multiline Text Editor dialog box to change the value for the dimension.

You will select Format and change the text orientation so that it is parallel to the dimension line.

Pick: **Format**

Use the dialog box that appears to orient the text parallel to the dimension line. To do so, pick on the image tile at the upper right of the dialog box until it is oriented as shown in Figure 7.33.

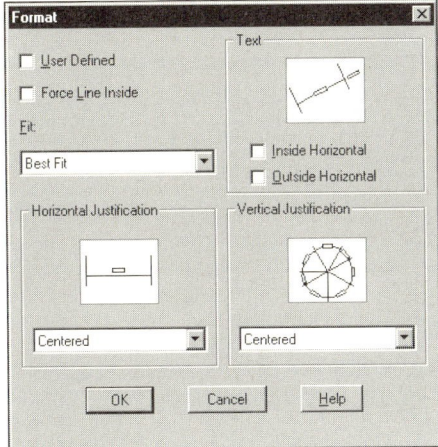

Figure 7.33

Pick: **OK**

Pick: **OK**

The position of the 0.50 vertical dimension changes so that it is now parallel to the dimension line, as shown in Figure 7.34. The dimension style for MECHANICAL has not changed; only that single dimension was overridden to produce a new appearance.

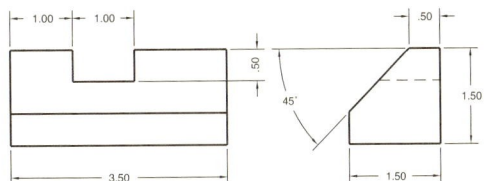

Figure 7.34

Updating Dimensions

When you change a dimension style, all the dimensions that you created with that style automatically take on the new appearance. (This method works only for dimensions created with associative dimensioning on.) You will change the arrow style to a new appearance for all the dimensions created with style MECHANICAL. Use the Dimension Styles dialog box to change the arrow style.

Pick: **Dimension Style icon**

The Dimension Styles dialog box appears on your screen. You will use the Geometry option. Be sure that the radio button for Parent is highlighted. Then,

Pick: **Geometry**

Pick: **Right Angle** *(in the Arrowheads section of the dialog box)*

Pick: **OK** *(to exit the Geometry dialog box)*

Pick: **Save**

Pick: **OK** *(to exit the Dimension Styles dialog box)*

The dimensions in the drawing update automatically to reflect this change. The arrows are now right-angled arrows instead of filled arrows. Your drawing should be similar to Figure 7.35.

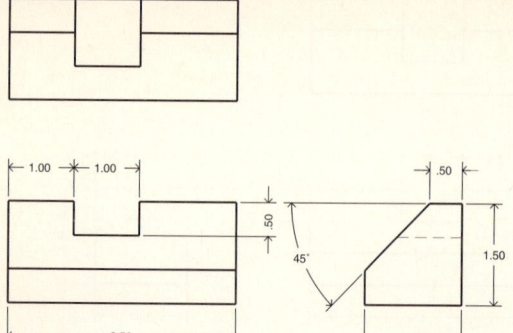

Figure 7.35

Now switch to paper space and use the Edit Text (Ddedit) command to edit the text in the title box to correct the date, name, and scale, as necessary. Title your drawing *ADAPTER*. Return to model space when you have finished and save the drawing on your own.

Close any open toolbars except the Draw and Modify toolbars, which should be docked to the edge of the graphics window, and exit AutoCAD. You have completed Tutorial 7.

> ■ **TIP** If you are not sure what dimension style and overrides you may have used when creating a particular dimension, use the List command and pick on the dimension. You will see a list of the dimension style and any overrides that were used for the dimension. ■

associative dimensioning
baseline dimensioning
chained (continued)
  dimensioning
dimension lines
dimension style

dimension values
dimension variables
  (dim vars)
dimensions
extension lines
extension line offset

image tile
inspection
leader line
manufacture
overall dimensions
tolerance

Angular Dimension
Baseline Dimension
Continue Dimension
Diameter Dimension

Dimension Status
Dimension Styles
Linear Dimension
Radius Dimension

Stretch
Xplode

EXERCISES

Draw and dimension the following shapes. The letter M after an exercise number means that the units are metric. Add a note to the drawing saying, METRIC: All dimensions are in millimeters for metric drawings. Specify a general tolerance for the drawing.

7.1M Stop Plate

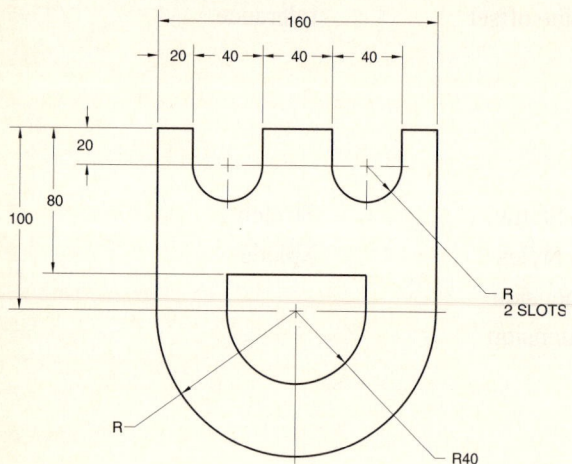

7.2M Hub

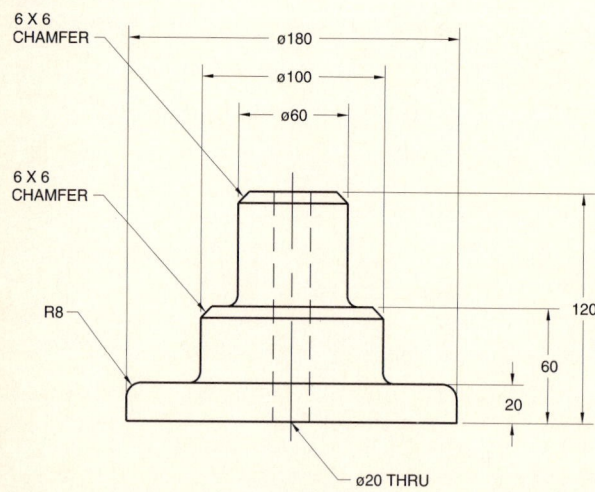

ALL FILLETS R6

⚙ 7.3 Guide Block

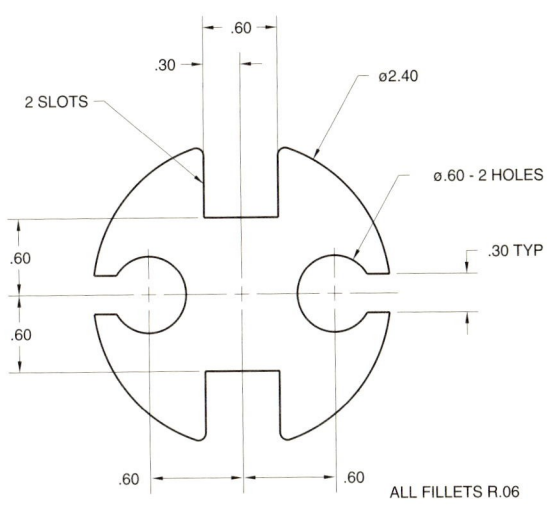

2 SLOTS

.60
.30

ø2.40

ø.60 - 2 HOLES

.30 TYP

.60

.60

.60

.60

ALL FILLETS R.06

⚙ 7.4 Angle Bracket

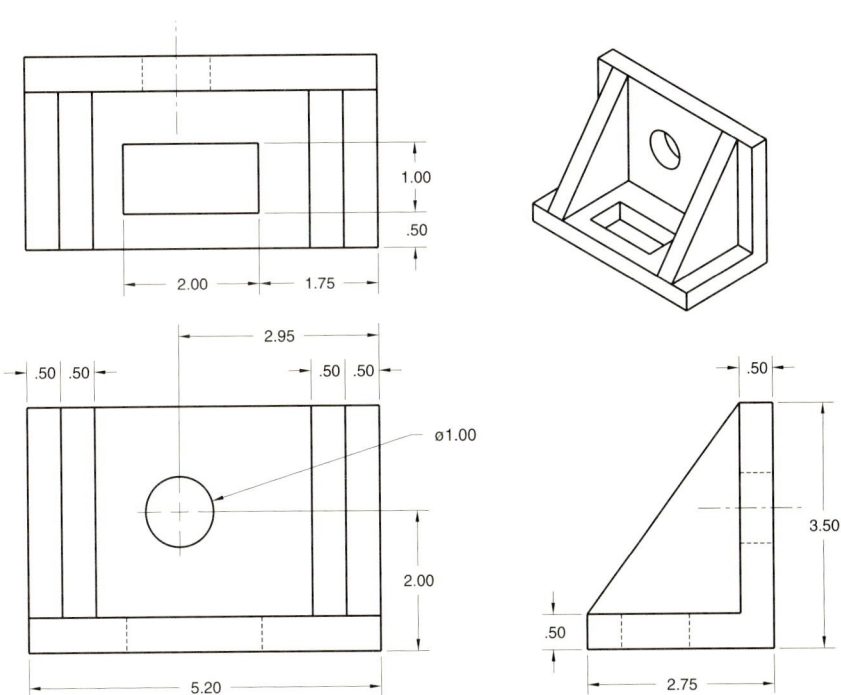

1.00

.50

2.00

1.75

2.95

.50 | .50

.50 | .50

ø1.00

2.00

5.20

.50

3.50

.50

2.75

7.5 Interchange

Draw and dimension this intersection, then mirror the circular interchange for all lanes

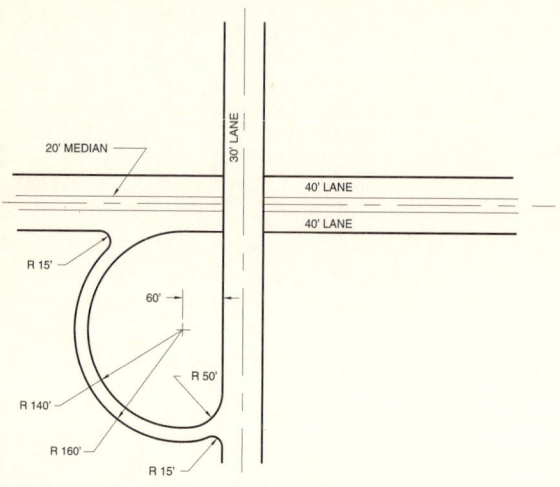

7.6 Bearing Box

Draw and dimension the following shape. Specify a general tolerance for the drawing.

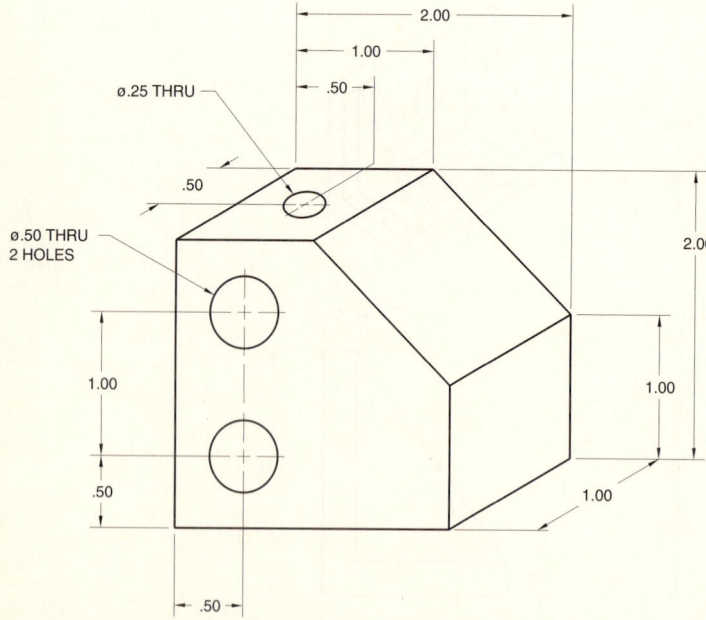

7.7 Plot Plan

Draw and dimension using the civil engineering dimensioning style shown, and find the missing dimensions.

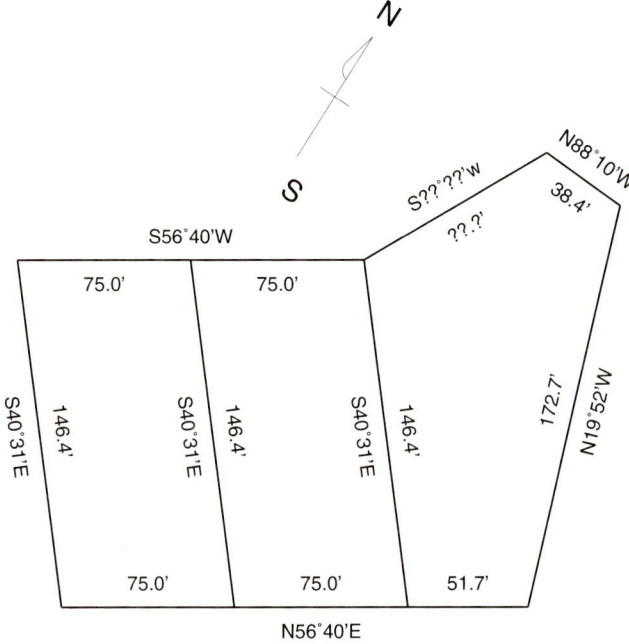

7.8 Hub

Draw and dimension the hub shown. Use 1 grid unit = 0.125" for a decimal inch drawing or 25 mm for a metric drawing.

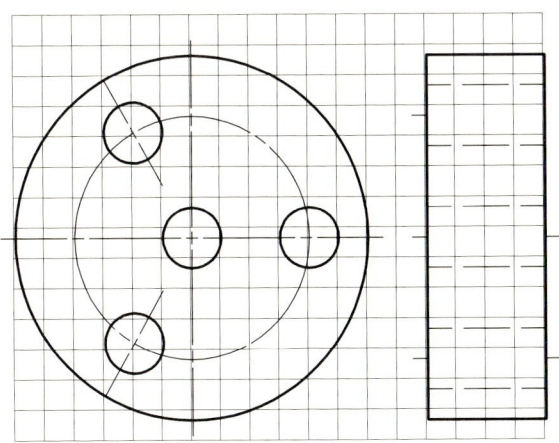

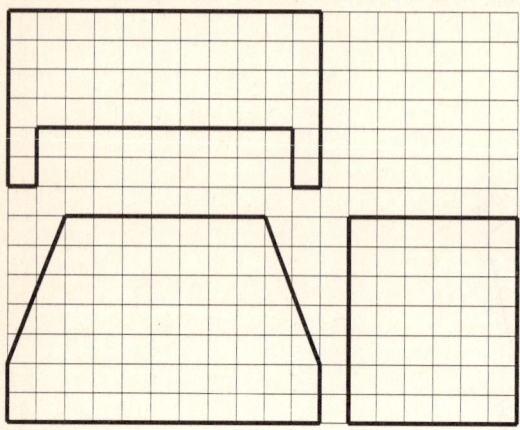

7.9 Support

7.10 Floor Plan

Use architectural units to draw and dimension the floor plan shown, making good use of layers.

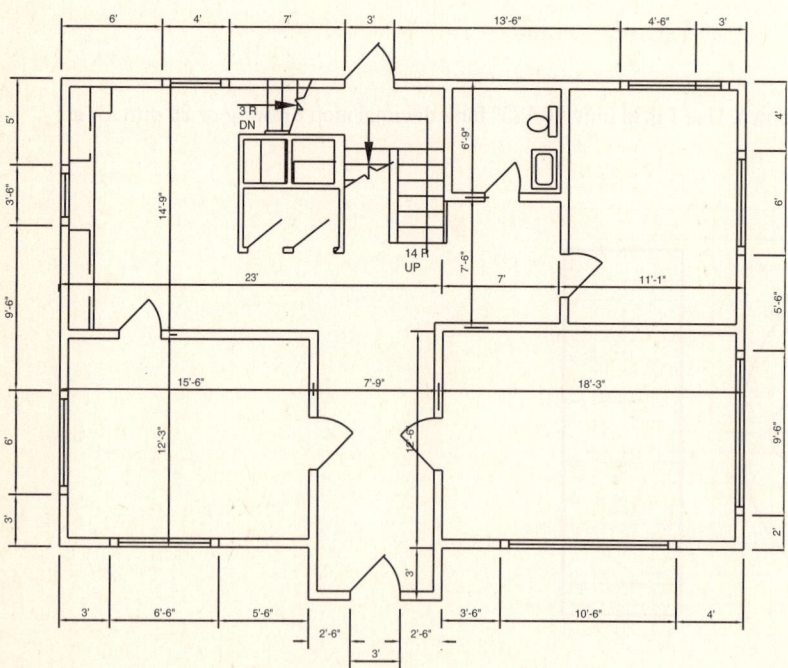

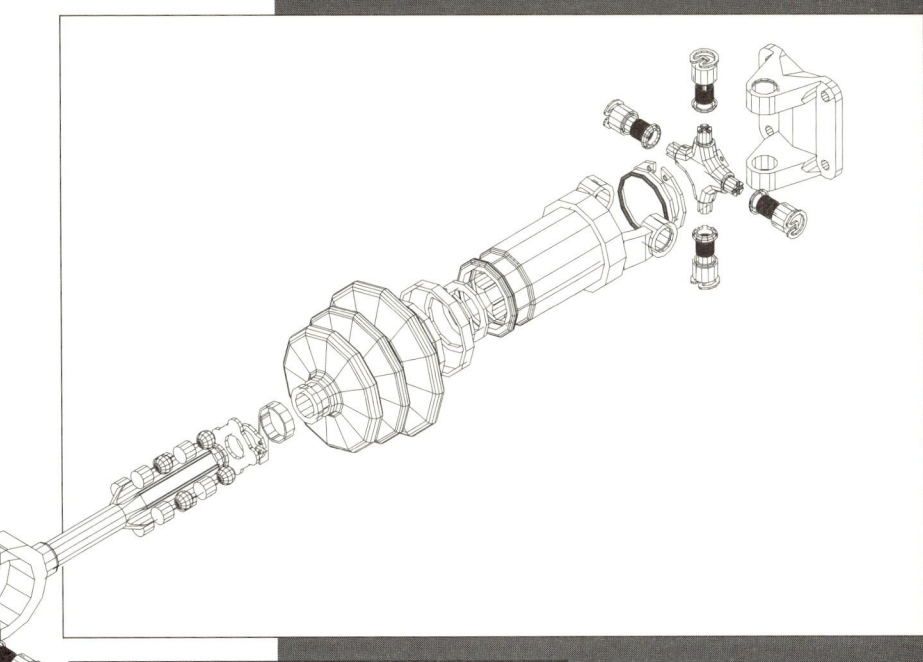

Using Blocks and Customizing Toolbars

Objectives

When you have finished this tutorial, you will be able to

1. Use the Block, Write Block, Insert Block, and Donut commands.

2. Create your own toolbar.

3. Add your block symbols to buttons.

4. Create a block library drawing.

5. Create a logic circuit diagram using blocks.

6. Create named groups.

Introduction

You can unify a collection of drawing objects into a single symbol by using AutoCAD's Block command. Using blocks can help you organize your drawing and make updating symbols easy. Blocks can have associated text objects called attributes, which you can even make invisible. You can extract this attribute information to an external database or spreadsheet for analysis or record keeping. You will learn more about using attributes in Tutorial 11. The use of blocks has many advantages. You can construct a drawing by assembling blocks that consist of small details. You can draw objects that appear often once and then insert them as needed, rather than drawing them repeatedly. Inserting blocks rather than copying basic objects results in smaller drawing files, which saves time in loading and regenerating drawings. Blocks can also be nested, so that one block is a part of another block.

In this tutorial you will create symbols used in drawing the electronic logic gates *Buffer, Inverter, And, Nand, Or,* and *Nor.* You will learn to use the Block and Insert Block commands to add symbols easily to your drawing. Using Write Block, you will make the symbols into separate drawings that can be added to any drawing. You will learn how to create a block library drawing, which you can use to insert a group of blocks into your current drawing. Finally, you will learn to create a custom toolbar to which you will add your block symbols.

You can create named groups in AutoCAD using the Object Group command. A *group* is a named selection set. You can also form *selectable groups*. When a group is selectable, picking on one of its members selects all the members of the group that are in the current space and not on locked or frozen layers. An object can belong to more than one group. Groups are another powerful way of organizing the information in your drawing. Some CAD programs use groups only to organize drawing information, instead of layers. In many ways groups are like blocks, but groups differ because of their special use for selecting objects by named sets.

One case in which you might use blocks and groups together is in an architectural drawing. You could create each item of furniture—a table, chair, etc.—as a separate block and then insert it multiple times into the drawing. You could use several different layers to draw each block. You could then name all the tables as a group called TABLES and all the chairs as a group called CHAIRS. The TABLES and CHAIRS groups could belong to a group named FURNITURE. Unlike blocks, which act as a single object, you can still move or modify items in groups individually, but you can also quickly select the group (using the Select Group selection method) and modify it as a unit.

Starting

Begin AutoCAD and, from the Start Up Dialog box, start a new drawing by using the drawing file *electr.dwg* provided with the data files as a template. On your own, save the drawing and name it *eblocks.dwg*. When you have finished, it should appear on your screen, as shown in Figure 8.1.

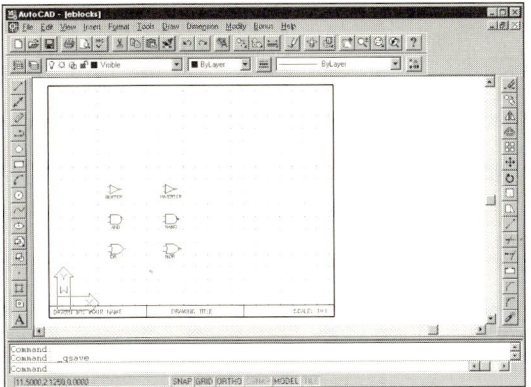

Figure 8.1

Figure 8.1 shows six symbols used in drawing electronic logic circuits: the Buffer, Inverter, And, Nand, Or, and Nor gates. You will make each symbol into a block.

Using the Block Command

A *block* is a set of objects formed into a compound object or symbol. You define a block from a set of objects in your current drawing. You specify the block name and then select the objects that you want to be part of the block. The *block name* is used whenever you insert the compound object (block) into the drawing. Each insertion of the block into the drawing, the *block reference*, can have its own scale factors and rotation. AutoCAD treats a block as a single object; you select the block for use with commands such as Move or Erase simply by picking any part of the block. You can explode blocks into their individual objects by using the Explode or Xplode commands, but doing so removes the blocks' special properties.

A block can be composed of objects that were drawn on several layers with several colors and linetypes. The layer, color, and linetype information of these objects is preserved in the block. Upon insertion, each object is drawn on

its original layer with its original color and linetype, regardless of the current drawing layer, object color, and object linetype. There are three exceptions to this rule.

1. Objects that were drawn on layer 0 are generated by using the current layer when the block is inserted, and they take on the characteristics of that layer.

2. Objects that were drawn with color BYBLOCK inherit the color of the block (either the current drawing color when they are inserted or the color of the layer when color is set by layer).

3. Objects that were drawn with linetype BYBLOCK inherit the linetype of the block (either the current linetype or the layer linetype, as with color).

> ■ *TIP* Block names within a drawing can be as many as 31 characters long and can contain letters, numbers, and the following special characters: $ (dollar sign), - (hyphen), and _ (underscore). AutoCAD converts all letters to uppercase. If a block with the name you choose already exists, AutoCAD tells you that it does. You can either exit without saving so that you do not lose the original block, or you can choose to redefine the block. ■

Blocks defined with the Block command are stored only in the current drawing and you can use them only in the drawing in which they were created. To use a block in another drawing, you need to use the Write Block command or the Windows Clipboard. You can use blocks in other drawings by copying vectors (graphical objects) to the Windows Clipboard and pasting them into a drawing, which results in a block. In addition, you can insert the drawing containing a block into another drawing. When you insert a drawing as you did in Tutorial 4 with the subdivision drawing, *subdivis.dwg*, any blocks it contains are then defined in the drawing in which it is inserted.

Unlike using the Windows Clipboard to copy a block or inserting a file containing a block, using the Write Block command saves the selected block or drawing objects as a separate drawing file. This block file can then be inserted into any drawing. You will learn more about Write Block later in this tutorial.

On your own, zoom up the area of the drawing showing the And gate so that your screen is similar to Figure 8.2.

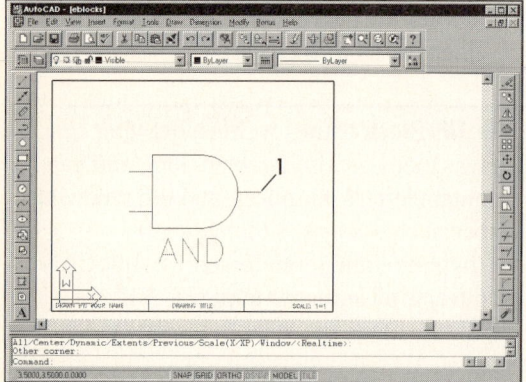

Figure 8.2

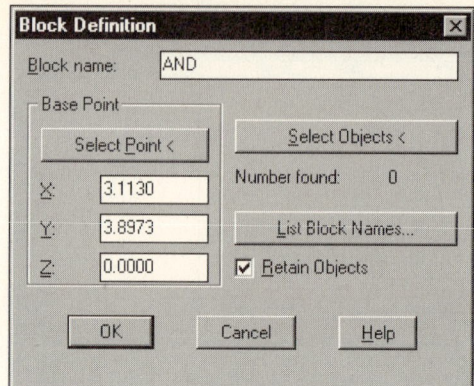

Figure 8.3

The dialog box allows you to select a base point, type in a block name, select objects to be used in that group, and view existing blocks. First you will type the block name, then select a base point, and finally select the objects to be used.

Type: **AND *(to the right of Block name)***

Pick: **Select Point *(located in the Base Point section of the dialog box)***

Be sure to choose a base point that will be useful when you insert the block in another drawing. If you do not select a base point, AutoCAD uses the default base point coordinates (0,0,0). Use Figure 8.2 to help in base point selection and AutoSnap to pick the endpoint.

Insertion base point: *(pick endpoint 1)*

When you return to the dialog box, you will choose to Select objects.

Pick: **Select Objects**

Select objects: *(use implied Crossing to select all of the objects making up the And gate, including the name AND)* ⏎

Pick: **OK *(when returned to the dialog box)***

You will use the Block icon on the Draw toolbar to create a block from the objects making up the And gate on the screen. Be sure to select Block and not Insert Block on the same toolbar. On your own, turn on running object snap endpoint. Refer to Figure 8.2 to help you select point 1.

Pick: **Block icon**

A Block Definition dialog box appears, similar to Figure 8.3.

You are returned to the drawing editor and now have a block of the And gate to be used in later drawings.

On your own, use Zoom Previous to restore the original view of the drawing, where you can see each symbol, and create a block of each of the other logic gates in the drawing. Remember to select the right-hand endpoint of the lead from each electronic logic gate (as you did for And) for the base point. Name the blocks NAND, OR, NOR, INVERT, and BUFFER.

To see how you can use blocks within the current drawing, you will insert the AND block you created.

Inserting a Block

 The Insert Block command inserts blocks or other drawings into your drawing. When inserting a block, you use a dialog box to specify the block name, the insertion point (the location for the base point you picked when you created the block), the scale, and the rotation for the block. A selection of 1 for the scale causes the block to remain the original size. The scaling factors for the X-, Y-, and Z-directions do not have to be the same. However, using the same size ensures that the inserted block has the same proportions, or aspect ratio, as the original. Blocks (even those with different X-, Y-, and Z-scaling factors) can be exploded back into their original objects with the Explode command. A rotation of 0° ensures the same orientation as the original. You can also use Insert Block by typing *INSERT* at the command prompt.

You will select Insert Block from the Block fly-out on the Draw toolbar and use the dialog box to insert the And gate into your drawing.

Pick: **Insert Block icon**

The Insert dialog box appears on your screen, as shown in Figure 8.4.

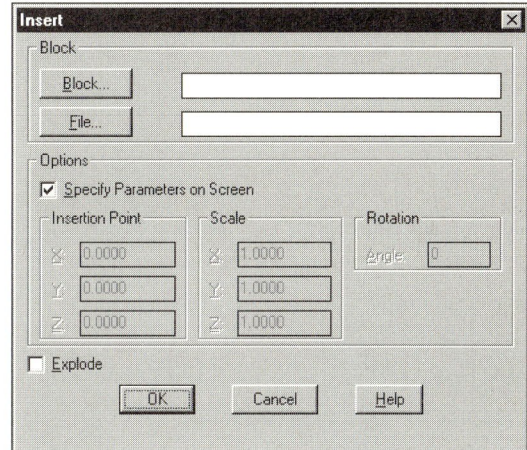

Figure 8.4

Select the Block button from the dialog box to bring up the Defined Blocks dialog box, which lets you select from the blocks defined in your drawing. The Defined Blocks dialog box looks like Figure 8.5.

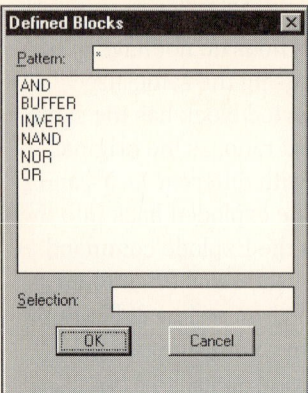

Figure 8.5

Pick: **AND**

Pick: **OK**

You return to the Insert dialog box. Be sure that the box to the left of Specify Parameters on Screen has a check in it. This feature allows you to pick the location for the block from your screen after you exit the dialog box. Otherwise, you must type in the exact coordinates for its location in the dialog box, as well as the rotation angle and scale. To exit the dialog box and begin specifying the insertion parameters on screen,

Pick: **OK**

The AND block appears, attached to the crosshairs by its endpoint, the base point you specified. Continue with the Insert command.

> Insertion point: *(pick a point anywhere to the right of the logic symbols)*
>
> X scale factor <1>/Corner/XYZ: ⏎
>
> Y scale factor (default = X): ⏎
>
> Rotation angle <0>: ⏎

The AND block is added to your drawing. On your own, try to erase one of the lines of the newly added block. Note that the entire block behaves as one object and is erased all together. Cancel the Erase command without erasing anything. In Tutorial 7, when you added associative dimensions to your drawing, they were also blocks. That's why they behaved as a single object, even though they were a collection of lines, polylines, and text. Before you can change individual pieces of a block, you must use the Explode command, as you did with Xplode in Tutorial 7. However, exploding takes away all of a block's special properties.

Using Explode

 AutoCAD treats blocks as a single object. The Explode command replaces a block reference with the simple objects comprising the block. Explode also turns 2D and 3D polylines, associative dimensions, and 3D mesh back into individual objects. When you explode a block, the resulting image on the screen is identical to the one you started with, except that the color and linetype of the objects may change. That can occur because properties such as the color and linetype of the block return to the settings determined by their original method of creation, either BYLAYER or the set color and linetype with which they were created. There will be no difference unless the objects were created in layer 0 or with the color and linetype set BYBLOCK, as specified in the list at the beginning of this tutorial.

You will use the Explode command to break the newly inserted AND block into individual objects for editing.

> *Pick:* **Explode icon**
>
> Select objects: *(pick any portion of the newly inserted AND block)*
>
> Select objects: ⏎

The block is now broken into its component objects. On your own, try to erase a line from the block. Note that you can erase one line at a time, rather than the whole block.

Using Write Block

The block you just created is only defined in the current drawing. You can use the Write Block command to make the block into a separate drawing for insertion into other drawings. A drawing file made with the Write Block command is the same as any other drawing file. You can call it up and edit it like any other drawing that you have created. You can use any AutoCAD drawing as a block and insert it in any other drawing, as you did in Tutorial 5. This is a very powerful feature for your use when creating drawings in AutoCAD.

■ *TIP* You can use Write Block to export part of any drawing to a separate file. This action can speed up your computer's response time by keeping file sizes smaller. When you are working in an extremely large file, but just need to work on a certain part of it, use Write Block to export that area to a new file. Work on the new, smaller file that contains only the part of the drawing you need. When done, use Insert Block to reinsert the finished work in the large drawing. You will have to explode it after inserting it before you can edit it. ■

The Write Block command involves two steps: Specify the new name for the drawing file that will be created on your default drive and then select the block. You can select the block in four ways:

1. Type the name of a block that you have created with the Block command in the current drawing at the command prompt and press ↵ . The objects comprising the specified block are written to the drawing file that you created in the first step.

2. Type an equals sign (=) and press ↵. You can use this shortcut when the block that you previously created with the Block command has the same name as the AutoCAD drawing file that you just specified.

3. Type an asterisk (*) and press ↵. These actions save the entire current drawing to the name specified as the Write Block file name (as the Save command does), except that unused named objects, layers, and other definitions are eliminated.

4. Press ↵. This null response is followed by prompts that allow you to specify the objects and the insertion base point, as in the Block command prompts. AutoCAD writes the selected objects to the drawing file name that you specified and deletes them from the current drawing. You can use the Oops command to retrieve them if necessary.

You will use the Write Block command to write the AND block that you created to a new AutoCAD drawing file called *and-gate.dwg*.

Command: **WBLOCK** ↵

If you configured AutoCAD to use dialog boxes for file operations (the default), the Create Drawing File dialog box shown in Figure 8.6 appears on the screen. (If not, set Filedia equal to 1 on your own and try again.)

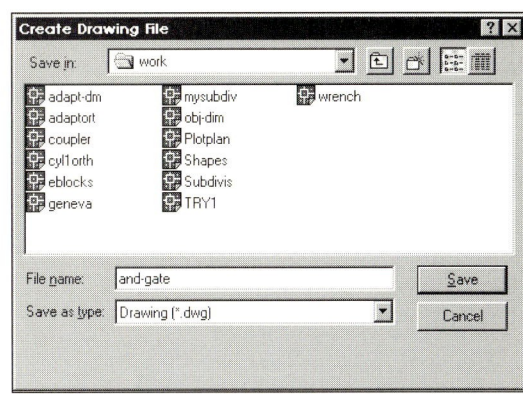

Figure 8.6

You will type *and-gate* for the new file that will be created. You do not need to include the file extension; the *.dwg* extension is automatically added. The name must have only eight characters to be a legal DOS file name. (Be sure to use the *c:\work* directory for the new file to keep your hard drive organized.)

Type: AND-GATE *(next to File name)*

When you have finished,

Pick: Save

■ **Warning:** Do not give the write block the same name as an existing drawing. If you do, AutoCAD will display the message *This file already exists. Replace existing file?* Select No unless you are certain that you no longer want the old drawing file. ■

Next, AutoCAD will prompt you for the block name. Type in the name of the AND block you created previously.

Block name: AND ⏎

The block that you created, named AND, has been saved to the new AutoCAD drawing file *and-gate.dwg*.

> ■ **TIP** Write Block does not save views, User Coordinate Systems, viewport configurations, unreferenced symbols (including block definitions), unused layers or linetypes, or text styles. This compresses your drawing file size by getting rid of unwanted overhead. ■

On your own, use the Write Block command to create separate drawing files for each of the other blocks that you created. Name them *or-gate, nor-gate, nnd-gate, inverter,* and *buffer*.

Creating Block Libraries

You can build block libraries of standard symbols, such as electronic, standard mechanical fastener, furniture, and landscaping symbols—and many other items that you might frequently use—for any application. You can also order block libraries from third-party sources. Many third-party vendors offer disks of standard shapes saved in block form.

To create a block library, start a new drawing with the name for the library. Insert each of the blocks into the library drawing. You can also add text below each block, indicating where the insertion point for each block is and the exact block name. Save the file.

To use the block library, insert the library file into the current drawing, using the Insert Block command and selecting File as the type of object to insert. Pick OK. At the prompt for the insertion point, pick Cancel. Now each of the blocks inserted into the library file is available in the current drawing. The reason is that, when you insert a drawing, all its block definitions (and groups, too) are inserted with it. Even though you cancel the Insert command, the blocks remain defined and you can insert each block you want to use from within the drawing.

Using block libraries helps keep your hard drive organized. Group like symbols together into a library file. Print out the drawing of it, showing the block insertion points and names of the blocks and hang it near your computer for quick access to block information. Doing so also helps inform coworkers about the blocks that you have created that they could be using.

Creating Custom Toolbars

One of the ways in which you can *customize* your AutoCAD software is by creating new toolbars and programming the existing toolbars so that they contain the commands you use most frequently. You can also reprogram the individual buttons on the AutoCAD toolbars. Next, you will create your own toolbar for the logic symbols that you created. You will program the buttons on the new toolbar to insert the logic symbols. To create a new toolbar,

Pick: **View, Toolbars**

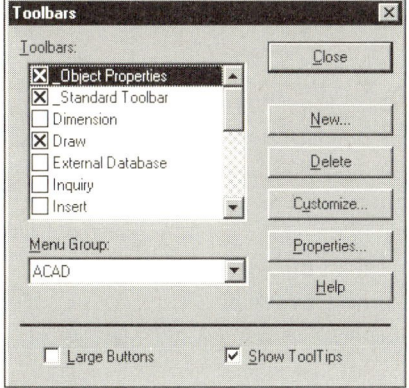

Figure 8.7

The Toolbars dialog box appears on your screen, as shown in Figure 8.7. You will pick New to create a new toolbar.

Pick: **New**

Figure 8.8 shows the New Toolbar dialog box.

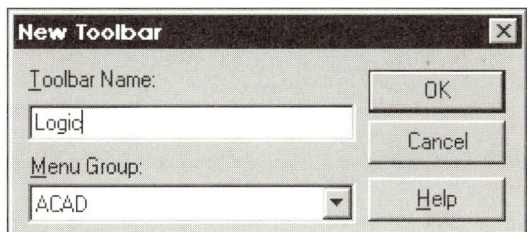

Figure 8.8

You will use the New Toolbar dialog box to type in the name of the toolbar that you want to create. You can also use it to select the menu group that you want to customize. You will add your toolbar to the ACAD menu group. Pick in the text box below Toolbar Name.

Type: **LOGIC**

Pick: **OK**

> ■ *TIP* The Toolbar dialog box and its associated dialog boxes are case-sensitive. What you type here will appear on your toolbar *exactly* as you have typed it. If you want your toolbar to look like those used by AutoCAD, use an initial capital letter, such as Logic, to name your toolbar. ■

You will see a small toolbar with the name Logic at the top appear on your screen. (The name Logic may be partially covered by the Windows close box.) It should be similar to Figure 8.9.

Figure 8.9

The Toolbars dialog box should still be on your screen. You will use the Customize selection to add buttons to your Logic toolbar.

Pick: **Customize**

Now the Customize Toolbars dialog box appears on your screen. You can use it to select standard button icons to add to the toolbar. The button icons are grouped by their functions under category headings. In the Categories area, use the list that appears and

Pick: **Custom**

When you have Custom selected, the dialog box appears as shown in Figure 8.10.

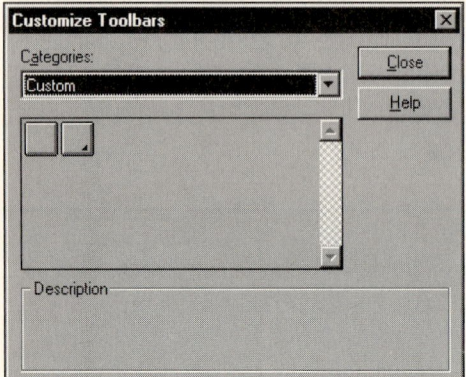

Figure 8.10

Note the two types of custom buttons: regular and flyout. To add a button to your toolbar, pick on the button icon you want to add and, holding the pick button down, drag a copy of that button to the toolbar. On your own, drag a regular button over to your Logic toolbar. Now drag a regular button to the toolbar three more times so that you have a total of four regular buttons on the toolbar. (Note that the Logic toolbar expands to accommodate the buttons as you add them.) Next, drag a flyout button to the toolbar. When you have finished adding the five buttons to the Logic toolbar,

Pick: **Close**

Now you are ready to add commands to your custom buttons. To edit the button properties, you will pick on the custom button that you want to modify, using the return button on your pointing device or mouse; this opens the Button Properties dialog box. If the Toolbars dialog box is closed, picking once with the return button on a toolbar button opens it.

Picking with the return button a second time on the toolbar button opens the Button Properties dialog box. On your own, pick using the return button on the leftmost regular button on the Logic toolbar to open the Button Properties dialog box shown in Figure 8.11.

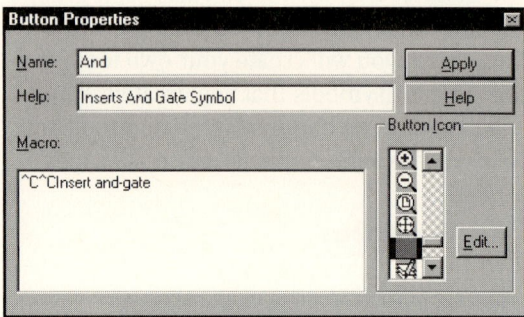

Figure 8.11

The input box to the right of Name allows you to specify the name of the button that you are creating. This name will appear as the tool tip for the button exactly as you type it. You use the input box to the right of Help to enter the help line for the button, which will appear in the status bar at the bottom of the screen. The area below the heading Macro lets you type in the command selection that you want AutoCAD to use when you pick the button. Do not add any spaces between ^C^C and the first command you type. AutoCAD uses a space to represent pressing ⏎ or SPACEBAR.

You will program these buttons to insert the blocks that you have created into your current drawing. When you pick the button, the command will use the Insert Block command automatically to insert the specified block.

> ■ **TIP** The block or file that you are inserting must be in the current path or you must tell AutoCAD where the block is located by adding the path name ahead of the block name. (You can change the current path in the Preferences dialog box.) The backslash has a special meaning in this dialog box: pause for user input. Therefore, when you add a path to a macro, you use the forward slash instead of the backslash to separate the directories (e.g., *c:/work/and-gate.dwg*). Finally, you can create a block library drawing and insert it into the drawing so that all the blocks are available in the current drawing prior to using the Logic toolbar that you just created. ■

Using Special Characters in Programmed Commands

There are some special characters that you can use in programming the toolbars. One of these is ^C. If you put ^C in front of the command name you are programming, it will cancel any unfinished command when you select its button from the toolbar. In programming menus and buttons, putting cancel twice before a command is often a good idea in case you have left a subprompt active (so that it would take two cancels to return to the command prompt). Putting cancel before the Insert command is useful because you cannot select this command during another command. Do not add cancel before the toggle modes or transparent commands (a ' must precede them) because, if you execute cancel before you use them, you can't use them during another command.

A space or ⏎ is automatically added to the end of every command you program for the buttons. This way the command is entered when you select the button. You can enter a string of commands, or a command and its options, by separating them with a space to act as ⏎. You can use the following special characters to program the toolbar buttons.

Character	Meaning
space	⏎
;	⏎
^Z	At the end of a line, suppresses the addition of a space to the end of a command string
+	At the end of a command string, allows it to continue to the next line
\	Pause for user input
/	Separates directories in path names, because \ (backslash) pauses for user input
^V	Changes current viewport
^B	Ctrl-B Snap mode toggle
^C	Ctrl-C Cancel
*^C^C	At the beginning of a line, automatically restarts command sequence (repeating)
^D	Ctrl-D Coordinates toggle
^E	Ctrl-E Isoplane toggle
^G	Ctrl-G Grid mode toggle
^O	Ctrl-O Ortho mode toggle
_	(Underscore) preceding a command or option, automatically translates AutoCAD keywords and command options for use with foreign-language versions

Pick: (in the input box to the right of Name)
Type: **AND**
Pick: (in the input box to the right of Help)
Type: **INSERTS AND GATE SYMBOL**
Pick: (to the right of ^C^C, below Macro)
Type: **INSERT AND-GATE**

■ *TIP* You may need to add the path to where your *and-gate* drawing is stored (i.e., ^C^CInsert c:/work/And-gate). ■

Pick: **Edit (to the right of the button icon pictures)**

The Button Editor dialog box appears on your screen. (If you double-click with the right button on an icon *when the Toolbars dialog box is not yet open*, the Toolbars dialog box opens and then the Button Editor dialog box opens.) The button editor works much like many paint-type drawing programs. Each little box in the button editor represents a pixel, or single dot, on your screen. You can select the drawing tool from the buttons at the top of the editor. The color of the drawing tool is selected from the colored buttons at the right of the editor. Use the button editor to draw a picture of the And gate, as shown in Figure 8.12. If you have trouble using the button editor, pick the Help button at the bottom right of the editor.

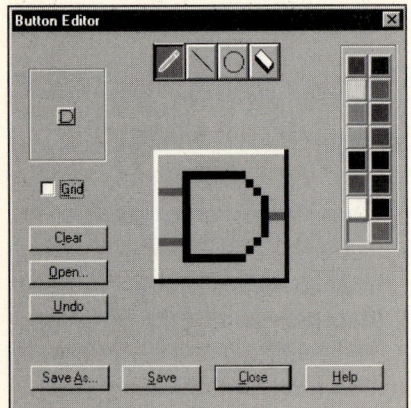

Figure 8.12

When you have finished drawing your button icon,

Pick: **Save As**

On your own, use the Save As dialog box that appears to name the icon *and* in the *c:\work* directory. (Again, you need not include the file extension; AutoCAD automatically adds the *.bmp* extension).

■ *TIP* If you want to save your *.bmp* files (or any other files) to a subdirectory in *c:\work* (or any other directory), you must add that subdirectory to your environment support path. Otherwise, AutoCAD won't be able to find your *.bmp* files upon loading. ■

To test an icon, all your dialog boxes must be closed; when you close them, AutoCAD updates the necessary files so that you can use your new icons. When you have finished using the button editor,

Pick: **Close**

Pick: **Apply (then close the Button Properties dialog box)**

Pick: **Close**

When you pick Apply, your selections are applied to the icon. You should now see the And gate icon that you made appear on your Logic toolbar.

All the dialog boxes should now be closed. On your own, test the new And button that you created on your toolbar.

Pick: **And icon**

The And gate should now be attached to the crosshairs, ready for insertion into your drawing. Position it as you like in your drawing and press the pick button. Press ⏎ for the remaining prompts for scale and rotation. The And gate should appear in your drawing.

Next, you will program the Nand gate button for the toolbar so that it will insert the NAND block when you pick it. Use the Button Properties dialog box by double-clicking the return button while the arrow cursor is positioned over the button second from the left on the Logic toolbar. Do so now, on your own.

The Button Properties dialog box appears on your screen. Type the entries listed below. (Note that there is no space between ^C^C and Insert.)

Name: **NAND**

Help: **INSERTS NAND GATE SYMBOL**

Macro: **^C^CINSERT NND-GATE**

On your own, pick on the And icon you created from the icons displayed in the Button Icon area of the dialog box and pick Edit. The button editor appears on your screen, with the icon for the And gate already showing. Pick the Save As button and save it to the new name *nand.bmp*.

■ *TIP* If you do not want to create the bit maps for each button, files have been provided with the data files. To use one, when the button editor is on your screen, pick Open and switch to the *c:\datafile* directory. The files are named *and.bmp*, *nand.bmp*, *or.bmp*, *nor.bmp*, *inverter.bmp*, and *buffer.bmp*. Once the appropriate file is selected, okayed, and appears in the button editor, pick Save As and save it to the *c:\work* directory. ■

Now use the button editor to change it so that it looks like a Nand gate, not an And gate, by adding pixels at the right end to look like the small circle on the Nand gate. (Making fine details is difficult; just a block at the end will look OK on the button.) When you have finished drawing it,

Pick: **Save**

Pick: **Close**

Pick: **Apply**

Now you will add the Or and Nor gates to the next two buttons. Use the Button Editor dialog box on your own to create button icons for them.

Name: **OR**

Help: **INSERTS OR GATE SYMBOL**

Macro: **^C^CINSERT OR-GATE**

Name: **NOR**

Help: **INSERTS NOR GATE SYMBOL**

Macro: **^C^CINSERT NOR-GATE**

When you have finished, close all the dialog boxes. On your own, test your new icons.

Creating the Inverter Flyout

Next, you will create a flyout for the Buffer and Inverter symbols. When you create a flyout, you essentially create a small toolbar that appears when the pick button is held down over that item. First, you will create a new toolbar named Inverter. Then you will add the buttons for the *Inverter* and *Buffer gates* to it. After finishing the Inverter toolbar, you will use it as a flyout on the Logic toolbar.

Pick: **View, Toolbars**

Pick: **New**

Type: **INVERTER** *(in the Toolbar Name text box)*

Pick: **OK**

The new toolbar appears on your screen without any buttons.

Pick: **Customize**

Use the Customize Toolbars dialog box to select the category Custom and add two regular blank buttons to the Inverter toolbar. If you need help adding the buttons, refer to adding the buttons to the Logic toolbar earlier in this tutorial. (Sometimes you may need to drag your toolbar to a location where you can reach it easily.) When you have added two buttons to the Inverter toolbar, close the Customize Toolbars dialog box on your own.

Now open the Button Properties dialog box using the Return button. Use the dialog box to make the following entries for the left button on the Inverter toolbar.

Name: INVERTER

Help: INSERTS INVERTER SYMBOL

Macro: ^C^CINSERT INVERTER

Now do the same for the right button on the Inverter toolbar.

Name: BUFFER

Help: INSERTS BUFFER SYMBOL

Macro: ^C^CINSERT BUFFER

Create your own button icons for the inverter and the buffer and save them to the appropriate *.bmp* names. When you are finished modifying the button properties, don't forget to pick Apply to apply these settings to the buttons. Close the dialog boxes on your own.

Now you are ready to add the Inverter toolbar as a flyout shown on the Logic toolbar.

Adding a Flyout

Double-click on the empty flyout button on the Logic toolbar using the return button. The Flyout Properties dialog box appears on your screen. Use it to make the selections shown in Figure 8.13. Be sure to turn on Show This Button's Icon at the bottom of the dialog box.

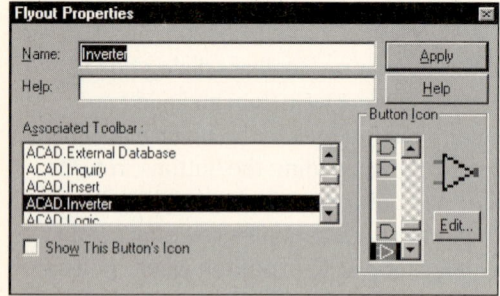

Figure 8.13

Pick: **Apply**

The Inverter flyout is added to the Logic toolbar.

On your own, close the Flyout Properties dialog box and then use the Customize selection from the Toolbars dialog box to add the Explode button to the Logic toolbar that you are creating. To add the Explode button, all you need do is pick the category Modify and drag a copy of the Explode button that appears over to your Logic toolbar. When you have finished, close all the dialog boxes and the Inverter toolbar.

> ■ **TIP** Remember, when there is no Close button in a dialog box, use the Windows close box in the upper right corner to close it. (In Windows NT, the close box is in the upper left corner.) ■

Editing Your Menu Files

When you have finished creating a custom toolbar and close the dialog box, AutoCAD writes to the *.mnr* and *.mns* menu files. It may not, however, write the correct *.bmp* files to your new buttons. To save the correct *.bmp* files to your new buttons you may need to edit your *.mns* file and copy the changes to your *.mnu* file. If your new toolbar does not show correctly, save your drawing now and exit AutoCAD on your own. (If you do not have difficulty with the toolbar buttons showing correctly, skip to Drawing the Half-Adder Circuit.)

Open your menu file in your word processor so you can edit it. Your menu file should be located in the AutoCAD R14 Support directory, or *c:\AutoCAD R14\support\acad.mns*.

Find the ***TOOLBARS section and within that section find **LOGIC. Below the Logic Toolbar entry should be **INVERTER. Edit these two sections so that they look like the following.

**LOGIC

ID_Logic_0	[_Toolbar("Logic", _Floating, _Show, 400, 50, 1)]
ID__0	[_Button("And gate", and.bmp, and.bmp)] ^C^CInsert and-gate
ID__1	[_Button("Nand gate", nand.bmp, nand.bmp)] ^C^CInsert nnd-gate
ID__2	[_Button("Or gate", or.bmp, or.bmp)] ^C^CInsert or-gate
ID__3	[_Button("Nor gate" nor.bmp, nor.bmp)] ^C^CInsert nor-gate
ID__4	[_Flyout("Inverter", inverter.bmp, inverter.bmp, _OwnIcon, ACAD.INVERTER)]
ID_Explode_0	[_Button("Explode", ICON_16_EXPLOD, ICON_32_EXPLOD, _OtherIcon, ACAD.TB_EXPLODE)]

**INVERTER

ID_Inverter_0	[_Toolbar("Inverter", _Floating, _Hide, 400, 50, 1)]
ID__5	[_Button("Inverter", Inverter.bmp, Inverter.bmp)] ^C^CInsert inverter
ID__6	[_Button("Buffer", Buffer.bmp, buffer.bmp)] ^C^CInsert buffer

ID_Logic_0 gives the information for the Logic Toolbar: that its name is Logic; that it's floating, not docked; and AutoCAD will open with it showing. ID_0 gives the information for the first icon, And gate. Its icon is *and.bmp* and the command it invokes is Insert and-gate. As you can see in the line defining the Explode icon, you can provide two different file names for the icons; the first is a 16-bit file and produces a small icon. In the Toolbars dialog box there is an option with which you can show large icons, the second file name in the menu file is the file used for the large icon. The large icons use *.bmp* files which have an image that is 32 pixels square. In the case of the icons you just created, if you chose to show large icons, your And gate, for instance, would appear in the upper left of the button, and AutoCAD would fill the rest of the button with some default pattern. To create *.bmp* files for use as large buttons, use a paint program such as Windows Paintbrush to create and save your own files. Make the image 32 pixels by 32 pixels.

Save your file *acad.mns* on your own. Copy the **LOGIC and **INVERTER sections of this file. Close *acad.mns* and Open *acad.mnu*. At the bottom of the ***TOOLBARS section, paste the **LOGIC and **INVERTER sections that you copied from *acad.mns*. Save *acad.mnu* on your own. Exit your word processor and start AutoCAD again.

Your Logic Toolbar should appear exactly as you left it, and you can use it to create your half-adder circuit. You can choose Hide from the Toolbar Properties dialog box or use the Windows control box to remove this toolbar from your screen at any time.

Drawing the Half-Adder Circuit

Now you have learned to customize the AutoCAD toolbars. You can create your own blocks and add them to other toolbars. You can also program your own commands onto toolbars and icons or modify existing toolbars and icons using the techniques that you have learned. Next, you will use the new toolbar that you created to draw a half-adder circuit with electronic logic symbols.

On your own, start a new drawing named *hfadder.dwg* from the template drawing *proto.dwt* provided with the data files. Drag the Logic toolbar to the upper right of the screen (or some other convenient location). Now switch to paper space and use the Edit Text command to change the text in the title bar to *HALF-ADDER CIRCUIT* and type your name. Specify NONE for the scale, as electronic diagrams usually aren't drawn to scale. Switch back to model space when you are finished.

> ■ *TIP* The Minsert command lets you insert a rectangular array of blocks. It can be useful when you're laying out drawings that have a pattern, such as rows of desks or rows of electronic components. The Divide and Measure commands also allow you to insert blocks. The Divide command divides an object that you select into the number of parts that you specify. You can either insert a point object where the divisions will be (as you did in Tutorial 5), or you can select the Block option and specify a block name. The Measure command works in much the same way, except that instead of specifying the number of divisions that you want to use, you specify the lengths of the divisions. Use the online Help facility to get more information about these commands. ■

On your own set the current layer to THIN and Snap to 0.1042 so that it will align with the spacing between the leads on the logic symbols that you created. Use the Logic toolbar on your own to insert the logic symbols, as shown in Figure 8.14. Use the Line and Arc

commands with Snap to Endpoint to connect the output from one logic circuit to the inputs of the next as shown.

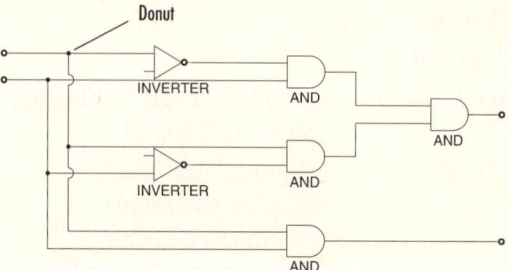

Figure 8.14

You will use the Donut command to make the filled and open circles for connection points.

Using Donut

The Donut command draws filled circles or concentric filled circles. See Figure 8.15.

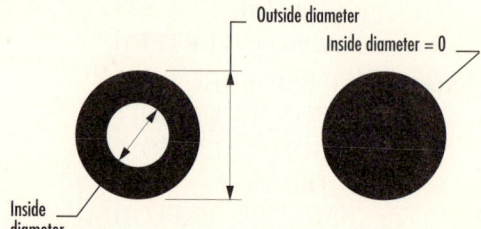

Figure 8.15

Found in the Draw menu, the Donut command requires numerical values for the inside and outside diameters of the concentric circles, as well as a location for the center point. If you enter a zero value for the inside diameter, a solid dot appears on the screen. The diameter of the dot will equal the stated outside diameter value. Next, you will draw some donuts off to the side of your logic circuit drawing.

Pick: **Draw, Donut**

Inside diameter<0.5000>: ⏎

Outside diameter<1.0000>: ⏎

Note that a doughnut-like shape is now moving with your cursor. The *Center of doughnut* prompt is repeated so that you can draw more than one donut. Select points on your screen.

Center of doughnut: *(select a point off to the side of your drawing)*

Center of doughnut: *(select another point)*

To exit the command when you have finished practicing drawing donuts of the same size, you will press ⏎.

Center of doughnut: ⏎

On your own, erase the donuts that you drew to the side. Restart the Donut command and set the inside diameter for the donut to 0. Set the outside diameter to 0.075. Draw the solid connection points for the logic circuit that you see in Figure 8.14 on your own.

Now add the open donuts at the ends of the lines. First, start the Donut command and set the inner diameter to 0.0625 and the outer diameter to 0.075. Create four donuts on the side of your drawing and then place them by using the Move command so that each donut is attached at its outside edge, not its center. Use Snap to Quadrant to select the base point on the donut and then Snap to Endpoint to select the endpoint of the line. When you have drawn and placed all the connection points, your drawing should look like that in Figure 8.14.

Freezing the TEXT Layer

The drawing *electr.dwg*, which you used to create the blocks for the logic gates, contained different layers for the text, lines, and shapes of the logic gates. You can freeze or turn off layers to control the visibility within parts of a block. You will freeze the text in the blocks because it is there to help identify the blocks while you are inserting them but would not ordinarily be shown on an electronic circuit diagram. The text in the blocks is on layer TEXT, as is the text in your title block. As you do not want to freeze the text in the title block, you will first move it onto a new layer.

On your own switch to paper space. Make a new layer, TEXT1, and then change all three portions of the title block text onto layer TEXT1.

Pick: **(Select the three portions of the title block)**

Pick: **Text1 (from the Layer control pull-down list on the Object Properties toolbar)**

Next, you will use Layer Control to freeze the layers that hold the text showing the name of the type of gate. Switch back to model space on your own.

Pick: **Layer Control icon**

Use the Layer Control list that appears on your screen to freeze layer TEXT. The names of the gates should disappear from the screen.

Save your drawing on your own before continuing.

Your drawing should look like Figure 8.16.

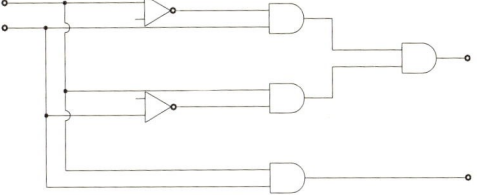

Figure 8.16

Creating Object Groups

Another powerful method of organizing your drawing is to use named groups. You can select these groups for use with other commands. Object Group is in the Tools menu.

Pick: **Tools, Object Group**

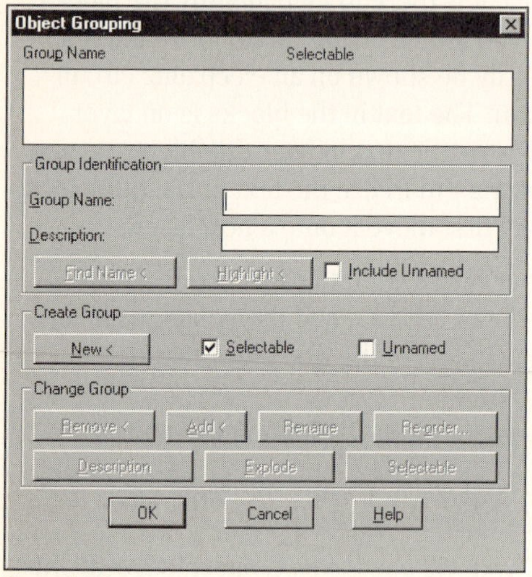

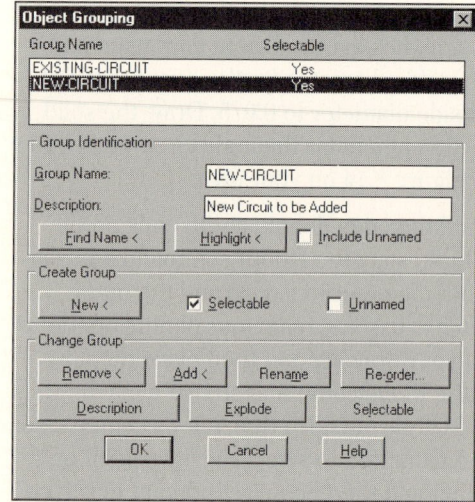

Figure 8.17

The Object Grouping dialog box appears on your screen, as shown in Figure 8.17. You will create a new group named EXISTING-CIRCUIT. For the existing circuit, you will select the upper Inverter and And gate and the two leads into each of them. Groups can be selectable, not selectable, named, or unnamed. When a group is selectable, the entire group becomes selected if you pick a member of the group when selecting objects. You can return to the dialog box and turn Selectable off at any time. Type the following information into the appropriate areas of the dialog box.

Group Name: **EXISTING-CIRCUIT**

Description: **Portions of the Existing Circuit**

Pick: **New**

Select objects: *(pick the upper inverter and upper And gate and the two leads to the left of each of them)*

Select objects: ⏎

The dialog box returns to your screen so that you can continue creating groups. You will group the remainder of the circuit and give it the group name NEW-CIRCUIT. Type the information into the dialog box, as shown in Figure 8.18. When you have finished, pick New. Select all the remaining objects to form the NEW-CIRCUIT group. Press ⏎ when you are done selecting. Pick OK to exit the dialog box.

Figure 8.18

Now, you are ready to try your groups with drawing commands. Remember, you can pick on any member of a selectable group to select the entire group. You will change the linetype property for the NEW-CIRCUIT group to a HIDDEN linetype.

Pick: **Properties icon**

Select objects: *(pick on one of the symbols from the NEW-CIRCUIT group)*

The entire group becomes selected, as shown in Figure 8.19.

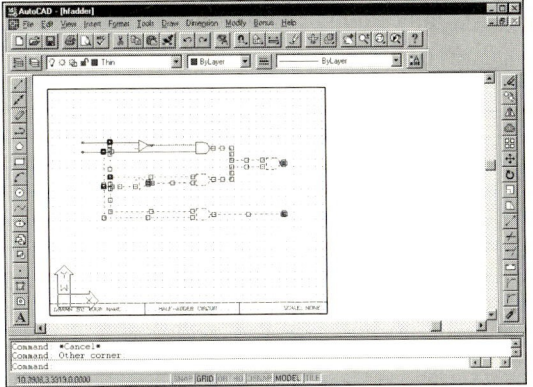

Figure 8.19

Select objects: ←

The Object Properties dialog box appears on your screen.

    Pick: **Linetype**

    Pick: **Hidden**

    Pick: **OK *(twice)***

The wiring diagram for the new portion of the circuit becomes dashed lines.

Use Undo from the Standard toolbar on your own to reverse this linetype change before you go on.

To complete your drawing, you will erase the unused leads from the inverters and reposition the remaining lead. To be able to select each object individually, return to the Object Grouping dialog box on your own and turn Selectable off.

Use the Explode command you added to your Logic toolbar to explode the inverters and erase the unused leads from them. Use the

Stretch command with hot grips to move the remaining lead so that it begins in the center of the inverter as shown in Figure 8.20. Save and plot your drawing.

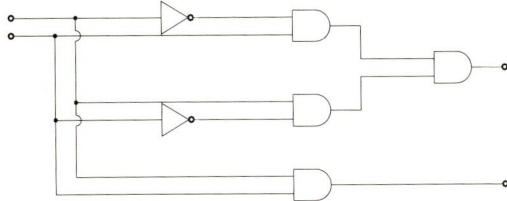

Figure 8.20

> ■ **TIP** You can quickly change blocks by redefining them. If you redefine a block, the appearance of all the blocks with that name will update. To redefine a block, make the changes you want to the block. Then use the Block command to save it as a block again, retaining the same file name. Insert the block again or type INSERT ← at the command prompt. When prompted for the block name, type the old block name followed by = and then the new block name. Do not include spaces because the spacebar has the same effect as ←. The appearance of all existing blocks that have the same name as the redefined block will update. ■

You have completed Tutorial 8. Save your file and close any toolbars that you may have open, except for the Draw and Modify toolbars. Exit AutoCAD.

KEY TERMS

And gate	Buffer gate	Nand gate
block	customize	Nor gate
block name	group	Or gate
block reference	Inverter gate	selectable group

KEY COMMANDS

Block	Object Group	Multiple Insert
Donut	Insert Block	Write Block
Explode		

8.1 Stress Test Circuit

Create this circuit board layout, using Grid and Snap set to 0.2. Use donuts and polylines of different widths to draw the circuit.

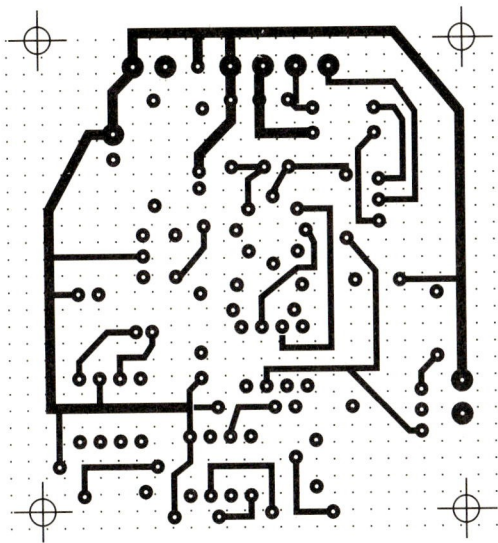

8.2M Decoder Logic Unit

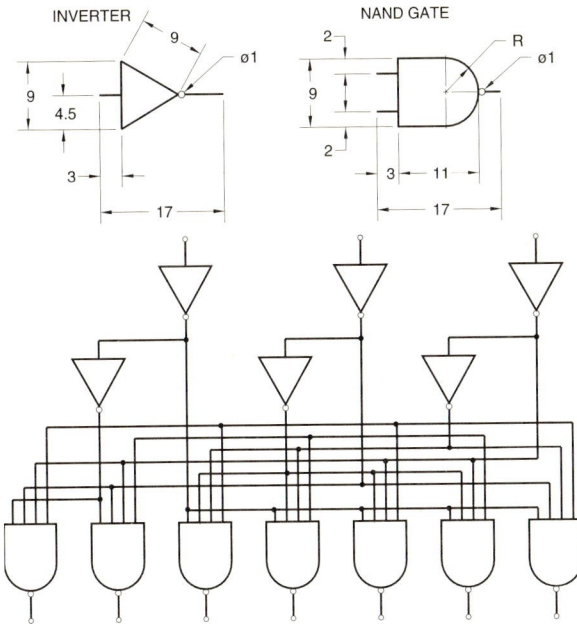

8.3 Floor Plan

Draw the floor plan shown for the first floor of a house. Create your own blocks for furniture and design a furniture layout for the house.

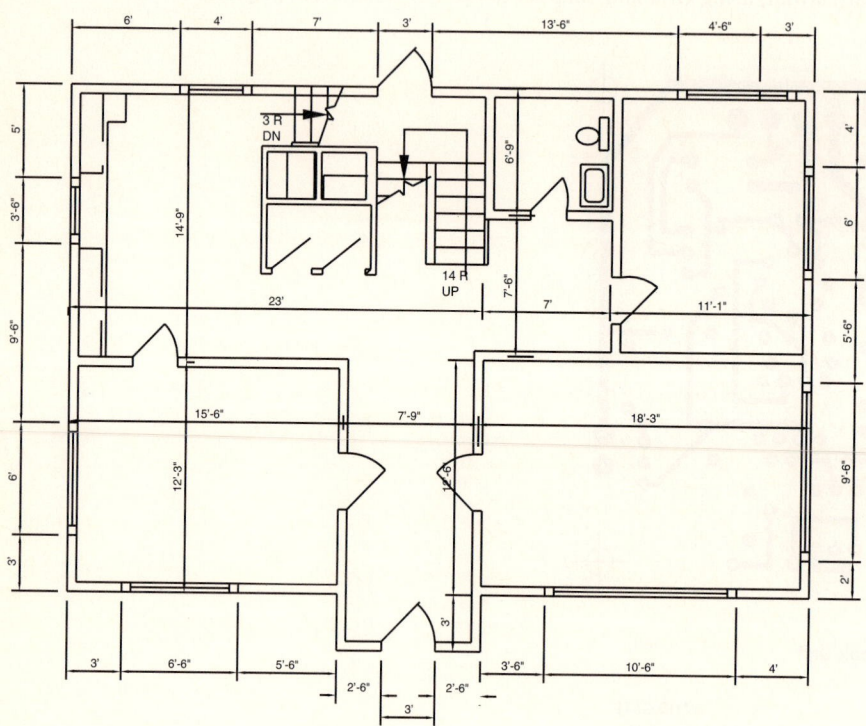

8.4M Arithmetic Logic Unit

Create blocks for the components according to the metric sizes shown. Use the blocks that you create to draw the circuit shown.

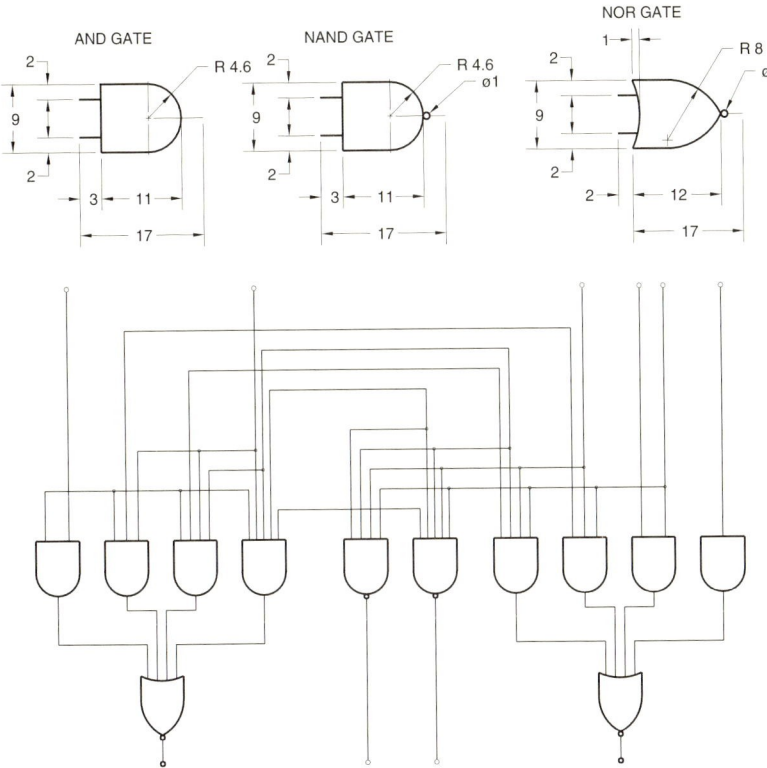

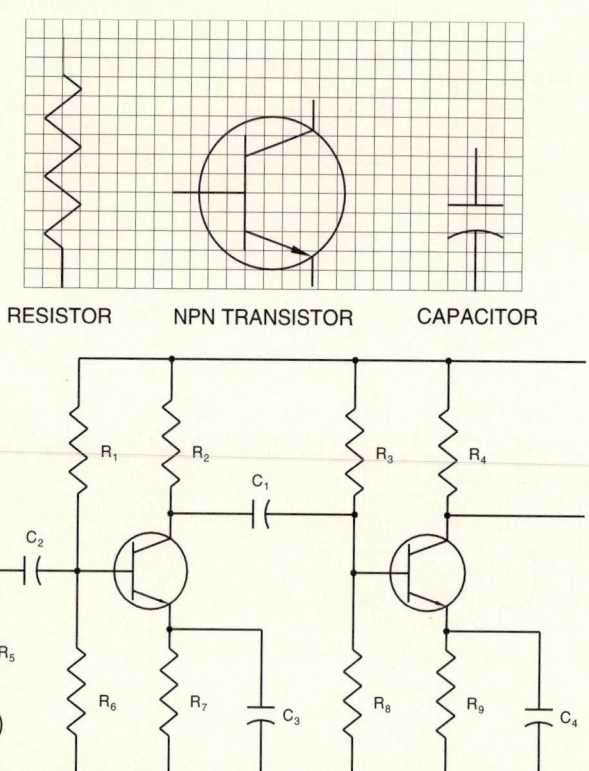 8.5 **Amplifier Circuit**

Use the grid at the top to draw and create blocks for the components shown. Each square equals –0.0625. Use the blocks that you create to draw the amplifier circuit.

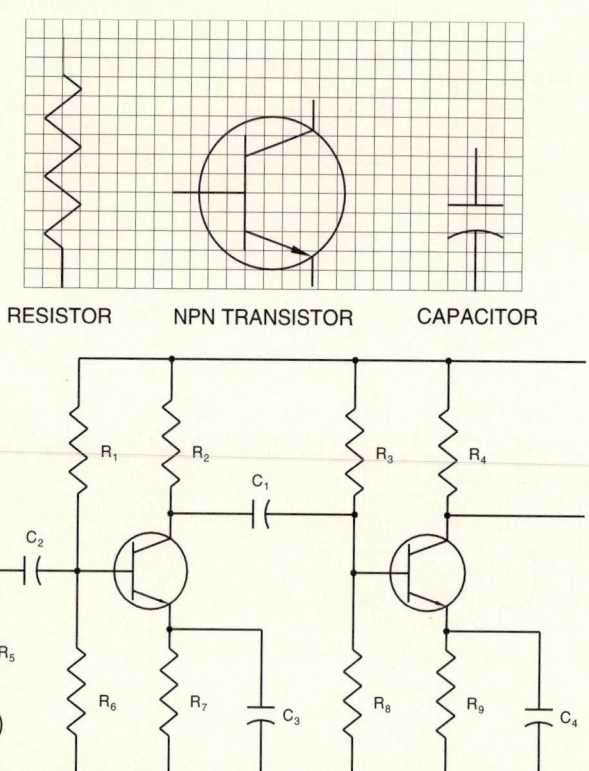

RESISTOR   NPN TRANSISTOR   CAPACITOR

8.6 Office Plan

Draw the office furniture shown and create blocks for DeskA, DeskB, ChairA, ChairB, and the window. Use the floor plan shown as a basis for creating your own office plan.

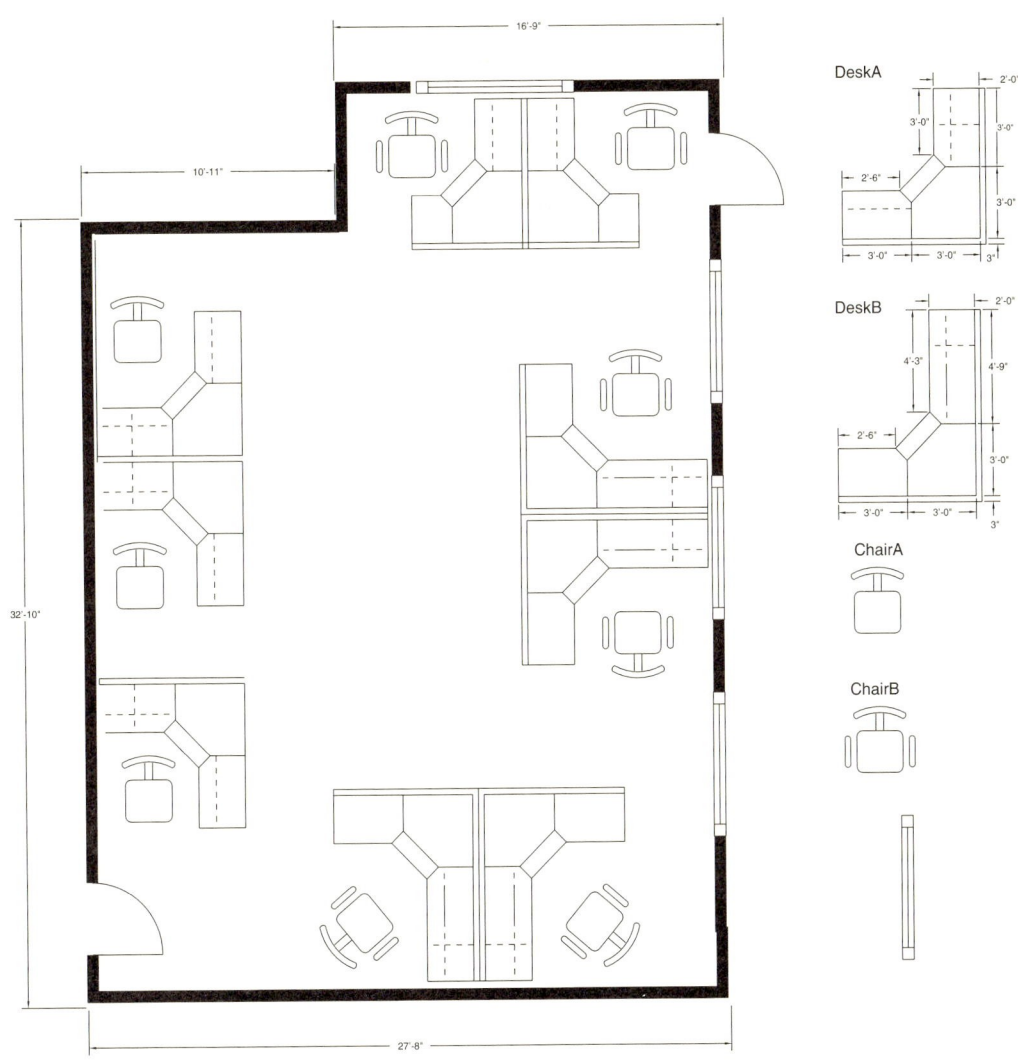

8.7 Piping Symbols

Draw the following piping symbols to scale; the grid shown is 0.2 inch (you do not need to show it on your drawing). Use polylines 0.03 wide. Make each symbol a separate block and label it. Insert all the blocks into a drawing named *pipelib.dwg*.

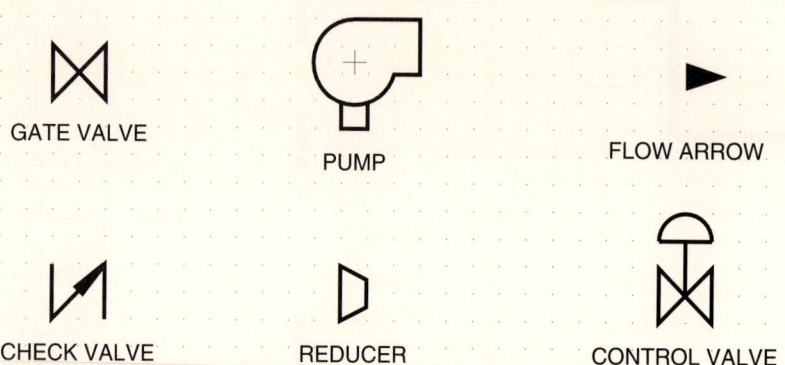

GATE VALVE

PUMP

FLOW ARROW

CHECK VALVE

REDUCER

CONTROL VALVE

PIPING SYMBOLS

8.8 Utilities Layout

Construct this representation of the utilities layout for a group of city blocks (grid spacing is 0.25). Using Polylines 0.05 wide, construct and label the three main symbols shown. Then scale the symbols down to 1/4 size, make each one a block, and insert them in the proper locations.

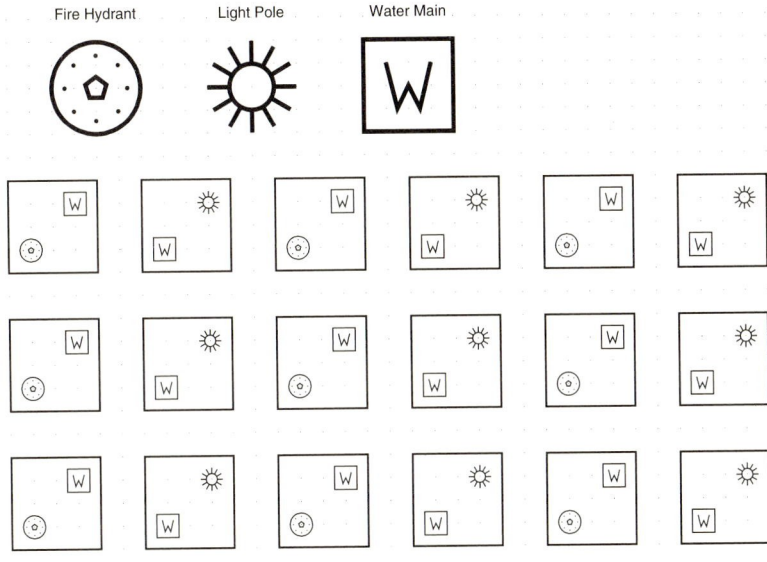

8.9 Mechanical Fasteners

Draw the hex head bolt, square head bolt, regular hex nut, and heavy hex nut according to the proportions shown. Make each into a block. Base the dimensions on a major diameter (D) of 1.00. This way when you insert the block you can easily determine the scaling factor that will be necessary to produce the correct major diameter. Insert each block into a drawing named *fastenrs.dwg*.

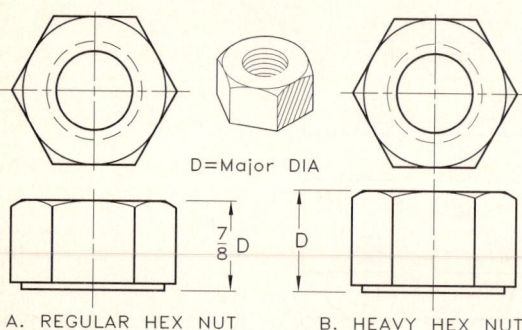

A. REGULAR HEX NUT   B. HEAVY HEX NUT

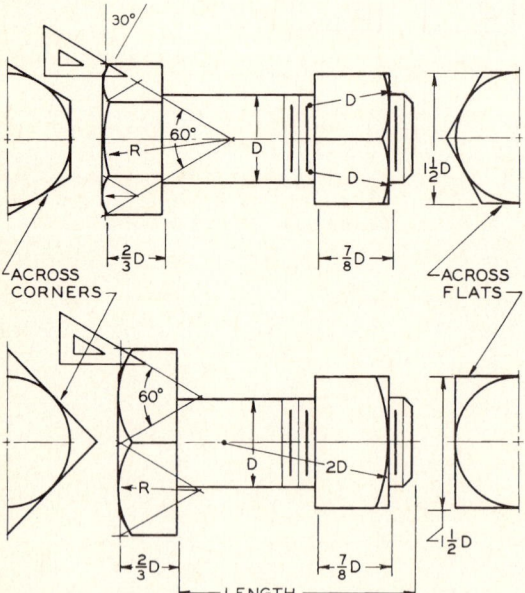

8.10 Wood Screws

Create blocks for each of the wood screws shown. The grid gives the proportions for drawing them. Make the diameter of each 1.00 so that you can easily determine the scale needed for inserting them at other sizes.

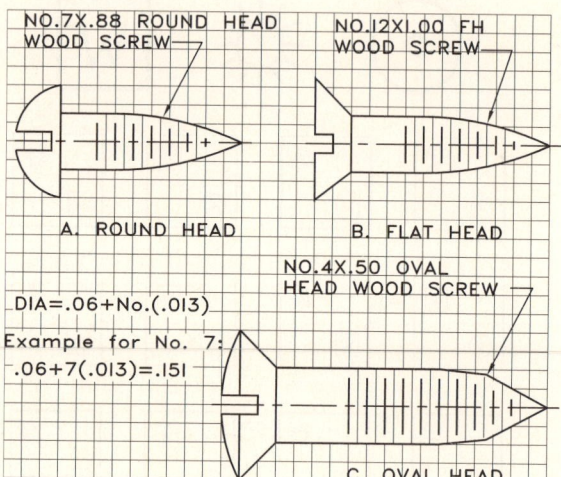

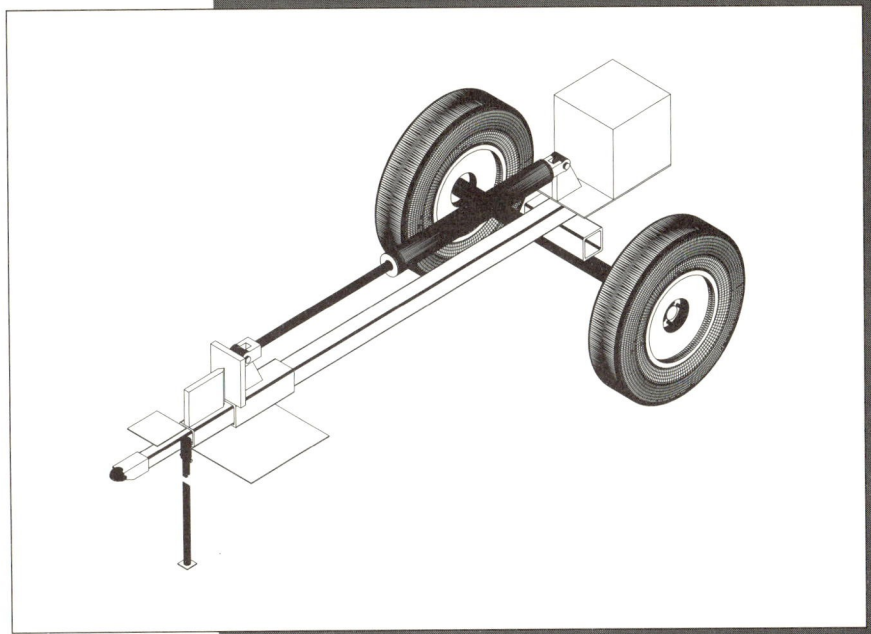

Introduction to Solid Modeling

Objectives

When you have completed this tutorial, you will be able to

1. Change the 3D viewpoint.

2. Work with multiple viewports.

3. Set individual limits, grid, and snap for each viewport.

4. Create and save User Coordinate Systems.

5. Set Isolines to control model appearance.

6. Create model geometry, using primitives, extrusion, and revolution.

7. Use Boolean operators to add, subtract, and intersect parts of your model.

8. Use region modeling.

9. Shade and plot your models.

10. Convert AME models.

Introduction

In Tutorials 1 through 8 you have learned how to use AutoCAD to model three-dimensional (3D) objects. When you create a multiview drawing, it is a model of the object that contains enough information within the various views to give the person interpreting it an understanding of the complete 3D shape. Your models have been composed of multiple two-dimensional (2D) views, which are used to convey the information. Now you will learn to use AutoCAD to create actual 3D models.

AutoCAD allows three types of 3D modeling: wireframe, surface, and solid modeling. *Wireframe modeling* uses 3D lines, arcs, circles, and other graphical objects to represent the edges and features of an object. It is called wireframe because it looks like a sculpture made from wires. *Surface modeling* takes 3D modeling one step further to add surfaces to the wireframe so that the model can be shaded and hidden lines removed. A surface model is like an empty shell: There is nothing to tell you how the inside behaves. A *solid model* is most like a real object because it represents not only the lines and surfaces, but also the volume contained within. In this tutorial you will first learn to create solid models and then surface models.

Using the computer, you can represent three dimensions in the drawing database; that is, you can create drawings by using X-, Y-, and Z-coordinates. Solid modeling is the term for creating an accurate 3D model of the drawing that goes beyond representing just the lines that form the edges of the surfaces. It describes a volume contained by the surfaces and edges making up the object. When you create a solid modeled drawing database, in some ways it is as if the actual object were stored inside the computer. Benefits of solid models include: They are easier to interpret; they can be rendered (shaded) so that someone unfamiliar with engineering drawing can visualize the object easily; 2D views can be generated directly from them; and their *mass properties* can be analyzed. The need to create a physical prototype of the object may even be eliminated.

Creating a solid model is somewhat like sculpting the part in clay. You can add and subtract material with *Boolean operators* and create parts by *revolution* and *extrusion*. Of course, when modeling with AutoCAD, you can be much more precise than when modeling with clay. At first, AutoCAD's solid modeling will look much like wireframe or surface modeling. The reason is that wireframe representation usually is used even in solid modeling and surface modeling to make many drawing operations quicker. When you want to see a more realistic representation, you can shade or hide the back lines of solid models and surface models.

Starting

Launch AutoCAD. Begin a new drawing on your own, based on the default template drawing *acad.dwt* provided in the AutoCAD software. Call your new drawing *block.dwg*.

When you are working in 3D, the grid is very useful. It can help you to relate visually to the current viewing direction and coordinate system. On your own, be sure that you are in model space and that the grid is turned on. The regularly spaced grid dots should appear every 0.5 unit.

3D Coordinate Systems

AutoCAD defines the model geometry using precise X-, Y-, and Z-coordinates in space called the *World Coordinate System* (*WCS*). The WCS is fixed and is used to store drawing geometry in both 2D and 3D in the database. Its default orientation on the screen is a horizontal X-axis, with positive values to the right, and a vertical Y-axis, with positive values above the X-axis. The Z-axis is perpendicular to the computer screen, with positive values in front of the screen. The default orientation of the axes is shown in Figure 9.1.

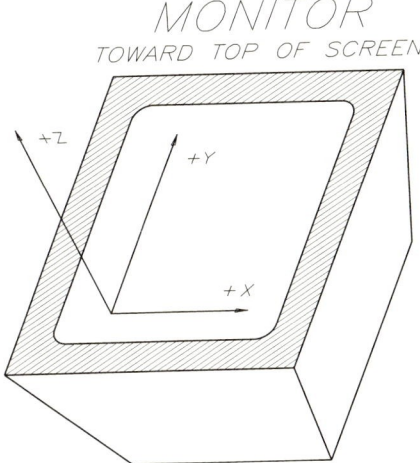

Figure 9.1

While creating 2D geometry, you have been using this default WCS. You have been looking straight down the Z-axis, so that a line in the Z-direction appears as a point. When you do not specify the Z-coordinate, AutoCAD uses the default elevation, which is zero. This method has made creating and saving 2D drawings easy.

Setting the Viewpoint

When creating 3D geometry, you will find it useful to establish a different direction, or several directions, from which to view the XY-plane, so that you can see the object's height along the Z-axis. You can do so with the Vpoint command and, in fact, can use it to view the

coordinate system from any direction you want. You will select Tripod from the Views menu bar to set the *viewpoint*.

Pick: **View, 3D Viewpoint, Tripod**

Rotate/<View point> <0.0000,0.0000,1.0000>:

An X-, Y-, and *Z-coordinate system locator* and *globe* appear on your screen, as shown in Figure 9.2.

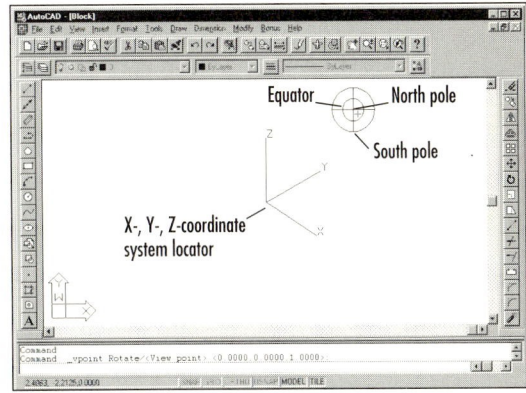

Figure 9.2

You will use the globe to select the viewing direction. You can think of the center point of the globe as the view looking straight down on the top of the XY-plane. The center circle is like the equator. Any point inside the center circle shows the view looking down on the object from the top. Points selected outside the center circle show the view looking up at the object from below. A horizontal line divides the globe into front and back. If you pick a point below the line, you are viewing the object from the front; above the line, and you are viewing the object from the back. Table 9.1 summarizes the use of the globe.

Table 9.1

	Inside of center circle	Outside of center circle
Above horizontal line	Top rear view	Bottom rear view
Below horizontal line	Top front view	Bottom front view

Move your pointing device so that the crosshairs are inside the center circle in the lower right-hand quadrant. When you have them positioned as shown in Figure 9.2, press the pick button to select the viewing direction. The location that you have selected produces a viewpoint showing the coordinate system from above and to the right.

Your screen should be similar to Figure 9.3, depending on the exact point you selected. Note that your view has changed. You are no longer looking straight down on the XY-plane, which is represented by the grid area. Your view now is from an angle.

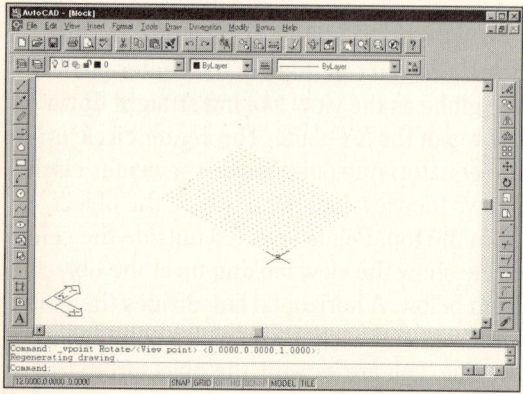

Figure 9.3

User Coordinate Systems

A *User Coordinate System (UCS)* is a set of X-, Y-, and Z-coordinates that you can define to help you create your 3D models. You can define your own UCS, which may have a different origin and rotation from the WCS and may be tilted at any angle with respect to it. UCSs are helpful because your mouse moves in only two dimensions, and UCSs let you orient the basic drawing coordinate system at any angle in the 3D model space of the drawing, allowing you to continue using your mouse or other pointing device for drawing. You can define any number of UCSs, give them names, and save them in a drawing; however, only one can be active at a time.

The UCS Icon

The UCS icon appears in the lower left of the screen. It will help you to orient yourself when looking at views of the object. The monitor screen is essentially flat, even if the object is a 3D solid model in the database, so only 2D views of it can be represented on the monitor. Because wireframe models look the same from front and back or from any two opposing viewpoints, keeping track of which view you are seeing on the screen is especially important.

The UCS icon is always drawn in the XY-plane of the current UCS. The arrows at the X- and Y-ends always point in the positive direction of the X- and Y-axes of the current UCS. A W appearing in the UCS icon tells you that the UCS is currently aligned with the WCS. The box in the lower corner of the icon indicates that you are viewing the UCS from above. A plus sign in the lower left of the icon indicates that the icon is positioned at the origin of the current UCS. Note the absence of a plus sign in the lower left of the UCS now. When you start a drawing from the *acad.dwt* template, the origin (0,0,0) of the drawing is in the lower left of the screen. If the UCS were at the origin, it would be partially out of the view; thus the default for the UCS icon is not at the origin. You can use the UCSIcon command to reposition the icon so that it is at the origin of the X-, Y-, and Z-coordinate system. When you do so, a plus sign appears in the icon.

A special symbol may appear instead of the UCS icon to indicate that the current viewing direction of the UCS is edgewise. (Think of the X- and Y-coordinate system of the UCS as a flat plane like a piece of paper; then, the viewing direction is set so that you are looking directly onto the edge of the paper.) In this case you can't use most of the drawing tools, so the icon appears as a box containing a broken pencil. Take special note of this icon.

The *perspective icon* replaces the UCS icon when perspective viewing is active; it appears as a cube drawn in perspective. Many commands are also limited when perspective viewing is in effect. The Dview command turns on perspective viewing. You will use the Dview command in Tutorial 13, when you create auxiliary views.

Figure 9.4 shows the different icons.

On your own, create three new layers. One layer will contain the solid model, another will contain viewports, and the third will contain the drawing's border. On your own, use the Layer Control dialog box to create the following layers:

MODEL	Magenta	Continuous
VPORT	White	Continuous
BORDER	White	Continuous

Set layer MODEL as the current layer.

Showing the Solids Toolbar

On your own, use the View selection from the menu bar and select the Toolbars option. From the list of toolbars that appears, pick to show the Solids toolbar. It will appear on your screen, as shown in Figure 9.5. Use the methods that you have learned to position it in an area of your screen where it will be accessible, but not in the way of your drawing.

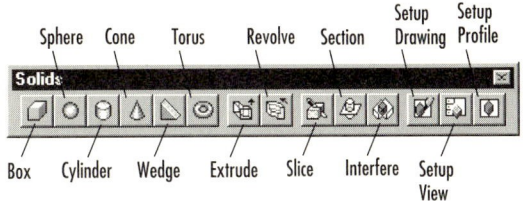

Figure 9.5

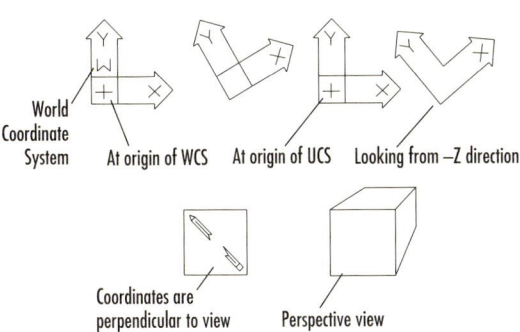

Figure 9.4

Creating an Object

AutoCAD provides three methods for creating model geometry: primitives, extrusion, and revolution. *Primitives* are basic shapes (e.g., boxes, cylinders, and cones) that can be joined to form more complex shapes. You will create a solid 3D box. Later in this tutorial you will learn how to join shapes by using Boolean operators and how to create other shapes with extrusion and revolution.

The Box command enables you to draw a rectangular prism solid model. You can draw a box by specifying the corners of its base and its height, its center and height, or its location and length, width, and height. Figure 9.6 shows the information you specify to draw a box using these methods. The default method of defining a box is to specify two corners across the diagonal in the XY-plane and the height in the Z-direction.

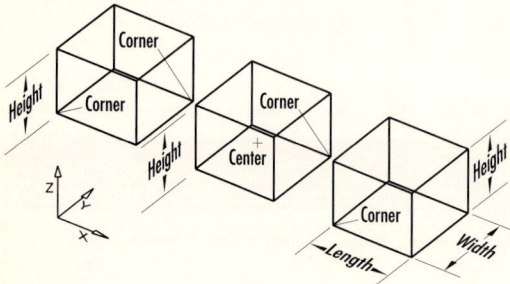

Figure 9.6

 You will use the Box selection on the Solids toolbar and draw a solid box by specifying the corners of the base and the height of the box.

> *Pick:* **Box icon**

The prompt for the box primitive appears. You will type in the coordinates.

Center/<Corner of box> <0,0,0>: **2,2,0** ⏎
Cube/Length/<other corner>: **8,6,0** ⏎
Height: **3** ⏎

Your screen should be similar to Figure 9.7.

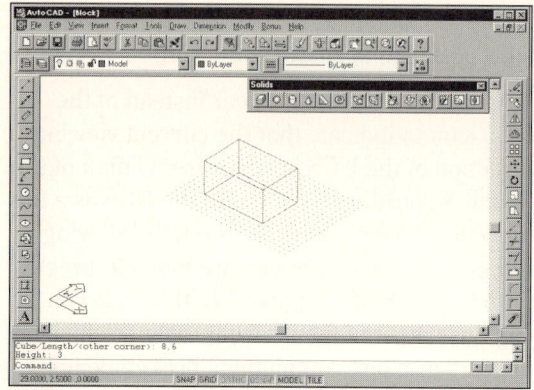

Figure 9.7

Creating Multiple Viewports

Now you have a 3D solid object on your screen. You will next create viewports to show several views of the object at the same time. This way you can create just one solid model and produce a drawing with the necessary 2D orthographic views directly from that 3D model. Having several views also makes creating the model easier. You have already used a single paper space viewport for printing and plotting in Tutorial 5. Now you will create four viewports and then change them so that they contain four different views of the model. Going back to the Rose Bowl analogy in Tutorial 5, you can think of the four separate viewports as four separate TV screens, each one of them showing the Rose Bowl game. Each TV set can show a different view of the game (like when the same event is on several different stations at once). Each cameraperson is looking at the game from a different angle, producing a different view on each TV set. Yet there is only one Rose Bowl being played.

Multiple viewports in AutoCAD work the same way. You can have many viewports, each of which contain a different view of the model, but there is only one model space and each object in it need only be created once.

On your own, set layer VPORT as the current layer before you continue.

Next, you will switch to paper space by selecting it from the View menu. When you do so, AutoCAD automatically sets Tilemode to 0 if needed. Once paper space has been enabled, do not change Tilemode. Doing so disables paper space and can become confusing when you use paper space viewports.

Pick: **View, Paper Space**

When you switch from model space to paper space, note the triangular paper space icon in the lower left of your screen. Recall from Tutorial 5 that paper space allows you to arrange views of your model, text, and other objects as you would on a sheet of paper. You can't see model space until you create a viewport in paper space (like turning on your TV). You use the Mview command to create viewports. You will pick it from View on the menu bar.

Pick: **View, Floating Viewports, 4 Viewports**

■ *TIP* Floating Viewports will be grayed out unless you are in paper space. ■

The Mview command is echoed in the command prompt area, and its prompts appear. You will place the viewports inside the area that you want to plot full size, centered on your page, as you did in Tutorial 5. You may find

that the following values work for your printer. If you used a different setting in Tutorial 5, apply the one that works for your printer.

Fit/<First Point>: **.25,.25** ⏎
Second point: **10.25,7.75** ⏎

Four viewports appear on your screen, each containing an identical view of the object. Now you will set the paper space limits to the size of the sheet of paper you will use when plotting. Each viewport, and paper space, can have its own Limits, Grid, and Snap settings.

The default limits in AutoCAD are set to 12, 9. Because the paper size that you will print on is only 11 x 8.5, the default limits are too large. It is a good practice to set the limits in paper space close to the paper size; that way you can see how much area you actually have on the sheet for drawing. Starting the limits at (0, 0) usually lets you position your plots more easily where you want them on the page when you are printing. Also, if you use Zoom All, the graphics window fills your screen, not a larger area (which would make your viewports smaller on the screen and harder to work in). While still in paper space,

Pick: **Format, Drawing Limits**

ON/OFF/<Lower left corner> <0.0000,0.0000>: ⏎
Upper right corner <12.0000,9.0000>: **10.5,8** ⏎

To fill the screen area with the new drawing limits, you will pick the Zoom All icon from the Zoom flyout on the Standard toolbar.

Pick: **Zoom All icon**

On your own, set layer MODEL as the current layer. Your screen should be similar to Figure 9.8.

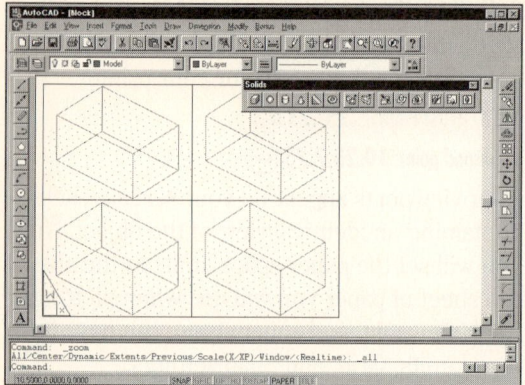

Figure 9.8

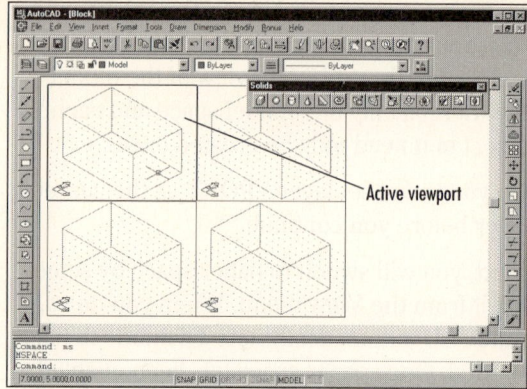

Figure 9.9

Next, you will change the viewpoint for each viewport so that together they show a top, front, right-side, and isometric view of the model.

Remember, when you are in paper space, you cannot change things that are in model space. The original box was created in model space, where you create and make changes to your model. To switch back to model space,

Command: MS ⏎

You know that you are in model space when the UCS icon reappears. It is now displayed in *each* viewport.

You will make the top left-hand viewport active and change it to show the view of the model looking straight down from the top.

Selecting the Active Viewport

To make the upper left viewport active, you will move the arrow cursor until it is positioned in the top left viewport and then press the pick button.

Pick: (anywhere in the upper left viewport)

The crosshairs will appear in the viewport, as shown in Figure 9.9, indicating that it is now the active viewport.

You can only draw in a viewport and pick points in it when it is active. After you create something, whatever you created is visible in all other viewports showing that area of the WCS, unless the layer that the object is on is frozen in another viewport. (Using the Vplayer command, you can freeze layers in specific viewports.) Keep in mind that, while you are using a certain viewport to *access* model space, there *is* only one model space. Thus you can start drawing something in one viewport and finish drawing it in another.

> ■ *TIP* You can press Ctrl-R to toggle the active viewport. This action is helpful when you are unable to pick in the viewport for some reason. ■

Changing the Viewpoint

Now you will change the viewpoint for the upper left-hand viewport so that it shows the top view of the object. By controlling the viewpoint with the Vpoint command, you can create a view from any direction. The numbers you enter in the Vpoint command are the X-, Y-, and Z-coordinates of a point that defines a *vector*, or directional line. The other point defining the vector is the origin (0,0,0). Your line of sight, or viewpoint, is defined by this vector, or imaginary line, toward the origin from the coordinates of the point you enter. The actual *size* of the number you enter does not matter—only the *direction* it establishes. Thus entering *2,2,0* is the same as entering *1,1,0*, as only the direction of sight is determined by this number. See Figure 9.10.

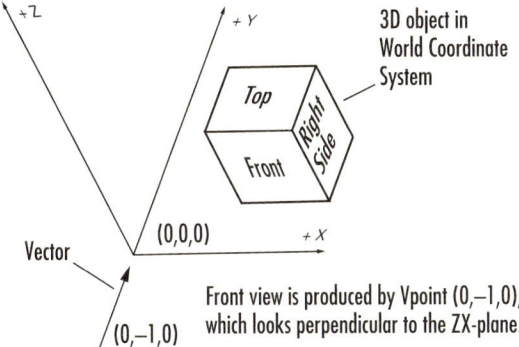

Figure 9.10

You can also select the Rotate option and establish the viewing direction by specifying the rotation angle in the XY-plane from the X-axis and then the rotation above the XY-plane for the viewport. See Figure 9.11. Experiment with these options on your own.

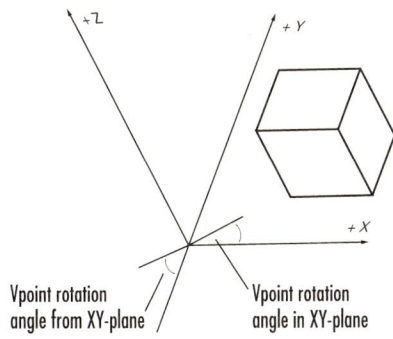

Figure 9.11

Using the Viewpoint Flyout

 The Viewpoint flyout on the Standard toolbar is shown in Figure 9.12. You can use it quickly to pick different viewing directions for the active viewport. The selections on the Viewpoint flyout use the Vpoint command to select commonly used viewing directions. You will use the flyout to pick the Top View icon.

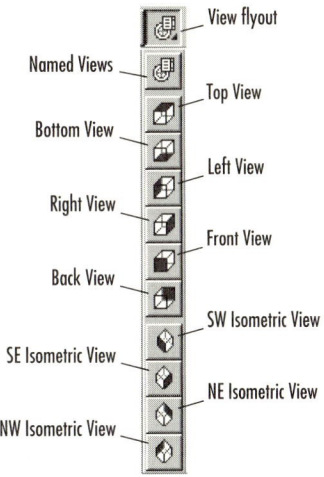

Figure 9.12

Pick: **Top View icon**

The view in the upper left viewport changes. Now you are looking straight down on top of the box. The top view of the box is too large to fit entirely within the view. Your screen should look like Figure 9.13, which shows the 3D box viewed from the top in the upper left viewport. Note that the view of the UCS icon is straight on now that you have changed the viewing direction.

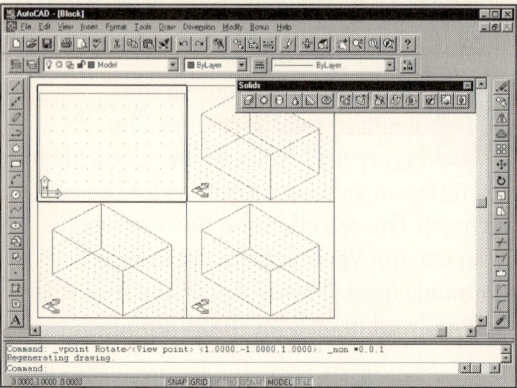

Figure 9.13

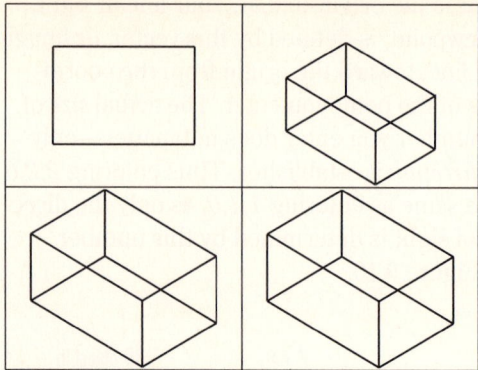

Figure 9.14

> ■ **TIP** You can turn on the Viewpoint toolbar by selecting View, Toolbars, Viewpoint from the menu bar; it contains the same selections as the View flyout. You may find it convenient to turn on this toolbar so that the selections for changing the view direction are always available on the screen and easy to pick. ■

 You will use the Zoom command with the Scale XP (meaning relative to paper space) option to move the viewing distance farther away so that the object appears smaller and will fit the viewport. You will type the alias for the Zoom command instead of using the icon.

Command: **Z** ⏎

All/Center/Dynamic/Extents/Previous/Scale(X/XP)/Window/
<Realtime>: **.5XP** ⏎

The top view of the box should fit into the viewport at half paper space scale. Your drawing should look like that in Figure 9.14.

Each viewport can contain its own setting for Grid, Snap, and Zoom. Now make the lower left viewport active by moving the arrow cursor into that viewport and pressing the pick button. The crosshairs appear in the lower left viewport.

 You will use the View flyout on the Standard toolbar to pick the Front View icon and change the view of the lower left viewport.

Pick: **Front View icon**

Now the view in this viewport is of the front of the object, as though you had taken the top view and tipped it 90° away from you. (Imagine that the original Y-axis is projecting straight into your monitor.) The object is again too large to fit in the viewport. Think of the distance of the object from the viewport or drawing screen as determined by the Zoom command. Use Zoom Scale XP to zoom out so that the entire view will fit the viewport.

Pick: **Zoom Scale icon**

Enter Scale factor: **.5XP** ⏎

Now you see the entire front view of the object in the viewport. Your screen should be similar to Figure 9.15.

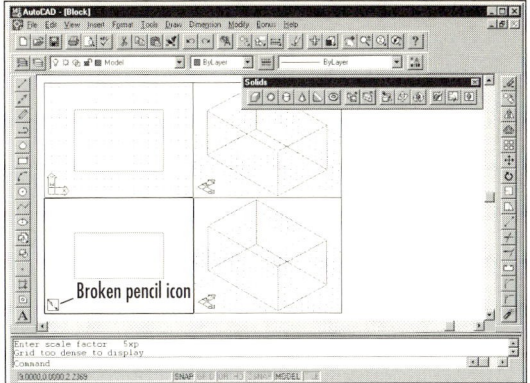

Figure 9.15

Note the broken pencil icon in the lower left corner. This icon tells you that your drawing's current XY-plane, or UCS, is parallel to the viewing direction. If you were to try to draw, you couldn't keep track of what you were creating. Every shape you drew would appear as a straight line because your view of the drawing surface would be edgewise. Think about what would happen if you looked edgewise at a piece of paper at eye level and then tried to draw on it. You may also have seen the message *Grid too dense to display* when you originally changed your viewpoint. If the coordinate system is viewed edge on, the grid is, of course, a solid line of dots, making it too dense to appear on the screen in a usable manner.

Creating User Coordinate Systems

You can have more than one coordinate system in AutoCAD. The object is stored in the WCS, which stays fixed, but you can use the UCS command to create a User Coordinate System oriented any way you want. The UCS command allows you to position a User Coordinate System anywhere with respect to the WCS. You can also use it to change the origin point for the coordinate system and to save and restore named coordinate systems.

 You will create a new UCS that is rotated 90° around the WCS X-axis, so that the XY-plane of the new UCS is parallel to the front viewing plane. Doing so will provide a coordinate system that will simplify modeling objects in the front view. AutoCAD uses the right-hand rule to determine the positive direction for rotation around an axis. To determine the direction of rotation, point your thumb in the positive direction of the axis in question; your fingers then will curl in the positive rotation direction. You can use the Standard toolbar UCS flyout shown in Figure 9.16 to pick the options of the UCS command quickly.

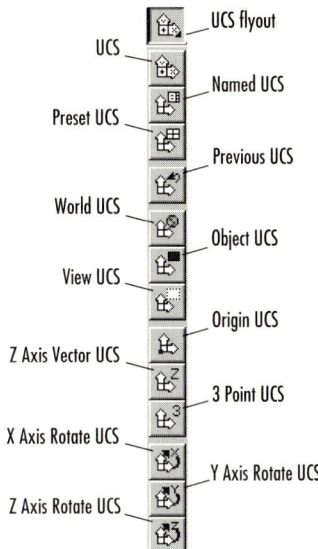

Figure 9.16

Pick: **X Axis Rotate UCS icon**

Rotation angle about X axis <0>: **90** ⏎

The grid in your drawing now appears parallel to the front view (lower left viewport). The broken pencil now appears in the top view (upper left viewport). Refer to Figure 9.17.

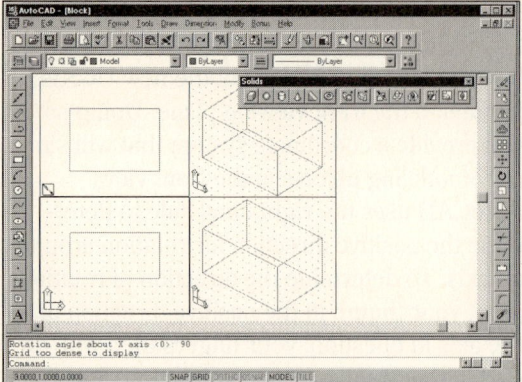

Figure 9.17

Now the XY-plane of the current coordinate system is aligned with the front view, but not the top view. You will save this UCS so that you can return to it later when you want to work in the front view. You will press ⊖ or the return button on your pointing device to repeat the UCS command, pick the Save option, and name the coordinate system FRONT because you see it in the front view. To restart the UCS command,

Command: ⊖

UCS Origin/ZAxis/3point/OBject/View/X/Y/Z/Prev/ Restore/Save/Del/?/<World>: **S** ⊖

?/Desired UCS name: **FRONT** ⊖

■ *TIP* The UCS command and the View command work well in conjunction with one another. After you have established a User Coordinate System, you can set up a view that aligns with it by using the Plan command and selecting the UCS option. This allows you to easily define a view that corresponds to any defined UCS. The View command lets you save a named view in the drawing. Use the View command to save the view with the same name as the UCS with which it is aligned. Save named views of any important viewing directions in your model. ■

Now that you have completed setting the viewing direction and UCS for the front view, make the lower right viewport active on your own. Now the crosshairs appear in the lower right viewport. Next, you will create a side view of the model in this viewport.

 You will use the Right View icon from the View flyout to select the viewing direction for this view.

Pick: **Right View icon**

The view changes to show a right-side view of the object. Again, the entire object will not fit in the viewport at the current zoom factor. You will use the Zoom Scale icon to zoom the view relative to paper space, as you did with the first two views.

Pick: **Zoom Scale icon** ⊖

Enter Scale factor: **.5XP** ⊖

Now the right-side view will fit the lower right viewport. But this viewport now has the broken pencil icon. You will create another UCS parallel to this view to make working in this viewport easy.

 The View UCS icon creates a UCS parallel to the viewing plane in the active viewport and at the origin of the Z-axis.

Pick: **View UCS icon**

The grid appears in the lower right viewport, and the broken pencil icon appears in the top and front views. Save this UCS so that you can return to it when you want to draw in the right-side view. Press ⊖ to repeat the previous UCS command.

Command: ⊖

Origin/ZAxis/3point/OBject/View/X/Y/Z/Prev/ Restore/Save/Del/?/<World>: **S** ⊖

?/Desired UCS name: **SIDE** ⊖

When you are finished, your screen should look like Figure 9.18.

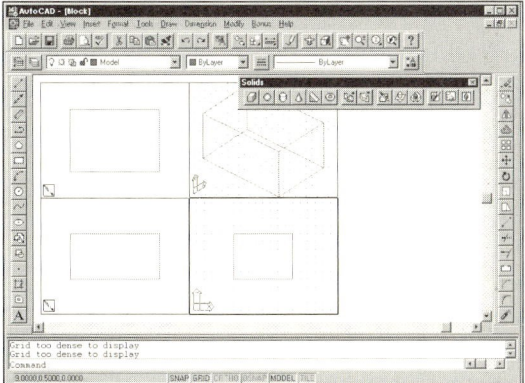

Figure 9.18

Make the upper left viewport active on your own before you continue. Restore the WCS to the top left viewport.

Command: **UCS** ⏎

Origin/ZAxis/3point/OBject/View/X/Y/Z/Prev/Restore/Save /Del/?/ <World>: ⏎ *(to accept the default of World)*

This action returns the drawing to the original WCS. Now the UCS icon shows the letter W and appears parallel in the top view. In

AutoCAD the default XY-plane is considered to be the *plan view* of the WCS. A plan view is basically a top view, which may sound familiar if you have worked with architectural drawings. Any time you want to restore the original coordinate system, set the UCS equal to the WCS with the command you just used.

Make the upper right viewport, in which the box appears as its 3D shape, active by moving the arrow cursor to this viewport and pressing the pick button. The crosshairs will appear in the viewport when you select it. On your own, pick the SE Isometric view from the Viewpoint flyout to change the view for this viewport. Use Zoom .8X to make the view 80% of the previous size so that it fills the upper right viewport.

In the next portion of the tutorial, you will familiarize yourself with the basic solid modeling primitives that you can use to create drawing geometry. You will add and subtract them with Boolean commands to create more complicated shapes.

Creating Cylinders

 Next, you will add a cylinder to the drawing. Later, you will turn the cylinder into a hole by using the Subtract command. You create cylinders by specifying the center of the circular shape in the XY-plane and the radius or diameter, then giving the height in the Z-direction of the current UCS. You can use the Baseplane option to change the height above the XY-plane that the circular shape is drawn in. The Elliptical option allows you to specify an elliptical shape instead of circular and then give the height.

The Center of other end option of the Cylinder command allows you to specify the center of the other end by picking or typing coordinates instead of giving the height.

> ■ *TIP* If you do not specify a Z-coordinate, it is assumed to be your current elevation, which is currently 0. ■

Pick: **Cylinder icon**

Elliptical/<center point> <0,0,0>: **4,4** ⏎

Diameter/<Radius>: **.375** ⏎

Center of other end/<Height>: **3** ⏎

Next, you will change the color of the cylinder. For many drawings the best approach is to make objects on separate layers and use the layer to determine the color, but that is not always the case with solid modeling. When you use a Boolean operator to join two objects, the result of that operation will always be on the current layer unless you change it later. Sometimes seeing features is easier if they are different colors, so you may want to set the color for an object before you join it with another object by using Boolean operations. To change the color of the cylinder, which you will use to create a hole,

Pick: (select the cylinder)

Pick: **Blue (from the Color Control pull-down list on the Object Properties toolbar)**

The cylinder should be changed to the new color. Your screen should look like Figure 9.19.

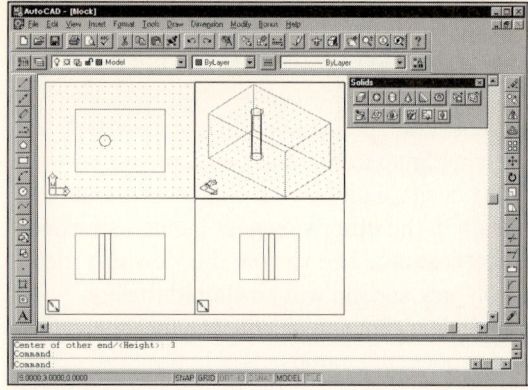

Figure 9.19

Now you have a box and a cylinder, each occupying the same space. Although they could not do so in the real world, in the drawing database these two objects are both occupying the volume inside the cylinder.

Setting Isolines

Before you continue, you will set the variable called Isolines. This variable controls the wireframe appearance of the cylinders, spheres, and tori (like a 3D donut) on the screen. You can set the value for Isolines between 4 and 2047, to control the number of tessellation lines used to represent rounded surfaces. Displayed on a curved surface, *tessellation lines* help you better visualize the surface. The number of tessellation lines that you set will be shown on the screen, representing the contoured surface of the shape. The higher the value for Isolines, the better the appearance of rounded wireframe shapes will be. The default setting of 4 looks poor but saves time in drawing. The highest setting may look best, but more time is needed for the calculations, especially for a complex drawing.

You will set the value to 12. You can change the setting for Isolines and regenerate the drawing at any time. You will use the Regenall command to regenerate all of the viewports to show the new setting for Isolines.

Command: **ISOLINES** ⏎

New value for ISOLINES <4>: **12** ⏎

Command: **REGENALL** ⏎

The cylinder in the upper right viewport should look similar to Figure 9.20.

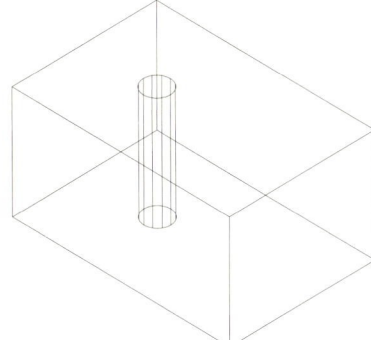

Figure 9.20

Next you will use the Boolean operator Subtract to remove the cylinder from the box so that it forms a hole.

Building Complex Solid Models

You can create complex solid models with Boolean operators, which find the *union* (addition), *difference* (subtraction), and *intersection* (common area) of two or more sets. These operations are named for Irish logician and mathematician George Boole, who formulated the basic principles of set theory. In AutoCAD the sets can be 2D areas (called *regions*) or they can be 3D solid models. Often *Venn diagrams* are used to represent sets and Boolean operations. Figure 9.21 will help you understand how the Union, Subtract, and Intersection commands work. The order in

which you select the objects is important only when you're subtracting (i.e., A subtract B is different from B subtract A).

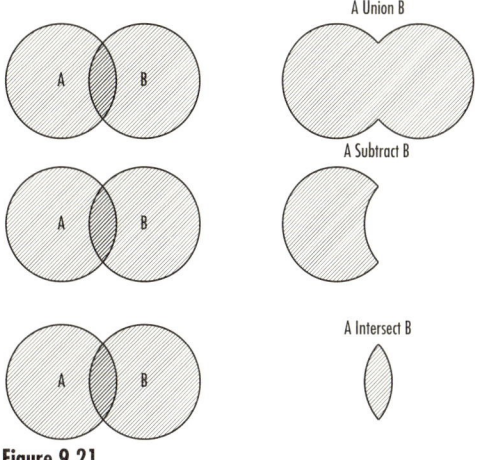

Figure 9.21

The Boolean operators are in the Modify menu, under Boolean.

You will use Subtract to remove the volume of the cylinder from the box, thereby forming a new solid model with a hole in it.

Pick: **Modify, Boolean, Subtract**

Select solids and regions to subtract from...

Select objects: **(pick the box and press ⏎)**

Select solids and regions to subtract...

Select objects: **(pick the cylinder and press ⏎)**

The resulting solid model is a rectangular prism with a hole through it. If you selected the items in the wrong order (i.e., picked the cylinder and subtracted the box from it) the result would be a null solid model or one that has no volume. If you picked both the box and the cylinder and then pressed ⏎, the command would not work unless you picked something else to subtract. If you make a mistake, you can back up by typing *U* ⏎ at the command prompt to use the Undo command and then try again.

Next, you will shade your model in the top right viewport.

Pick: **View, Render, Render**

The Render dialog box appears on your screen, as shown in Figure 9.22.

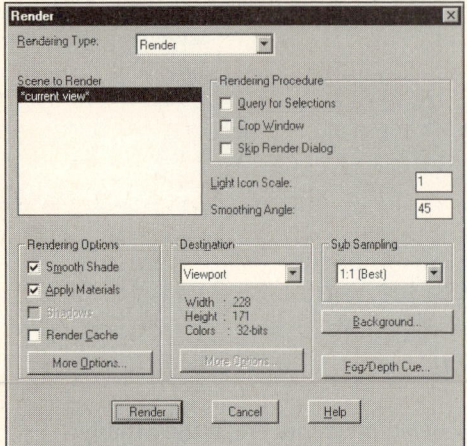

Figure 9.22

You will accept the defaults, using Render as the type of rendering to perform. You will learn more about using Render and its features in Tutorial 15.

Pick: **Render**

The model in the active viewport becomes shaded. Note that you can now see that the cylinder has formed a hole in the block. The color inside the hole is blue and the rest of the block is magenta.

Your drawing in the upper right viewport should look like that in Figure 9.23.

Figure 9.23

■ *TIP* The Facetres variable adjusts the smoothness of shaded objects and objects from which hidden lines have been removed. Its value can range from 0.5 (the default) to 10. Viewres controls the number of straight segments used to draw circles and arcs on your screen. Viewres can be set between 1 and 20000. You can type these commands at the command prompt and set their values higher to improve the appearance of rendered objects. ■

You will regenerate your drawing to eliminate the shading so that you can continue to work on it by typing *REGEN* at the command prompt on your own. (You can't select objects when they are shaded.)

Command: REGEN ←

Save your drawing, *block.dwg*, on your own before continuing. Saving periodically will prevent you from losing your drawing in the event of a power failure or other hardware problem. Also, saving your work after you've completed a major step and before you go on to the next step is useful. That way, if you want to return to the preceding step, you can open the previous version of the drawing, discarding the changes that you have made.

Creating Wedges

Now you will add a Wedge primitive to the drawing and then subtract it from the main block. The Wedge command has two options: Corner (the default) and Center. Wedge Corner first draws the rectangular base of the wedge, prompting you for two points that define the diagonal of the rectangular base. The height you enter starts at the first point specified for the base and gets smaller in the X direction toward the second point. The Center option asks you to specify the center point of the wedge you want to draw. As with other solid primitive commands, you can use the Baseplane option to select a different height, or Z level, at which to draw.

Pick: **Wedge icon**

Center/<Corner of Wedge> <0,0,0>: **8,6** ⏎

Cube/Length/<other corner>: **6,2** ⏎

Height: **3** ⏎

Now you will use the Boolean operators to subtract the wedge from the object.

Pick: **Modify, Boolean, Subtract**

Select solids and regions to subtract from...

Select objects: *(pick the box with the hole and press ⏎)*

Select solids and regions to subtract...

Select objects: *(pick the wedge and press ⏎)*

With the wedge subtracted, your drawing should look like that in Figure 9.24.

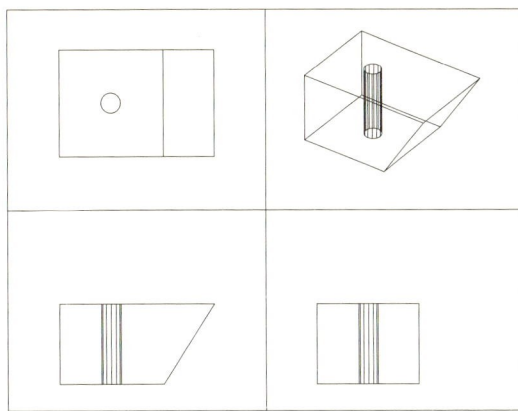

Figure 9.24

Next, you will use Hide to remove hidden lines from the upper right view. Be sure that the upper right viewport is active. Then,

Command: **HIDE** ⏎

The 3D object will appear on your screen with the hidden lines removed. The shape in the upper right viewport should look like that in Figure 9.25.

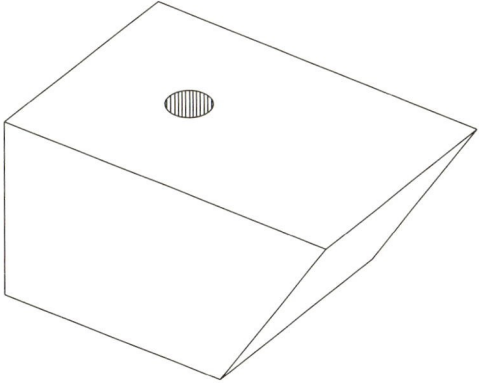

Figure 9.25

Regenerate your drawing display from the drawing database so that you can continue to work with wireframe.

Command: **REGEN** ⏎

Creating Cones

 The cone primitive creates a solid cone, defined by a circular or elliptical base and tapering to a point perpendicular to the base. It is similar to the cylinder primitive that you have already used. The circular or elliptical base of the cone is always created in the XY-plane of the current UCS. The height is along the Z-axis of the current UCS.

Now you will create a cone using the cone primitive and then subtract it from the block to make a countersink for the hole. It will have a circular base and a negative height. Specifying a negative height causes the cone to be drawn in the opposite direction from a positive height (i.e., in the negative Z-direction).

Pick: **Cone icon**

Elliptical/<center point> <0,0,0>: **4,4,3** ⏎

Diameter/<Radius>: **.625** ⏎

Apex/<Height>: **–.75** ⏎

Use the Boolean operator to subtract the cone from the block to make a countersink for the hole.

Pick: **Modify, Boolean, Subtract**

Select solids and regions to subtract from...

Select objects: *(pick the block and press* ⏎*)*

Select solids and regions to subtract...

Select objects: *(pick the cone and press* ⏎*)*

Next, shade the object in the upper right viewport using the Render command as you did before.

Pick: **View, Render, Render**

Pick: **Render**

The upper right viewport now displays a view of the shaded object. Your screen should look like Figure 9.26.

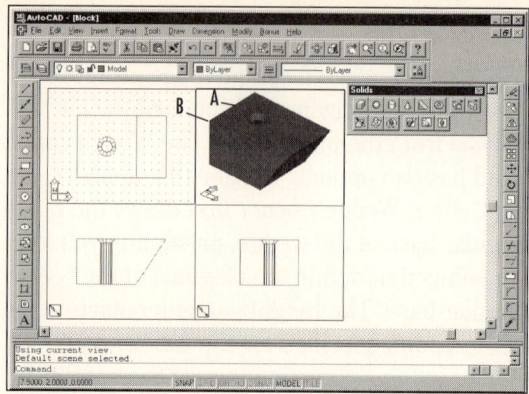

Figure 9.26

Next, you will use the cylinder primitive to add a rounded surface to the top of the block. On your own, use Regenall to regenerate the views.

Selecting a Named Coordinate System

 First you will restore the User Coordinate System SIDE that you saved earlier in this tutorial for drawing in the side view. Doing so will simplify creating objects parallel to the side view. You will use the Standard toolbar to pick the UCS flyout selection for Named UCS.

Pick: **Named UCS icon**

The UCS Control dialog box appears on your screen, as shown in Figure 9.27. You will use it to select the UCS for the side view.

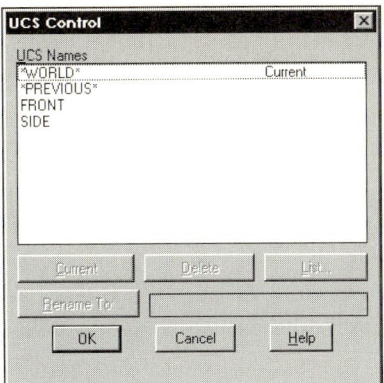

Figure 9.27

Pick: **SIDE**

Pick: **Current**

Pick: **OK**

The grid appears in the side view, and the broken pencil icon appears in the top and front views. In the isometric view in the upper right viewport, the grid changes to a different angle, as it now aligns with the side, not the top, of the object.

On your own, verify that the upper right viewport is active.

Next, you will use the Cylinder icon and choose to draw a cylinder by specifying its center. On your own, be sure that OSNAP is turned on, with midpoint and endpoint selected, to aid in object selection. Refer to Figure 9.26 for the points to select.

Pick: **Cylinder icon**

Elliptical/<center point> <0,0,0>: *(pick the back top line, labeled A)*

Diameter/<Radius>: *(pick the endpoint of the edge, labeled B)*

Center of other end/<Height>: **1** ⏎

Your drawing should look like the one shown in Figure 9.28.

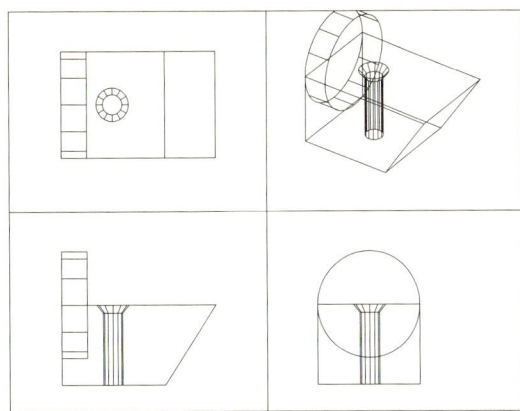

Figure 9.28

Next, you will use the Boolean operator Union to join the cylinder to the block.

Pick: **Modify, Boolean, Union**

Select objects: *(pick the new cylinder and the block)*

Select objects: ⏎

When the objects are joined, your drawing should be similar to Figure 9.29.

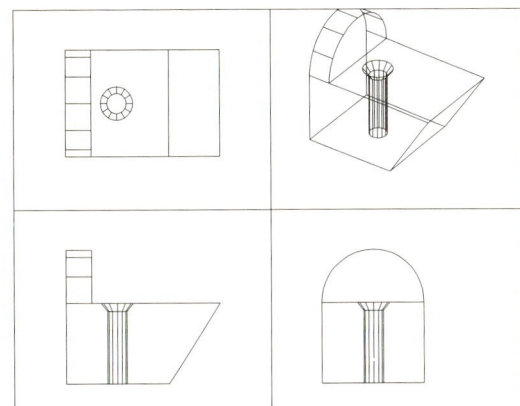

Figure 9.29

On your own, use the Pan command to pan the view of the object in the upper right viewport so that the entire object fits the available space.

Save your completed *block.dwg* drawing. Double-click the OSNAP toggle on the Status bar to turn it off.

> ■ *TIP* When you are picking on the solid model, you may have trouble at first deciding which surface to select because each wire of the wireframe represents not one surface, but the intersection of two surfaces. Many commands have the Next option to allow you to move surface by surface until the one you want to select is highlighted. If the surface you want to select is not highlighted, use the Next option until it is. When the surface you want to select is highlighted, press ↵. ■

Plotting Solid Models from Paper Space

Next, you will plot your multiview drawing of the block. Change to paper space on your own by typing *PS* ↵ at the command prompt. When you have done so, the paper space icon replaces the UCS icon.

You will use the Mview command with the Hideplot option so that when you plot your drawing from paper space, the back surface lines will be automatically removed. When selecting the objects for the Hideplot option of the Mview command, remember that you are selecting which *viewports* to show with hidden lines removed. To select a viewport, pick on its border, not inside it.

Command: **MVIEW** ↵

ON/OFF/Hideplot/Fit/2/3/4/Restore/<First Point>: **H** ↵

ON/OFF: **ON** ↵

Select objects: *(pick on the border of the upper right viewport)*

Select objects: ↵

Nothing noticeable happens when you finish the command. However, when you plot your drawing, the back surface lines will automatically be removed from this viewport.

On your own, make BORDER the current layer and freeze layer VPORT. This way the viewport borders won't print on your drawing. Use the Line command to draw a border around all the viewports while you are in paper space. Add a title strip and notes if you want to, as you did in Tutorial 5.

Be sure that you are in paper space and plot the drawing limits at a scale of 1=1. The views that you have drawn should be exactly half-size on the finished plot because the Zoom XP scale factor was set to 0.5. Your plotted drawing should be similar to that in Figure 9.30.

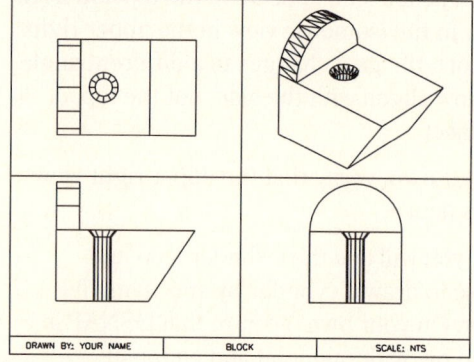

Figure 9.30

Saving Your Multiview Setup as a Template

Switch your drawing back to model space.

The UCS icons are restored to the viewports.

On your own, thaw layer VPORT and set layer MODEL as the current layer in the drawing.

Now you will erase the object and save the basic settings to use as a template drawing when creating new 3D drawings. Type the alias for the Erase command at the prompt.

Command: **E** ⏎

Select objects: *(pick the solid block object you have drawn)*

Select objects: ⏎

It is erased from all viewports.

 Next, you will restore the World Coordinate System, using the World UCS icon from the UCS flyout.

Pick: **World UCS icon**

Use Saveas to save this drawing to file name *soltemplate.dwt*.

■ *Warning:* Do NOT explode 3D models after you have created them, as it coverts them back to 3D lines and surfaces. Once you've exploded a 3D solid, you can't rejoin the resulting objects to form the 3D object again. If you need to make changes to a solid and can't do so with the techniques used in these tutorials, recreating the 3D solid is the best option. ■

Creating Solid Models with Extrude and Revolve

Next, you will learn how to create new solid objects, using the extrusion and revolution methods. Start a new drawing, using the New icon from the Standard toolbar. Use the drawing *solpro_d.dwg* provided with your data files as a template or the one that you just created, *soltemplate.dwt*. Call the new drawing *extrusn.dwg*.

Your screen should be similar to Figure 9.31.

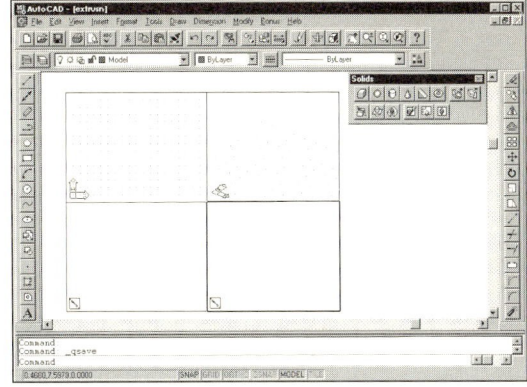

Figure 9.31

Because you started your new drawing from the solid modeling template, the viewports already exist. The UCSs called FRONT and SIDE that you saved previously are still available with the UCS Restore option. Layer MODEL should be the current layer.

On your own, pick in the upper left viewport to make it active. Verify that Grid and Snap are active and Snap is set to .5. You will draw the I shape shown in Figure 9.32, using the 2D Polyline command. The coordinates have been provided to make selecting the points easier for you.

Pick: **Polyline icon**

From point: *(pick point A)*

Current line-width is 0.0000

Arc/Close/Halfwidth/Length/Undo/Width/<Endpoint of line>: *(pick points B-L in order)*

Arc/Close/Halfwidth/Length/Undo/Width/<Endpoint of line>: **C** ⏎

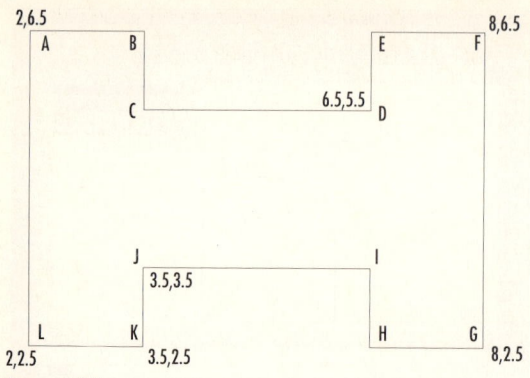

Figure 9.32

You will set the fillet radius to 0.25 and then use the Fillet command's Polyline option to round all the corners of the polyline.

Command: **FILLET** ⏎

Polyline/Radius/Trim/<Select first object>: **R** ⏎

Enter fillet radius <0.0000>: **.25** ⏎

Command: ⏎

FILLET Polyline/Radius/Trim/<Select first object>: **P** ⏎

Select 2D polyline: *(pick the polyline you just created)*

12 lines were filleted

> ■ *TIP* If you did not close your object by typing *C*, you will get the message *11 lines were filleted.* In this case, undo the last two commands and redraw your I shape, closing the polyline with the Close option. ■

On your own, use the Snap grid to draw the two circles shown in Figure 9.33.

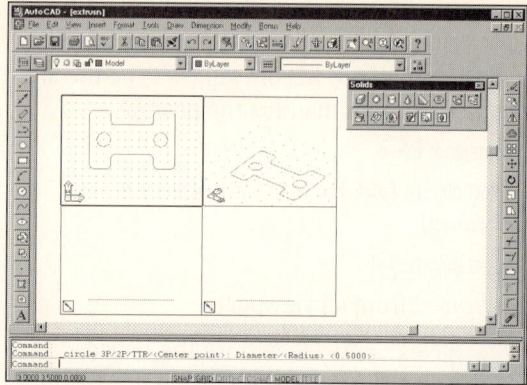

Figure 9.33

Region Modeling

You can also use the Boolean operators with the closed 2D shapes made by circles, ellipses, and closed polylines. You can convert closed 2D shapes to regions, which essentially are 2D solid models, or areas. Use the Region icon from the Draw toolbar. Region modeling and extrusion can be combined effectively to create complex shapes. Turn Snap off on your own to make selecting easier.

Pick: **Region icon**

Select objects: *(use implied Crossing to pick the polyline and both circles; then press ⏎)*

3 loops extracted

3 regions created

Now you are ready to subtract the two circles from the polyline region to form holes in the area.

Pick: **Modify, Boolean, Subtract**

Select solids and regions to subtract from...

Select objects: *(pick the polyline region and press ⏎)*

Select solids and regions to subtract...

Select objects: *(pick the circles and press ⏎)*

Nothing noticeable happens; however, now there is only one region, which has two holes in it. You can confirm this result by selecting the object. When you do so, note that the circles become highlighted as part of the object's outline.

Extruding a Shape

Now you can extrude this shape to create a 3D object. Extrusion is the process of forcing material through a shaped opening to create a long strip that has the shape of the opening. AutoCAD's extrusion command works similarly to form the shape. Closed 2D shapes such as splines, ellipses, circles, donuts, polygons, regions, and polylines can be given a height, or extruded.

Polylines must have at least three vertices in order to be extruded. You can specify a taper angle if you want the top of the extrusion to be a different size than the bottom. You can use the Path option to extrude a shape along a path curve that you have previously drawn. Path curves can be lines, splines, arcs, elliptical arcs, polylines, circles, or ellipses. You may encounter problems when trying to extrude along path curves with a large amount of curvature and where the resulting solid would overlap itself. The Extrude icon appears on the Solids toolbar.

Pick: **Extrude icon**

Select objects: (*pick the region and press* ⏎)

Path/<Height of Extrusion>: **2** ⏎

Extrusion taper angle <0>: ⏎

Now render the object in the upper right viewport. The solid object on your screen should look like that in Figure 9.34.

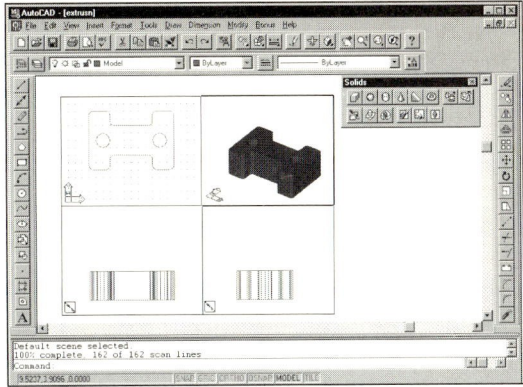

Figure 9.34

> ■ *TIP* Using the Pedit command to join objects into a closed polyline is an effective way to create 2D objects to extrude. ■

On your own, save your drawing, *extrusn.dwg,* and start a new drawing from the template provided with your data files, *solpro_d.dwg.* Call your new drawing *revolutn.dwg.*

Creating Solid Models by Revolution

Creating a solid model by revolution is similar in some ways to creating an extrusion. You can use it to sweep a closed 2D shape about a circular path to create a symmetrical solid model that is basically circular in cross section.

Use the 2D Polyline command on your own to draw the closed shape shown in Figure 9.35.

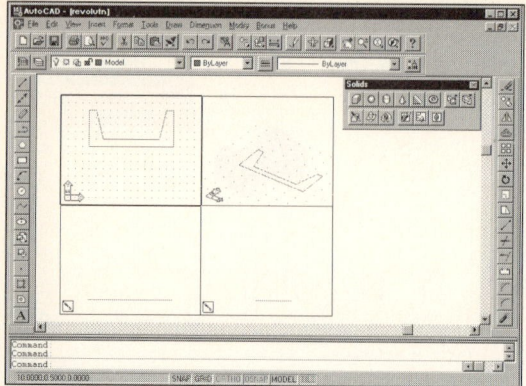

Figure 9.35

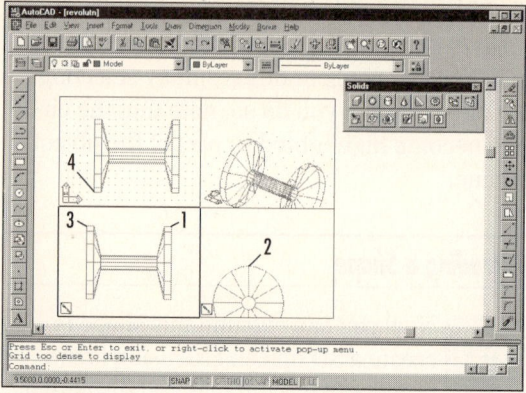

Figure 9.36

Now you will revolve the polyline about an axis to create a solid model. You don't need to draw the axis line; you can specify it by two points. The Revolve icon is on the Solids toolbar.

Pick: **Revolve icon**

Select objects: *(pick the polyline and press ⏎)*

Axis of revolution - Object/X/Y/<Start point of axis>:
   (pick one endpoint of the bottom line)

<End point of axis>: *(pick the other endpoint)*

Angle of revolution <full circle>: ⏎

Next, you will use the Pan Realtime command and Mvsetup to help align the views in the lower left, lower right, and upper left viewports. On your own, be sure that OSNAP is turned on, with quadrant selected, and pick to make the lower left viewport active. Then,

Pick: **Pan Realtime icon**

Move the solid until it is centered in the lower left viewport.

Your screen should now be similar to Figure 9.36.

Next, you will type the command Mvsetup and align the lower right and upper left views relative to the lower left viewport using the OSNAP quadrant. Use Figure 9.36 to help in making the selections.

Type: **MVSETUP ⏎**

Align/Create/Scale Viewports/Options/Title Block/Undo: **A ⏎**

Angled/Horizontal/Vertical alignment/Rotate view/
   Undo: **H ⏎**

Basepoint: *(select point 1, using the AutoSnap quadrant marker to help)*

Other point: *(select the lower right viewport and point 2, using the AutoSnap quadrant marker)*

The view in the lower right viewport should align with the view in the lower left viewport relative to the points you selected. Now you will repeat these steps for the upper left viewport.

Angled/Horizontal/Vertical alignment/Rotate view/
   Undo: **V ⏎** *(if this option line is not active, then press the return button to reactivate MVSETUP and choose align again)*

Basepoint: *(select point 3 using the AutoSnap quadrant marker to help)*

Other point: *(select the lower right viewport and point 4, using the AutoSnap quadrant marker to help)*

When you have finished aligning the view in the upper left viewport, use the Pan Realtime command again in the upper right viewport so that the view looks similar to Figure 9.37.

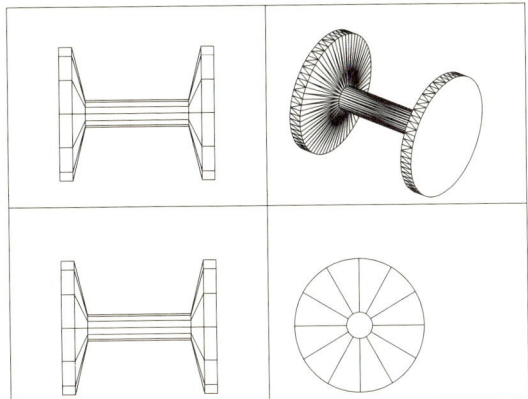

Figure 9.37

Save your drawing and then pick the Save As command to save the drawing with a new name. Call your new drawing *intsct.dwg*. On your own, erase the solid model that you just made with the Revolve command.

Using the Boolean Operator Intersection

Like the Union and Subtract Boolean operators you learned to use earlier in the tutorial, Intersection lets you create complex shapes from simpler shapes. Intersection finds only the area common to the two or more solid models or regions that you have selected. Next, you will create the shape shown in Figure 9.38 by creating two solid models and finding their intersection.

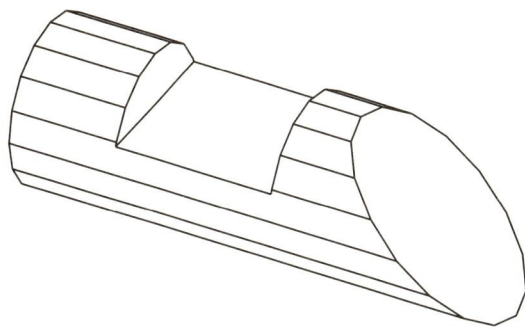

Figure 9.38

First, you will create the shape in the front view of a surface that you will extrude to create the angled face and notch. On your own, make the lower left viewport active and restore the UCS FRONT.

The grid appears in the lower left viewport, parallel to the front view. Use the Polyline command to create a polyline that defines the shape of the object in the front view, as shown in Figure 9.39.

Pick: **Polyline icon**

From point: **2.5,0** ↵

Current line-width is 0.0000

Arc/Close/Halfwidth/Length/Undo/Width/
    <Endpoint of line>: **2.5,1.5** ↵

Arc/Close/Halfwidth/Length/Undo/Width/
    <Endpoint of line>: **3.5,1.5** ↵

Arc/Close/Halfwidth/Length/Undo/Width/
    <Endpoint of line>: **3.5,1** ↵

Arc/Close/Halfwidth/Length/Undo/Width/
    <Endpoint of line>: **5,1** ↵

Arc/Close/Halfwidth/Length/Undo/Width/
    <Endpoint of line>: **5,1.5** ↵

Arc/Close/Halfwidth/Length/Undo/Width/
    <Endpoint of line>: **5.5,1.5** ↵

Arc/Close/Halfwidth/Length/Undo/Width/
    <Endpoint of line>: **7,0** ↵

Arc/Close/Halfwidth/Length/Undo/Width/
    <Endpoint of line>: **C** ↵

After you've drawn the shape of the object in the front view, you may need to use the Pan Realtime and Mvsetup commands so that you can see it in the top and side views and align them with the front view.

When you have finished, your drawing will look like the one shown in Figure 9.39.

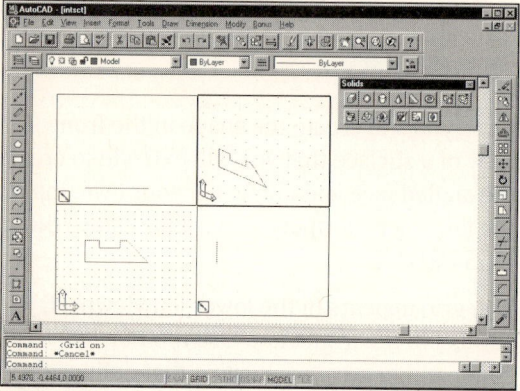

Figure 9.39

Now extrude this shape to form a solid model.

Pick: **Extrude icon**

Select objects: *(pick the polyline)* ⏎
Path/Height of extrusion: **−3** ⏎
Extrusion taper angle <0>: ⏎

The solid model shown in Figure 9.40 has been created.

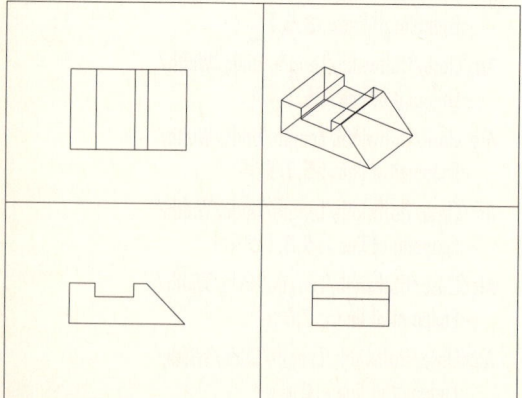

Figure 9.40

Next, you will switch to the side view to create the circular shape of the object, as shown in Figure 9.41.

On your own, pick in the lower right viewport to make it current. Restore UCS SIDE to activate the grid and coordinates parallel to the side view.

Pick: **Circle icon**

Circle 3P/2P/TTR/<center point>: **1.5,0.75** ⏎
Diameter/<Radius>: **0.75** ⏎

In the top view, front view, and isometric view of your entire object, the circle is drawn a distance away from the previously drawn solid model. You will use the Extrude command to elongate the circle into a cylinder.

Pick: **Extrude icon**

Select objects: *(pick the circle)* ⏎
Path/Height of extrusion: **10** ⏎
Extrusion taper angle <0>: ⏎

The isometric view of your drawing in the upper right viewport should look like that in Figure 9.41.

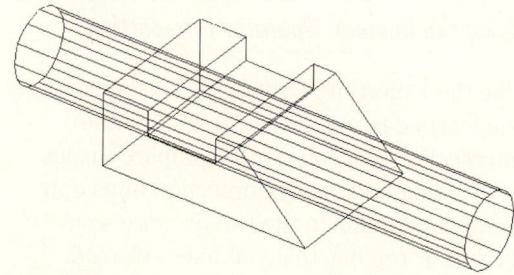

Figure 9.41

Now you are ready to use the Intersection command to create a new solid model from the overlapping portions of the two solid models that you have drawn.

Pick: **Modify, Boolean, Intersect**

Select objects: *(pick the cylinder and the extruded polyline)* ⏎

The solid model is updated. Your drawing should look like Figure 9.42.

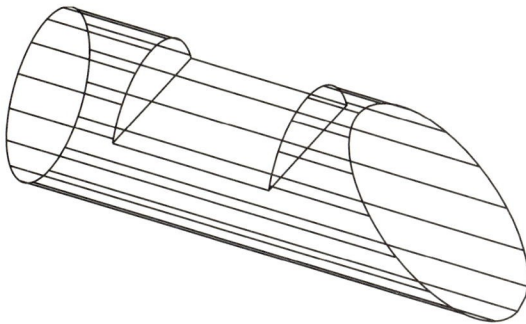

Figure 9.42

Save your *intsct.dwg* drawing.

> ■ *TIP* The Delobj variable controls whether the 2D object is automatically deleted after being extruded or revolved. A value of 1 means delete the object that was used to create the other object (the circle that created the cylinder, for example); a value of 0 means do not delete the object. The default setting in AutoCAD is 1 (delete). ■

On your own, experiment by typing *DELOBJ* at the command prompt and changing the setting to 0. Create a polyline and extrude it in one direction using a taper angle. The original polyline will remain. Select it again and extrude it in the other direction using the same taper angle. This way you can create a surface that tapers in both directions.

On your own, hide the Solids toolbar by picking on its close box.

Converting AME Solid Models

The Ameconvert command allows you to convert solid models created with AutoCAD's AME 2 or 2.1 solid modeler to ACIS solid models used in AutoCAD Release 14. To convert an AME model, type *AMECONVERT* ⏎ at the command prompt and then select the objects to convert. AutoCAD ignores objects that are not AME 2 or 2.1 solid models. You may notice a difference in the models after they've been converted. The ACIS modeler used by AutoCAD Release 14 is more accurate and may interpret some surfaces on the AME model as not meeting exactly or as holes not going all the way through the model.

The ACIS modeler has several advantages over the AME modeler. The ACIS models and mass property calculations performed on them are more accurate than the AME models and calculations. ACIS models can be created with more complex fillets and chamfers and with extrusion along a path curve to easily create a wider variety of shapes. The AME models had the advantage of Constructive Solid Geometry (CSG) storage, so the individual primitives could be edited even after being joined in Boolean operations. The ACIS models are stored in the drawing database with Boundary Representation (B-Rep), and therefore the primitives are not available for editing after being joined with Boolean operations.

Creating Surface Models

You can also use AutoCAD to create surface models. AutoCAD's surface models are composed of a faceted polygonal mesh that approximates curved surfaces. Surface modeling is more difficult to use than solid modeling and provides less information about the object. The reason is that only the surfaces and not the interior volumes of the object are described in the drawing database. However, surface modeling is well suited to applications such as modeling 3D terrain for civil engineering applications.

In general you should not mix solid modeling, surface modeling, and wireframe modeling, as you can't edit them in the same ways to create a cohesive single structure. Select the single method that is best for your application.

You can convert solid models to surface models by using the Explode command. Surface models can be likewise converted to wireframe models. However, because they do not contain the same information, you can't go from wireframe to surface models or from surface models to solid models.

Next, you will show the Surface toolbar. On your own, use the View menu bar selection and

Pick: **Toolbars, Surfaces *(then close the Toolbars dialog box)***

The Surfaces toolbar appears on the screen as shown in Figure 9.43.

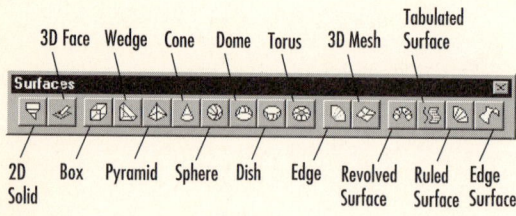

Figure 9.43

The Surface toolbar contains selections for 2D Solid, 3D Face, Box, Wedge, Pyramid, Cone, Sphere, Dome, Dish, Torus, Edge, 3D Mesh, Revolved Surface, Tabulated Surface, Ruled Surface, and Edge Surface. Many of the selections are similar to the selections on the solids toolbar. However, the objects created with surface modeling are like empty shells. They do not contain information about the volume and mass properties of the object, as solid models do. An important consideration is that you cannot use Boolean operators to join surface models. You must edit the mesh that creates surface models differently.

Because using Box, Wedge, Pyramid, Cone, Sphere, Dome, Dish, Torus, Revolved Surface, and Tabulated Surface is basically similar to the method you used earlier in the tutorial to create solid models, they will not be covered here. Instead, on your own, refer to AutoCAD's Help command for further information about creating these shapes.

The 2D Solid selection lets you create a solid surface using four points. When this surface is parallel to the screen, it is filled with a solid color.

The 3D Face selection lets you add a three- or four-sided surface anywhere in your drawing. It can be used to create surfaces on top of wireframe drawings.

The Edge command lets you change the visibility of an edge of a 3D face, which you created with the previous command. You can use it to hide edges that join with other 3D faces or to hide back edges that may not hide correctly when you use the Hide command.

In this tutorial, you will learn to create a 3D Mesh, an Edge Surface, and a Ruled Surface. Understanding how to create these objects will allow you to create a wider variety of shapes that are especially useful for modeling 3D terrain, such as mountainous topography.

On your own, start a new drawing from the template drawing *contrdat.dwg*, which was provided with the data files. Name your new drawing *surf1.dwg*. When you have finished, your screen should look like Figure 9.44.

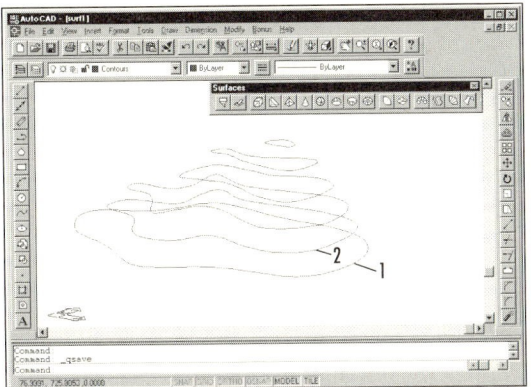

Figure 9.44

On the screen are 2D splines representing contours (lines of equal elevation on a contour map). The splines have been moved along the Z-axis so that they are at different heights. The bottom line is at Z 200, the next at Z 220, the next at Z 240, and so on. You can create these contour lines on your own by drawing splines (or polylines) through your data points, joining points of equal elevation. Then use Move to relocate the resulting spline. Specify a base point of (0,0,0) and a displacement of (0,0,200) (or whatever your elevation may be). If needed, change the viewpoint and use Zoom so that you can see the lines clearly.

On your own, make layer MESH current.

Creating a Ruled Surface

Using the Ruled Surface selection from the Surfaces toolbar, you can create a surface mesh between two graphical objects. The

objects can be lines, points, arcs, circles, ellipses, elliptical arcs, 2D polylines, 3D polylines, or splines. The objects can either be both open (like lines) or both closed (like circles). Points can be connected to either open or closed objects.

Controlling Mesh Density

The mesh comprising a surface is defined in terms of a *matrix* of M and N vertices; M and N specify the column and row locations of vertices. Two system variables control the density of the mesh for creating surfaces. Surftab1 controls the density of the mesh in the M direction, and Surftab2 controls the density of the mesh in the N direction. The larger the value for Surtab1 and Surftab2, the more tightly the generated mesh will fit the initial objects selected. As with other commands, increasing the density of the mesh increases the time for calculation and display of the object.

The Surftab1 variable controls the density of the mesh for Ruled Surfaces because they are always a 2 x Surftab1+1 mesh. To set the value for Surftab1,

Command: **SURFTAB1** ⏎

New value for SURFTAB1 <6>: **40** ⏎

 Next, you will create a ruled surface between the two bottom splines. Use the Surfaces toolbar to select Ruled Surface. Refer to Figure 9.44 for the lines to select.

Pick: **Ruled Surface icon**

Select first defining curve: *(pick 1)*

Select second defining curve: *(pick 2)*

The ruled surface will be added between the two splines as shown in Figure 9.45.

Figure 9.45

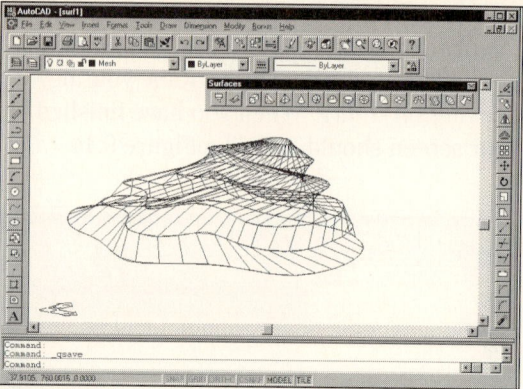

Figure 9.46

> ■ *TIP* When you are picking the defining curves, pick in a similar location on each curve. If you pick on the opposite end of the second curve, the generated mesh will be intersecting. ■

On your own, repeat the command for Ruled Surfaces between curves 2 and 3, 3 and 4, 4 and 5, and 5 and 6 shown in Figure 9.45. You may need to zoom in so that you can pick the red splines. (Remember, Zoom is transparent and can be used during the command.) If you pick on the blue mesh, you will not be able to use it as a defining curve. When you have finished, the drawing on your screen will look like Figure 9.46. Note that for the upper curves, half the mesh for the ruled surface is connected to each curve. To create an accurate model, sometimes you must change Surftab1 to vary the mesh refinement to avoid so much interpolation.

On your own, freeze layer CONTOURS to remove the original splines from the screen and save your drawing.

Start a new drawing on your own from the template *edgsurf.dwg*. Name your new drawing *surf2.dwg*. On your own, make MESH the current layer. Your screen should look like Figure 9.47.

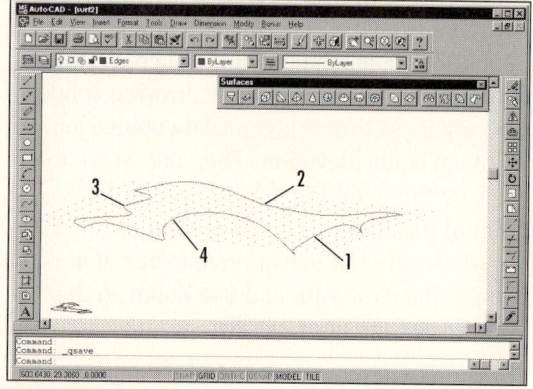

Figure 9.47

You will use the four 3D splines shown on the screen to define an edge surface.

Creating an Edge Surface

The Edge Surface selection from the Surfaces toolbar creates a mesh defined by four edges you select. The mesh density for an edge surface is controlled by the Surftab1 and the Surftab2 variables. The Edge Surface selection creates a Coons patch mesh. A *Coons patch mesh* is a *bicubic surface* interpolated between the four edges. First, you will set the values for Surftab1 and Surftab2 to control the mesh density. Then you will select Edge Surface from the Surfaces toolbar. Refer to Figure 9.47 for the splines to select.

Command: **SURFTAB1** ⏎
New value for SURFTAB1 <6>: **20** ⏎
Command: **SURFTAB2** ⏎
New value for SURFTAB2 <6>: **20** ⏎
Pick: **Edge Surface icon**
Select edge 1: *(pick 1)*
Select edge 2: *(pick 2)*
Select edge 3: *(pick 3)*
Select edge 4: *(pick 4)*

The edge surface appears in your drawing as shown in Figure 9.48.

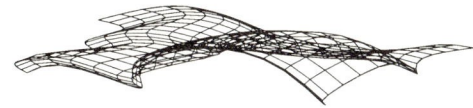

Figure 9.48

Save your drawing on your own.

Next, you will define a rectangular mesh. You will erase the objects from the screen and use Save As to give the drawing a new name.

Pick: **Erase icon**
Select objects: **ALL** ⏎
Select objects: ⏎

All the objects are erased from the screen. On your own, use the Save As command to save the drawing with the name *mesh*. The name *mesh* appears in the title bar at the top of the current drawing.

Creating 3D Mesh

The 3D Mesh command lets you construct rectangular polygon meshes by entering the X-, Y-, and Z-coordinates of the points in the mesh. First, you define the number of columns (M) and then the number of rows (N) for the mesh matrix. The values for M and N must fall between 2 and 256. After you define M and N, AutoCAD prompts you to type the values for the vertices defining the mesh. You can often use a Lisp program or script in conjunction with the 3DMesh command to automate creation of the mesh.

Pick: **3D Mesh icon**
Mesh M size: **3** ⏎
Mesh N size: **3** ⏎
Vertex (0, 0): **0,0,0** ⏎
Vertex (0, 1): **15,0,1** ⏎
Vertex (0, 2): **30,5,2** ⏎
Vertex (1, 0): **0,15,0** ⏎
Vertex (1, 1): **18,18,3** ⏎
Vertex (1, 2): **30,15,0** ⏎
Vertex (2, 0): **0,30,2** ⏎
Vertex (2, 1): **15,30,0** ⏎
Vertex (2, 2): **30,34,3** ⏎

The mesh appears in your drawing. You will change the size of your drawing so that you can see it in its entirety. You will use Zoom Extents to do this.

 You will change your view so that you can see the mesh clearly in your drawing.

Pick: **Zoom Extents icon**

Pick: **SW Isometric View icon**

Your drawing should appear similar to Figure 9.49, which has the coordinates of the vertices noted.

Now you know how to establish viewports and viewing directions and create and join the basic shapes used in solid modeling. With these tools you can create a variety of complicated shapes. In Tutorial 10, you will learn how to change solid models further, as well as to plot them. Practice creating shapes and working with the User Coordinate Systems on your own.

You've now completed Tutorial 9. Save your drawing and close toolbars before exiting AutoCAD.

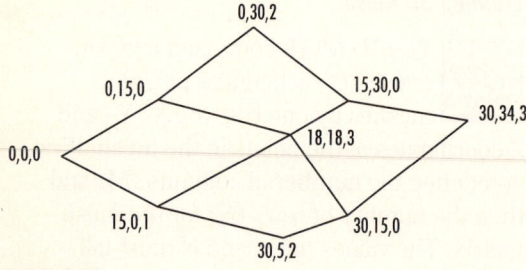

Figure 9.49

bicubic surface
Boolean operators
Coons patch mesh
coordinate system locator
difference
extrusion
globe
intersection
mass properties

matrix
perspective icon
plan view
primitives
regions
revolution
solid modeling
surface modeling
tessellation lines

union
User Coordinate System (UCS)
vector
Venn diagrams
viewpoint
wireframe modeling
World Coordinate System (WCS)

KEY COMMANDS

3DMesh
Box Corner
Cylinder
Delobj
Edge Surface
Extrude
Facetres
Front View
Hide
Intersection
Isolines
Mview

Named UCS
Paper Space
Plan
Regenerate All
Region
Render
Revolve
Right View
Ruled Surface
Southwest Isometric View
Subtract
Surftab1

Surftab2
Top View
Union
UCS
Viewres
View UCS
Vpoint
Wedge Center
Wedge Corner
World UCS
Zoom Extents

EXERCISES

Use solid modeling to create the parts shown according to the specified dimensions.

 9.1 Connector

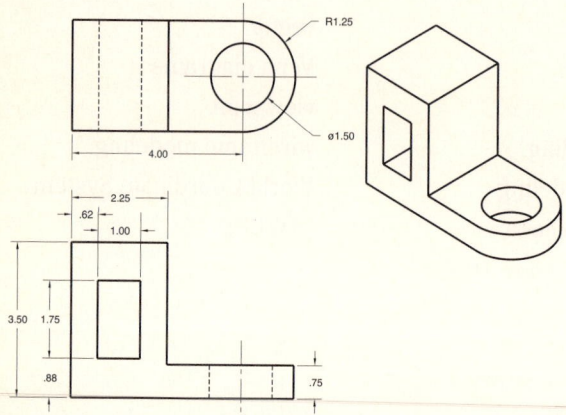

 9.2 Angle Link

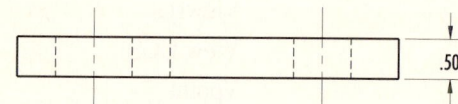

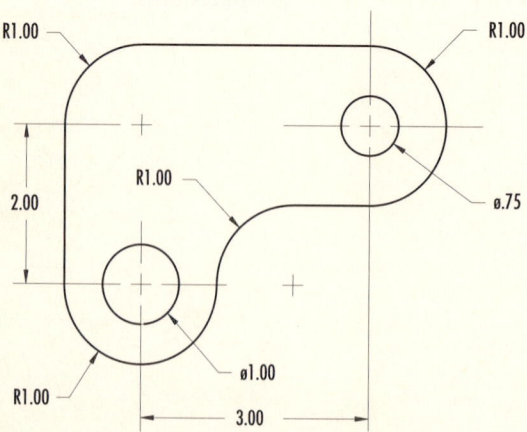

9.3 Support Base

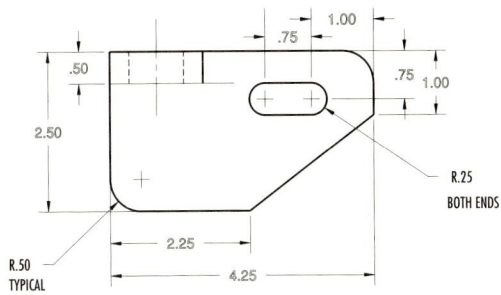

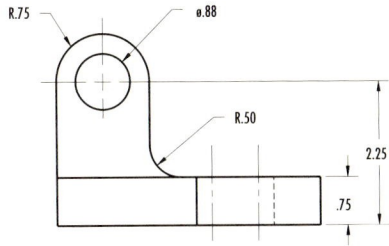

 ## 9.4 Chess Piece

Create the rook chess piece body by revolving a polyline. Use Subtract to remove box primitives to form the cutouts in the tower. Add an octagon for the base. Extrude it to a height of 0.15 and use a taper angle of 15°. (Use your *soltemplate.dwt* as the template from which you start the rook.)

Polyline used for revolution

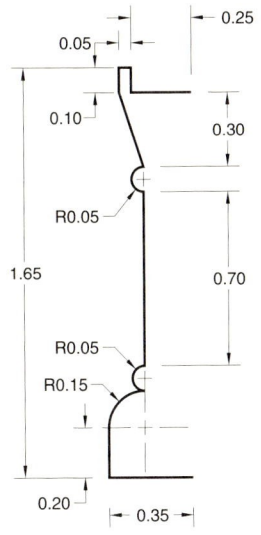

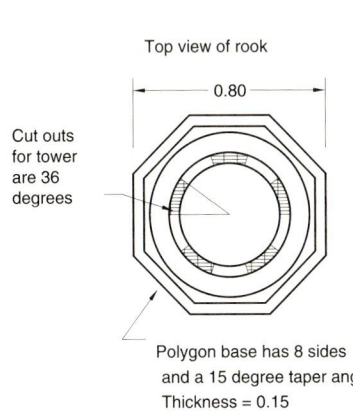

Top view of rook

Cut outs for tower are 36 degrees

Polygon base has 8 sides and a 15 degree taper angle. Thickness = 0.15

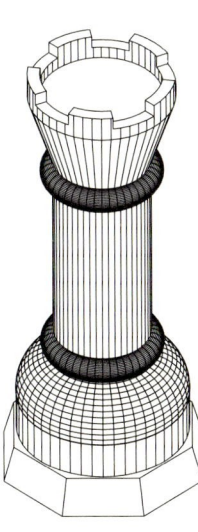

Draw the following shapes by using solid modeling techniques. The letter M after an exercise number means that the units are in millimeters.

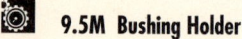 **9.5M Bushing Holder**

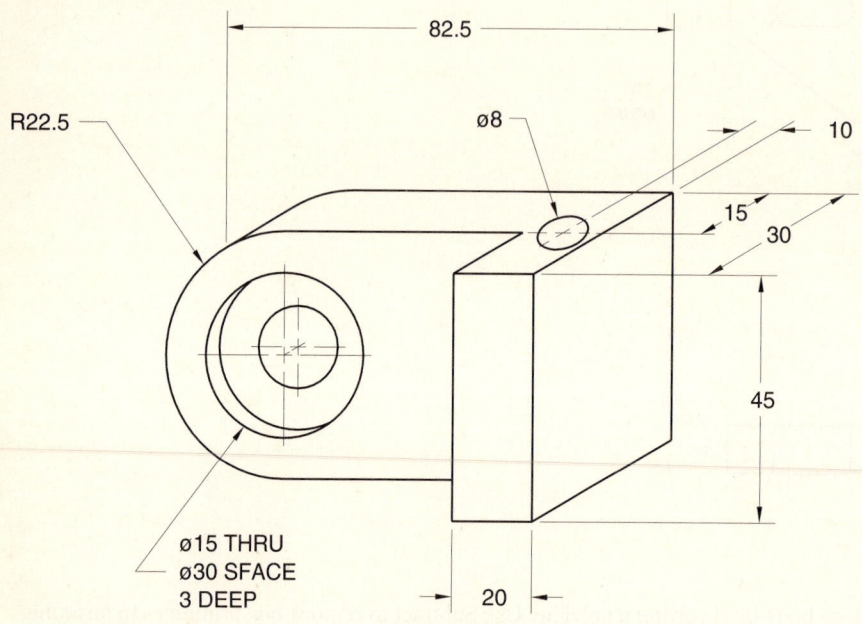

R22.5

82.5

ø8

10

15

30

45

20

ø15 THRU
ø30 SFACE
3 DEEP

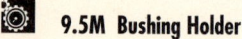 **9.6 Shaft Support**

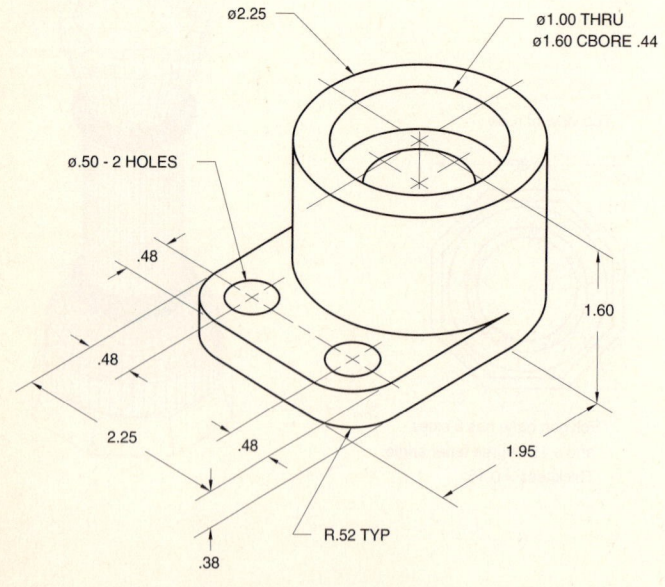

ø2.25

ø1.00 THRU
ø1.60 CBORE .44

ø.50 - 2 HOLES

.48

.48

2.25

.48

1.60

1.95

.38

R.52 TYP

9.7 Balcony

Design a balcony like the one shown here, or a more complex one, using the solid modeling techniques you have learned. Use "two-by-fours" (which actually measure 1.5" x 3.5") for vertical pieces. Use Divide to ensure equal spacing.

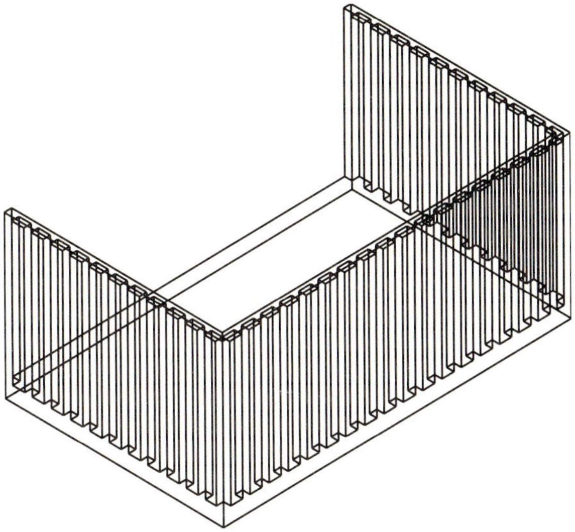

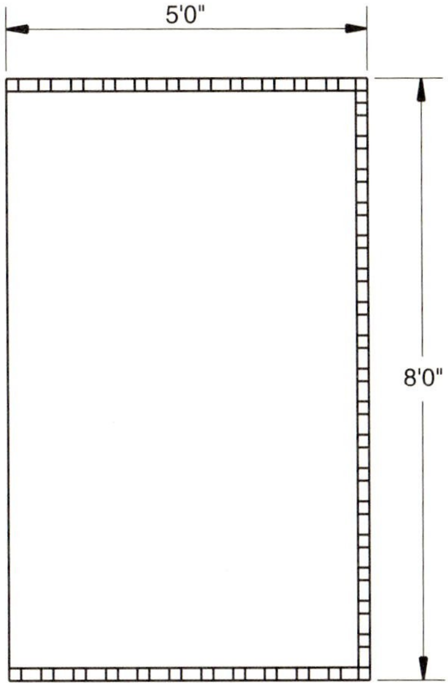

5'0"

8'0"

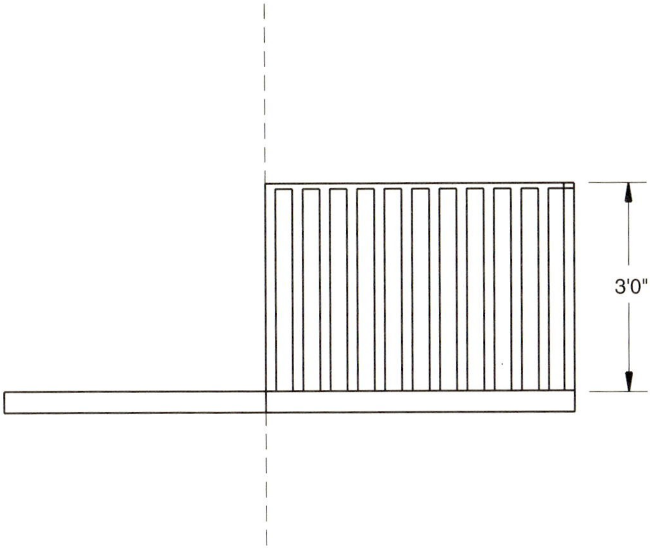

3'0"

9.8 Bridge

Create the bridge shown. First, determine the size of the bridge that you want and the area to be left open with the arc. Use extrusion to create the arc. Add the structure of the bridge, making sure that the supports are evenly spaced and of sufficient height. (You can enter one-half of the supports and then use Mirror.)

Box primitive

Extruded ARCs

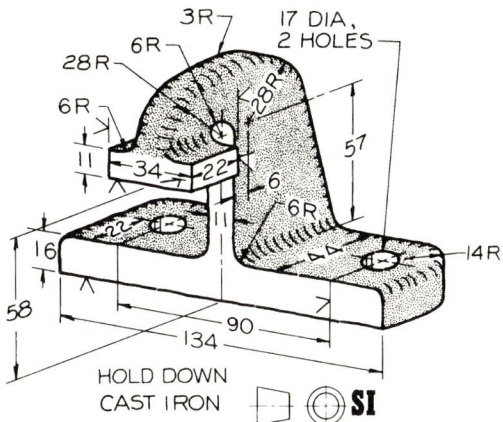

3R
17 DIA,
2 HOLES
6R
28R
28R
57
6R
11
34 — 22
6
6R
16
22
11
6R
14R
58
134 — 90

HOLD DOWN
CAST IRON **SI**

9.10 Rock Plan

Use the information in the file *rockplan.pts* to create the plan and profile drawings for the railroad bed as shown below. The file has the information in the following order: point number, northing (the y-coordinate to the north from the reference point), easting (the x-coordinate to the east from the last reference point), elevation, and description. Start by locating each point at the northing and easting given. Extrapolate between elevations and draw polylines for the contour lines at 1' intervals. Make the contour lines for 95' and 100' wide, so that they stand out. Create a 3D mesh using the information in the file *rockplan.pts* modeling the terrain shown.

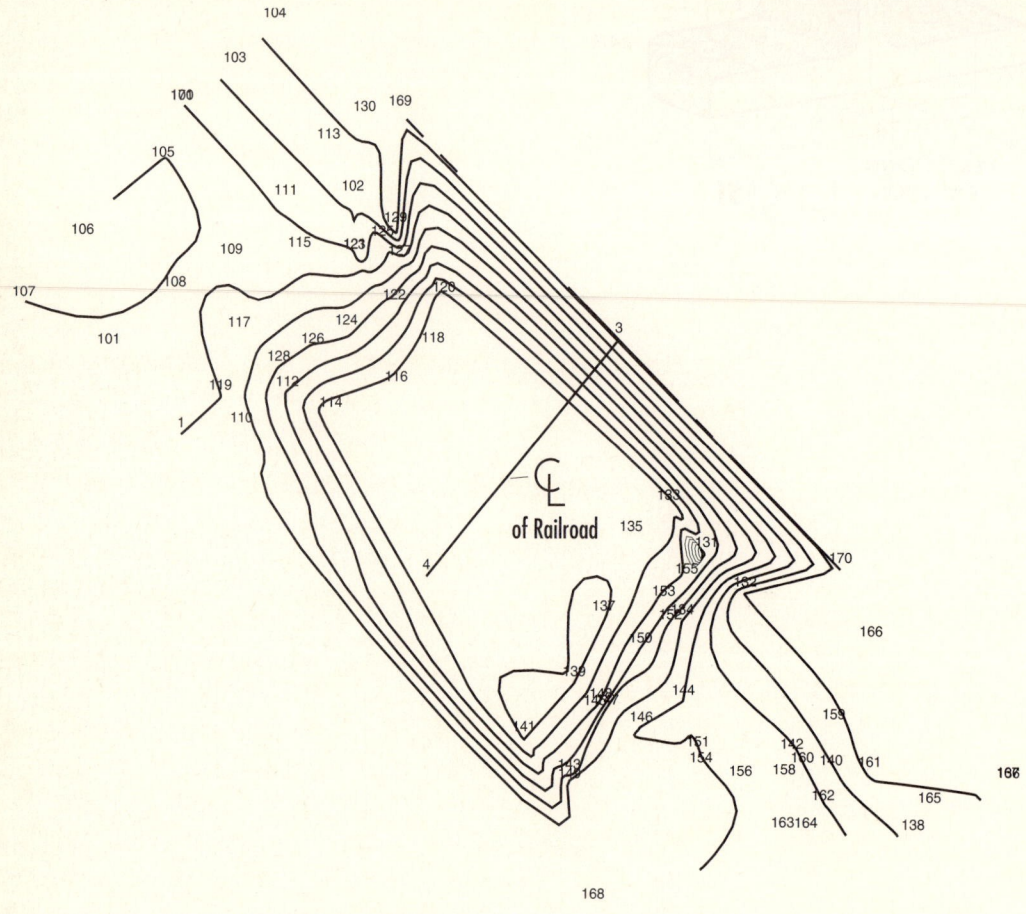

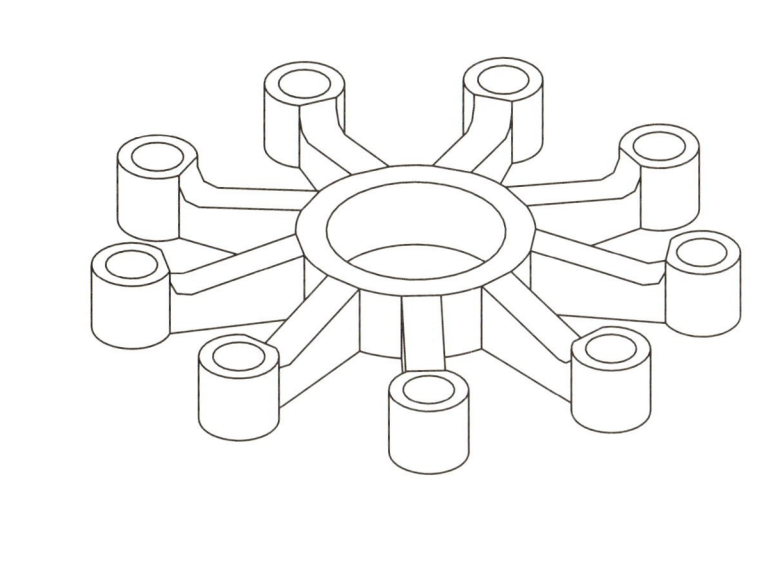

TUTORIAL 10

Changing and Plotting Solid Models

Objectives

When you have completed this tutorial, you will be able to

1. Add a rounded fillet between two surfaces.

2. Add an angled surface or chamfer to your model.

3. Remove a portion from a solid.

4. List the solid information and tree structure.

5. Plot 2D views of the model with hidden lines shown correctly.

6. Control visibility of layers within each viewport.

Introduction

You have learned how to create solid models and use multiple viewports. In this tutorial you will learn to use the solid modeling editing commands to make changes and create a larger variety of shapes. You can create solid models for many engineering drawing needs. From these models you can directly generate two-dimensional (2D) orthographic views that correctly show hidden lines. You will learn a method for controlling layer visibility that allows you to add centerlines so that they appear in a single view. You can also use this method for adding dimensions that appear only in a single viewport. In future tutorials you will learn how to apply solid modeling to create many different standard types of engineering drawings.

Starting

Launch AutoCAD. From the Start Up dialog box, open the existing drawing *block.dwg*, which you created in Tutorial 9. Or you can open the drawing *Solblock.dwg* from the data files provided as a template and save it as a new drawing. Your screen should be similar to Figure 10.1.

Open the Solids toolbar on your own and position it to the right of the drawing so it will be available for use in this tutorial.

■ **TIP** As you work through this tutorial, you may need to redraw or regenerate the drawing to eliminate partially shown lines or shading. Use Redraw to eliminate partially shown lines in a single viewport. Use RedrawAll to redraw all the viewports at the same time. The Redraw commands refresh the screen from the current display file. The Regen commands recalculate the display file from the drawing database and refresh the screen. Use Regen to regenerate the current viewport or RegenAll to regenerate all the viewports at once. To make selecting easier, remember the 'Zoom Window command. Transparent zooming allows you to zoom a view during another command. You can use it to enlarge an area to make selecting easier. To return the area to the original size, always use 'Zoom Previous, not 'Zoom All. When using transparent Zoom, don't forget to add the apostrophe in front of the command, select from the toolbar, or select from View on the menu bar to use the transparent version of the command. ■

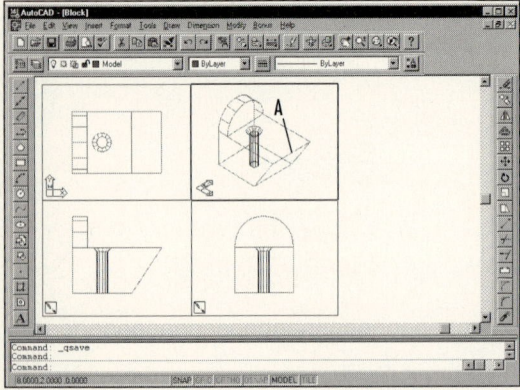

Figure 10.1

Using Fillet on Solids

AutoCAD's Fillet command lets you add concave or convex rounded surfaces between plane or cylindrical surfaces on an existing solid model. You have already used it to create rounded corners between 2D objects. Now you will use the command to create a rounded edge for the front, angled surface of the object. Refer to Figure 10.1 for your selection of point A.

Pick: **Fillet icon**

Polyline/Radius/Trim <Select first object>: *(pick on line A)*

Enter radius <0.0000>: *.5* ↩

Chain/Radius <Select Edge>: ↩

Your drawing with the rounded edge added should look like Figure 10.2.

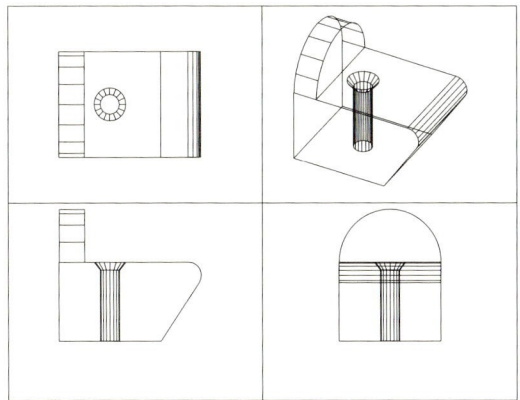

Figure 10.2

Next, you will undo the fillet and then use the Chamfer command to add an angled surface.

Command: **U** ↩

The fillet that you added to your drawing is eliminated.

Using Chamfer on Solids

The Chamfer command works on solid models in a way similar to the Fillet command, except that it adds an angled surface instead of a rounded one. Its Loop option allows you to add a chamfer to all edges of a base surface at once.

Pick: **Chamfer icon**

Polyline/Distance/Angle/Trim/Method/
    <Select first line>: **D** ↩

Enter first chamfer distance<0.0000>: *.75* ↩

Enter second chamfer distance<0.7500>: ↩

Command: ↩

Polyline/Distance/Angle/Trim/Method/<Select first line>:
    (pick line A in Figure 10.1)

Select base surface/Next/ <OK>: *(if the top surface of the object is highlighted, press* ↩*; if not, type N so that the next surface becomes highlighted)*

Enter base surface distance<0.7500>: ↩

Enter other surface distance<0.7500>: ↩

Loop/Select edge: *(pick line A)* ↩

Loop/Select edge: ↩

Your screen should look like Figure 10.3.

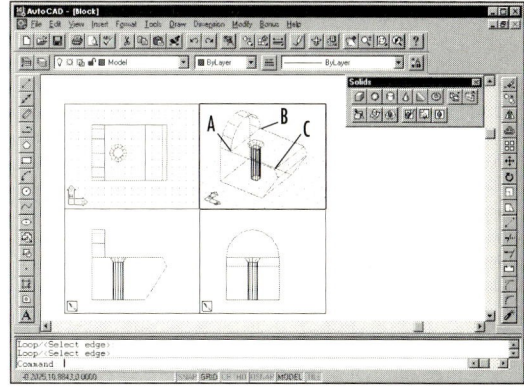

Figure 10.3

On your own, undo the Chamfer command before continuing.

Using Slice

Using the Slice command, you can cut a solid using a specified cutting plane. The cutting plane does not have to be an existing drawing entity. You can specify the *cutting plane* in several ways.

Option	Function
Object	Select an existing planar entity
Zaxis	Specify a point for the origin and a point that gives the direction of the Z-axis of the cutting plane
View	Align a plane parallel to the current view and through a point
XY, YZ, ZX	Choose a plane parallel to the XY-, YZ-, or ZX-coordinate planes and through a point
3points	Choose three points to define a plane

The options for the cutting plane, also called the *slicing plane*, are the same as those presented whenever the CP, or *construction plane*, prompt appears. When you have specified the cutting plane, AutoCAD prompts you to pick a point on the side of the plane where you want the object to remain. The portion of the object on the other side is then deleted. You can choose the Both sides option so that the object is cut into two pieces, with neither side being deleted.

In this example you will use the 3points option to slice the existing object.

You will use Slice to separate the rounded top surface from the block itself. On your own, be sure that running object snap Endpoint is on. Refer to Figure 10.3 for your selections.

Pick: **Slice icon**

Select objects: *(select the block)*

Select objects: ⏎

Slicing plane by Object/Zaxis/View/XY/YZ/ZX/<3points>: ⏎

1st point on plane: *(pick endpoint A using AutoSnap)*

2nd point on plane: *(pick endpoint B using AutoSnap)*

3rd point on plane: *(pick endpoint C using AutoSnap)*

Both sides/<Point on desired side of the plane>: **B** ⏎

Now if you select the half-cylinder that used to be the top surface of the block, it will be separate from the block. The grips on the cylinder would appear separate from the block, as shown in Figure 10.4.

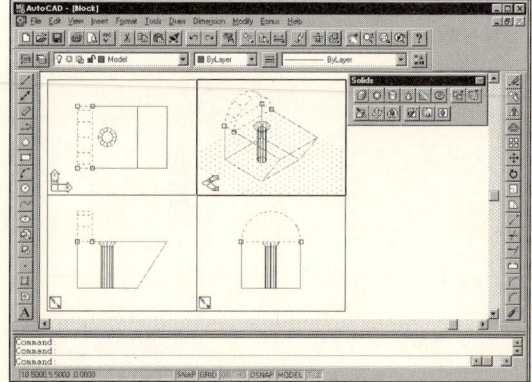

Figure 10.4

Next, you will move the half-cylinder up one unit in the Z-direction and then create a new rectangular piece beneath it.

> ■ *TIP* Many of the editing commands that you learned in the 2D tutorials are still effective in solid modeling. Don't be afraid to try them when they would be useful. ■

Pick: **Move icon**

Select objects: *(pick on the half-cylinder that you separated from the block, unless it is already selected)*

Select objects: ⏎

Base point or displacement: **0,0,1** ⏎

Second point of displacement: ⏎

The cylinder is moved upward 1 unit, as shown in Figure 10.5.

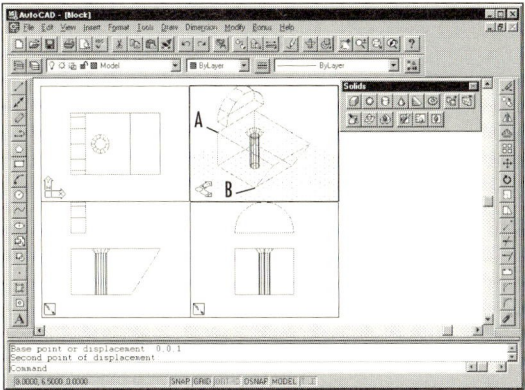

Figure 10.5

Now you will create two rectangular boxes. One will be immediately below the cylinder and the second will be where you subtracted a wedge in Tutorial 9. The WCS should still be active. Use Figure 10.5 for your selections.

Pick: **Box icon**

Center/<Corner of box> <0,0,0>: *(pick corner A in Figure 10.5, using the AutoSnap endpoint marker)*

Cube/Length/<other corner>: **L** ⏎

Length: **1** ⏎

Width: **4** ⏎

Height: **1** ⏎

Command: ⏎

Center/<Corner of box> <0,0,0>: *(pick corner B in Figure 10.5, using the AutoSnap endpoint marker)*

Cube/Length/<other corner>: **L** ⏎

Length: **2** ⏎

Width: **4** ⏎

Height: **3** ⏎

The boxes should appear on your screen as shown in Figure 10.6.

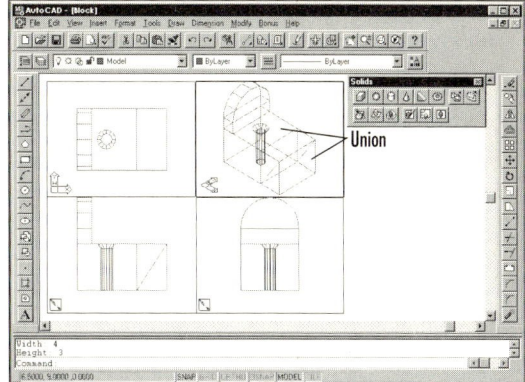

Figure 10.6

Now you will use Union to join the block and the second box you created (where you subtracted the wedge in Tutorial 9).

Pick: **Modify, Boolean, Union**

Select objects: *(used implied crossing to select ONLY the block and the second box, using Figure 10.6 for reference)*

Select objects: ⏎

The two objects should now be joined, forming one object similar to that in Figure 10.7.

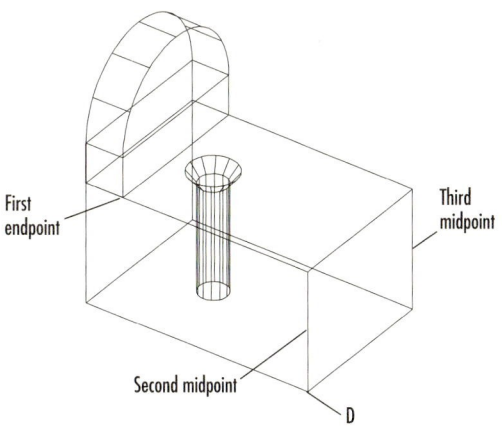

Figure 10.7

Now you will use the 3points Slice option again to slice an angled portion from the block. Refer to Figure 10.7 for the three points.

Pick: **Slice icon**

Select objects: *(select the block object)*

Select objects: ⏎

Slicing plane by Object/Zaxis/View/XY/YZ/ZX/<3points>: *(target the endpoint labeled first endpoint, using the AutoSnap endpoint marker)*

2nd point on plane: *(target the front vertical edge labeled Second midpoint, using the AutoSnap midpoint marker)*

3rd point on plane: *(target the front vertical edge labeled Third midpoint, using the AutoSnap midpoint marker)*

Both sides/<Point on the desired side of the plane>: *(pick on the bottom left corner, labeled D)*

The upper portion of the block is cut off, leaving an angled surface, as shown in Figure 10.8.

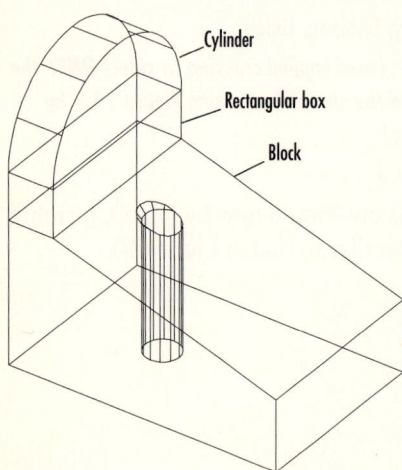

Figure 10.8

Now use the Union command to join the three parts.

Pick: **Modify, Boolean, Union**

Select objects: *(use implied crossing to select all three objects)*

Select objects: ⏎

Your drawing should look like that in Figure 10.9.

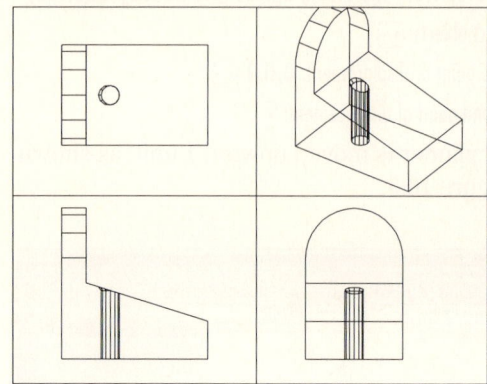

Figure 10.9

Note that the object is not centered well inside the upper right viewport. You will use the Zoom Scale command to view the object from farther away.

Be sure that the upper right viewport is active before continuing.

Pick: **Zoom Scale icon**

Enter scale factor: **.4XP** ⏎

The view should fit within the viewport now, and the drawing should appear as shown in Figure 10.10.

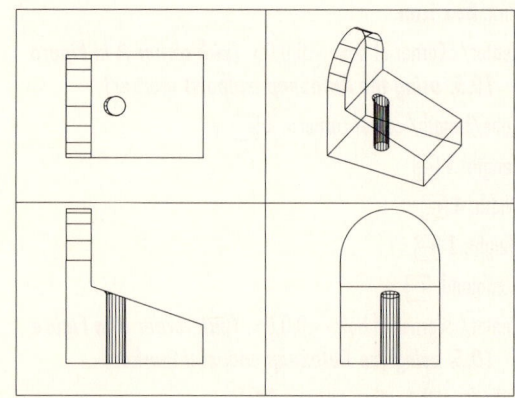

Figure 10.10

The views in the other three viewports should line up. If they don't appear to, then use the command MVSetup and Align, as you did in Tutorial 9.

Next, you will add a 0.375 deep hole of diameter 0.5, perpendicular to the angled surface.

To create a hole perpendicular to the angled surface, you will create a User Coordinate System that aligns with the angled surface. To do so, you will use the UCS 3 point command to locate the origin and X- and Y-axes.

The upper right viewport should already be active. OSNAP should still be active, with Endpoint turned on, so that you can select accurately during the following commands. Refer to Figure 10.11 as you pick points.

Pick: **3 Point UCS icon**

Origin point <0,0,0>: *(target the lower left corner of the angled surface, point 1 in Figure 10.11)*

Point on positive portion of the X axis <9.0000, 2.0000, 1.5000>: *(target the lower right corner of the angled surface, point 2)*

Point on positive-Y portion of the UCS XY plane <7.0000, 2.0000, 1.5000>: *(target the upper left corner of the angled surface, point 3)*

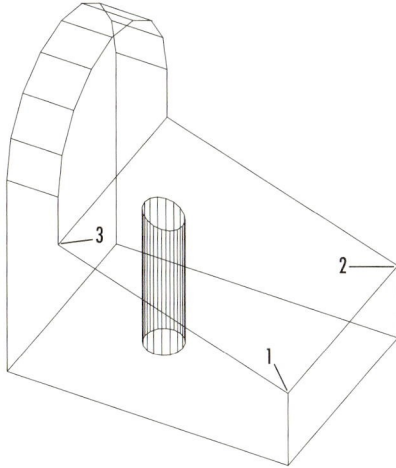

Figure 10.11

The grid in all the views changes so that it now aligns with the angled surface.

■ *TIP* Always keeping the grid turned on helps you see how the UCS is aligned. You can also look at the UCS icon, but often the grid is more noticeable. ■

Now use the Plan command to align your view so that you are looking straight down the Z-axis. The resulting view shows the XY-plane of the new UCS that you created in the previous step. The Plan command will fit the drawing extents to the viewport.

Command: **PLAN** ↵

<Current UCS>/UCS/World: ↵

The view in the upper right viewport should now be directly perpendicular to the angled surface on the object. Your drawing should look like that in Figure 10.12.

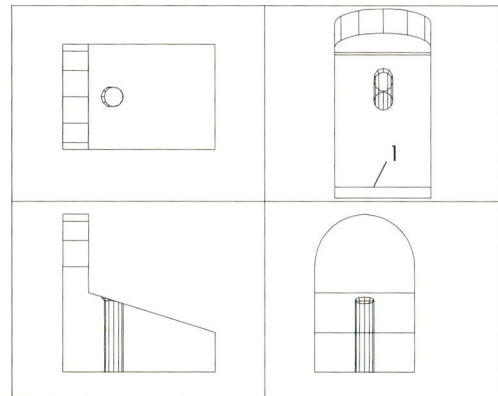

Figure 10.12

Save the view and the UCS so that you can restore them later if you want to.

Pick: **UCS icon**

Origin/ZAxis/3point/Object/View/X/Y/Z/Prev/Restore/Save /Del/?/<World>: **S** ↵

?/Desired UCS name: **Angle** ↵

Command: **VIEW** ⏎

?/Delete/Restore/Save/Window: **S** ⏎

View name to save: **Angle** ⏎

> ■ *TIP* AutoCAD has a built-in geometry calculator that lets you specify points that would be impossible to specify by other drawing means. Whenever you come to a prompt requesting input of a point, real number, integer, or vector, you can use the transparent command 'CAL. 'CAL lets you write an expression in infix notation, such as 1+2*sin(30), at the prompt and the resultant value is used. You can also use OSNAP functions in combination with the geometry calculator. The geometry calculator is useful in creating complex models. ■

Creating the Hole

You will use the cylinder primitive and then subtract it to create the hole.

You will also use the Snap From icon to aid in the location of the cylinder. Remember that the Snap From icon is on the Object Snap flyout. On your own, open the Solids toolbar if it is not already active and make sure that the running object snap Midpoint is active. Use Figure 10.12 for your selection point.

Pick: **Cylinder icon**

Elliptical/<Center point><0,0,0>: *(pick the Snap From icon)*

Base point: *(select point 1, using AutoSnap)*

<Offset>: **1** ⏎

Diameter/<Radius>: **D** ⏎

Diameter: **0.5** ⏎

Center of other end/<Height>: **−.375** ⏎

Subtract the cylinder to form a hole in the object.

Pick: **Modify, Boolean, Subtract**

Select objects: *(pick the large block)*

Select objects: ⏎

Select solids and regions to subtract...

Select objects: *(pick the short cylinder you just created)*

Select objects: ⏎

Now you will use the World UCS icon to change back to World Coordinates.

Pick: **World UCS icon**

Reset the viewpoint for the upper right viewport with the Vpoint command.

Command: **VPOINT** ⏎

Rotate/<View point><0.2873, 0.0000,
 0.9578>: **1.5,−3, 2.4** ⏎

A view similar to the original view should be restored. Name this view 3D, using the View command.

Command: **VIEW** ⏎

?/Delete/Restore/Save/Window: **S** ⏎

View name to save: **3D** ⏎

On your own, use the Zoom Scale icon with a scale factor of .4XP to make the image fit better within the viewport. Your drawing should now look like the one shown in Figure 10.13.

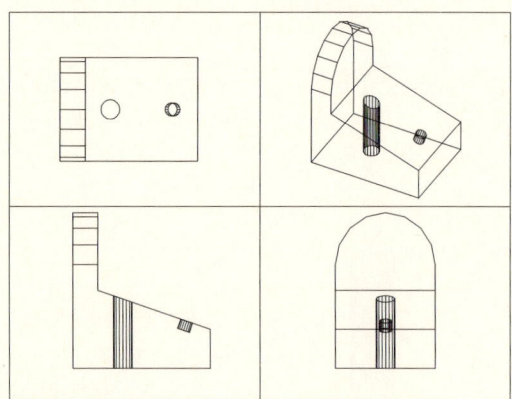

Figure 10.13

Use the Save As selection from the File menu to save the drawing. Name your drawing *anglblok*.

Plotting with Hidden Lines

 To generate views that plot correctly using hidden lines, you can use the Setup Profile (SOLPROF) command to create a 2D projection of each view. AutoCAD automatically generates separate layers for hidden and visible lines for each viewport. The tangential edges, or multiple lines, that make up circular features can also be automatically deleted. To profile the object in each viewport, first make the upper right viewport active and verify that the Solids toolbar is turned on. Next,

Pick: **Setup Profile icon**

Select objects: *(pick on an edge of the object in the upper right viewport)*

Select objects: ⏎

Display hidden profile lines on a separate layer? <Y>: ⏎

Project profile lines onto a plane? <Y>: ⏎

Delete tangential edges? <Y>: ⏎

One solid selected.

You will see lines drawn over the original colored lines that designate the edges of your model in the current viewport. The default color for the profile lines is white (which appears black on a light background).

Next, pick the upper left viewport to make it active.

Then repeat the process to create a 2D profile of the object in the upper left viewport.

Pick: **Setup Profile icon**

Select objects: *(pick on an edge of the object in the upper left viewport)*

Select objects: ⏎

Display hidden profile lines on a separate layer? <Y>: ⏎

Project profile lines onto a plane? <Y>: ⏎

Delete tangential edges? <Y>: ⏎

One solid selected.

Again, lines appear over the original colored lines in this viewport.

On your own, repeat this process for the two remaining viewports.

When you have completed profiling the object in each viewport, you will see only white lines (which appear black on a light background) and colored tessellation lines on the screen. The Setup Profile command creates new layers for the lines of the profile, and the default color setting is white (or black). Because you responded "Yes" to the prompt for displaying hidden profile lines on a new layer, two new types of layers exist in the drawing—one for visible profile lines and one for hidden profile lines.

AutoCAD automatically names the layers for visible profile lines. The layer names for profiled visible lines start with the letters PV-. The layer names for profiled hidden lines start with the letters PH-. The remaining portion of the layer name is based on the handle number of the viewport that the object was profiled in.

To get the handle number for each viewport, switch to paper space and use the List command. Later you will use this information selectively to freeze and thaw the profiled lines for the desired layers.

On your own, change the current setting to paper space.

Note that the Paper Space icon, a triangle, is now displayed in the lower left corner of your screen. Next, you will list the information for the viewports. To select a viewport, you must pick on its border and not just somewhere inside the viewport.

Pick: **List icon**

Select objects: **(pick the border of the upper right viewport)**

Select objects: ⏎

The information shown in Figure 10.14 is displayed on your screen.

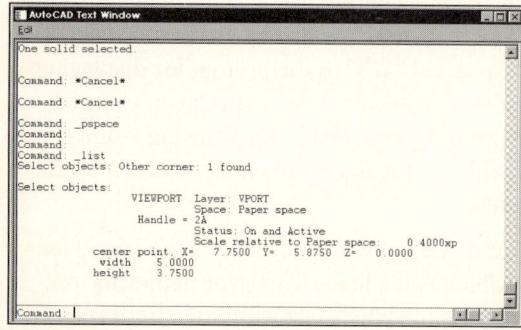

Figure 10.14

You will write the handle number for the upper right viewport in the space provided. The handle numbers begin with hexadecimal number 1, which is assigned to the first object you create in your drawing, and then proceed

with hexadecimal numbers for each additional entity. (Hexadecimal representation for decimal number 10 is A; 11 is B; and so on.) The handle number that you see in your drawing may be different from that listed in Figure 10.14, depending on how you created each object in the drawing.

Pick: **(the close box to remove the text window)**

Pick: **List icon**

Select objects: **(pick the border of the upper left viewport)**

Select objects: ⏎

The information and handle number for the upper left viewport are listed on the text screen. Write the handle number for the upper viewports in the space provided. Then write the handle numbers for the lower left and lower right viewports on your own.

Upper Right Viewport Handle Number:

Upper Left Viewport Handle Number:

Lower Left Viewport Handle Number:

Lower Right Viewport Handle Number:

You will need to keep track of these numbers in order to freeze and thaw profile layers in any particular viewport.

Loading the Linetypes

You must load the linetypes before they can be used in the next steps.

On your own, return to model space and use the Linetype icon on the Object Properties toolbar to load the Hidden and Center linetypes for use in the drawing. If you need help, refer to Tutorial 4 for details on loading linetypes.

Setting the Linetypes and Colors for Generated Layers

 Now you will use the Layer & Linetype Properties dialog box to set the color and linetype for the new layers.

Pick: **Layers icon**

The Layer & Linetype Properties dialog box should appear on your screen. Note that eight more layers are listed in the dialog box than were previously. These are the generated layers for the profiles that AutoCAD created when you used the Setup Profile, or SOLPROF, command. You will use the dialog box to set the color and linetype of the generated hidden line layers. Remember to hold down the Shift key in order to select more than one layer name at a time.

Pick: **(each of the layer names that start with PH-)**

Pick: **Continuous (under the heading Linetype, across from one of the selected layers)**

Pick: **Hidden**

Pick: **OK (to exit the Select Linetype dialog box)**

Pick: **(the Color box across from one of the selected layers)**

Pick: **Blue**

Pick: **OK (to exit the Select Color dialog box)**

Pick: **OK (to exit the Layer & Linetype Properties dialog box)**

You have finished setting the color blue and linetype HIDDEN for the generated hidden line layers. Now you will freeze the generated visible line layers. Each layer name begins with the letters PV- and ends with the hexadecimal number for the viewport in which the profile was created.

Using the techniques that you learned earlier in this tutorial, you will freeze all layers starting with a PV- using the Layer Control pull-down list on the Object Properties toolbar. On your own, perform that task so that the pull-down list appears as shown in Figure 10.15

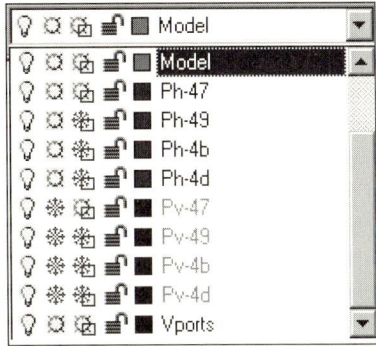

Figure 10.15

The linetype and color for those layers are already set to CONTINUOUS and white, so you do not need to change them.

The PH- and PV- layers are frozen in the current viewport, which AutoCAD does automatically for you when generating the layers; otherwise, each 2D profile would appear in every viewport, creating a lot of confusion.

On your own, set layer 0 as the current layer and freeze the layer MODEL so that you will not see the original solid model that you created underneath the hidden lines. Doing so will make selecting specific lines easier as you continue to work on your drawing.

When you return to your drawing, only blue hidden lines and the viewport borders should be visible on the screen, as shown in Figure 10.16.

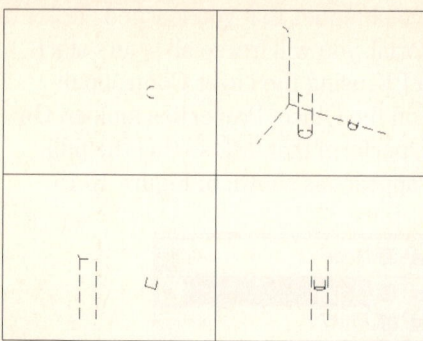

Figure 10.16

Eliminating Duplicate Hidden Lines

When printing or plotting your drawing, it is not good practice to double-plot lines. Plotting visible lines over hidden lines in the drawing would cause these lines to appear darker than normal. To eliminate the unwanted hidden lines, you will explode the blocks in which they were created and then erase the unwanted lines. Recall that you shouldn't explode the actual model, as you can't rejoin the separate objects to form the model again. To ensure that this does not happen, you have frozen the layer MODEL on which you created the actual model. The lines remaining on your screen are the profile lines.

 Before you can erase the unwanted hidden lines, you must use the Explode icon to turn the block of hidden lines back into single entities.

Pick: **Explode icon**

Select objects: *(pick on the hidden lines in the upper right viewport)*

Select objects: ⏎

The lines should now be separate entities, and you can select any one of them. On your own, explode the block of the hidden lines in each of the other viewports.

Now you can erase the outer hidden lines and any others over which visible lines would fall. You will erase all of the lines in the isometric view except those showing the depth of the blind hole. Hidden lines are not usually shown in that view unless they are needed to show the depth of a hole or some other feature that cannot be inferred without them. In an isometric view, holes are assumed to be through holes unless the depth is indicated in the drawing by hidden lines, a visible back surface showing the depth, or a note.

When you have finished, your drawing should look like that in Figure 10.17.

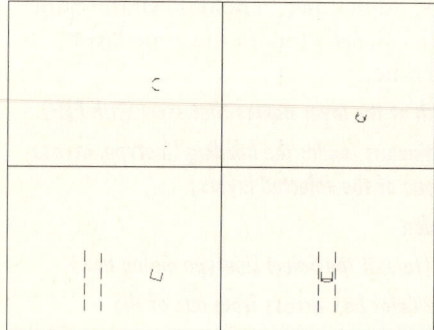

Figure 10.17

Now you will thaw the profile visible layers. On your own, use the Layer Control pull-down list on the Object Properties toolbar to do so.

Pick on the lower left viewport to make it active before continuing.

Next, you will restore the UCS for the front view.

Pick: **Named UCS icon**

The UCS Control dialog box appears on your screen. Select the UCS for the front view and then make it current, as shown in Figure 10.18.

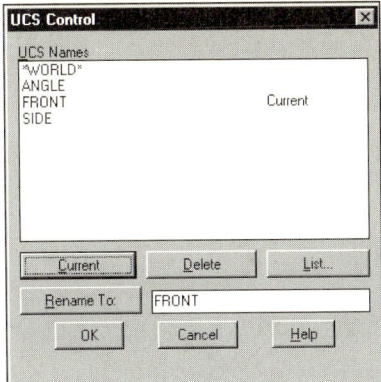

Figure 10.18

When you have restored the UCS for the front view, the grid will appear in the lower left viewport. Now you are ready to draw the centerline for the front view.

Your screen should look like Figure 10.19.

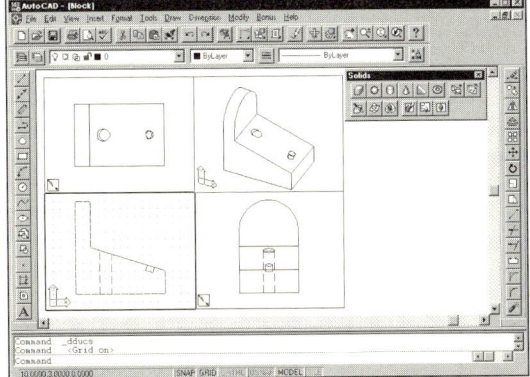

Figure 10.19

Adding the Centerlines

Next you will draw the centerlines for the front view. First you will create a new layer named CL-FRONT for the front view centerlines, using the Layers icon.

Pick: **Layers icon**

Pick: **New**

Type: **CL-FRONT**

Pick: **(select the Color box to the right of the new layer)**

Pick: **Green (from within the Select Color dialog box)**

Pick: **OK (to exit the Select Color dialog box)**

Pick: **(select the linetype CONTINUOUS to the right of the new layer)**

Pick: **CENTER (on your own, load CENTER if it has not been loaded yet)**

Pick: **OK (to exit the Select Linetype dialog box)**

Pick: **OK (to exit the Layer & Linetype Properties dialog box)**

On your own, be sure that Snap is turned on and use Snap and the Line icon to draw a centerline for the through hole in the front view. When you have finished, your drawing should be similar to Figure 10.20.

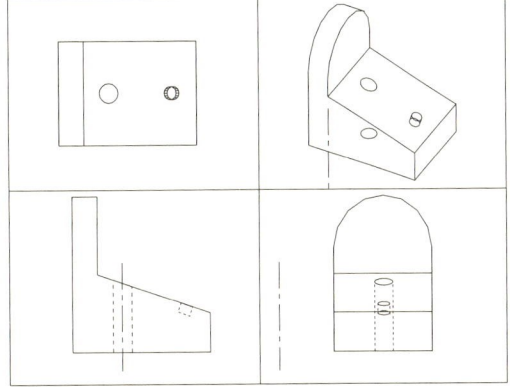

Figure 10.20

What happened? The line that you drew appeared in the other viewports. Anything you add in model space will be visible in all the views unless you tell AutoCAD otherwise. You can use Freeze/Thaw in the Current Viewport selections in the Layer Control pull-down list to turn off the layers in specific viewports.

Using Freeze/Thaw in Current Viewport

When you add lines such as centerlines and dimensions to 3D drawings, you usually have to create layers for each viewport. Using the Freeze/Thaw in Current Viewport selections, you can turn off the visibility of the layers in the viewports in which you do not want the lines to show. AutoCAD did this for you with the generated layers for profile visible and profile hidden. You must control the visibility of layers that you create unless you want them to show in every viewport. Imagine putting dimensions in the front view and having them show up edgewise in the top and side views.

The Freeze/Thaw in Current Viewport selection is shown in Figure 10.21.

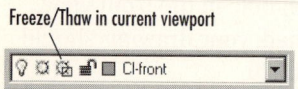

Freeze/Thaw in current viewport

Figure 10.21

Be sure that the lower right viewport is active before continuing.

> *Pick: (select the Layer Control pull-down list from the Object Properties toolbar)*

> *Pick: (the Freeze/Thaw in Current Viewport selection, located to the left of layer CL-FRONT)*

Now the centerline you drew should be visible only in the front and isometric views (you can not see it in the Top view because it is a line perpendicular to the page). Now you will activate the upper right viewport.

> *Pick: (select the Layer Control pull-down list from the Object Properties toolbar)*

> *Pick: (the Freeze/Thaw in Current Viewport selection, located to the left of layer CL-FRONT)*

The centerline you drew should now be visible only in the front view.

Next, you will use the Layer & Linetype Properties dialog box to create layers for the top and side view centerlines. Then you will set their visibility and draw the appropriate centerlines. Draw the centerline for the smaller hole before continuing.

On your own, pick the upper left viewport to make it active.

> *Pick: **Layers icon***

The Layer & Linetype Properties dialog box appears on your screen. Use the techniques that you have learned to create two new layers: CL-SIDE and CL-TOP. Set their linetypes to CENTER and colors to green. When you have finished creating the layers, you are ready to control their visibility.

> *Pick: **Details (located at the bottom of the dialog box)***

The Layer & Linetype Properties dialog box contains a Details section, as shown in Figure 10.22.

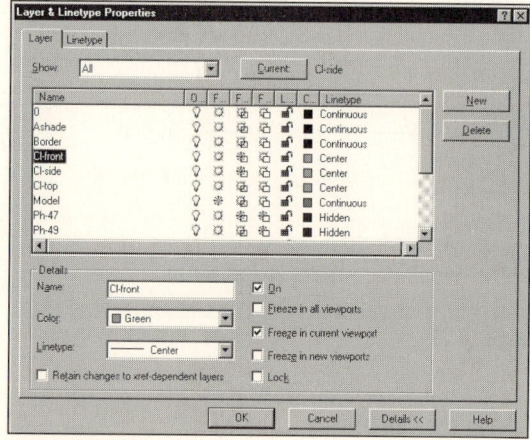

Figure 10.22

The options in the Details section allow you to make the same changes that you did by picking the icons on the Layer Control list.

You have already made the top view the active, or current, viewport. You will freeze the layers that you created so that your list looks like the one shown. Layer CL-FRONT should already be frozen for the current viewport because you froze it previously with the Layer Control pull-down list. You will freeze CL-SIDE in the current viewport.

Pick: (either the Freeze in Current Viewport selection in the Details section or the sun with a small viewport)

The sun with the small viewport becomes a snowflake with a small viewport, signifying that the layer has been frozen in the current viewport. Before exiting, make CL-SIDE the current layer.

Pick: OK (to exit the dialog box)

On your own, restore the UCS for the side view. Make the lower right viewport current and draw the side view centerlines. The lines will not appear in the top view because the current layer is frozen in that viewport.

On your own, make the lower left viewport (containing the front view) current and then return to the Layer & Linetype Properties dialog box. This time, use the dialog box to freeze layers CL-SIDE and CL-TOP in the lower left viewport. Make layer CL-TOP current. When you have finished, pick OK to exit the dialog box.

Next, use the UCS command to restore the WCS. Draw the top view centerlines.

Make the lower right viewport active and use the Layer Control pull-down list to freeze all of the centerline layers in that viewport except CL-SIDE. Then make the isometric viewport active and freeze all the centerline layers in that viewport.

Your drawing should be similar to the one in Figure 10.23. You can also create dimensions that appear in only one viewport by controlling layer visibility as you did for the centerlines.

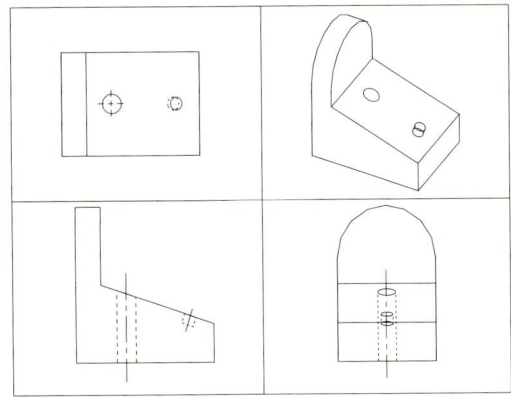

Figure 10.23

Plotting

Now you are ready to plot your drawing. Switch to paper space for plotting.

Adjust the LTSCALE so that the centerlines and hidden lines in the drawing appear at an appropriate scale.

Command: LTSCALE ↵
New scale factor <1.000>: .5 ↵

> ■ *TIP* Try various values until you are pleased with the results. ■

Usually, multiview engineering drawings do not have border lines between the views. This is why you created the layer VPORTS when you created the viewport boundaries. Now you will freeze this layer so that the views of your drawing plot correctly, with no lines dividing the views.

Pick: (select the Layer Control pull-down list)
Pick: (freeze the layer VPORTS)

The border lines disappear from the screen.

On your own, use Save As to save your drawing to the file name *angl_plt.dwg* before your plot. Use the Plot/Print icon to plot your drawing. You have now completed Tutorial 10.

10.1 Angled Support

Create this figure as a 3D solid model. Create necessary views and plot with correct hidden lines and centerlines.

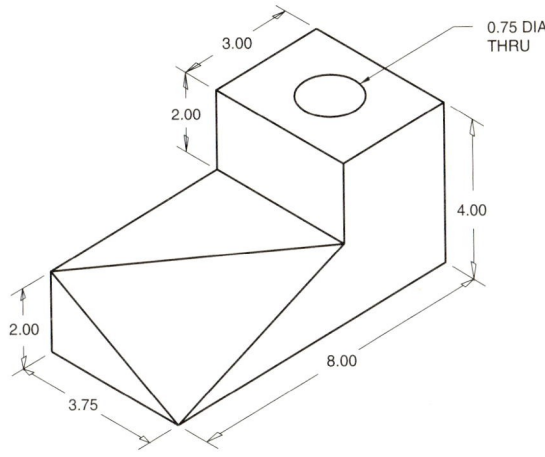

10.2 Step Model

Construct a solid model of this object. Grid spacing is 0.25 inch. Use a viewpoint of (3,–3,1) in the upper right viewport.

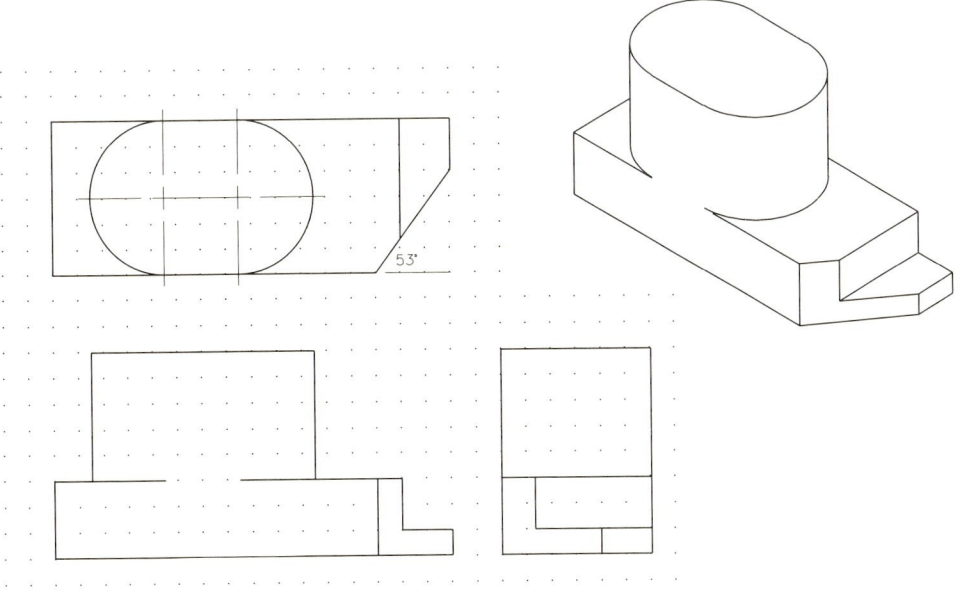

⬚ 10.3 Rod Holder

Create solid models for the objects shown. Show the necessary orthographic views in the correct viewports. Use SOLPROF to generate hidden lines.

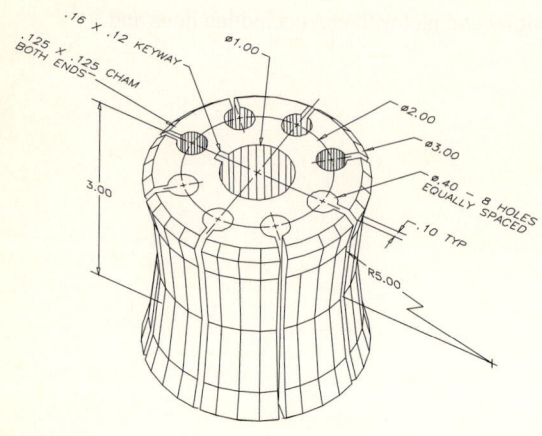

⬚ 10.4 Gear

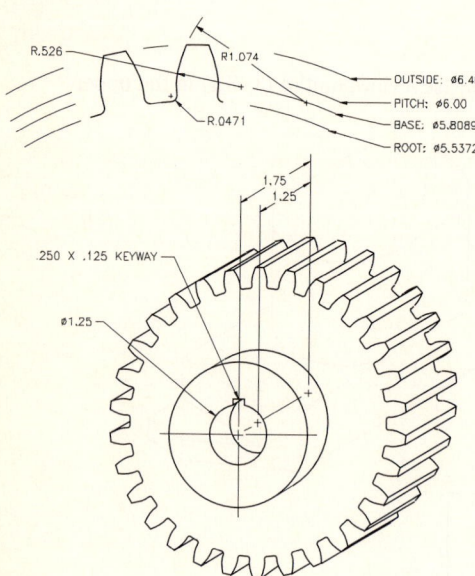

10.5 Router Guide

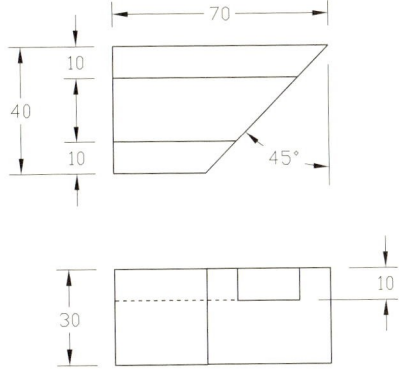

10.6 Slide Support

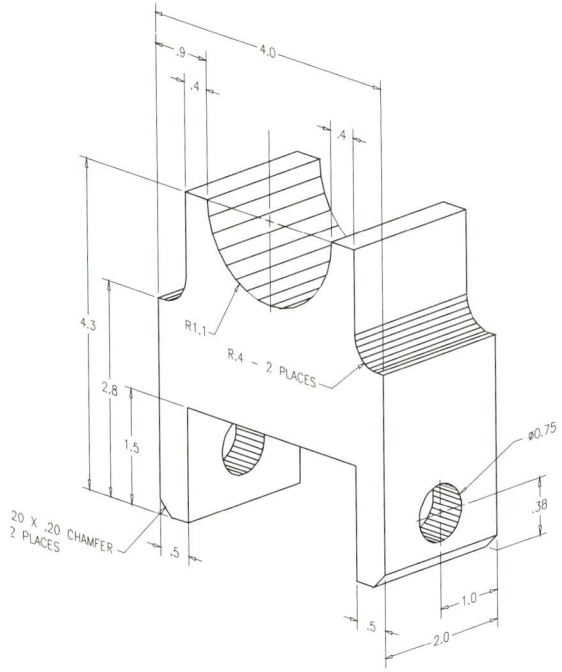

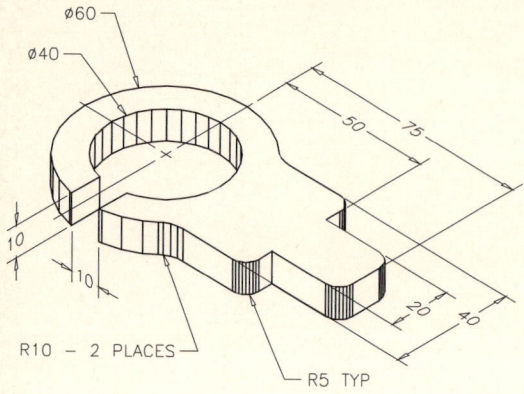

Ø60
Ø40
50
75
10
10
20
40
R10 — 2 PLACES
R5 TYP

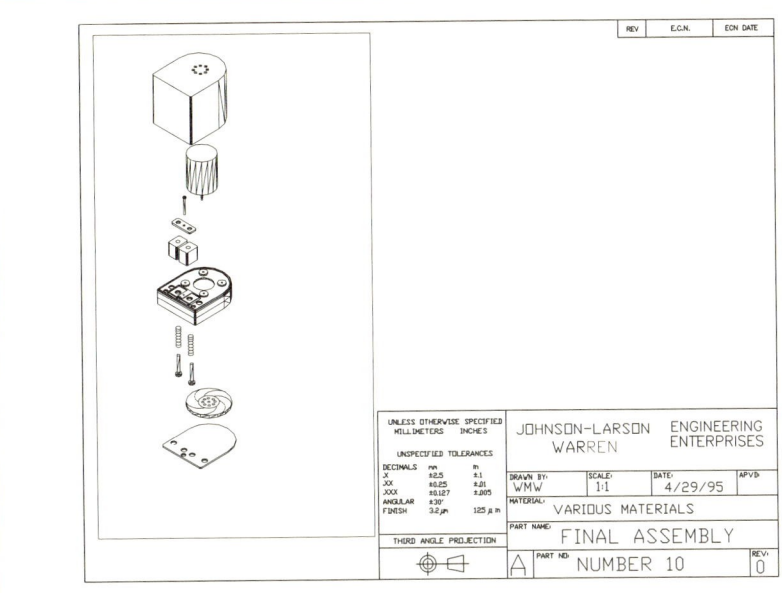

TUTORIAL 11

Creating Assembly Drawings from Solid Models

Objectives

When you have completed this tutorial, you will be able to

1. Create an assembly drawing from solid models.

2. Use external references.

3. Check for interference.

4. Analyze the mass properties of solid models.

5. Create an exploded isometric view.

6. Create and insert ball tags.

7. Create and extract attributes.

8. Create a parts list.

9. Use Windows Object Linking and Embedding (OLE).

Introduction

In previous tutorials you created solid models of objects and projected 2D views to create dimensioned part drawings. To make a set of working drawings using solid modeling, you will create an assembly drawing of a clamp, similar to the one shown in Figure 11.1. The parts for the assembly drawing have been created as solid models and are included with the data files. You will proceed directly to creating the assembly drawing. The purpose of an assembly drawing is to show how the parts go together, not to describe completely the shape and size of each part. Assembly drawings usually do not show dimensions or hidden lines.

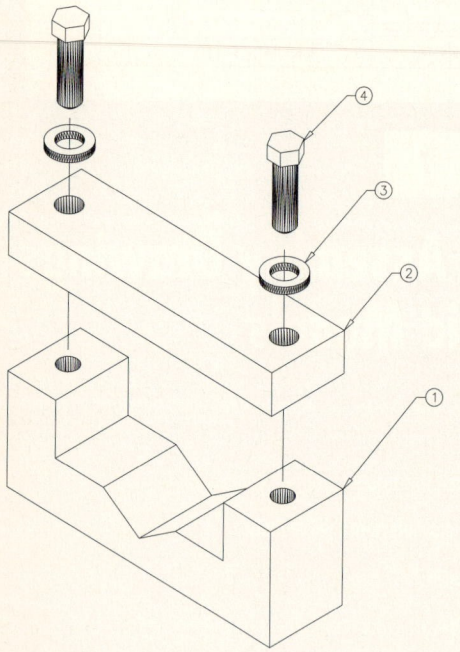

Figure 11.1

Attaching Solid Models

In this tutorial you will learn how to use the External Reference commands to attach part drawings to create an assembly drawing. When you use external references to attach a part drawing to the assembly drawing, AutoCAD automatically updates the assembly drawing whenever you make a change to the original part drawing.

You will also learn more about how to use the Insert command, as you did in Tutorial 8 when inserting the electronic logic symbols. Objects added to a drawing with the Insert command are included in the new drawing, but if you update the original drawing, AutoCAD doesn't automatically update the new drawing.

The parts in Figures 11.2, 11.3, 11.4, and 11.5 are the parts provided with the data files. The insertion points are identified as 0,0 in the drawings.

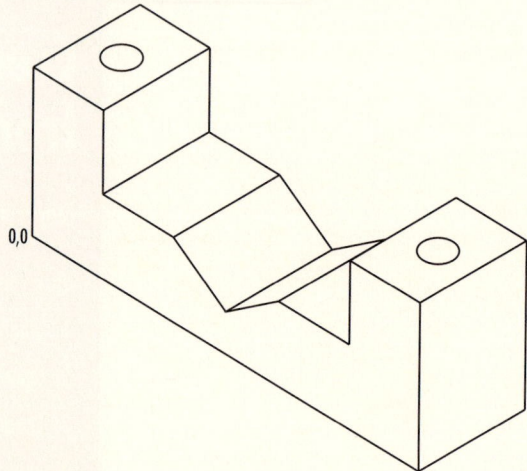

Figure 11.2

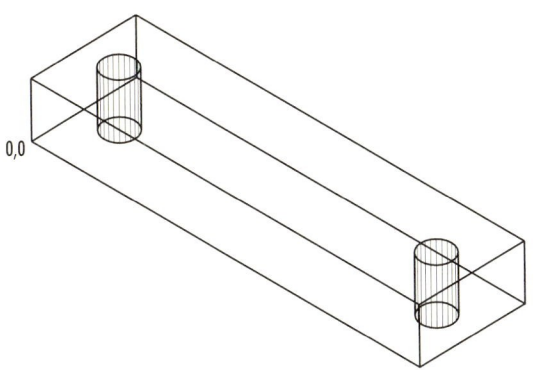

Figure 11.3

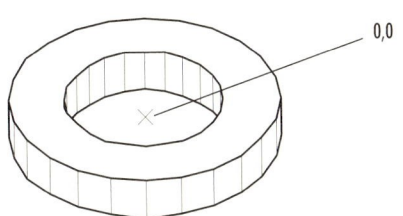

Figure 11.4

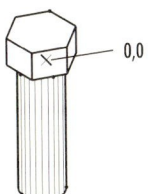

Figure 11.5

Starting

Start AutoCAD and begin a new drawing on your own using the drawing called *prot-iso.dwg* (provided with the data files) as a template. Name the drawing *asmb-sol.dwg* on your own.

The AutoCAD drawing editor appears on your screen with the settings from a template drawing for 3D isometric views. On your own, turn on the Reference toolbar, shown in Figure 11.6, by selecting it from the Toolbars dialog box, which can be opened from the View menu.

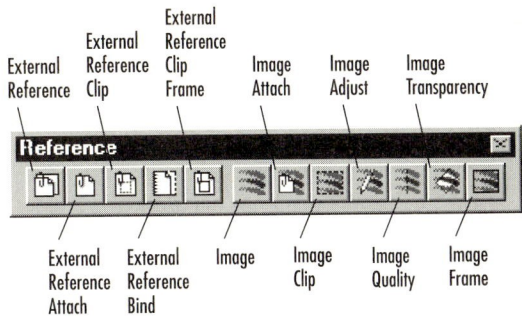

Figure 11.6

Position the toolbar in an area of the screen where it will be out of your way while you are drawing. Remember, you can move it to a different location at any time. Your screen should be similar to Figure 11.7.

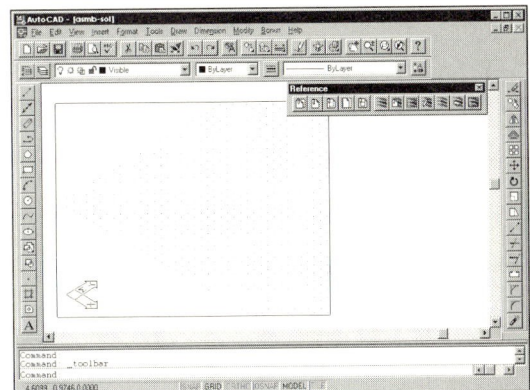

Figure 11.7

Using External References

 The External Reference command lets you use another drawing without adding it to your current drawing. The advantage of this is that if you make a change to the referenced (original) drawing, any other drawing to which it is attached updates automatically the next time you open that drawing. As with blocks, you can attach external references anywhere in the drawing; you can also scale and rotate them. You can quickly select the External Reference command options from the Reference toolbar. As with any other AutoCAD command, you can also type the command name, XREF, and select the option you want. You will begin your drawing by attaching the base of the clamp to your assembly drawing as an external reference, using the External Reference icon from the Reference toolbar.

Pick: **External Reference icon**

The External Reference dialog box appears, as shown in Figure 11.8.

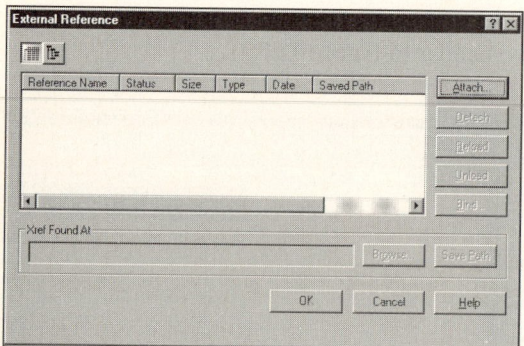

Figure 11.8

Pick: **Attach**

The Select file to attach dialog box appears on your screen. On your own, use this dialog box to select the file *base-3d.dwg*, which is in the *c:\datafile* directory.

When you have selected the *base-3d.dwg* file,

Pick: **Open**

The Attach Xref dialog box appears, similar to Figure 11.9.

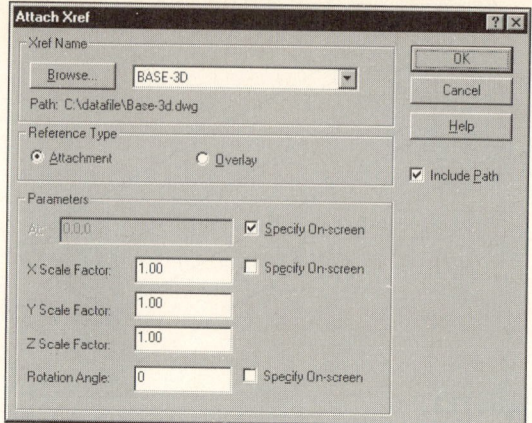

Figure 11.9

The options available are for Xref Name, Reference Type, and Parameters. Xrefs can be used as an attachment or used as an overlay. If Overlay is selected, when the drawing containing the xref overlay is inserted into a new drawing, the xref will NOT be carried along. When Attachment is selected, the xref'd file comes along with it.

Under the Parameters portion of the dialog box, deselect the Specify On-screen (remove the check) option next to the basepoint (At) section.

Type: **0,0,0 (next to At, in the Parameters portion of the dialog box)**

Pick: **OK**

You will set the Isolines variable to a larger value to produce more tessellation lines and improve the appearance of the rounded surfaces.

Command: **ISOLINES** ⏎
New value for Isolines <4>: **12** ⏎
Command: **REGEN** ⏎

The base should now appear in your drawing. On your own, use the Zoom command with the XP option to set the paper space zoom factor to 0.5. When you have finished these steps, your screen should be similar to Figure 11.10.

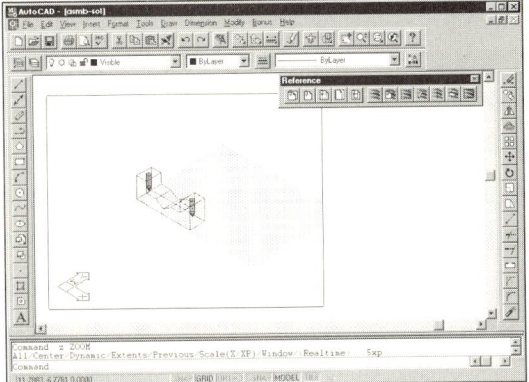

Figure 11.10

You can't modify drawings attached with the External Reference command (as *base-3d.dwg* was) in the new drawing, nor can you analyze the mass properties of a referenced object.

You can use the object snaps to locate new objects in relation to the referenced object's geometry.

Layers and External References

When you attach a drawing to another drawing by using External Reference Attach, all its subordinate features (e.g., layers, linetypes, colors, blocks, and dimensioning styles) are attached with it. You can control the visibility of the layers attached with the referenced drawing, but you cannot modify anything or create any new objects on these layers. The Visretain system variable controls the extent to which you can make changes in the visibility, color, and linetype. If Visretain is set to 0, any changes that you make in the new drawing apply only to the

current drawing session and will be discarded when you end it or reload or attach the original drawing.

Use Layer Control from the Object Properties toolbar to pull down the list of available layers. Scroll to the top of the list, if necessary, to see the result of attaching the base as an external reference. The list should be similar to that in Figure 11.11.

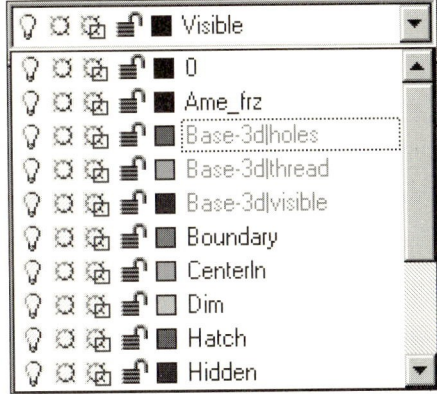

Figure 11.11

Note that the layer names attached with the external reference have the name of the external reference drawing in front of them.

Pick: **(to freeze layer BASE-3D|VISIBLE)**

The base externally referenced to the drawing disappears from the screen.

Pick: **(to thaw layer BASE-3D|VISIBLE)**

You can also control which portions of an external reference are visible by using the External Reference Clip command. This command creates a paper space viewport that you can use as a window to select the portion of the external reference that you want to be visible in the drawing. On your own, use the Help command to obtain more information about the Xrefclip command.

Inserting the Solid Part Drawings

Next, you will insert the solid model drawing *cover-3d.dwg* into the assembly drawing as a block. You will use the Insert command, as you did in Tutorial 8 when inserting the electronic logic symbols. On your own, verify that OSNAP is active, with Endpoint and Center selected. Use the Draw toolbar to select the icon for Insert Block.

> *Pick:* **Insert Block icon**
>
> *Pick:* **File**

Use the Select Drawing File dialog box that appears to select the drawing *cover-3d.dwg* from the directory containing the data files provided. When you have selected the file,

> *Pick:* **Open *(to exit the Select Drawing File dialog box)***
>
> *Pick:* **OK *(to exit the Insert dialog box)***

You return to the command prompt for the remaining prompts.

> Insertion point: *(pick anywhere in the drawing editor, using Figure 11.12 as a reference)*
>
> X scale factor <1> /Corner/XYZ: ⏎
>
> Y scale factor (default=X): ⏎
>
> Rotation angle <0>: ⏎

The solid model of the clamp cover appears in your drawing, as shown in Figure 11.12.

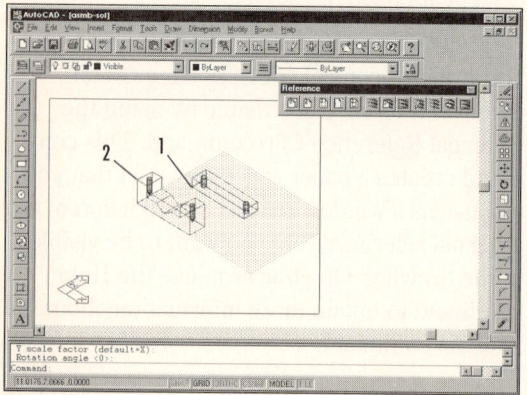

Figure 11.12

You will now use the Align command to align the edge of the cover with the edge of the base.

> Command: **ALIGN** ⏎
>
> Select objects: *(select the cover that you just inserted)*
>
> Select objects: ⏎
>
> Specify 1st source point: *(use AutoSnap Endpoint to select point 1 in Figure 11.12)*
>
> Specify 1st destination point: *(use AutoSnap Endpoint to select point 2 in Figure 11.12)*
>
> Specify 2nd source point: ⏎

The solid model of the clamp cover moves to align with the base, as shown in Figure 11.13. Save your drawing on your own.

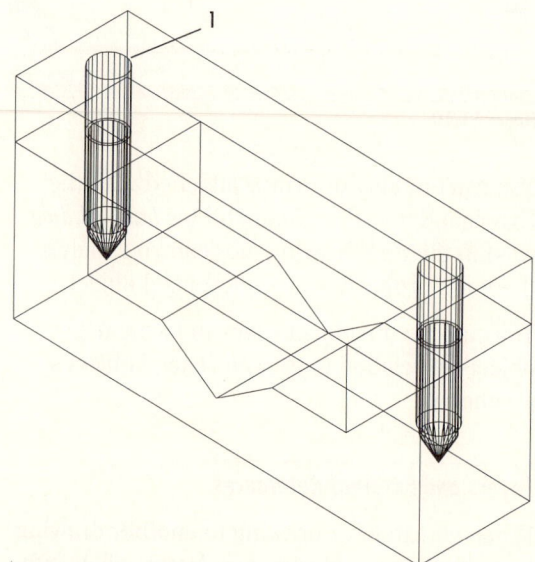

Figure 11.13

Now you are ready to insert the washers into the assembly. You will select the Attach button from the External Reference dialog box. On your own, be sure that Snap is off and use Zoom Window to enlarge the view to make selection easier.

Pick: **External Reference icon**

Pick: **Attach *(within the External Reference dialog box)***

This time select the drawing *washr-3d.dwg,* being sure to select the correct drive and directory. Pick Open to exit the Select File to Attach dialog box. You return to the Attach Xref dialog box where you will now select Specify On-screen for the basepoint selection in the Parameters portion of the dialog box.

Pick: **Specify On-screen *(select the check box to the right of the basepoint (At) section within Parameters)***

Pick: **OK *(to exit the Attach Xref dialog box)***

Insertion point: **(*target the top edge of hole 1 in Figure 11.13, using the AutoSnap Center marker*)**

Insert the washer for the right side of the part on your own.

The following message appears: *Xref WASHR-3D has already been defined. Using existing definition.* If you knew that drawing *washr-3d.dwg* had changed and you wanted to update it without having to exit the drawing, you could use the Reload option of the Xref command. Reload is especially useful if you are working on a networked system and sharing files with other members of a design team. At any point you can use the Reload option to update the definition of the referenced objects in your drawing. In this case, you want to insert another copy of the same drawing, so you can just ignore this message. (You could also use the Copy command to create the other washer.)

Once you have inserted the washers, your drawing should look like that in Figure 11.14.

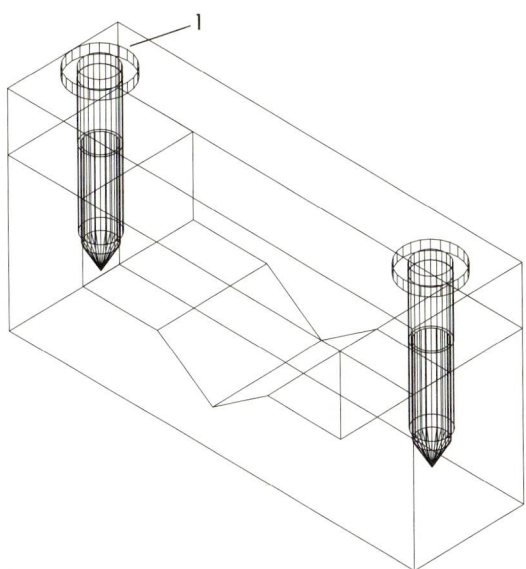

Figure 11.14

Next, you will add the two screws to the assembly to finish inserting the parts that you will need. You will use Insert, not External Reference, so that you can use other AutoCAD features—for example, checking interference—later in this tutorial.

Pick: **Insert Block icon**

On your own, use the Insert dialog box that appears on your screen to select *screw-3d.dwg* from the data files provided. Close the Select Drawing File dialog box by picking Open after selecting the correct file. Be sure that Specify Parameters on Screen in the Insert dialog box is enabled and pick OK to return to the command prompt for the remaining selections.

Insertion point: **(*pick the top edge of the washer labeled 1 in Figure 11.14*)**

X scale factor <1> /Corner/XYZ: ⏎

Y scale factor (default=X): ⏎

Rotation angle <0>: ⏎

The screw should appear in your drawing. Repeat this process on your own for the second screw. When you repeat the process, you don't need to select the file again; you can insert another instance of the same block by selecting it from the list in the Insert dialog box.

Now your drawing should depict the parts in completely assembled form, as shown in Figure 11.15.

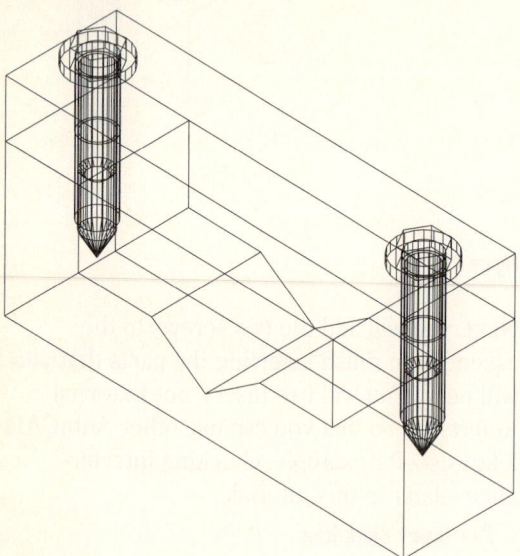

Figure 11.15

Checking for Interference

 When you are designing a device, you often want to know whether the parts will fit correctly. You can use AutoCAD's Interfere command to determine whether two parts overlap. If they do, Interfere creates a new solid showing where they do. You can use this command to determine whether the objects fit together as intended.

On your own, load the Solids toolbar and place it where it does not interfere with either the Reference toolbar or the drawing.

You will determine whether the screws in the assembly will fit the holes in the cover. The Interfere command will only analyze solids, but the objects that you've inserted in the drawing are blocks. To return the cover and screws to solids, you will explode them. When you explode an object, doing so removes one level of grouping at a time, so exploding the block will return it to a solid. If you explode the solid, it will become a surface model. On your own, list the cover and two screws, which are block references. To explode the cover and screws,

Pick: **Explode icon**

Select objects: *(pick the cover and the two screws)*

Select objects: ⏎

The cover and two screws are exploded. Now list them again on your own; they should now be listed as 3dsolid objects. Next you will determine whether the screws interfere with the cover.

> ■ *TIP* You may need to use Zoom Window to select the screw. ■

From the Solids toolbar,

Pick: **Interfere icon**

Select the first set of solids:

Select objects: *(pick on the cover)*

Select objects: ⏎

Select the second set of solids:

Select objects: *(pick on the left-hand screw)*

Select objects: ⏎

Comparing 1 solid against 1 solid.

Solids do not interfere.

Because the solids do not interfere, you know that the screw on the left fits through the cover. If the solids did interfere, you would see a message similar to

Interfering solids: 2 Interfering pairs: 1 Create interference solids ? <N>:.

You can type *Y* ⏎ if you want AutoCAD to create a new solid on the current layer from the overlap between the interfering solids.

Determining Mass Properties

 AutoCAD enables you to inquire about the mass properties of an object. One of the advantages of the solid modeler used in Release 14 is the accuracy with which volumes and mass properties are calculated. Pick the Mass Properties icon from the Inquiry/List flyout on the Standard toolbar. Like Interfere, this command works only on solids, not on external references.

Pick: **Mass Properties icon**

Select objects: *(pick the clamp cover)*

Select objects: ⏎

The AutoCAD text window opens displaying the mass property information, as shown in Figure 11.16.

```
AutoCAD Text Window                                    _ □ X
Edit
Command: _massprop
Select objects: 1 found

Select objects:

-------------- SOLIDS   ---------------

Mass:                     5.9594
Volume:                   5.9594
Bounding box:         X:  0.0000   --  5.5000
                      Y:  0.0000   --  1.5000
                      Z:  2.0000   --  2.7500
Centroid:             X:  2.7500
                      Y:  0.7500
                      Z:  2.3750
Moments of inertia:   X: 38.4038
                      Y: 93.4026
                      Z: 64.0179
Products of inertia: XY: 12.2913
                     YZ: 10.6152
                     ZX: 38.9225
Radii of gyration:    X:  2.5385
                      Y:  3.9589
                      Z:  3.2775

Press ENTER to continue
```

Figure 11.16

Close the AutoCAD text window on your own. AutoCAD provides the opportunity to write the mass properties information to a file. Once you have closed the text window, you will see the prompt,

Press ENTER to continue: ⏎

Write to a file ? <N>: ⏎

Creating an Exploded Isometric View

Often used to show how parts are assembled, exploded views show the parts removed from their assembled positions yet still aligned, as though the assembly had exploded. You will create an exploded view from your assembly drawing by moving the parts away from each other. Because the parts still should align with their assembled positions, you will move them only along one axis. You will use the Move icon on the Modify toolbar to move the screws from their assembled positions.

Pick: **Move icon**

Select objects: *(pick the two screws)* ⏎

Base point or displacement: **0,0,0** ⏎

Second point of displacement: **0,0,5** ⏎

The screws move up in the drawing and off the view. You will use the Zoom Extents command to fit the entire drawing on the screen.

Pick: **Zoom Extents icon**

Your drawing should be similar to that in Figure 11.17.

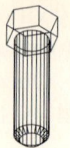

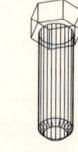

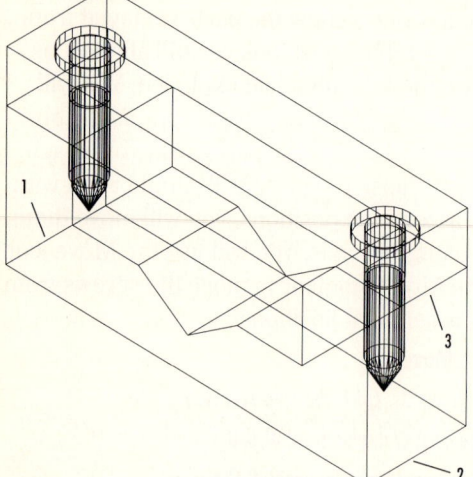

Figure 11.17

You will continue to use the Move command and relocate the washers, keeping them aligned with their present position but three units above, along the Z-axis.

Pick: **Move icon**

Select objects: **(pick the two washers)** ⏎
Base point or displacement: **0,0,0** ⏎
Second point of displacement: **0,0,3** ⏎

The washers align below the screws and all remain aligned with the holes in the object. Next, move the clamp cover up two units, along the Z-axis.

Command: ⏎
Select objects: **(pick the cover)** ⏎
Base point or displacement: **0,0,0** ⏎
Second point of displacement: **0,0,2** ⏎

When you create an exploded view, you should add thin lines to show how the parts assemble.

> ■ *TIP* Another very useful technique in aligning 3D objects in an assembly drawing is to use the Tracking command. You can Attach or move the solids with reference to snapped locations on other solids. ■

On your own, create a new layer called ALIGN that is color cyan and linetype CONTINUOUS. Set it as the current layer.

 To create lines in the same plane as the objects' centers, you need to define a UCS that aligns through the middle of the objects. Zoom in, if necessary, to pick the correct lines on the part. Also be sure that OSNAP is turned on with Midpoint selected.

Pick: **3 Point UCS icon**

Origin point <0,0,0>: **(pick the middle of line 1 in Figure 11.17)**

Point on positive portion of the X axis <1.0000,0.7500,0.0000>: **(pick the middle of line 2)**

Point on positive-Y portion of the UCS XY plane <1.0000,0.7500,0.0000>: **(pick the middle of line 3)**

The UCS icon changes to align with the plane through the middle of the parts. If it does not, repeat the command and try again. On your own, verify that Snap is on and draw the lines, indicating how the parts align, by picking points on the Snap. Save your drawing before you continue.

Creating a Block for the Ball Tags

Ball tags identify the parts in the assembly. They are made up of the item number in the assembly enclosed in a circle, hence the name ball tags. You can easily add ball tags to the drawing by making a simple block with one visible attribute and several invisible attributes. An *attribute* is basically text information that you can associate with a block. It is an effective method of adding information to the drawing. You can also extract attribute information from the drawing and import it into a database or word processing program for other uses.

Figure 11.18 shows ball tags like the ones you will create.

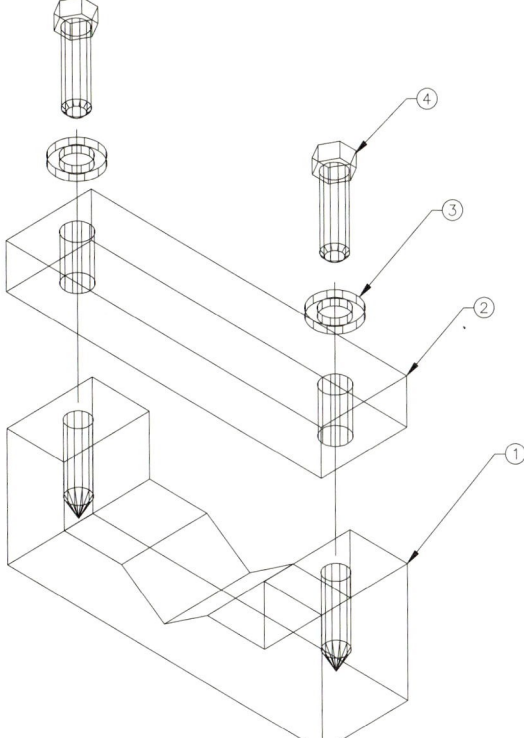

Figure 11.18

On your own, set layer TEXT as the current drawing layer. Select the View UCS to align the UCS with the view so that you can create text that appears straight on.

You will draw the circle first. Then you will add the attributes and convert the circle and attribute objects into a block.

Pick: **Circle icon**

On your own, draw a circle of diameter 0.25 above and to the right of the assembly drawing.

Defining Attributes

The Define Attributes command is on the Draw menu. You can use Define Attributes to create special attribute text objects.

Pick: **Draw, Block, Define Attributes**

The Attribute Definition dialog box appears on your screen. You will use it to make the selections shown in Figure 11.19.

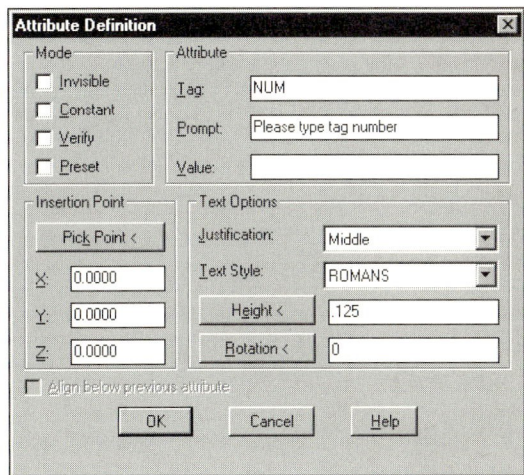

Figure 11.19

Attribute Modes

Attributes can have the following special modes, or properties.

Mode	Definition
Invisible	The attribute does not appear in the drawing, but can still be used for purposes such as extracting to a database.
Constant	The value of the attribute is set at the beginning of its definition, instead of AutoCAD prompting you for a value on insertion of the block.
Verified	You can make sure that the value is correct after typing the value.
Preset	You can change the attribute later, but you will not be prompted for the value when you insert it.

You will now use the Attribute Definition dialog box to add the number in the center of the ball tag as a visible attribute. You will also create attributes for the other information commonly shown in the parts list, such as the part name, part number, material, and quantity, as invisible attributes.

Leave each of the Mode boxes unchecked to indicate that you want this attribute to be visible, variable, not verified, and typed in, not preset.

Attribute Tag

The *attribute tag* is a variable name that is replaced with the value that you type when prompted as you insert the attribute block. The tag appears when the block is exploded or before the attribute is made into a block. You will use NUM for the attribute tag. On your own, type *NUM* in the Tag text box.

Attribute Prompt

The *attribute prompt* is the prompt that appears in the command area when you insert the block into a drawing. Make your prompt descriptive, so it clearly identifies the information to be typed and how it should be typed. For instance, *Enter the date (dd/mm/yy)* is a prompt that specifies not only what to enter—the date—but also the format—numeric two-digit format, with day first, then month, and then year.

On your own, type *Please type tag number* in the Prompt text box.

> ■ *TIP* The prompt appears with a colon (:) after it to indicate that some entry is expected. AutoCAD includes the colon automatically; you do not need to type a colon in the dialog box. ■

Default Attribute Value

The Value text box defines a default value for the attribute. Leaving the box to the right of Value empty results in no default value for the tag. As every tag number will be different, there is no advantage to having a default value for the ball tag attribute. If the prompt is often answered with a particular response, that response would be a good choice for the default value.

> ■ *TIP* You could create a title block for your drawings that would use attributes. As part of the attribute for the drafter's name, you would make the attribute prompt *Enter drafter's name*. In the Value text box, you would put your own name as the default value. When you inserted the block into a drawing, the default value after the prompt within the AutoCAD default angle brackets would be your own name. You could press ⏎ to accept the default rather than having to type in your name each time. ■

Attribute Text Options

Attributes and regular text use the same types of options. You can center, fit, and align attributes just like any other text. Select the Middle option from the Justification pull-down menu. Doing so will cause the middle of the text to appear in the center of the ball tag when you type it.

The text style romans is the default, which is acceptable. You do not need to change the Text Style setting.

Set the text height to 0.125.

Leave the rotation set at 0° for this attribute. You do not want the attribute rotated at an angle in the drawing.

> ■ *TIP* Clicking on the Height or Rotation buttons lets you set the height or rotation, either by picking from the screen or by typing at the command prompt. ■

Attribute Insertion Point

You can use the Pick Point button in the Insertion Point section to select the insertion point on your screen, or you can type the X-, Y-, and Z-coordinates into the appropriate boxes. Be sure that OSNAP is on with Center selected.

Pick: **Pick Point**

Your drawing returns to the screen.

Start point: **(pick on the edge of the circle)**

AutoCAD returns you to the Attribute Definition dialog box.

Pick: **OK**

The tag NUM is centered in the circle. The circle and the attribute tag should look like Figure 11.20. If the attribute is not in the center of the circle, use the Move command to center it in the circle on your own.

Figure 11.20

Next you will make the attributes for part name (description), part number, material, and quantity. These attributes will be invisible, to avoid making the drawing look crowded, but they can still be extracted for use in a spreadsheet or database.

To restart the Define Attribute command,

> Press: ⏎

The Attribute Definition dialog box appears on your screen. On your own, select the invisible mode. When you select it, a check appears in the box to its left. Continue to use the dialog box to create an invisible attribute with the following settings.

> Tag: **PART**
>
> Prompt: **Please type part description**
>
> Value: *(leave Value blank so that there is no default value)*
>
> Justification: **Left**
>
> Text Style: **ROMANS**
>
> Height: **.125**
>
> Rotation: **0**

When you have entered this information into the dialog box, you are ready to select the location in the drawing for your invisible attribute.

> Pick: **Pick Point**

Your drawing returns to the screen.

> Start point: *(pick a point to the right of the ball tag circle)*
>
> Pick: **OK *(to exit the Attribute Definition dialog box)***

The tag PART should appear at the location you selected.

On your own, use this method to create the following three invisible attributes. Choose the box to the left of Align below previous attribute to locate each new attribute below the previous attributes.

> Tag: **MATL**
>
> Prompt: **Please type material**
>
> Value: **Cast Iron**
>
> Justification: **Left**
>
> Text Style: **ROMANS**
>
> Height: **.125**
>
> Rotation: **0**
>
> Tag: **QTY**
>
> Prompt: **Please type quantity required**
>
> Value: *(leave blank)*
>
> Justification: **Left**
>
> Text Style: **ROMANS**
>
> Height: **.125**
>
> Rotation: **0**
>
> Tag: **PARTNO**
>
> Prompt: **Please type part number, if any**
>
> Value: *(leave blank)*
>
> Justification: **Left**
>
> Text Style: **ROMANS**
>
> Height: **.125**
>
> Rotation: **0**

When you have finished, each of the tags should appear in the drawing.

Defining the Block

Now you will convert the circle and all the attributes into a block. (Remember, first you create the shapes for the block and the attributes, and then you define them as a block.)

> Pick: **Block icon**
>
> Block name (or ?): **BALLTAG** ⏎
>
> Insertion base point: *(pick Snap to Quadrant icon)*
>
> _qua of: *(pick on the left side of the circle to select its left quadrant point)*
>
> Select objects: *(use implied Windowing to pick the circle and the tag names NUM, PART, MATL, QTY, PARTNO)* ⏎

The circle and the tag names disappear from your screen. Now you can insert the necessary ball tags into the drawing, using the block you have made.

> ■ **TIP** You may want to use the Write Block command (Wblock) so that the block is available for other drawings. Remember, without Write Block, the block is defined only in the current drawing. ■

Inserting the Ball Tags

On your own, turn on the Dimension toolbar.

 You will use the Leader icon on the Dimension toolbar to draw the arrow and the lines to the ball tag. Then, you will insert the block BALLTAG to add the circle and tags.

Before continuing, be sure that OSNAP Nearest is on.

You will use Figure 11.18 as your guide as you insert the ball tags.

On your own, verify that Ortho is off. Turn off the Reference toolbar.

> ■ **TIP** Because block BALLTAG has its insertion point on the left, you want to draw the leaders toward the right. ■

Pick: **Leader icon**

From point: *(pick a point on the right edge of part 1, the base, using the AutoSnap nearest marker)*

To point: *(pick a point that is above and to the right of part 1)*

To point (Format/Annotation/Undo)<Annotation>: *(Turn on Ortho and pick a point that is about 0.125 to the right)*

To point (Format/Annotation/Undo)<Annotation>: ⏎
Annotation (or press ENTER for options): ⏎
Tolerance/Copy/Block/None/<Mtext>: **B** ⏎
Block name (or ?): **BALLTAG** ⏎
Insertion point: *(pick the endpoint of the leader)*
X scale factor <1>/Corner/XYZ: ⏎
Y scale factor (default=X): ⏎
Rotation angle <0>: ⏎
Please type tag number: **1** ⏎
Please type part number, if any: **ADD1** ⏎
Please type quantity required: **1** ⏎
Please type material<Cast Iron>: **STEEL** ⏎
Please type part description: **BASE** ⏎

> ■ **TIP** Remember, the space bar does not act as ⏎ when you are entering text. Because text may include spaces, when you are entering text commands press ⏎ to end the command. ■

The circle and number 1 appear on your screen. Your prompts for attribute information may have appeared in a different order. If you made a mistake, you can correct it in the next section.

Continue to create the leader lines for parts 2 (the cover), 3 (the washers), and 4 (the hex head) on your own. The leader can start anywhere on the edge of the part. Try to pick points and leader line angles that place the ball tags in the drawing where they are accessible and easy to read. When adding the ball tags, refer to the table on the next page showing the information for each part.

Item	Name	Material	Quantity	Number
1	Base	Steel	1	ADD1
2	Clamp Top	1020ST	1	ADD2
3 .	438 x.750 x.125 Flat Washer	Stock	2	ADD3
4	.375-16 UNC x.125 Hex Head	Stock	2	ADD4

When completed, your drawing should look similar to Figure 11.18. Save your drawing before you continue.

Changing an Attribute Value

The Edit Attribute command is very useful for changing an attribute value. If you mistyped a value or you want to change an existing attribute value, select Attribute from the Modify menu.

Pick: **Modify, Object, Attribute, Single**

Select block: **(pick on one of the ball tags)**

The Edit Attributes dialog box appears, as shown in Figure 11.21. (Your attributes may appear in a different order.)

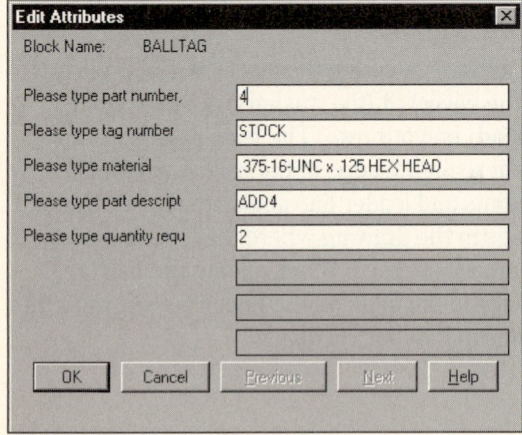

Figure 11.21

It contains the attribute prompts and a text box to the right of each prompt with the value you entered. To change a value, pick its text box and type the new value. Pick OK to exit the dialog box.

Changing an Externally Referenced Drawing

The primary difference between blocks and externally referenced drawings is that external references are not really added to the current drawing. A *pointer* is established to the original drawing (external reference) that you attached. If you change the original drawing, the change is also made in any drawing to which it is attached as an external reference. You will try this feature by opening drawing *base-3d.dwg* and filleting its exterior corners. On your own, open the file *base-3d.dwg*.

The drawing of the base appears on your screen. On your own, use the commands that you have learned to add a fillet of radius 0.25 to all four exterior corners of the base. Save the changes to the drawing *base-3d.dwg* so that they will occur in your assembly drawing.

Now reopen the assembly drawing *asmb-sol.dwg* and observe that the changes you made to *base-3d.dwg* appear in that drawing also.

The change to the base has been made in the assembly drawing, as shown in Figure 11.22. When you opened this drawing and loaded the newly changed external referenced drawing *base-3d.dwg* into *asmb-sol.dwg*, AutoCAD automatically updated the assembly.

On your own, verify that *asmb-sol.dwg* is your current drawing before continuing. Its name should appear in the title bar near the top of the screen.

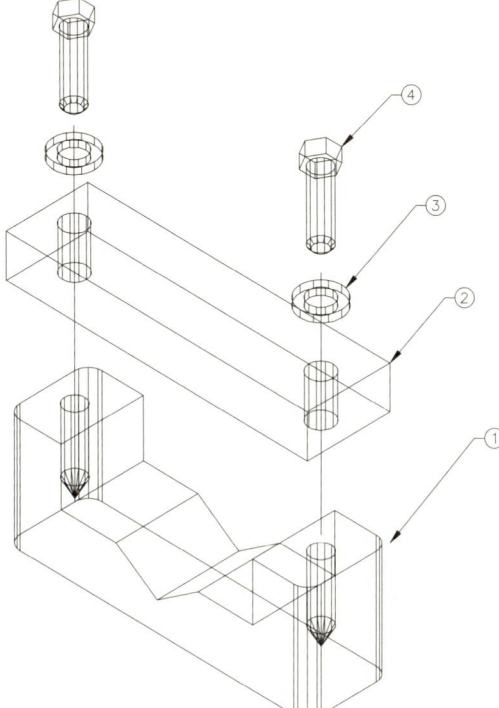

Figure 11.22

■ *TIP* You must carefully keep track of your drawings when using external references. If you did not intend to change the assembly but made a change to its externally referenced drawing, the drawing will change. If you want to change an externally referenced drawing but not the assembly, use Save As and save the changed drawing to a new file name. If you move an externally referenced drawing to a different directory, you will have to Reload the external reference and supply the new path in order for AutoCAD to find it. ■

Creating the Parts List

The next task is to create a parts list for your drawing. The required headings for the *parts list* are Item, Description, Material, Quantity, and Part No. The item number is the number that appears in the ball tag for the part on your assembly drawing. Sometimes a parts list is created on a separate sheet; but often in a small assembly drawing, such as the clamp assembly, it is included in the drawing. This method is preferable because having both in one drawing keeps the parts list from getting separated from the drawing that it should accompany. The parts list is usually positioned near the title block or in the upper right corner of the drawing.

There is no standard format for a parts list; each company may have its own standard. You can loosely base dimensions for a parts list on one of the formats recommended in the MIL-15 Technical Drawing Standards or in the ANSI Y14-2M Standards for text sizes and note locations in technical drawings.

Extracting Attribute Information

You will extract the attribute information that you created and then import it into a spreadsheet or word processing program to format. Afterward you can reinsert it in the drawing by using Windows Object Linking and Embedding (OLE). You will type the dynamic dialog command for attribute extraction, Ddattext.

Command: **DDATTEXT** ⏎

The Attribute Extraction dialog box appears on your screen, as shown in Figure 11.23. You can use it to set the format of the attribute information that you will extract.

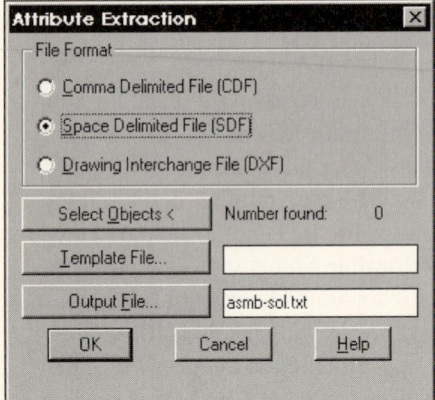

Figure 11.23

You can extract attributes to three different file formats:

Comma Delimited File (CDF) creates a file containing one record for each block reference, with the individual attributes separated by commas. The character fields in the record are enclosed in single quotation marks.

Space Delimited File (SDF) creates a file containing one record for each block reference. The fields of each record are a fixed width; no commas or other separators are used.

Drawing Interchange File (DXF) creates a file like the one used to exchange drawings between systems (DXF), except that it has only block reference, attribute, and end-of-sequence objects. This format does not require the use of a template file.

On your own, be sure the space delimited format is enabled. The radio button to the left of Space Delimited File (SDF) should be filled in.

The *template file* tells AutoCAD how to structure the extracted attribute information. You will use the template file provided with the data files, *ext.txt*. If you want to see the structure of the template file, use a text editor to open *ext.txt*. It is a text file that matches the block name, X- and Y-coordinates, and attribute names with the types and lengths of the fields to which the information will be extracted. The field information begins with either C (for character) or N (for numeric), indicating the type of field, followed by a three-digit number indicating the field length and three more places indicating the number of decimal places; for example, N007001 or C020000.

On your own, use the Attribute Extraction dialog box to select the template file *ext.txt*. Pick the Template File button. Use the dialog box that appears to select *ext.txt* as the template file. You will need to change to the *c:\datafile* directory in order to see *ext.txt* listed.

The *output file* has the same prefix, *asmb-sol*, as the AutoCAD file in which you are working, with the file extension *.txt*. Leave it set to the default name, *asmb-sol.txt*.

Next, you will select the attributes to extract from your drawing. You will pick the Select Objects button to return to your drawing for the selections.

Pick: **Select Objects**

Select objects: **(pick all four ball tags)** ⏎

Pick: **OK (to exit the Attribute Extraction dialog box)**

The message, *4 records in extract file*, will appear in the command window. The file *asmb-sol.txt* contains four lines.

Using Your Word Processing Program

On your own, minimize AutoCAD and open the word processing program you use on your system.

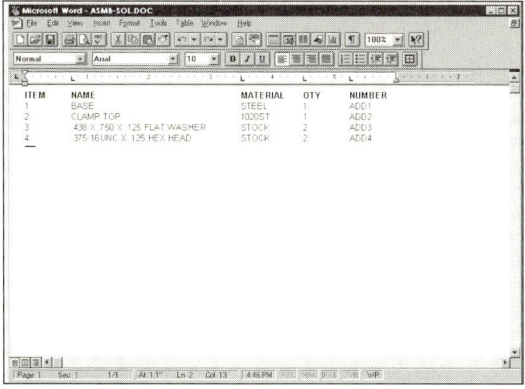

Figure 11.24

On your own, use your knowledge of your spreadsheet or word processing program to open or insert the extracted file, *asmb-sol.txt*, and clean up the formatting. Figure 11.24 shows an example of the file open and formatted with Microsoft Word. The titles for the parts list have been added as the top row. On your own, delete the XY-coordinate columns

because you do not need them for the parts list. (The coordinates for the attributes can be useful in a number of ways for inserting information or drawing objects back into your drawing.) Save the changes you made to *asmb-sol.txt*.

In your Windows spreadsheet or word processor, highlight the parts list text on your own. Use the spreadsheet or word processor's Edit, Copy command sequence to copy the selection to the Windows Clipboard. Minimize the word processing program and return to AutoCAD.

Pasting from the Clipboard

AutoCAD's drawing editor should be on your screen, showing the assembly drawing created earlier in this tutorial. Be sure that you have completed the previous step of highlighting and copying the selection from your spreadsheet to the Clipboard. Then,

Pick: **Edit, Paste**

The selection that you copied to the Windows Clipboard from the spreadsheet appears in your graphics window. On your own resize and position the pasted text until it is similar to that in Figure 11.25. Resize the pasted item by picking on the dark boxes that appear at the corners of the selection. The cursor changes to a double arrow when positioned over the boxes, and you can drag those boxes to move the object's outline. When you have the cursor positioned over the object but not at the edges where you can resize the selection, the four-way arrow cursor will appear. Position the pasted selection by picking on it when you see the four-way arrow cursor and hold the pick button down while dragging it to a new location.

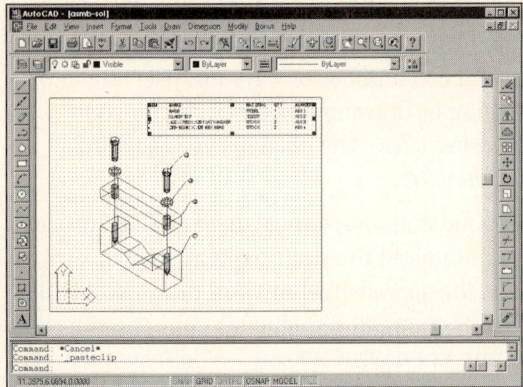

Figure 11.25

If you cannot extract the attributes, use any Windows-compatible word processor or spreadsheet to create the parts list. Once you have typed the information for the parts list, highlight all the text information that you have entered. Select the Edit menu, Copy option to copy the selection to the Windows Clipboard. Switch back to the AutoCAD program and use the Edit menu, Paste option to add the parts list to your AutoCAD drawing.

On your own, use the hot grips with Move, if necessary, to locate the parts list in the upper right portion of the drawing. Save your drawing.

Using Windows Object Linking and Embedding

Object Linking and Embedding (OLE) is a Windows feature that you can use to transfer information between Windows-based applications. With OLE, you can use several different forms of information inside a single file. Objects that can be embedded in Windows applications include spreadsheets, drawings, charts, and text. For example, you can embed the text for the parts list, which you previously exported and formatted, in your AutoCAD drawing. Embedding a linked object is similar to attaching an external reference drawing in AutoCAD. When you edit the original source document, the changes automatically appear in the document or drawing where it is embedded. The reason is that the data that actually make up the linked object reside only in the source document. (Pasting a copied selection into the drawing is similar to inserting a block.) To embed a Windows object into your AutoCAD drawing, you would use the AutoCAD menu bar.

Pick: **Insert, OLE Object**

The Insert Object dialog box appears on your screen, as shown in Figure 11.26.

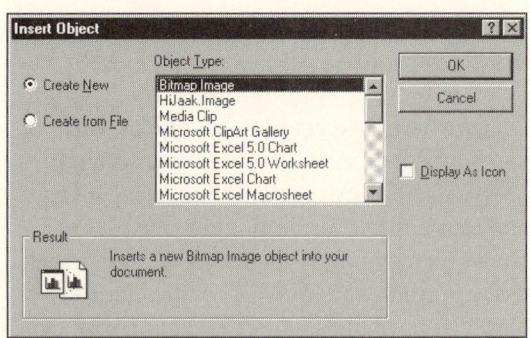

Figure 11.26

Use the scroll bars to scroll down the list of object types available on your system. Selecting an object type from the list automatically launches the application with which that object type is associated. For example, if you pick MS Word 6.0 Document, MS Word 6.0 will automatically start. You can then open the file that you previously saved or create a new file if you want. When you are finished using the application (e.g., MS Word 6.0) you would pick File, Close and Return to Object from the menu bar.

Pick: **Cancel** *(to cancel the Insert Object dialog box)*

Experiment on your own to embed objects in your AutoCAD drawing and embed AutoCAD drawings in your other documents. When objects are embedded in a destination application, double-clicking on them automatically starts the source application in which the object was created.

On your own, switch to paper space and plot your drawing. Close any spreadsheet or word processing programs that you may have left open.

■ *TIP* If you are going to send a drawing to someone on a disk, you must be sure to send any externally referenced drawings too. Otherwise the recipient will not be able to open the referenced portions of the drawing. A good way to avoid this problem is to use the External Reference command with the Bind option at the command prompt to link the two drawings. Doing so makes the referenced drawing a part of the drawing to which it is attached. Also, it prevents you from unintentionally deleting the original drawing. However, once you have done so, the external reference will no longer update. To use the External Reference Bind option, after you have attached the drawing, pick the External Reference icon, select the referenced objects you want to bind, select the Bind option, and then choose the Bind type (whether it be Bind or Insert). One drawback to binding the files is that the file size for the assembly drawing will be larger. ■

Hide any extra toolbars you may have on your screen. Leave the Draw and Modify toolbars showing. Exit AutoCAD. You have completed Tutorial 11.

EXERCISES

11.1 Clamping Block

Prepare assembly and detail drawings, and a parts list based on the information below using solid modeling techniques. Create solid models of Parts 1 and 2. Join Part 2 to Part 1 using two .500–3UNC x 2.50 LONG HEX-HEAD BOLTS. Include .500–13UNC NUTS on each bolt. Locate a .625 x 1.250 x.125 WASHER under the head of each bolt, between Parts 1 and 2, and between Part 1 and the NUT. (Each bolt will have 3 washers and a nut.)

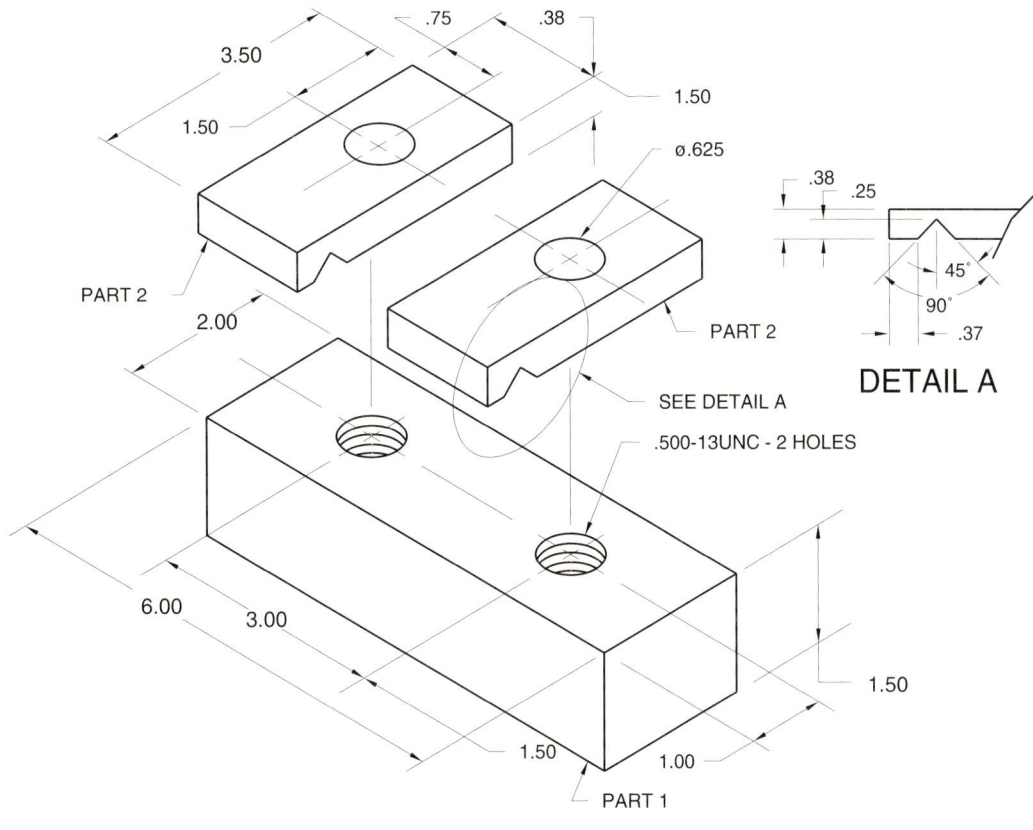

DETAIL A

⚙ 11.2 Pressure Assembly

Prepare assembly and detail drawings, and a parts list of the pressure assembly using solid modeling techniques. Create solid models of the BASE, GASKET, and COVER shown below. Use four .375–16UNC x 1.50 HEX-HEAD SCREWS to join the COVER, GASKET, and BASE. The GASKET assembles between the BASE and COVER.

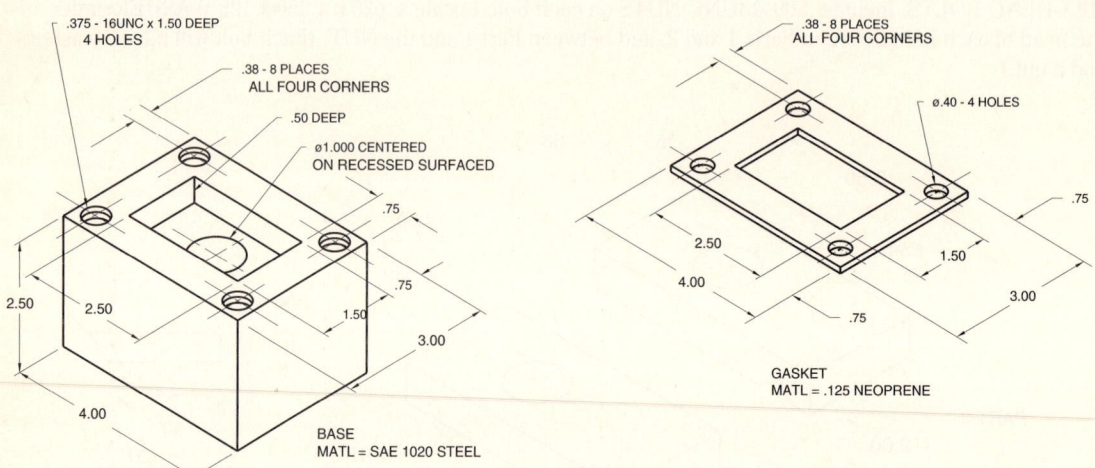

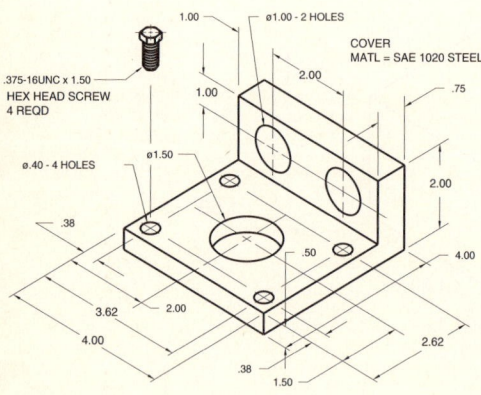

⬡ 11.3 Hub Assembly

Create an assembly drawing for the parts shown. Use 8 bolts with nuts to assemble the parts.

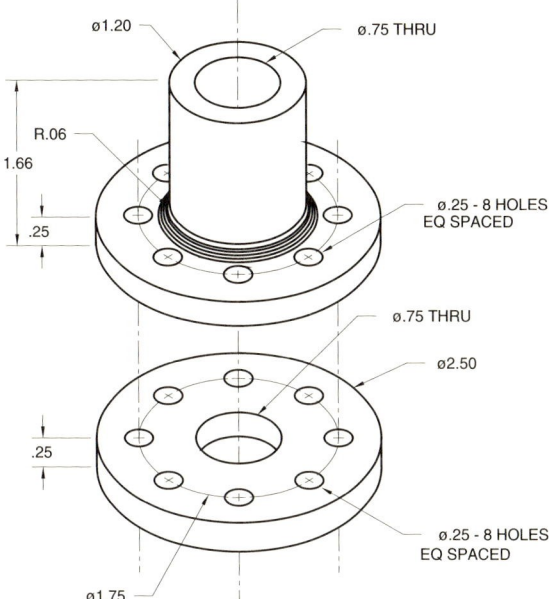

ø1.20

ø.75 THRU

R.06

1.66

.25

ø.25 - 8 HOLES
EQ SPACED

ø.75 THRU

ø2.50

.25

ø.25 - 8 HOLES
EQ SPACED

ø1.75

11.4 Turbine Housing Assembly

Create detail drawings for the two parts shown below and save each as a Wblock. Insert each block to form an assembly.

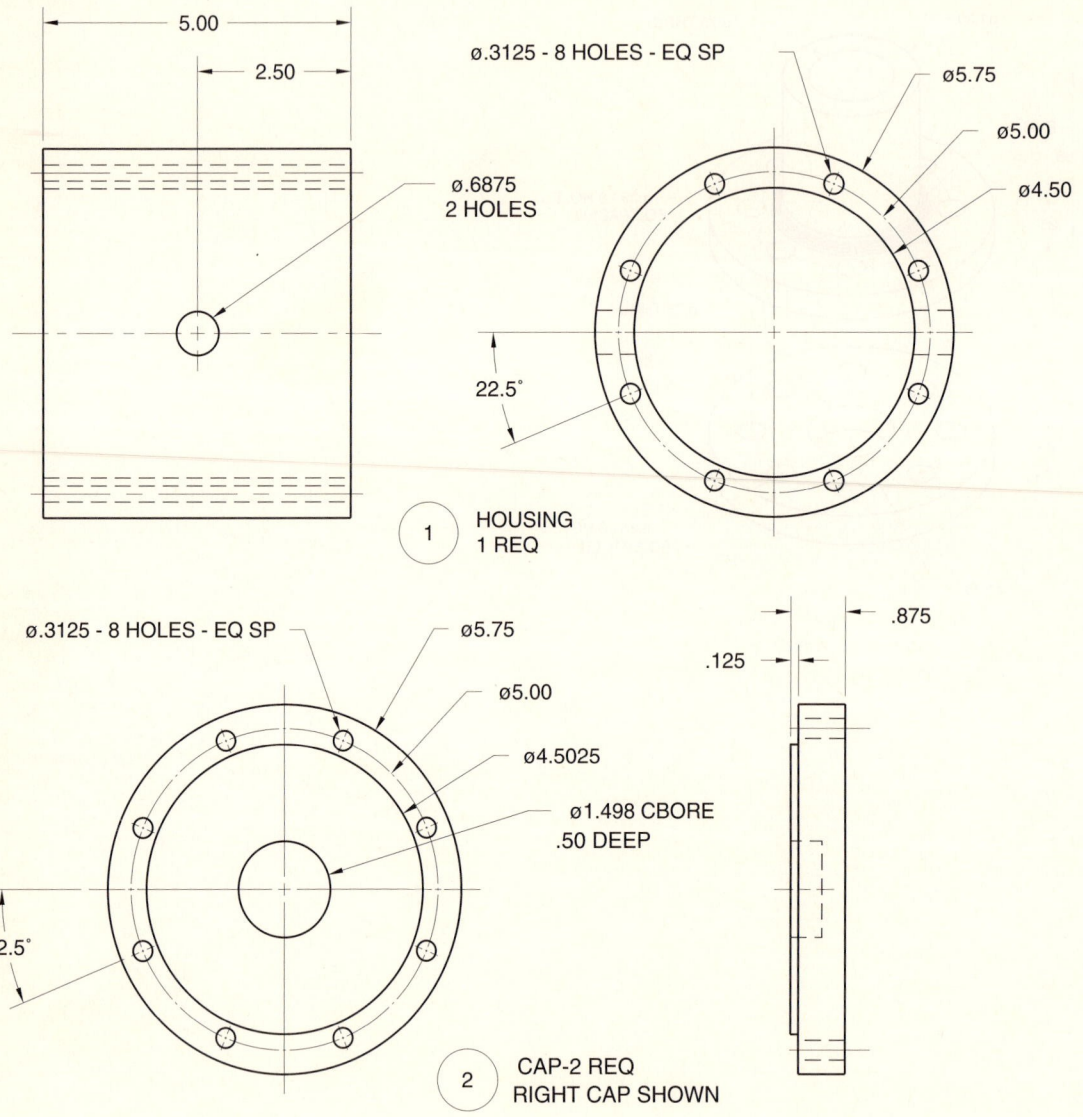

5.00

2.50

ø.6875
2 HOLES

ø.3125 - 8 HOLES - EQ SP

ø5.75

ø5.00

ø4.50

22.5°

1 HOUSING
1 REQ

ø.3125 - 8 HOLES - EQ SP

ø5.75

ø5.00

ø4.5025

ø1.498 CBORE
.50 DEEP

.875

.125

22.5°

2 CAP-2 REQ
RIGHT CAP SHOWN

Create an assembly drawing for the parts shown, using solid modeling techniques. Experiment with different materials and analyze mass properties.

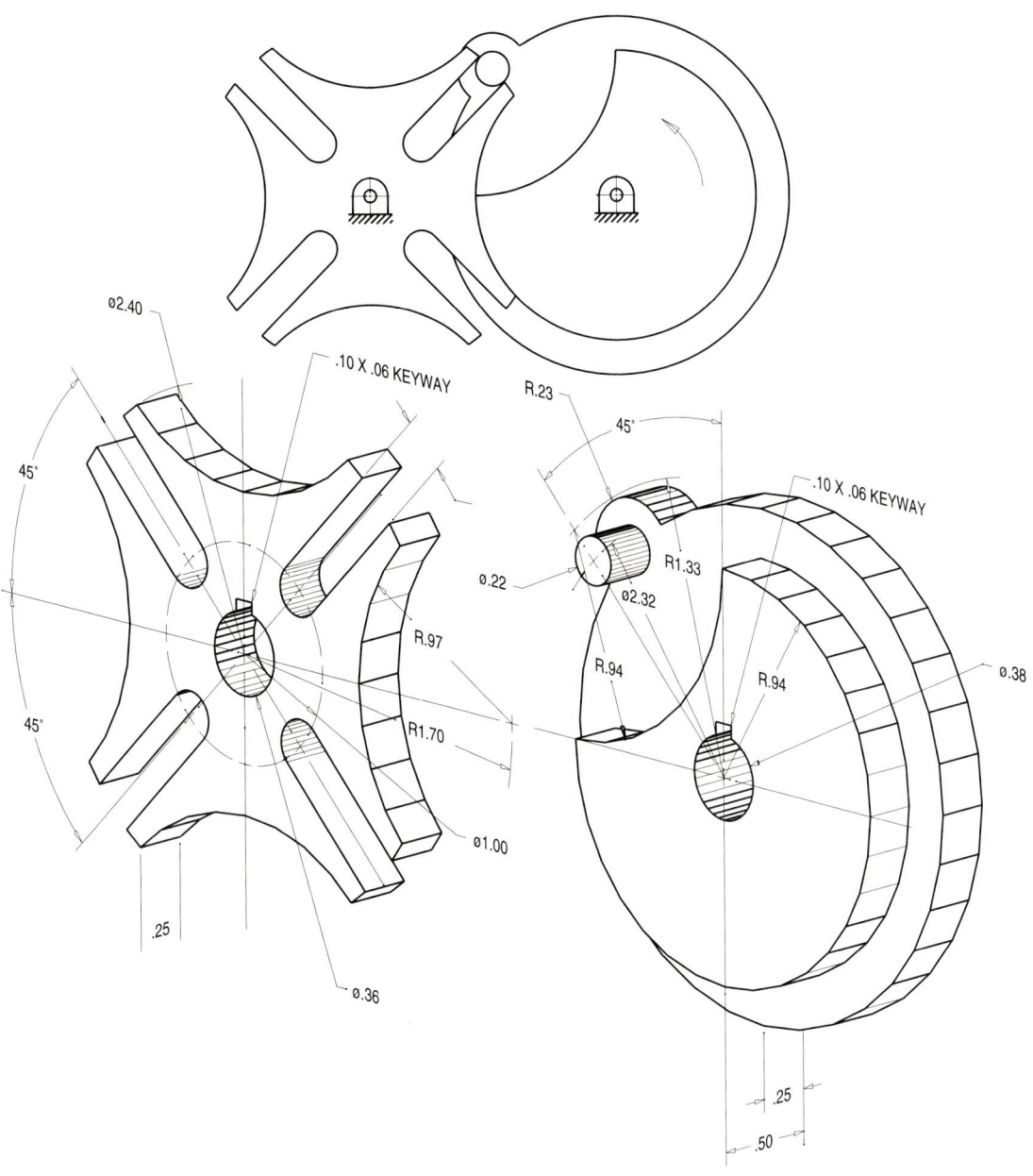

11.6 Staircase

Use solid modeling to generate this staircase assembly detail.

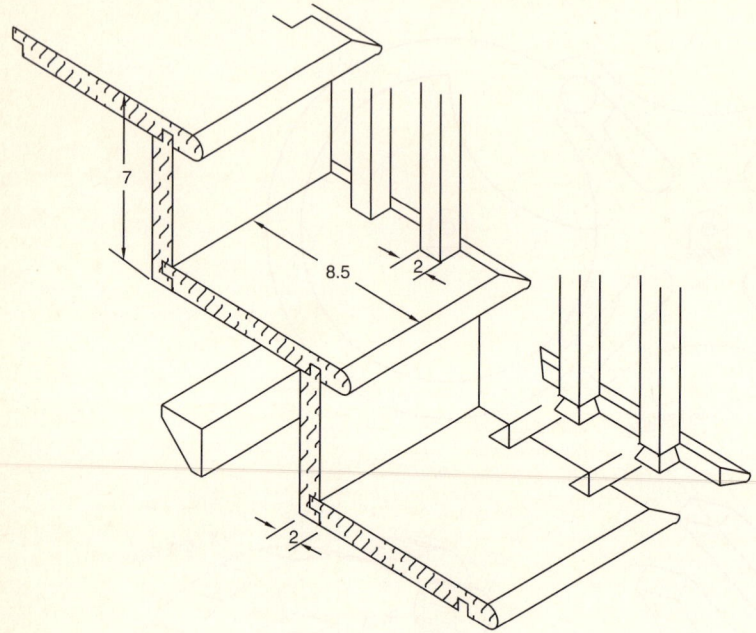

11.7M Universal Joint

Create an assembly drawing for the parts shown, using solid modeling techniques. Analyze the mass properties of the parts.

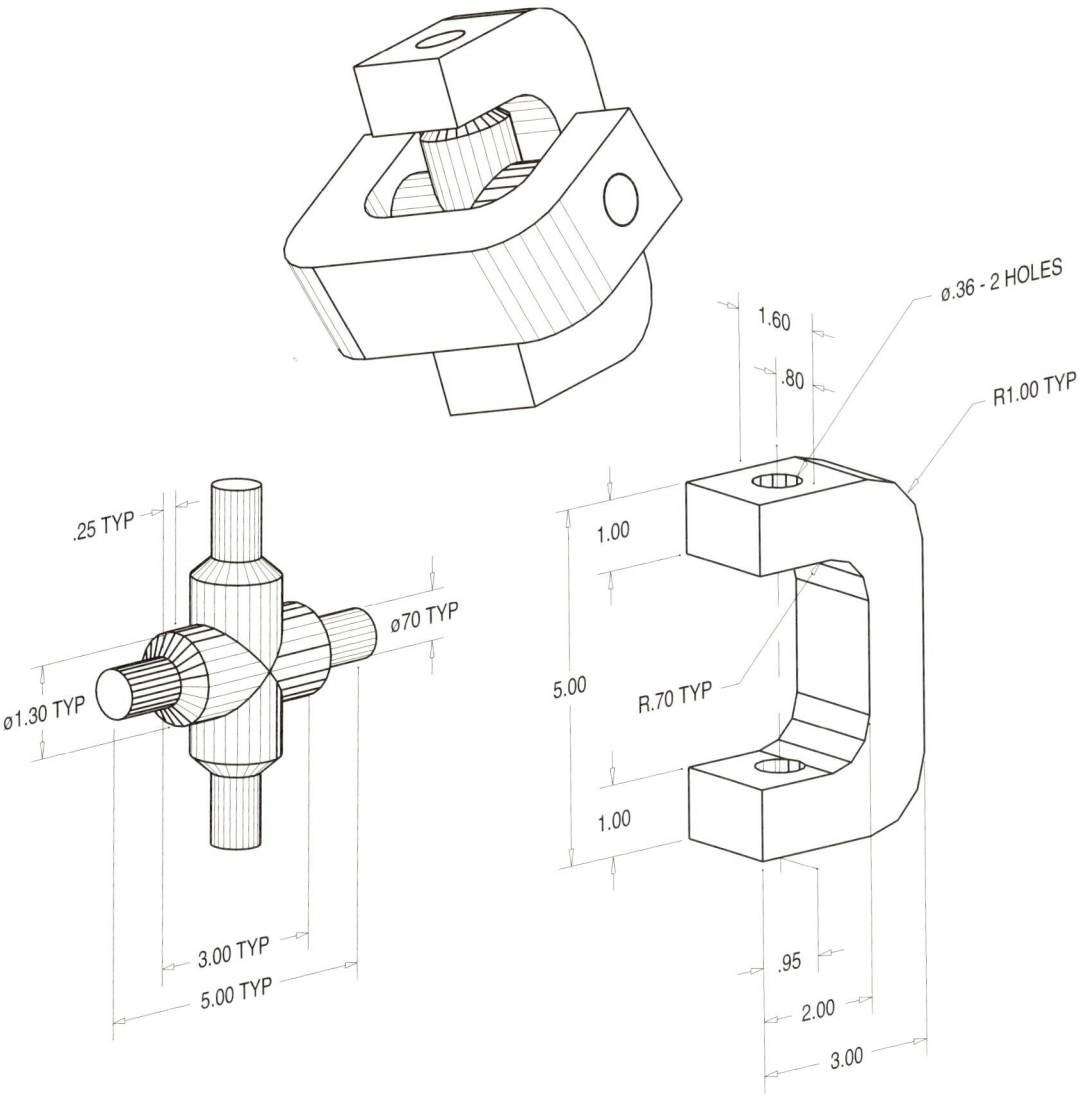

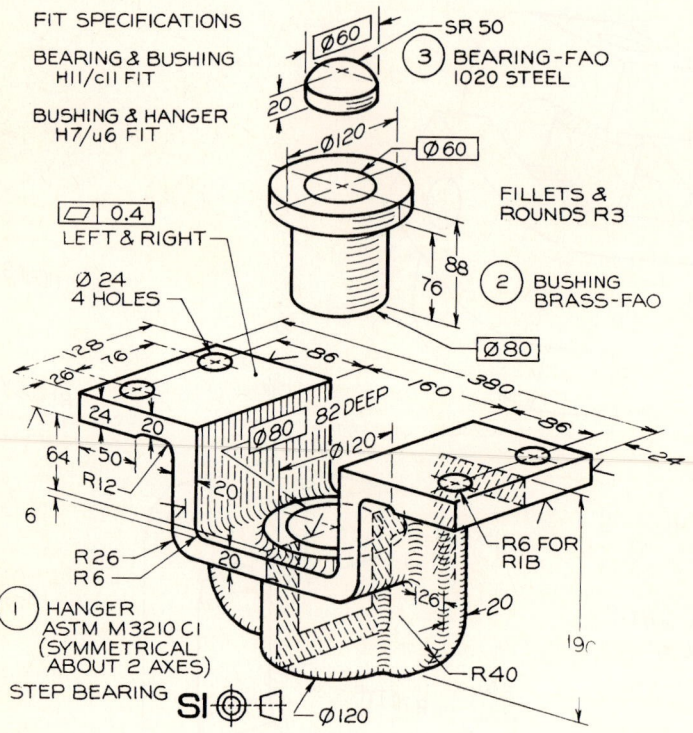

11.8M Step Bearing

Use solid modeling to create an assembly drawing of this step bearing.

11.9M Tensioner

Create an assembly drawing of this tensioner using solid modeling.

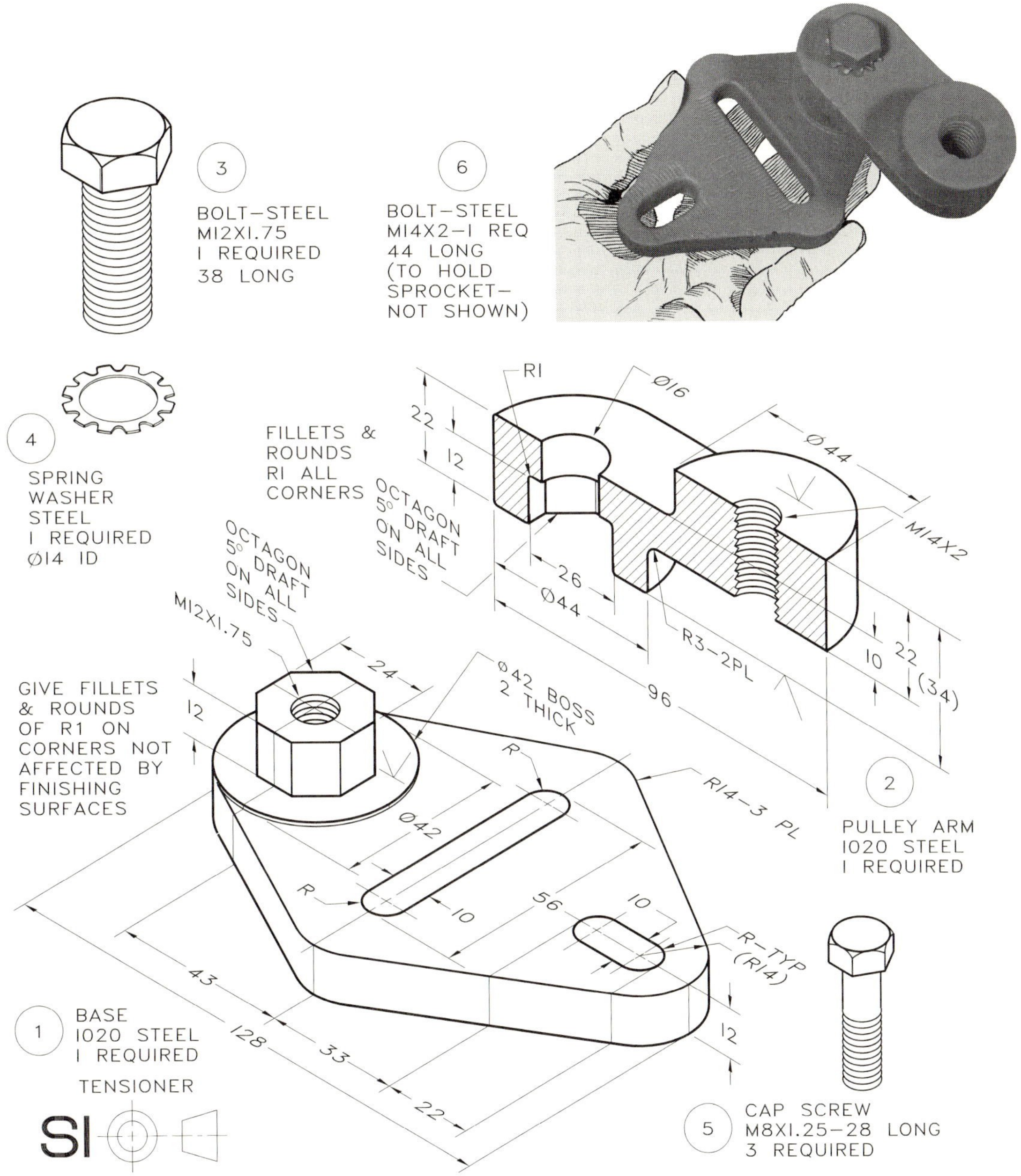

3 — BOLT—STEEL M12X1.75 I REQUIRED 38 LONG

6 — BOLT—STEEL M14X2—I REQ 44 LONG (TO HOLD SPROCKET— NOT SHOWN)

4 — SPRING WASHER STEEL I REQUIRED Ø14 ID

FILLETS & ROUNDS RI ALL CORNERS

OCTAGON 5° DRAFT ON ALL SIDES

RI

Ø16

Ø44

M14X2

22

12

26

Ø44

R3—2PL

10

22

(34)

96

2 — PULLEY ARM 1020 STEEL I REQUIRED

OCTAGON 5° DRAFT ON ALL SIDES

M12X1.75

24

12

GIVE FILLETS & ROUNDS OF R1 ON CORNERS NOT AFFECTED BY FINISHING SURFACES

Ø42 BOSS 2 THICK

Ø42

R

R

10

R14—3 PL

56

10

R—TYP (R14)

12

43

128

33

22

1 — BASE 1020 STEEL I REQUIRED

TENSIONER

SI

5 — CAP SCREW M8X1.25—28 LONG 3 REQUIRED

Make an assembly drawing of this trailer hitch using solid modeling.

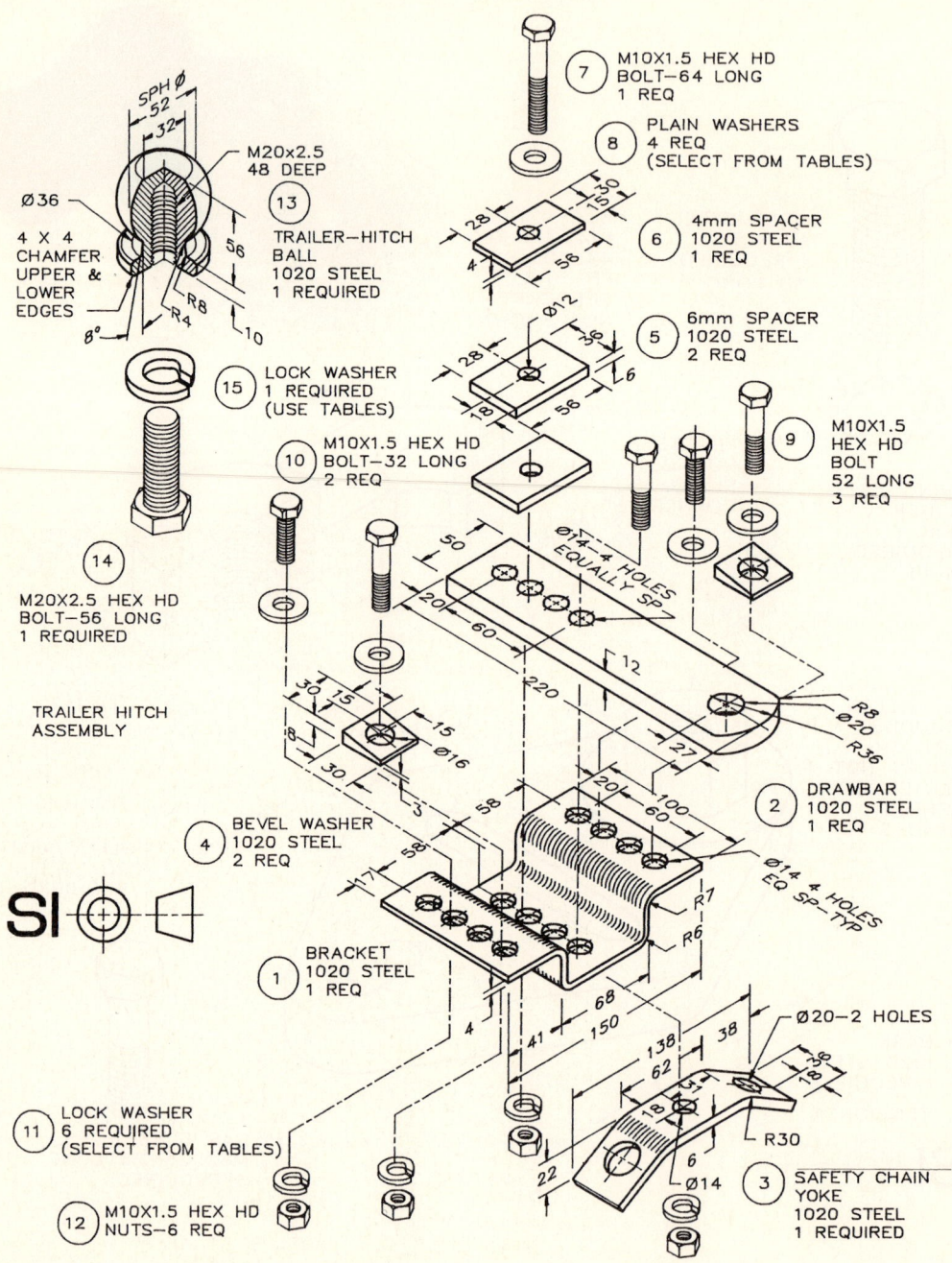

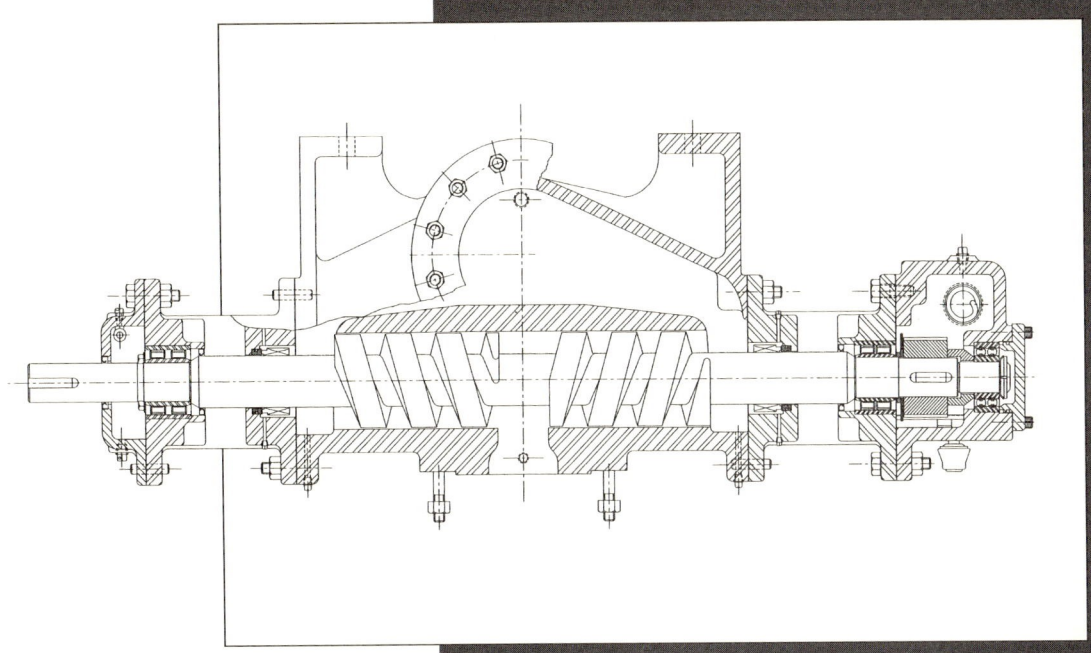

TUTORIAL 12

Creating Section Views Using 2D and Solid Modeling

Objectives

When you have completed this tutorial, you will be able to

1. Show the internal surfaces of an object, using 2D section views.

2. Locate and draw cutting plane lines and section lines on appropriate layers.

3. Hatch an area with a pattern.

4. Edit associative hatching.

5. Section and slice a solid model.

6. Control layer visibility within a viewport.

Introduction

In this tutorial you will learn how to draw section views. A *section view* is a special type of orthographic view used to show the internal structure of an object. Essentially, it shows what you would see if a portion of the object were cut away and you were to look at the remaining part. Section views are often used when the normal orthographic views contain so many hidden lines that they are confusing and difficult to interpret. If you are creating your drawings with solid modeling, you can generate a section view directly from the solid model. In this tutorial you will learn to use both two-dimensional (2D) and solid modeling to create section views.

Section View Conventions

Figure 12.1 shows the front and side views of a circular object. The side view contains many hidden lines and is somewhat difficult to interpret.

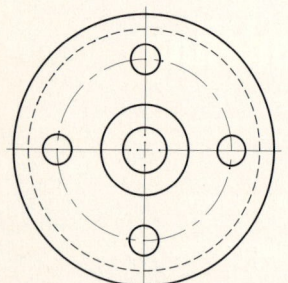 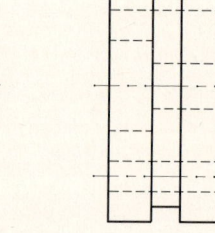

Figure 12.1

Figure 12.2 shows a pictorial drawing of the same object cut in half along its vertical centerline. This type of drawing is called a *pictorial full section*.

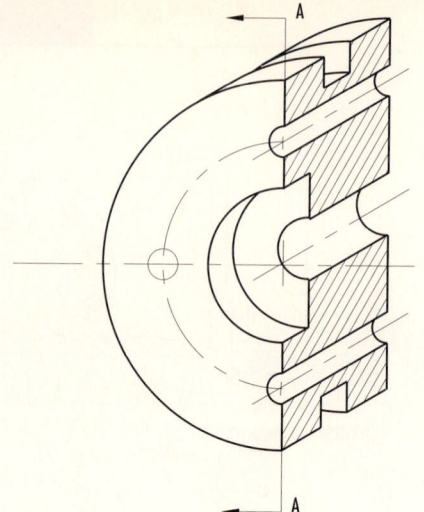

Figure 12.2

In many drafting applications, an area (such as the portion of an object that has been cut to show a section view) is filled with a pattern to make the drawing easier to interpret. The pattern can help differentiate components of a three-dimensional (3D) object, or it can indicate the material comprising the object. You can accomplish this *crosshatching* or pattern filling by using AutoCAD's Boundary Hatch command.

In Figure 12.2, crosshatching shows the solid portions of the object, that is, where material was cut to make the section view.

Figure 12.3 shows front and section views of the same circular object. A *cutting plane line*, line A–A, defines where the sectional cut should be made and indicates the direction in which you should view the object to produce the section view.

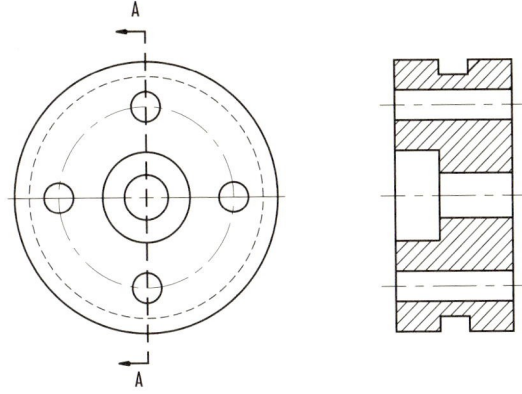

Figure 12.3

Compare the side view in Figure 12.1 with the section view in Figure 12.3 and the 3D section view shown in Figure 12.2. Figures 12.2 and 12.3 are easier to interpret than the front and side views shown in Figure 12.1.

Figure 12.4 shows a different object with a cutting plane line and the appropriate section view.

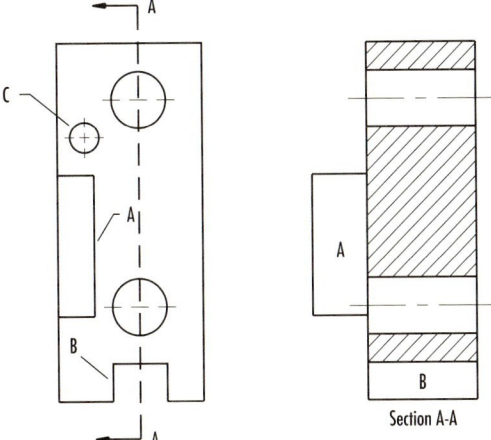

Section A-A

Figure 12.4

Crosshatching

Crosshatching is used to show where material has been cut. Note in Figure 12.4 that the surfaces labeled A and B are not crosshatched. Only material that has been cut by the cutting plane is crosshatched. Surfaces A and B are shown in section A–A because they are visible after the cut has been made, but they are not hatched because they were not cut by the cutting plane.

A section view shows what you would see if you looked at the object in the direction the arrows on the cutting plane line point, with all the material behind the arrows removed. When creating a section view, you should ignore the portion of the object that will be cut off. Draw the remaining portion, toward which the arrows on the cutting plane line point. Remember, once the object has been cut, some features that were previously hidden will be visible and should be shown. The hole labeled C in the side view of Figure 12.4 is in the remaining portion of the object, but is not visible, so it is not shown in the section view. The purpose of a section view is to show the internal structure without the confusion of hidden lines. You should use hidden lines only if the object would be misunderstood if the lines were not included. (You could use an offset cutting plane line—which bends 90° to pass through features that are not all in the same plane—to show hole C in the section views, if you want to. Refer to any standard engineering graphics text to review conventions for cutting plane lines and types of section views.)

Starting

Start AutoCAD and, on your own, use the data file *cast2.dwg* that came with your software as a template file to begin a new drawing. Save the drawing as *bossect.dwg*.

Creating 2D Section Views

The top, front, and side orthographic views of a casting are displayed on your screen. The horizontal centerlines have been erased in the top view. Your screen should look like Figure 12.5.

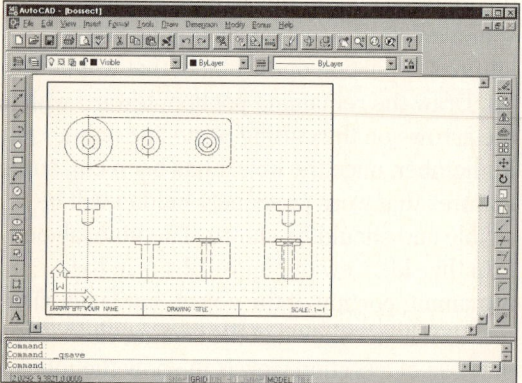

Figure 12.5

Often one or more of the orthographic views are replaced with section views. You will use the front orthographic view to draw a front section view. You will use the side view to draw a side section view through the boss.

Cutting Plane Lines

You can draw a cutting plane line by using one of two different linetypes: long dashes (DASHED) or long lines with two short dashes followed by another long line (PHANTOM). You should use only one type of cutting plane line in any one drawing. You will make dashed cutting plane lines by creating a new layer with the linetype DASHED. Linetype DASHED has longer dashes than linetype HIDDEN. You should use a different color for the cutting plane layer to help distinguish it from the other layers and linetypes. Also, cutting plane lines are drawn with a thick pen during plotting. Because using different object colors allows for different pens to be selected for plotting, you should use different colors for thick cutting plane lines and thin lines.

Creating Dashed Lines

Create a new layer for the cutting plane lines on your own. Name the new layer CUTTING_PLANE. Assign it the color white (which will appear black on your screen if you are using a white background) and linetype DASHED. Set layer CUTTING_PLANE as the current layer. Be sure that Snap is turned on and set to 0.25.

Pick: **Line icon**

On your own, add a horizontal dashed cutting plane line in the top view of the boss that passes through the center of the holes in the top view, that is, between points A–A in Figure 12.6. Then add a vertical dashed cutting plane line through the center of the top view of the boss line between points B–B. The lines should extend beyond the edges of the object by at least 0.5 unit.

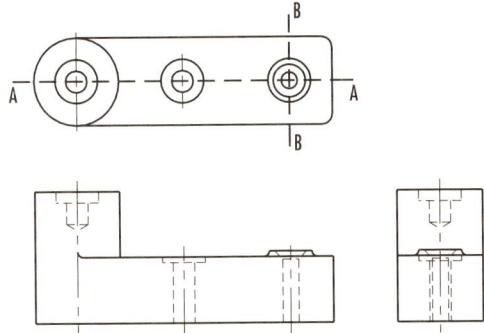

Figure 12.6

The cutting plane lines shown in Figure 12.6 should now appear in your drawing on layer CUTTING_PLANE. Save your drawing on your own.

Drawing the Leaders

As you learned in Tutorial 11, the Leader command lets you construct complex leaders. You will use the Leader command to add the arrowed line segments and the identifying letters for the cutting plane lines. Because the arrowed line segments are part of the cutting plane lines, they should be drawn on layer CUTTING_PLANE.

You will start the leader 0.5 unit to the left of the top end of cutting plane line B–B.

> ■ *TIP* The Leader command will not draw the arrows if the first line segment of the leader is too short. For AutoCAD to draw an arrow, the first line segment must be longer than the current arrowhead size. You can set the arrowhead size with the dimensioning variables that you learned to use in Tutorial 7. ■

On your own, turn on the Dimension toolbar that you used in Tutorial 7 before you continue. Use what you have learned about floating toolbars to position it on your screen so that it is out of the way, but still handy for picking commands.

On your own, verify that Snap is set to 0.25 before you draw the leader line. Use Ortho to assist you in drawing straight lines. Also be sure that OSNAP is turned on, with Endpoint selected.

Pick: **Leader icon**

From point: *(select a point 0.5 unit to the left of, and even with, the top end of cutting plane line B–B)*

To point: *(pick the top end of the cutting plane line B–B)*

To point (Format/Annotation/Undo)<Annotation>: ⏎

Annotation (or press ENTER for options): **B** ⏎

MText: ⏎

> ■ *TIP* Remember, after typing text you must press ⏎ on the keyboard. The return button on your pointing device does not work for entering text. ■

Repeat these steps on your own to create the leader for the bottom of cutting plane line B–B.

> ■ *TIP* You can type *U* at the *To point* prompt to remove the last leader segment drawn if you make an error. ■

For cutting plane line A–A, you will end the command after the first segment of the line has been drawn and use an option of the Leader command for not adding an annotation. Then you will use the Dtext command to place the text labeling cutting plane line A–A.

Leader Command Options

The Leader command provides the following options once you have drawn the leader line: Format, Annotation, and Undo. The Format option has further options that allow you to select the shape for the leader line: Splined, Straight, Arrow, and None. The Splined and Straight options let you choose splined lines (so that you can make a curved leader) or straight lines. The Arrow and None options let you choose to draw an arrow at the end of the leader or draw a leader with no arrow.

The Annotation option also has further selections: Tolerance, Copy, Block, None, and Mtext. The default option is Mtext; you have used this option to create text at the end of the leader line, and you can also use it to create multiple lines of text. The Tolerance selection lets you add a geometric tolerance feature control frame to the end of the leader line. (You will use this option in Tutorial 14 when you learn about advanced dimensioning features.) The Copy option lets you pick on an existing feature control frame, block, or text object and copy it to the end of the new leader line. Choosing Block lets you select a block to be added to the end of the leader, as you did in Tutorial 11. The None option lets you draw a leader line with no text added to the end. You will use the None option to end the Leader command when you draw the leaders for cutting plane line A–A.

Pick: **Leader icon**

From point: *(select a point 0.5 unit above, and even with, the left endpoint of cutting plane line A–A)*

To point: *(pick the left endpoint of cutting plane line A–A)*

To point (Format/Annotation/Undo)<Annotation>: ⏎

Annotation (or press ENTER for options): ⏎

Tolerance/Copy/Block/None/<Mtext>: **N** ⏎

Command: ⏎ *(to restart the Leader command)*

From point: *(pick a point 0.5 unit above, and even with, the right endpoint of cutting plane line A–A)*

To point: *(pick the right endpoint of cutting plane line A–A)*

To point (Format/Annotation/Undo)<Annotation>: ⏎

Annotation (or press ENTER for options): ⏎

Tolerance/Copy/Block/None/<Mtext>: **N** ⏎

Type: **DTEXT** ⏎

Justify/Style/<Start point>: **J** ⏎

Align/Fit/Center/Middle/Right/TL/TC/TR/ML/MC/MR/BL/BC/BR: **M** ⏎

Middle point: *(pick on the snap point below the left leader for cutting plane line–A)*

Height: <0.2000>: ⏎

Rotation angle <0>: ⏎

Text: **A** ⏎

Text: *(pick a point below the end of the right leader for cutting plane line A–A)*

Text: **A** ⏎

Text: ⏎

When completed, your leader lines should look like the ones shown in Figure 12.7.

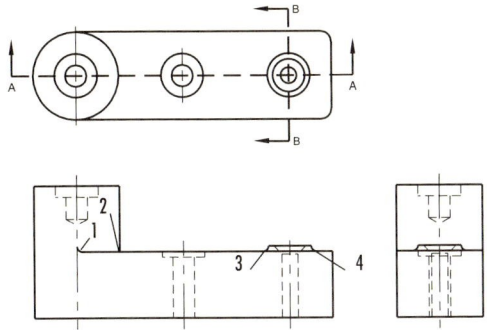

Figure 12.7

> ■ *TIP* You can easily insert arrowheads at the end of a cutting plane line by making your own arrowhead block. A quick way to make a block for arrowheads is to explode a leader line and make a block from its arrowhead. You can also use a Polyline with a zero starting width and larger ending width to create an arrowhead from scratch. ■

When you have finished creating the cutting plane lines as shown in Figure 12.7, turn off the Dimension toolbar on your own by clicking its Windows close box.

Adjusting Linetype Scale

As you have learned in previous tutorials, you can change the general linetype scaling factor to improve the overall appearance of the cutting plane lines by adjusting their scale. You will type *LTSCALE* at the command prompt to activate the Linetype Scale command.

Command: **LTSCALE** ⏎

New scale factor <0.7500>: **.5** ⏎

Observe how the overall lines are affected. You can continue to try different values until the lines suit your particular needs.

Next, you will remove the unnecessary lines from the front view. When you draw a section view, you do not want to see the lines that represent intersections and surfaces on the outside of the object.

On your own, turn Snap off and use the Erase command to remove the *runout* (or curve that represents the blending of two surfaces) labeled 1 in Figure 12.7. Use Trim to remove the portion of the line to the left of point 2. Use the Erase command to remove the solid line from point 3 to point 4, which defines the surface in front of the boss. Use Zoom Window to help identify the points. Your drawing should be similar to that in Figure 12.8 when you are finished.

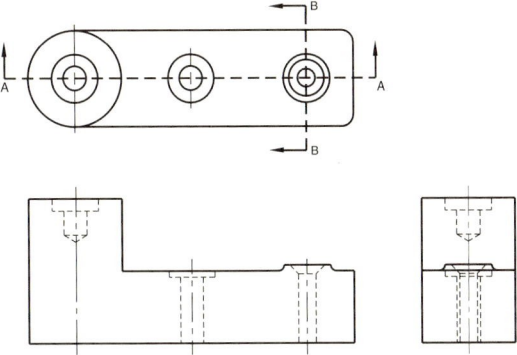

Figure 12.8

Changing Properties

 When you section the drawing, the interior details become visible. You will use the Properties selection to change the layer of the hidden lines to VISIBLE. Pick the Properties icon from the Object Properties toolbar.

Pick: **Properties icon**

Select objects: **(pick all of the hidden lines in the front view)**

Select objects: ⏎

The Change Properties dialog box appears on your screen. You will use it to pick the Layer button and then use the Select Layer dialog box to set the new layer for the selected lines to layer VISIBLE.

Pick: **Layer**

Pick: **VISIBLE**

Pick: **OK *(to exit the Select Layer dialog box)***

Pick: **OK *(to exit the Change Properties dialog box)***

Now all the hidden lines in the front view have been changed to layer VISIBLE. Your screen should now be similar to Figure 12.9.

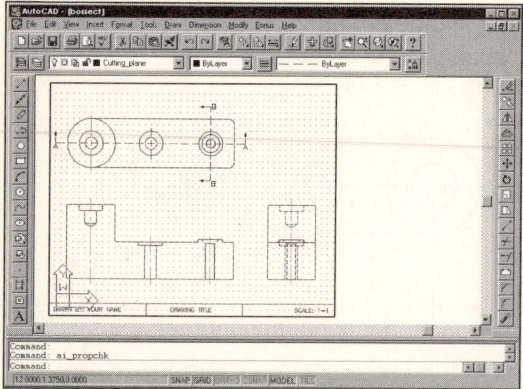

Figure 12.9

If you have zoomed your drawing, use Zoom Previous to return to the original size. You are almost ready to add the hatching to the drawing. The hatching will fill the areas of the drawing where the solid object was cut by the cutting plane.

Hatching should be on its own layer and should be plotted with a thin pen. Having the hatch on its own layer is useful so that you can freeze it if you do not want it displayed while you are working on the drawing. In the template drawing *cast2.dwg*, a separate layer for the hatching has already been created. Its color is red, and its linetype is CONTINUOUS.

Use Layer Control from the Object Properties toolbar to set HATCH as the current layer on your own.

Using Boundary Hatch

 Hatching comprises a series of lines, dashes, or dots that form a pattern. The Boundary Hatch command fills an area with a pattern. It creates *associative hatching*, which means that, if the area selected as the hatch boundary is changed, the hatching is automatically updated to fill the new area. Associative hatching is the default for the Boundary Hatch command. If you update the selected boundary so that it no longer forms a closed area, the hatching will no longer be associative. Boundary Hatch helps you choose the boundary for the pattern by selecting from its various boundary selection options. To activate Boundary Hatch, select the Hatch icon from the Draw toolbar.

Pick: **Hatch icon**

The Boundary Hatch dialog box appears on your screen, as shown in Figure 12.10. The checked box near the bottom right of the dialog box indicates that associative hatching is turned on.

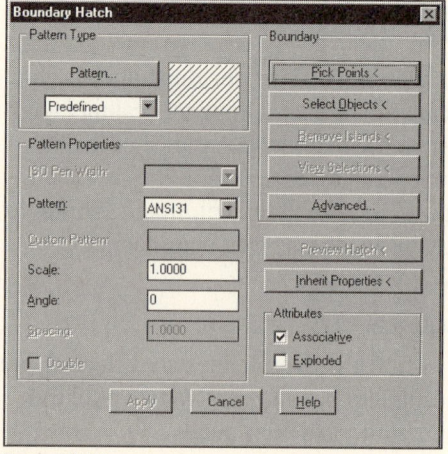

Figure 12.10

Selecting the Hatch Pattern

You can select predefined hatch patterns by pulling down the list in the Pattern Properties area. ANSI31 is the default hatch pattern. It consists of continuous lines angled at 45°, spaced about 1/16" apart, and is typically used to show cast iron materials. It is also used as a general pattern when you do not want to specify a material by the hatch pattern. To select a hatch pattern, pick on the name ANSI31 shown to the right of the word Pattern in the Pattern Properties area of the dialog box.

Pick: **ANSI31** *(to the right of Pattern)*

The list of predefined hatch patterns pulls down, as shown in Figure 12.11.

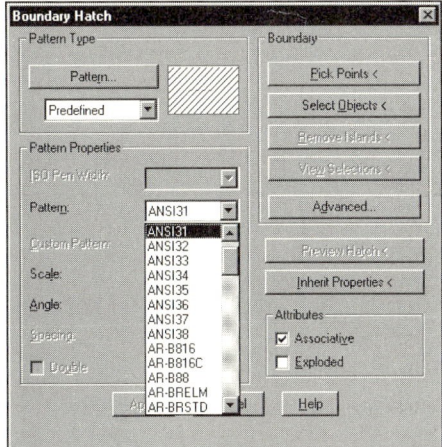

Figure 12.11

You can also select a hatch pattern by picking on the active image tile in the Pattern Type area of the dialog box to scroll through the available patterns. Refer to the discussion of image tiles in Tutorial 7 if you need help using them to select a pattern.

Pick: **(on the image tile to the right of Predefined)**

When you pick on the image tile, the hatch pattern changes. The name of the pattern appearing in the image tile is shown to the right of Pattern where the name ANSI31 used to

appear. After you have looked at some of the hatch patterns, return the selection to ANSI31 on your own.

In addition to Predefined hatch patterns, you can use *custom hatch patterns* that you previously created and stored in the file *acad.pat* or another *.pat* file of your own making. You can also apply *user-defined hatch patterns,* which you can create with the Boundary Hatch dialog box by changing the Pattern Type selection to Custom and specifying the angle and spacing for the hatching using the lower left portion of the dialog box. You can make user-defined hatch patterns more complex by setting the linetype for the layer on which you will create the hatch. Try these features on your own. Remember to refer to AutoCAD's online help for more information about any of the commands described in these tutorials.

Scaling the Hatch Pattern

You can set the scale for the hatch pattern by using the Boundary Hatch dialog box. Doing so determines the spacing for the hatched lines in your drawing by scaling the entire hatch pattern. The hatch pattern is similar in some ways to a linetype. Hatch patterns are stored in an external file, with the spacing set at some default value. Usually the default scale of 1.0000 is the correct size for drawings that will be plotted full scale, resulting in hatched lines that are about 1/16" apart. If the views will be plotted at a smaller scale, half-size for instance, you will need to increase the spacing for the hatch by setting the scale to a larger value so that the lines of the hatch are not as close together. For this drawing, you will not need to change the scale of the hatch pattern.

■ *Warning:* Be cautious when specifying a smaller scale for the hatch pattern because if the hatch lines are very close together, AutoCAD will take a long time to calculate all of them and may even run out of space on your hard disk, causing your system to crash. ■

Angling the Hatch Pattern

The Angle area of the Boundary Hatch dialog box allows you to specify a rotation angle for the hatch pattern. The default is 0°, or no rotation of the pattern. In the hatch pattern you selected, ANSI31, the lines are already at an angle of 45°. Any angle you choose in the Angle area of the dialog box will be added to the chosen pattern; for example, if you added a rotation angle of 10° to your current pattern, the lines would be drawn at an angle of 55°. Leave the default value of 0° for now.

Selecting the Hatch Boundary

The Boundary area of the Boundary Hatch dialog box lets you select among the different methods that you can use to specify the area to hatch. Pick Points allows you to pick inside a closed area to be hatched; Select Objects lets you select drawing objects forming a closed boundary for the hatching. Picking the Advanced button opens the Advanced Options dialog box, which lets you select more options for the creation of the hatch boundary.

Pick: **Advanced**

Advanced Options

You can create boundaries for hatching as one of two different types of objects: polylines or regions (recall from Tutorial 9 that a region is a 2D enclosed area). When determining the hatch boundary, AutoCAD analyzes all the objects on the screen. This step can be time-consuming if you have a large drawing data-base. To define a different area to be considered for the hatch boundary, you would pick the Make New Boundary Set button. You would then return to your drawing, where you could select the objects to form the new set. Using implied Window works well for this procedure.

You can use the Style area of the Advanced Options dialog box to change the way islands inside the hatch area are treated. Observe the image tile as you explore the options. The selection Normal causes islands inside the hatched area to be alternately hatched and skipped, starting at the outer area. Choosing Outer causes only the outer area to be hatched; any islands inside are left unhatched. You can also choose Ignore; then, the entire inside is hatched, regardless of the structure. When you are finished observing the effect of these selections on the image tile, be sure to leave this area of the dialog box set to Normal.

Turning off Island Detection (removing the check from this box) allows you to select the boundary by *ray casting*. Using the defaults, ray casting finds the nearest closed boundary in any direction. For quicker results, you can change the direction from which the ray casting is started by picking on Nearest and using the list that appears to select a different direction. Leave Island Detection turned on for now.

Pick: **OK (to exit the Advanced Options dialog box)**

Next, you will specify Pick Points as the method for selecting the area to fill with the hatch pattern. You will return to your drawing to select points *inside* the areas that you want to have hatched. (To use the Select Objects method, you would return to your drawing and pick *on the objects* that form the boundary.) AutoCAD will determine the boundary around the areas you picked. (If there are islands inside the area that you did not want hatched, you would use Remove Islands to pick them after you have picked the area to be hatched.)

When you are finished selecting, you will press ⏎ to tell AutoCAD that you are finished selecting and want to continue with the Boundary Hatch command. You can use any standard AutoCAD selection method to select the objects. Refer to Figure 12.12 as you make selections.

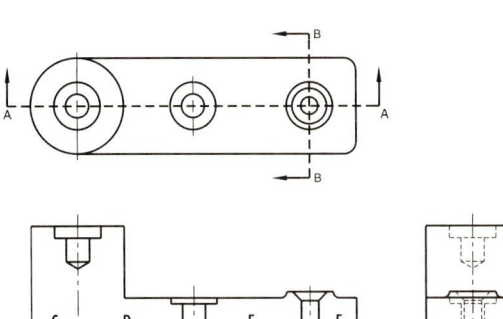

Figure 12.12

Pick: **Pick Points**

You are returned to your drawing screen and the following command prompt appears.

Select internal point: *(pick a point inside the area labeled C)*

The perimeter of the area that you have selected is highlighted, and you again see the prompt:

Select internal point: *(pick a point inside the area labeled D)*

Select internal point: *(pick inside the area labeled E)*

Select internal point: *(pick inside the area labeled F)*

Select internal point: ⏎

You return to the Boundary Hatch dialog box. Some of the choices that were previously grayed (so that you couldn't select them) are now available.

■ *TIP* Turn Grid off before previewing, as it can make reading the hatch pattern difficult. ■

Pick: **Preview Hatch**

You return to your drawing screen with the hatch showing so that you can confirm that the areas you selected are correct. You will pick the Continue box when you are ready to return to the Boundary Hatch dialog box. If the area was not hatched correctly, make the necessary changes and then preview the hatch again. When what you see is correct, you will pick Apply to apply the hatch. (Your screen may not show the preview perfectly, but if the areas are generally correct, they will probably hatch correctly.) If the hatch appeared correctly,

Pick: **Continue *(to return to the Boundary Hatch dialog box)***

Pick: **Apply *(in the lower left of the dialog box)***

The area that you selected becomes hatched in your drawing. Your drawing should now look like the one in Figure 12.13.

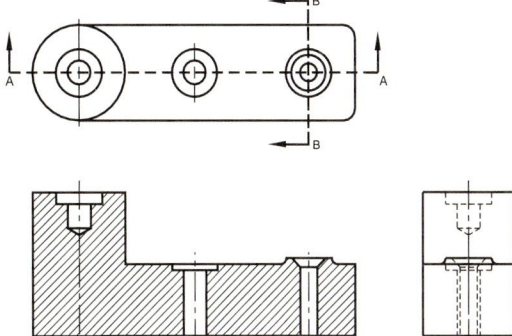

Figure 12.13

Adjusting Your Viewing Area

To draw section B–B, you will need to increase the space available to draw the side view on your screen. Use Zoom to zoom out so that more drawing area is shown in the viewport.

Command: **Z** ⏎

All/Center/Dynamic/Extents/Previous/Scale(X/XP)/Window/ <Realtime>: **.75XP** ⏎

Your drawing appears smaller, so more of it fits into the viewport.

Next, you will use the Pan Realtime command to move the view to the left in the viewport to make more room to the right of the existing views.

Pick: **Pan Realtime icon** *(move the image until it is close to the left hand side of the viewport)*

Pick: *(with the right mouse button)*

Pick: **Exit**

Now you have space to create section B–B to the right of the top view. You will use the Rotate icon from the Modify toolbar to reorient the side view so that you can align it with the top view, as shown in Figure 12.14. On your own, verify that Snap is on.

Pick: **Rotate icon**

Select objects: *(select all of the objects in the side view with implied Crossing)*

Select objects: ↵

Base point: *(pick the top left corner of the side view)*

<Rotation angle>/Reference: **90** ↵

> ■ *TIP* You can also use the hot grips to rotate. Use implied Crossing to select all the objects you want to rotate. Once you've selected them and you see all the blue hot grips, pick a base grip that will be the base point in the Rotate command. After selecting the base grip, you will see the Stretch command echoed in the command prompt area. Press the right mouse button to activate the pop-up menu and select Rotate. Follow the prompts that appear on your screen to rotate the objects in the side view 90°. ■

Now you will use the Move command to align the rotated side view with the top view.

Pick: **Move icon**

Select objects: **P** *(for previous selection set)* ↵

Select objects: ↵

Base point or displacement: *(pick the upper left corner of the rotated object)*

Second point of displacement: *(pick a point on the snap that lines up with the top line in the top view)*

Your drawing should now look like that in Figure 12.14.

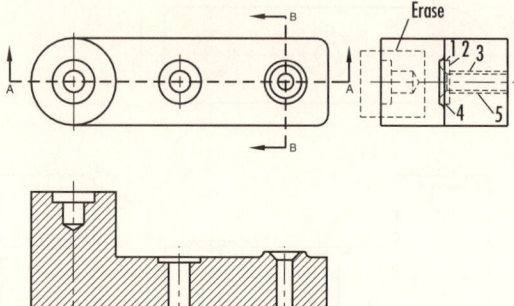

Figure 12.14

On your own, turn Snap off and use the Erase command to remove the unnecessary hidden lines (those for the counterbored hole and lines 1 through 5) from the rotated object. Refer to Figure 12.14 to help you make the selections. Then, on your own, change the hidden lines for the countersunk hole in the rotated side view to layer VISIBLE, as this hole is the only one that will show in the section view.

> ■ *TIP* You may need to use Zoom Window to enlarge the area on the screen. When you are finished erasing and changing the lines, use Zoom Previous to return the area to its original size. ■

Your screen should be similar to Figure 12.15 when you are finished.

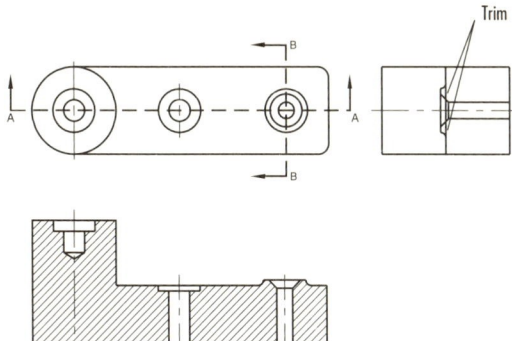

Figure 12.15

Next, use the Trim command on your own to remove the visible line that crosses the outer edge of the boss, as shown in Figure 12.15. When you are finished trimming, your drawing should look like that in Figure 12.16.

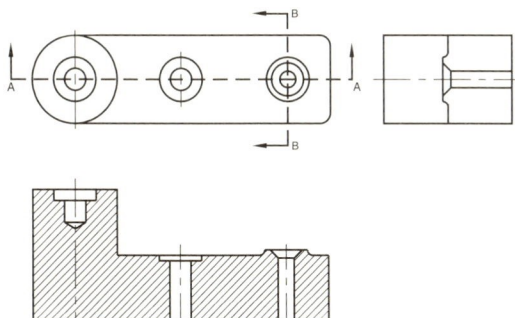

Figure 12.16

Now you are ready to add the hatching.

Pick: **Hatch icon**

In the Boundary Hatch dialog box, the hatch pattern ANSI31 should already be selected and shown at the top as the current pattern.

Pick: **Pick Points**

Select inside the areas you want to have hatched and press ↵ to return to the Boundary Hatch dialog box when you have finished. To be sure that the hatching shows correctly,

Pick: **Preview Hatch**

Pick: **Continue** *(to return to the dialog box)*

If the hatching appears the way you want it to,

Pick: **Apply**

When you are through, the hatching in your drawing should look like that on the screen in Figure 12.17.

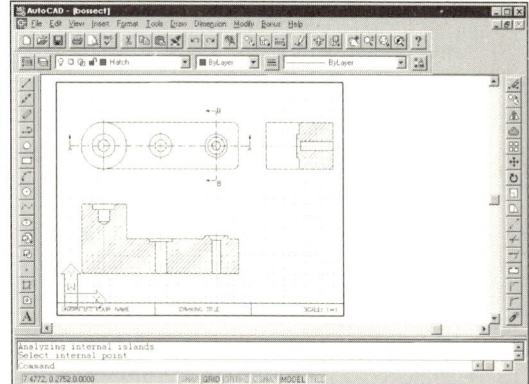

Figure 12.17

■ *TIP* Although you won't always be instructed to do so, remember to save your work frequently while creating drawings. AutoCAD has an autosave feature that saves automatically to a file named *auto.sv$*. You set the time interval for the autosave feature with the system variable Savetime. The default is to save every 120 minutes. You would not want to lose two hours of work, so you should set AutoCAD to save automatically at shorter intervals. To make such a change, type *SAVETIME* at the command prompt and enter the new number in minutes. However, do not rely on automatic saves. Save your file whenever you have made a major change that you want to keep or about every 10 minutes otherwise. ■

Plot your drawing on your own. Before you plot, be sure to switch to paper space. If you need help plotting, refer to Tutorial 3.

Solid Fill

A new feature in AutoCAD Release 14 is the Solid fill for hatching. Instead of patterns for a hatch, you can fill an area with a solid color. Use solid fill sparingly in your drawings because it is time consuming to plot, and some printers and plotters are not capable of showing it correctly. On your own, change the layer to DIM in order to use the color cyan.

Pick: **Rectangle icon**

Draw a small rectangle to the right of your front section view. Then,

Pick: **Hatch icon**

The Boundary Hatch dialogue box appears.

Pick: **Solid (from within the Pattern pull-down list)**

Pick: **Pick Points**

Select internal point: **(pick a point within the new rectangle box)**

Select internal point: ⏎

Pick: **Apply**

The rectangle is filled with a solid color, similar to Figure 12.18. The color is associated with the current layer DIM, so it should be cyan.

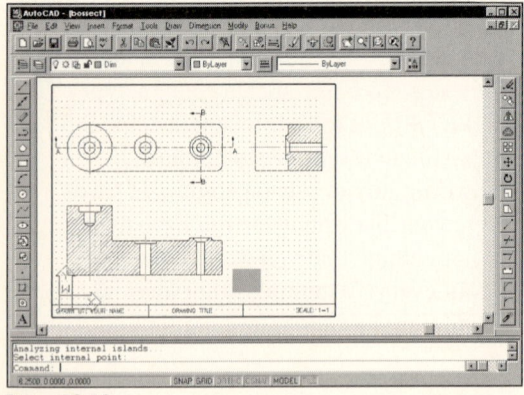

Figure 12.18

Draworder

 Another new feature in Release 14 is the Draworder icon. Draworder specifies which object takes precedence if two or more objects are over the top of one another in a view. On your own, turn on the toolbar Modify II.

On your own, move the rectangle a bit to the left so that it covers the front section view. When you have finished, your screen should look like Figure 12.19.

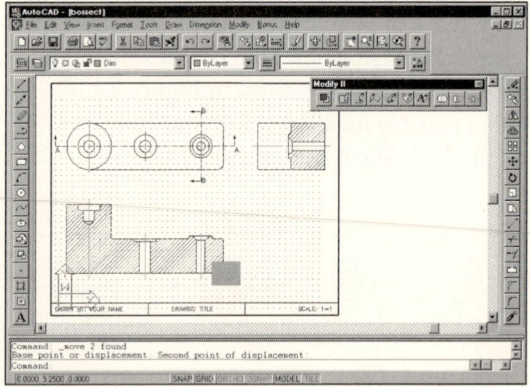

Figure 12.19

Pick: **Draworder icon**

Select objects: **(use implied Crossing to select the hatch solid and the rectangle)**

Select objects: ⏎

Above object/Under object/Front/<Back>: **U** ⏎

Select reference object: **(use implied Crossing to select the section hatch pattern and block)**

The Section hatch should now be drawn over the solid cyan hatch color. Use Zoom Window for a closer view of the overlay. Your screen should now look like Figure 12.20.

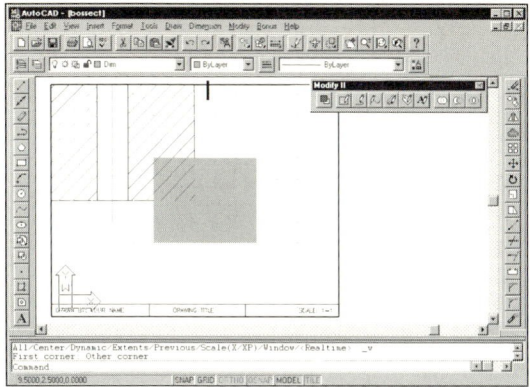

Figure 12.20

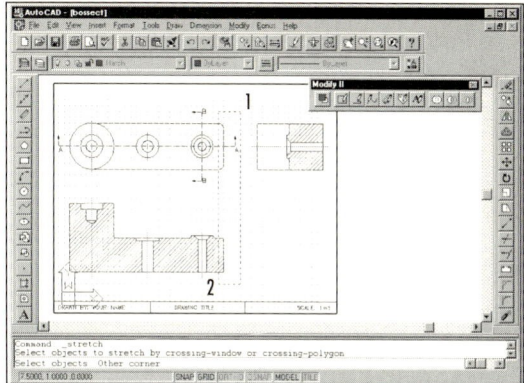

Figure 12.21

On your own, practice with the other Draworder options: Above object, Front, and Back. When you are done, return the Zoom to normal and erase the rectangle and solid Hatch, then change the current layer back to HATCH.

Changing Associative Hatches

The hatching created with the Boundary Hatch command is associative. In other words, it automatically updates so that it continues to fill the boundary when you edit the boundary by stretching, moving, rotating, arraying, scaling, or mirroring. You can also use grips to modify hatching that you've inserted into the drawing.

Next, you will stretch the top and front views of the drawing; the hatching automatically updates to fill the new boundary. Refer to Figure 12.21 for the points to select to create a crossing box. Be sure that you are in model space and that Snap is turned on.

Pick: **Stretch icon**

Select objects to stretch by crossing-window or crossing-polygon...

Select objects: *(pick point 1)*

Select objects: *(pick point 2)*

Select objects: ⏎

Base point or displacement: *(pick on the snap at the lower right corner of the front view)*

Second point of displacement: *(turn Snap off and pick to the right of the previous location)*

The front and top views of the drawing are stretched to the right. Your screen should be similar to Figure 12.22.

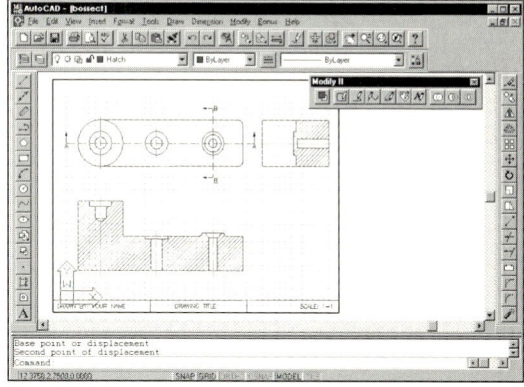

Figure 12.22

 The objects selected by the crossing box are stretched, as shown in Figure 12.22. The hatching updates to fill the new boundary. You will now use the Edit Hatch icon on the Modify II toolbar to change the angle of the hatching. Because the angled edge of the countersunk hole is drawn at 45°, you will change the angle of the hatch pattern so that the hatch is not parallel to features in the drawing. (Having the hatch run parallel or perpendicular to a major drawing feature is not good engineering drawing practice.) Figure 12.23 shows the Modify II toolbar with the Edit Hatch icon.

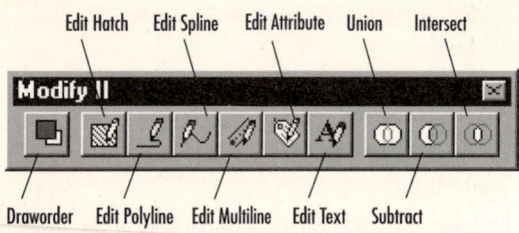

Figure 12.23

Pick: **Edit Hatch icon**

Select hatch object: *(pick the hatching in the front view)*

The Hatchedit dialog box shown in Figure 12.24 appears on your screen. Its appearance is identical to the Boundary Hatch dialog box. You can use this dialog box to edit hatching already added to your drawing.

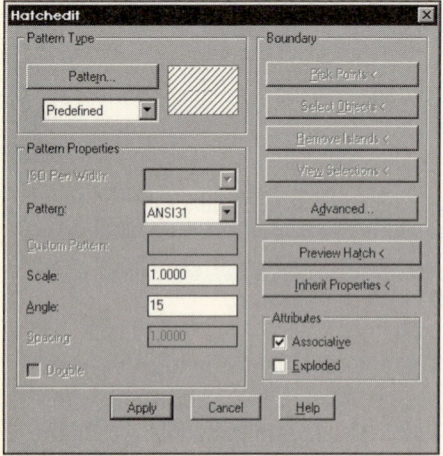

Figure 12.24

Type: **15** *(next to the Angle entry)*

Pick: **Apply**

The hatching updates so that it is angled an additional 15°, as shown in Figure 12.25.

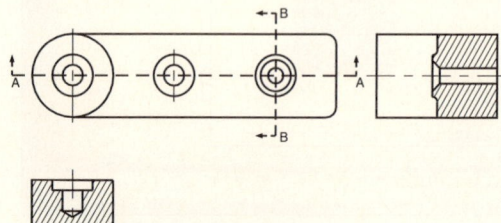
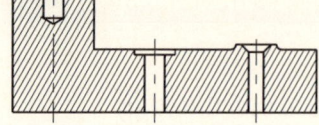

Figure 12.25

You will restart the Edit Hatch command by pressing ↵ at the command prompt.

Command: ↵

Select hatch object: *(pick the hatching in the side view)*

Again the Hatchedit dialog box returns to your screen. Type in a 15 for the hatch angle on your own, and when you have finished,

Pick: **Apply**

The final result of changing the hatch angles is shown in Figure 12.26.

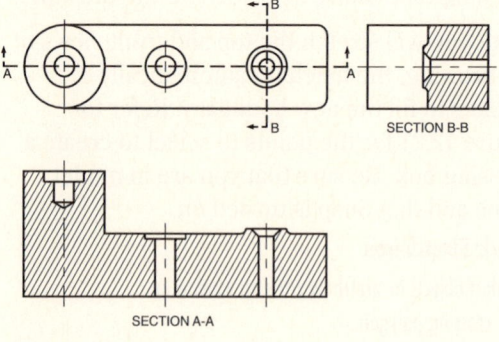

Figure 12.26

On your own, set layer TEXT current and use Dtext to label the section views SECTION A–A and SECTION B–B. Place the text directly below the section views, as shown in Figure 12.26. Save your *bossect.dwg* drawing.

■ *TIP* You can create open, irregular hatch boundaries by typing the Hatch command at the command prompt. Irregular hatch boundaries are useful for hatching large areas where you do not want to fill the entire area—for example, representing earth around a foundation on an architectural drawing—or in mechanical drawings to hatch only the perimeter of a large shape instead of filling in the entire area. Typing *HATCH* ↵ at the command prompt will start the Hatch command, rather than the Boundary Hatch command. Select the default pattern and specify the default scale and angle for the pattern. Then at the *Select objects* prompt, press ↵. Doing so allows you to specify points to form a boundary for the hatch. You will see the question *Retain polyline? <N>*. Accept the default to indicate that AutoCAD is not to retain a polyline boundary formed by the points you will pick. AutoCAD then prompts you to pick points. Type *C* ↵ to use the close option to return to the first point when you are done selecting and end the command. The area that you selected will be filled with the pattern. ■

You have completed the 2D portion of this tutorial. On your own, turn off the Modify II toolbar if it is still on.

Next, you will learn how to create section views automatically from a solid model, using the Section command.

Creating Sections from a Solid Model

This time you will start your drawing from a solid model template provided with the data files. It is similar to the object that you created in Tutorial 9.

Pick: **New**

Use the datafile *solblk2.dwg* as a template for your new drawing and save it as *blksect.dwg*.

A drawing similar to the one you created and modified in Tutorials 9 and 10 appears on your screen, as shown in Figure 12.27.

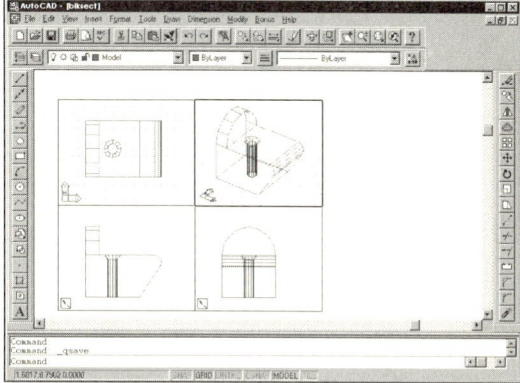

Figure 12.27

You will use SOLVIEW to create a section view. As the section view created will take the place of the lower left viewport image, you will erase it for now. On your own, change the drawing to paper space.

Pick: **Erase icon**

Select objects: *(use a window to select the objects in the lower left viewport)*

Select objects: ↵

The front view should disappear, as in Figure 12.28.

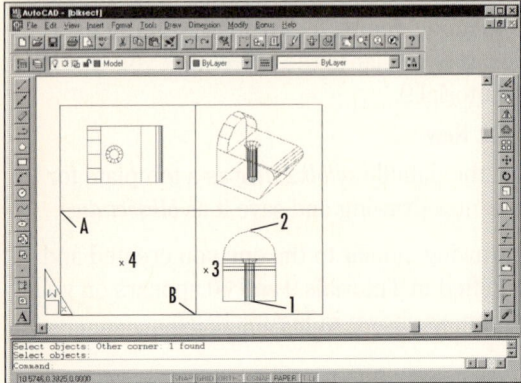

Figure 12.28

On your own, change back to model space, make the lower right viewport active, and change the UCS to Side. Now you are ready to create a section view.

> ■ **TIP** If you want to select in a viewport that is not active, press Ctrl-R to toggle the active viewport during commands. ■

Using Solview

The SOLVIEW and SOLDRAW commands can generate section views and auxiliary views quite easily. When doing a section view, you must specify the cutting plane, the side to view from, the view scaling factor, where to place the section view, and the size of your paperspace viewport. Then you use the SOLDRAW command to view the section hatch. AutoCAD automatically generates four new layers for the Section view and freezes them where necessary. Use Figure 12.28 to aid in selecting points. On your own, make sure that Object Snap Midpoint is on, and the lower right viewport is still active.

Command: **SOLVIEW** ←
UCS/Ortho/Auxiliary/Section/<eXit>: **S** ←
Cutting Plane's 1st point: *(pick the AutoSnap Midpoint at point 1)*
Cutting Plane's 2nd point: *(pick the AutoSnap Midpoint at point 2)*
Side to view from: *(pick near point 3)*
Enter view scale<0.5000>: ←
View center: *(pick near point 4)*
View center: ←
Clip first corner: *(pick Intersection A, using the Intersection snap from the Object Snap flyout)*
Clip other corner: *(pick Intersection B, using the Intersection snap from the Object Snap flyout)*
View name: **SECTION** ←
UCS/Ortho/Auxiliary/Section/<eXit>: **X** ←

Your screen should be similar to Figure 12.29.

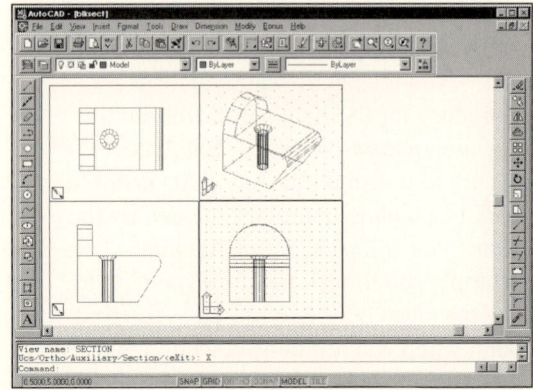

Figure 12.29

Note that the section hatch pattern is not shown in the front view. You will use SOLDRAW to show the hatch.

Command: **SOLDRAW** ←
Select objects: *(use a window to select the objects in the lower left viewport)*
Select objects: ←
One solid selected.

The section hatch should have appeared in the lower left viewport on the screen, similar to Figure 12.30.

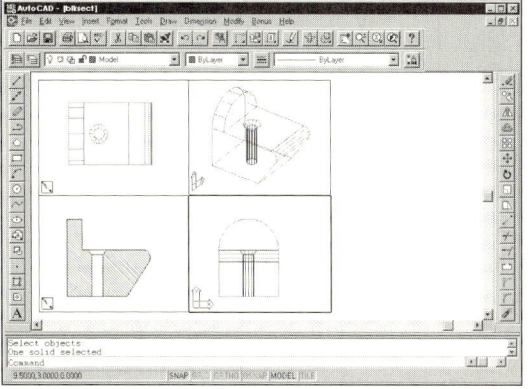

Figure 12.30

If you look at the Layer Control pull-down list, you will see that the SOLVIEW command has created a number of new layers with the proper visibilities set for each viewport. On your own, change the color for layer SECTION1-HAT to red.

Solid Profile

As you learned in Tutorial 10, Setup Profile, or the SOLPROF command, eliminates lines on the back surfaces of the object from the screen and inserts hidden lines where necessary in the drawing. Next, you will use the SOLPROF command to generate 2D profiles in the upper left and lower right viewports. On your own, activate the upper left viewport.

Command: **SOLPROF** ↵

Select objects: *(pick on an edge of the object in the upper right viewport)*

Select objects: ↵

Display hidden profile lines on a separate layer? <Y>: ↵

Project profile lines onto a plane? <Y>: ↵

Delete tangential edges? <Y>: ↵

One solid selected.

Make the lower right viewport active before restarting the command.

Command: ↵

SOLPROF

Select objects: *(pick on an edge of the object in the lower right viewport)*

Select objects: ↵

Display hidden profile lines on a separate layer? <Y>: ↵

Project profile lines onto a plane? <Y>: ↵

Delete tangential edges? <Y>: ↵

One solid selected.

On your own, change the linetype and color of the PH- layers to HIDDEN and blue, respectively. Then using techniques used earlier in this manual, freeze layer MODEL *in the upper left and lower right viewports only.* (Remember to use the Freeze/Thaw in Current Viewport selections on the Layer Control list). Black visible lines created by the SOLPROF command should be visible, with blue hidden lines as shown in Figure 12.31.

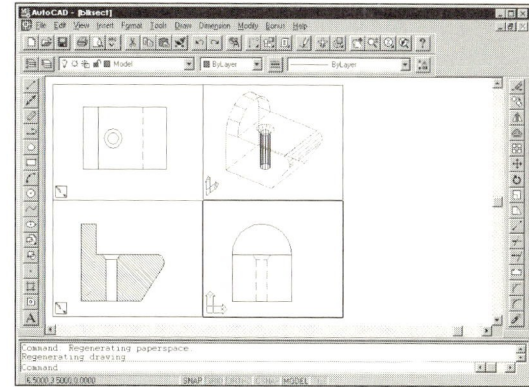

Figure 12.31

On your own, you may add a cutting plane line to the top view to show where the section was taken.

Next you will create an isometric section view using the Section and Slice commands. Save your drawing before continuing.

Using Section and Slice

You will start a new drawing, using the same prototype previously used.

Pick: **New**

Use the datafile drawing *solblk2.dwg* as a template and save it as *blksect2.dwg*.

A drawing similar to the one you just used appears on your screen as shown in Figure 12.32.

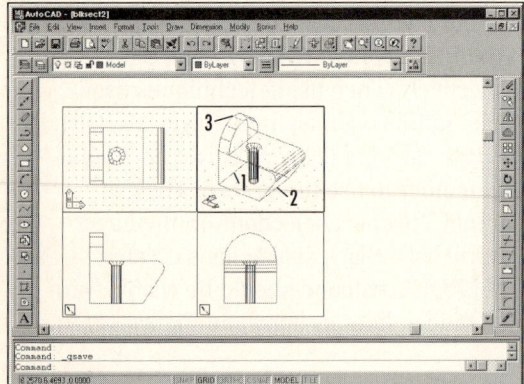

Figure 12.32

On your own, create layer HATCH. Assign it color red and linetype CONTINUOUS. Make it the current layer. AutoCAD places the region it generates during the Section command on the current layer. Turn on the Midpoint running object snap on your own, so that it is available for the next command.

You will use the Midpoint running object snap to create a UCS through the center of the part. You will be typing the UCS command at the command prompt. Refer to Figure 12.32 to make your selections.

Pick: (the upper right viewport to make it active)

Command: **UCS** ⏎

Origin/ZAxis/3point/OBject/View/X/Y/Z/Prev/Restore/Save/Del/?/<World>: **3** ⏎

Origin point <0,0,0>: *(pick near point 1)*

Point on positive portion of the X axis<3.0000,4.0000,0.0000>: *(pick near point 2)*

Point on the positive-Y portion of the UCS Xyplane<2.0000,5.0000,0.0000>: *(pick near point 3)*

You will type *UCSICON* to move the UCS icon to the origin of the newly created User Coordinate System. This helps you see whether the correct coordinate system has been selected.

Command: **UCSICON** ⏎

ON/OFF/All/Noorigin/ORigin <ON>: **OR** ⏎

The UCS icon should appear at the center of the part in the upper right view. Your screen should look like Figure 12.33. If it does not, repeat the preceding steps, being careful to select the midpoints indicated in Figure 12.32.

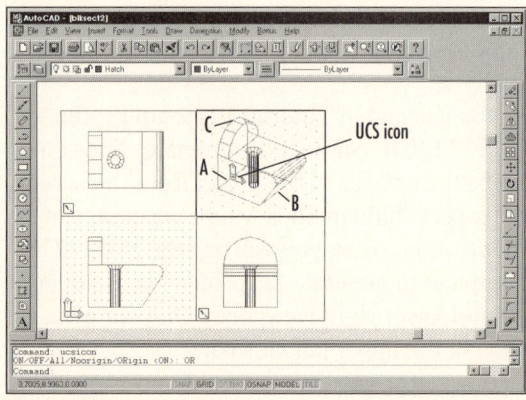

Figure 12.33

Use the View pull-down menu and select Toolbars to show the Solids toolbar. On your own, close the Toolbars dialog box and make sure that the upper right viewport is still active.

Using Section

 The Section command generates a region (a closed 2D area) from a solid and a cutting plane that you specify. The region is inserted on the current layer and at the location of the cross section you specified by selecting the cutting plane. You can pick the Section icon from the Solids toolbar.

Pick: **Section icon**

Select objects: *(pick the object displayed in the upper right viewport)*

Select objects: ⏎

Sectioning plane by Object/Zaxis/View/XY/YZ/ZX/ <3points>: ⏎

You can specify the cutting plane by choosing:

Object   aligns the cutting plane with a circle, ellipse, arc, 2D spline, or 2D polyline that you select.

Zaxis   prompts you for two points that determine the edge view of an XY plane. The cutting plane is the Z axis plane through the first point you select and is normal (perpendicular) to the line defined by the two points selected.

View   aligns the cutting plane parallel to the current viewing plane through the point you specify.

XY, YZ, or ZX   aligns the cutting plane parallel to the specified UCS plane through the point you specify.

3 points   prompts you for three points, through which the desired cutting plane passes.

You will specify three points to define the cutting plane, which will pass through the middle of the object. The Midpoint running object snap, which you turned on earlier, will select the midpoint of the lines you pick. Refer to Figure 12.33 for selections.

1st point on plane: *(pick line A)*

2nd point on plane: *(pick line B)*

3rd point on plane: *(pick line C)*

The Section command creates a region or a block in your drawing. You should see it appear in your drawing, as shown in Figure 12.34.

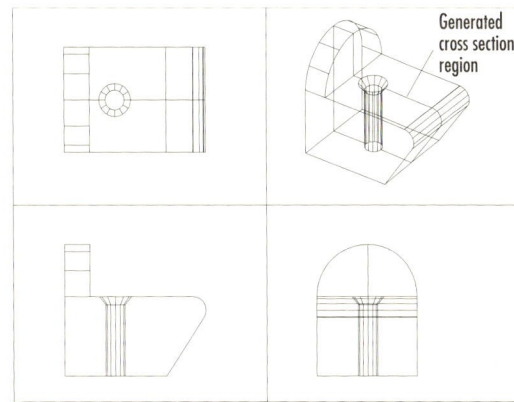

Generated cross section region

Figure 12.34

Next, use the Boundary Hatch command to hatch the section boundary that you created.

Pick: **Hatch icon**

The Boundary Hatch dialog box appears on your screen. On your own, choose the button for Select Objects to define the boundary. You will return to your drawing screen. Pick on the boundary that you created with the Section command and press ⏎ to return to the dialog box. Preview the hatching to make sure that the correct area will be hatched. When you are finished,

Pick: **Apply *(to apply the hatch and exit the Boundary Hatch dialog box)***

The hatching appears in your drawing, as shown in Figure 12.35.

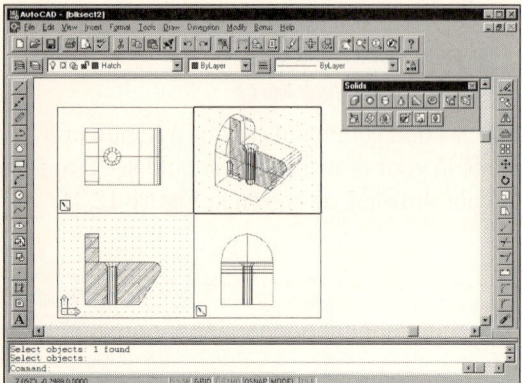

Figure 12.35

The hatching should only be drawn in the section view (in this case, the front view) according to good engineering drawing practice, but at this point it is still visible in every view. (It is on edge in the top and side views, so it just appears as a single line through the center of the part.) You will control the visibility for layer HATCH so that the hatching only shows in the front view.

> ■ *TIP* Remember that you can use Draworder if you want to make the hatch behind or in front of the object outline. ■

Controlling Layer Visibility

You will use Layer Control from the Object Properties toolbar to freeze the hatching in all of the viewports except the lower left one. First you will freeze layer HATCH within individual viewports.

On your own, make the top left viewport current.

> Pick: *(on layer name HATCH on the Object Properties toolbar)*

The list of layers pulls down from the current layer name, as shown in Figure 12.36.

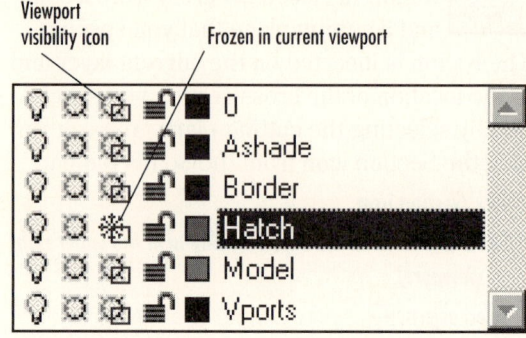

Figure 12.36

Pick on the viewport visibility icon to the left of layer name HATCH. It should change in appearance from a sun with a little viewport to its lower right to a snowflake with a little viewport. Pick anywhere in the graphics window for the selection to take effect. The line representing the edge view of the hatching should no longer be visible in the upper left viewport. Repeat this process on your own to freeze the hatching in the lower right viewport containing the side view. When you have finished, your screen should look like Figure 12.37.

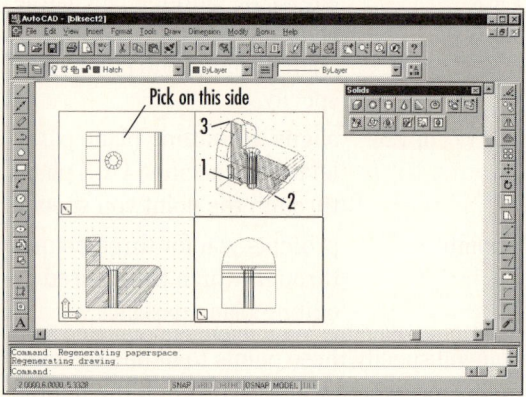

Figure 12.37

Using Slice

The Slice command lets you remove a portion of a solid by specifying a *slicing plane* similar to the cutting plane in the Section command. You can use the Section command in conjunction with the Slice command to create pictorial section views from solid models. Next, you will slice the model and cut away the front half. Refer to Figure 12.37 for the points to select. (They will be the same points you selected last time to define the section.) Make sure your isometric view in the upper right viewport is active on your own before you continue. You can pick the Slice icon from the Solids toolbar.

Pick: **Slice icon**

Select objects: *(pick the object displayed in the upper right viewport)*

Select objects: ⏎

Slicing plane by Object/Zaxis/View/XY/YZ/ZX/ <3points>: ⏎

The methods by which you select the cutting plane for a section are very similar to the methods for using the Slice command. You will specify three points to define the cutting plane, which will pass through the middle of the object. The Midpoint running object snap will select the midpoint of the lines you pick. Refer to Figure 12.37 for the lines to select.

1st point on plane: *(pick line 1)*

2nd point on plane: *(pick line 2)*

3rd point on plane: *(pick line 3)*

Both sides/<Point on desired side of the plane>: *(pick once in the top view to activate it and then near the point shown)*

Using Hide

As you saw in Tutorial 9, the Hide command eliminates lines from the back surfaces of the object from the screen so that the object appears correctly in the viewport. Next you will hide the back surface lines with the Hide command. On your own, make the upper right viewport active. Since you do not have the Render toolbar open, you will type *HIDE* at the prompt.

Command: **HIDE** ⏎

On your own, hide the remaining three viewports. When you are finished, your screen should look like Figure 12.38.

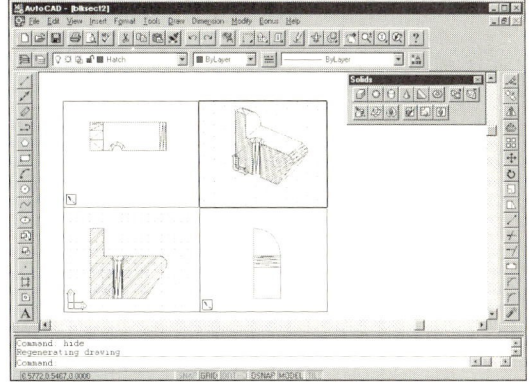

Figure 12.38

On your own, plot the upper right viewport from model space, if you want. Select the Hide Lines box in the Plot Configuration dialog box. To plot from paper space with hidden lines removed, use the Mview command and select the Hideplot option, and pick the viewports in which you want back surface lines removed when plotting.

Turn off the Midpoint running object snap on your own and close the Solids toolbar.

Save your drawing and exit AutoCAD.

You have completed Tutorial 12.

KEY TERMS

associative hatching	cutting plane line	runout
crosshatching	pictorial full section	section view
custom hatch patterns	ray casting	user-defined hatch pattern

KEY COMMANDS

Boundary Hatch	Hatch	Section
Edit Hatch	Savetime	Slice

Redraw the given front view and replace the given right side view with a section view. Use 2D methods or solid modeling. Assume that the vertical centerline of the front view is the cutting plane line for the section view. The letter M after an exercise means that the given dimensions are in millimeters.

12.1 High Pressure Seal

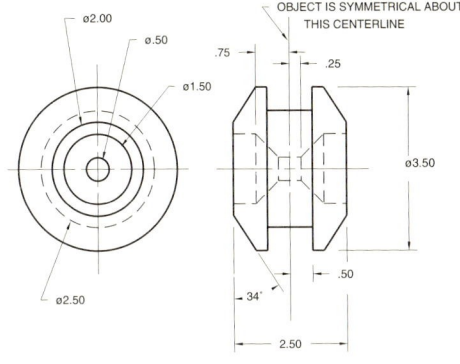

OBJECT IS SYMMETRICAL ABOUT THIS CENTERLINE

12.2M Pulley

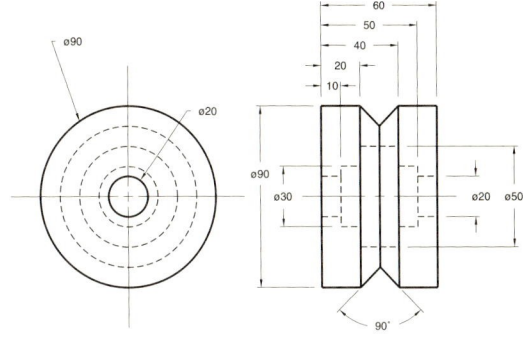

12.3 Valve Cover

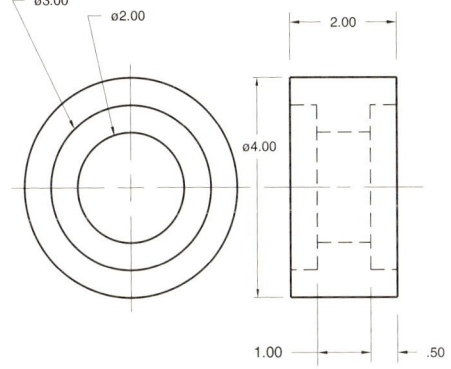

12.4 Double Cylinder

Use 2D techniques to draw this object; the grid shown is 0.25 inches. Show the front view as a half section and show the corresponding cutting plane in the top view. On the same views, completely dimension the object to the nearest 0.01 inches. Include any notes that are necessary.

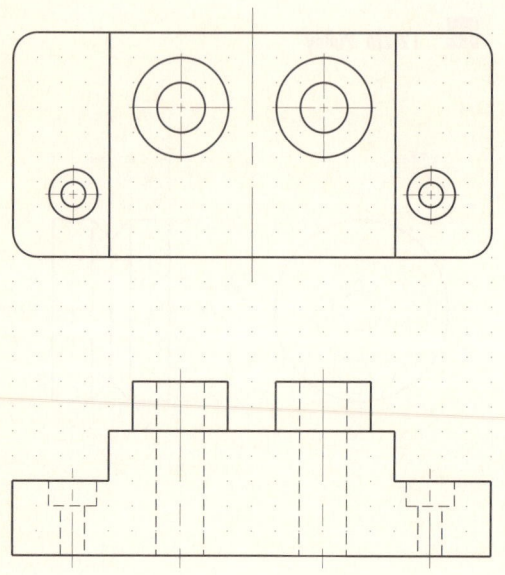

12.5 Plate

Use 2D to construct the top view with the cutting plane shown and the sectioned front view.

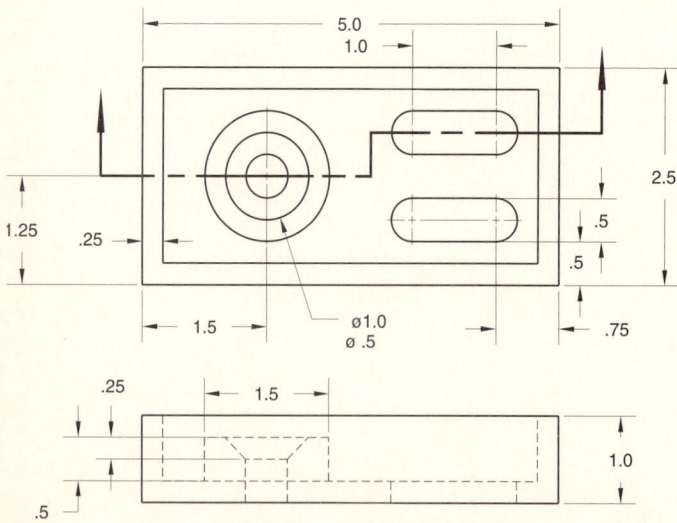

12.6 Support Bracket

Create a solid model of the object shown, then produce a plan view and full section front view. Note the alternate units in brackets; these are the dimensions in millimeters for the part.

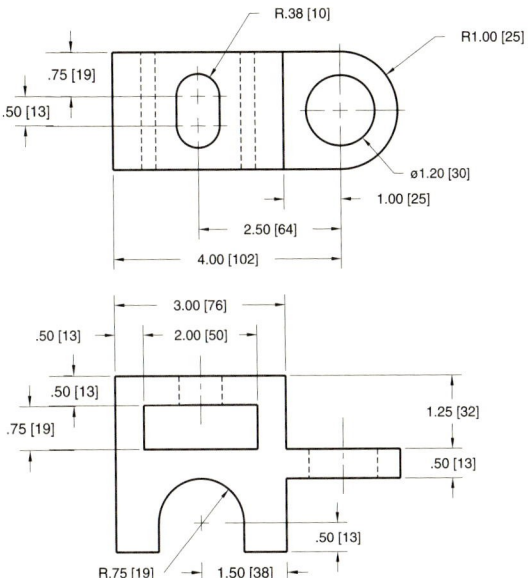

12.7 Shaft Support

Redraw the given front view and create a section view. Assume that the vertical centerline of the shaft in the front view is the cutting plane line for the section view.

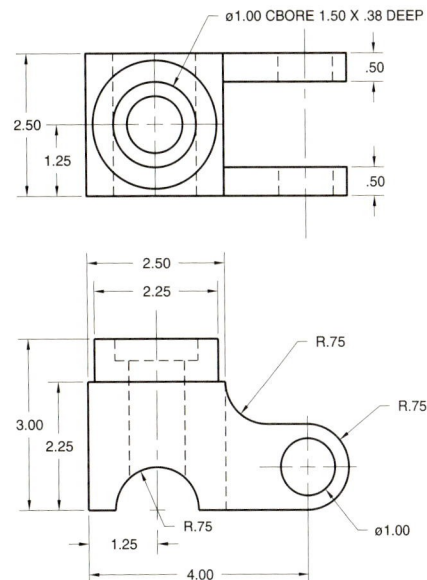

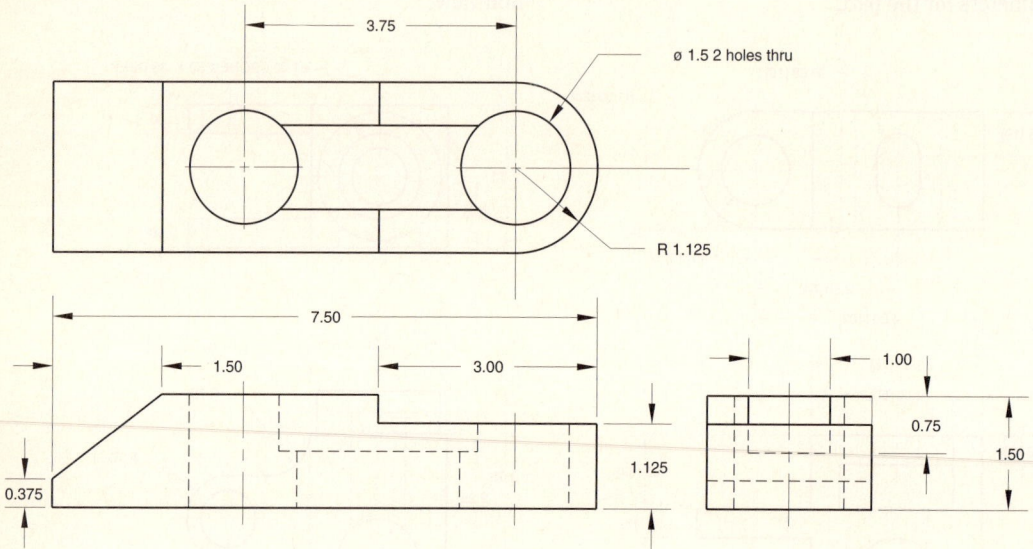

12.8 Wedge

Make a solid model based on the given views, then produce a plan view and a full-section front view. Align the sectioned front view with the plan view. Hidden lines and/or cutting planes are not needed in the plan view.

12.9 Turned Down Slab Detail

Use 2D methods to create the drawings shown. Use Boundary Hatch to fill the area with the pattern for concrete. Type the Hatch command and create the irregular areas to fill with patterns for sand and earth.

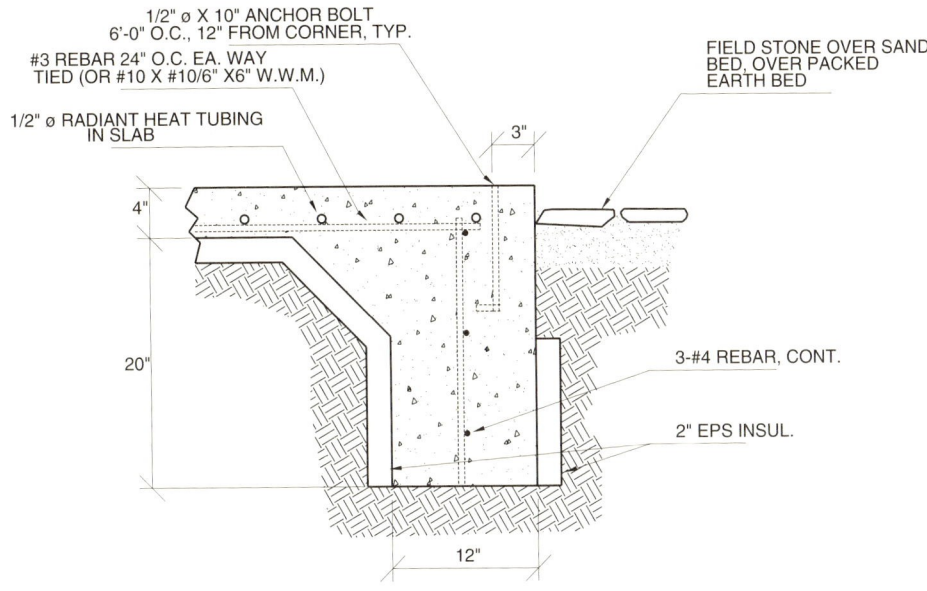

1/2" ø X 10" ANCHOR BOLT
6'-0" O.C., 12" FROM CORNER, TYP.

#3 REBAR 24" O.C. EA. WAY
TIED (OR #10 X #10/6" X6" W.W.M.)

1/2" ø RADIANT HEAT TUBING
IN SLAB

FIELD STONE OVER SAND
BED, OVER PACKED
EARTH BED

3"

4"

20"

3-#4 REBAR, CONT.

2" EPS INSUL.

12"

12.10M Pulley Arm

Use solid modeling to create a pictorial view like the one shown.

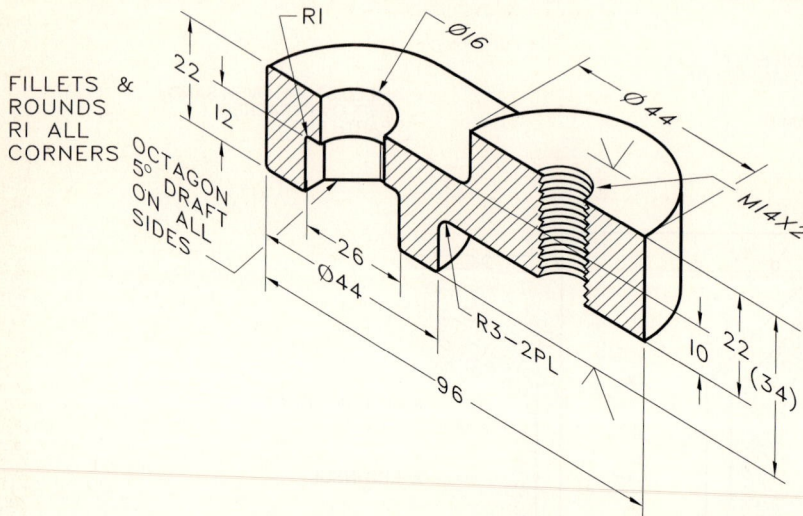

FILLETS &
ROUNDS
RI ALL
CORNERS

OCTAGON
5° DRAFT
ON ALL
SIDES

RI

Ø16

Ø44

M14X2

22

12

26

Ø44

R3-2PL

96

10

22

(34)

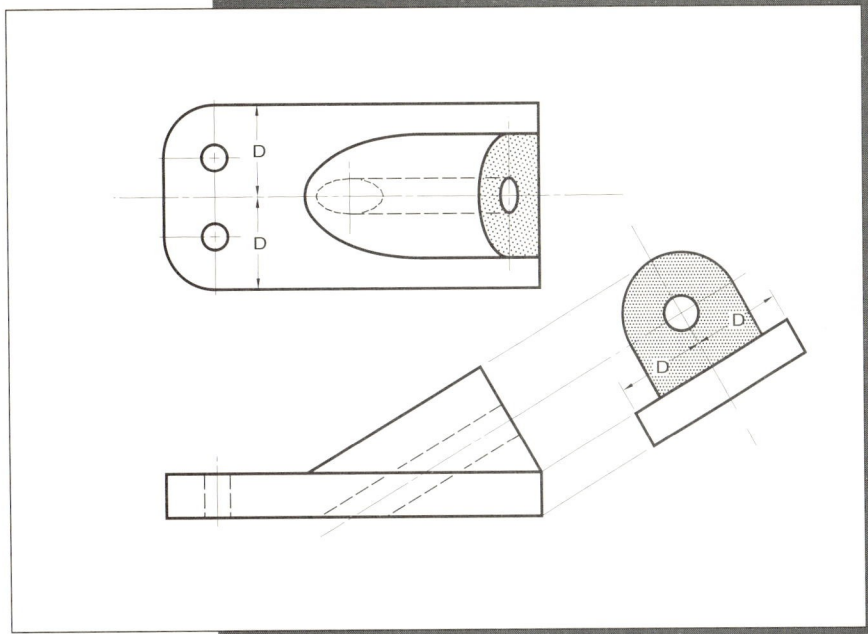

Creating Auxiliary Views with 2D and Solid Models

Objectives

When you have completed this tutorial, you will be able to

1. Draw auxiliary views, using 2D projection.

2. Set up a UCS to help create a 2D auxiliary view.

3. Rotate the snap to help create a 2D auxiliary view.

4. Draw auxiliary views of curved surfaces.

5. Create an auxiliary view of a 3D solid model.

Introduction

An *auxiliary view* is an orthographic view of the object that has a different line of sight from the six *basic views* (front, top, right-side, rear, bottom, and left-side). Auxiliary views are most commonly used to show the true size of a slanted or oblique surface. Slanted and oblique surfaces are *foreshortened* in the basic views because they are tipped away from the viewing plane, causing their projected size in the view to be smaller than their actual size. An auxiliary view is drawn with its line of sight perpendicular to the surface, showing the true size and shape of a slanted or oblique surface.

In this tutorial, you will learn to create auxiliary views, using 2D methods and 3D solid modeling. You can project 2D auxiliary views from the basic orthographic views by rotating the snap or by aligning a new User Coordinate System. You can generate auxiliary views directly from a 3D solid model by changing the viewpoint to show the true size of the surface.

Auxiliary Views

Figure 13.1 shows three views of an object and an auxiliary view.

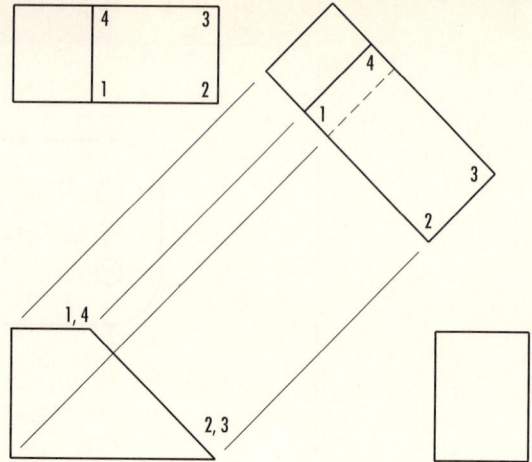

Figure 13.1

The surface defined by vertices 1–2–3–4 is inclined. It is perpendicular to the front viewing plane and shows on edge in the front view. Both the top and side views show foreshortened views of the surface; that is, neither shows the true shape and size of that surface of the object.

To project an auxiliary view showing the true size of surface 1–2–3–4, you create a new view by drawing projection lines perpendicular to the edge view of the surface, in this case the angled line in the front view. You transfer the width of the surface, measured from a *reference surface*, to the auxiliary view to complete the projection. Note that, as with the front, top, and side views, the auxiliary view shows the entire object as viewed with the line of sight perpendicular to that surface, causing the base of the object to show as a hidden line.

Starting

Start AutoCAD and on your own start a new drawing using the drawing *adapt4.dwg*, provided with the data files, as a template. Save your drawing as *auxil1.dwg*.

AutoCAD's drawing editor shows the drawing of the orthographic views for the adapter on your screen, as shown in Figure 13.2.

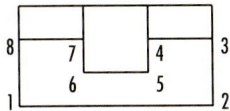

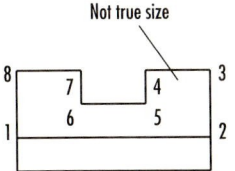
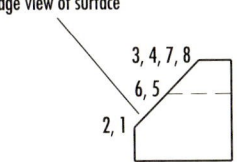

Figure 13.2

You will create an auxiliary view showing the true size of the inclined surface. The object is shown in Figure 13.2, with the surface labeled 1–2–3–4–5–6–7–8.

Drawing Auxiliary Views Using 2D

You can create the auxiliary view by using information from any two of the views shown, as any two views with a 90° relationship provide all the principal dimensions. For this tutorial, you will erase the top view and use the front and side views to project the auxiliary view.

Pick: **Erase icon**

Select objects: *(use implied Windowing to select the top view)*

Select objects: ↵

On your own, use Redraw as necessary. Your screen will look like Figure 13.3 once you have erased the top view.

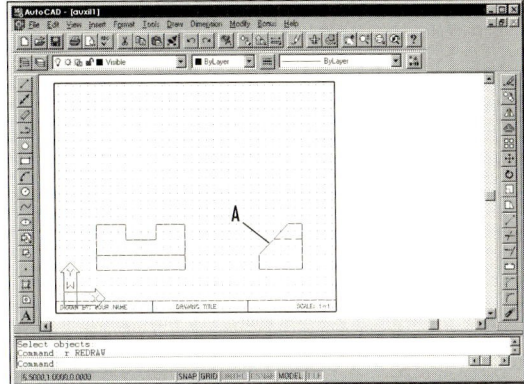

Figure 13.3

Next, you will draw projection lines perpendicular to surface 1–2–3–4–5–6–7–8 in the side view. You will create a new UCS and align it with the edge view to help you project these lines easily.

Aligning the UCS with the Angled Surface

 You will set the UCS so that it is aligned with the angled surface. You will use the Object option to align the new coordinate system. The Object option prompts you for the existing drawing object with which you want to align the UCS. You will align the UCS with the angled line in the side view. You will pick the Object UCS icon from the UCS flyout on the Standard toolbar.

Pick: **Object UCS icon**

Select object to align UCS: *(pick the angled line in the side view, identified as A in Figure 13.3)*

When you have finished aligning the UCS, your screen should look like Figure 13.4. Note that the crosshairs and grid, as well as the UCS icon, now align with the angled surface.

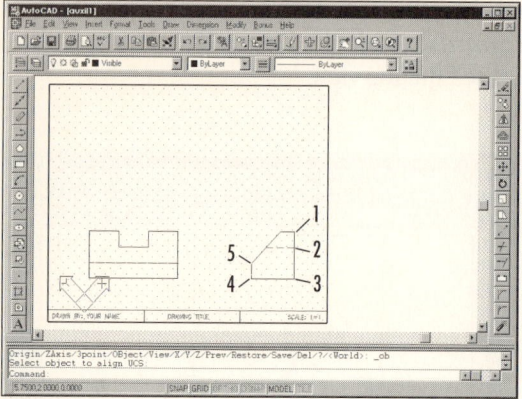

Figure 13.4

On your own, set layer PROJECTION as current and turn on OSNAP with Intersection selected. Be sure that Ortho is on and that Snap is off.

Now you will draw rays from each intersection of the object in the side view into the open area of the drawing above the front view. You will pick Ray from the Draw menu; if you need help creating rays, refer to Tutorial 4.

Pick: **Draw, Ray**

From point: *(pick intersection 1 in Figure 13.4)*

Through point: *(with Ortho on, pick a point above and to the left of the side view)*

Through point: ⏎

On your own, restart the Ray command and draw projection lines from points 2 through 5 in the left-side view. Then use the Line command to draw a *reference line* perpendicular to the projection lines. The location of this line can be anywhere along the projection lines, but it should extend about 0.5 unit beyond the top and bottom projection lines.

Your drawing should look like that in Figure 13.5.

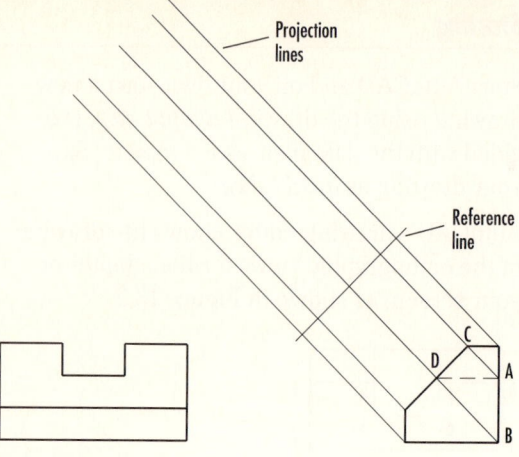

Figure 13.5

Note that the back edge of the slot (labeled A) and the lower right corner (labeled B) in the side view align with the projection lines through the corners of the angled surface (labeled C and D) to produce a single projection line. If points C and D did not align on the projection lines that you created, you would need to project them.

The width of the object is known to be 3. Next, you will use the Offset command to draw a line 3 units from and parallel to the reference line.

> ■ *TIP* If you did not know the dimensions, you could use the Distance command (available from the Inquiry flyout on the Object Properties toolbar) with the Intersection object snap in the front view to determine the distance. ■

Pick: **Offset icon**

Offset distance or Through <Through>: **3** ⏎

Select object to offset: *(pick the reference line, as shown in Figure 13.5)*

Side to offset? *(pick a point anywhere to the left of the reference line)*

Select object to offset: ⏎

Your drawing should look like the one in Figure 13.6.

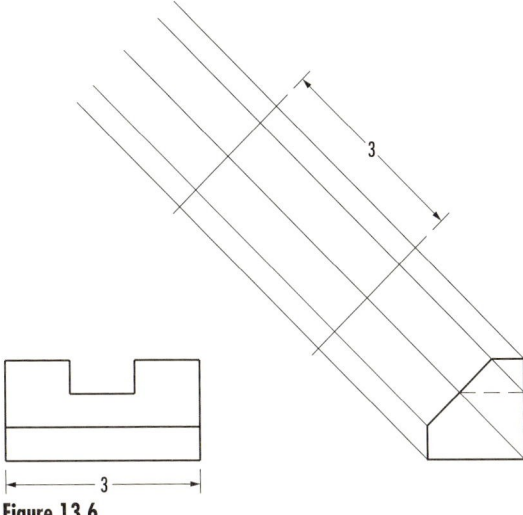

Figure 13.6

Add the width of slot on your own by using Offset to create two more parallel lines, 1 unit apart. Your drawing should look like that in Figure 13.7.

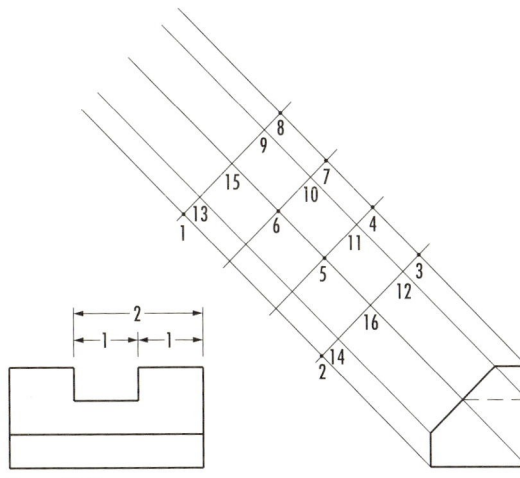

Figure 13.7

On your own, set layer VISIBLE as the current layer and verify that the Intersection running object snap is on before you continue. Make sure you see the AutoSnap Intersection marker when selecting points in the following steps.

Next, you will draw the visible lines in the drawing over the top of the projection lines. Be sure to draw the correct shape of the object as it will appear in the auxiliary view. Refer to Figure 13.7 for your selections.

> *Pick:* **Line icon**
>
> From point: **(pick the intersection of the projection lines at point 1)**
>
> To point: **(pick point 2)**
>
> To point: **(pick point 3)**
>
> To point: **(pick point 4)**
>
> To point: **(pick point 5)**
>
> To point: **(pick point 6)**
>
> To point: **(pick point 7)**
>
> To point: **(pick point 8)**
>
> To point: **C** ⏎

Now you will restart the Line command and draw the short line segments where the two short top surfaces of the object intersect the slanted front surface:

> Command: ⏎
>
> From point: **(pick point 9)**
>
> To point: **(pick point 10)**
>
> To point: ⏎
>
> Command: ⏎
>
> From point: **(pick point 11)**
>
> To point: **(pick point 12)**
>
> To point: ⏎

Next, you will draw the line representing the back edge of the slot in the auxiliary view.

> Command: ⏎
>
> From point: **(pick point 10)**
>
> To point: **(pick point 11)**
>
> To point: ⏎

Now you will add the line showing the intersection of the vertical front surface with the angled surface.

Command: ⏎

From point: *(pick point 13)*

To point: *(pick point 14)*

To point: ⏎

On your own, set layer HIDDEN_LINES as the current layer.

Next, you will draw the hidden line showing the intersection of the bottom surface and the back surface. It will be hidden from the line of sight.

Command: **LINE** ⏎

From point: *(pick point 15)*

To point: *(pick point 16)*

To point: ⏎

Freeze layer PROJECTION on your own. Use the Trim command to remove the portion of the hidden line that coincides with the visible line from point 5 to point 6 before you continue.

When you have finished, your screen should look like Figure 13.8.

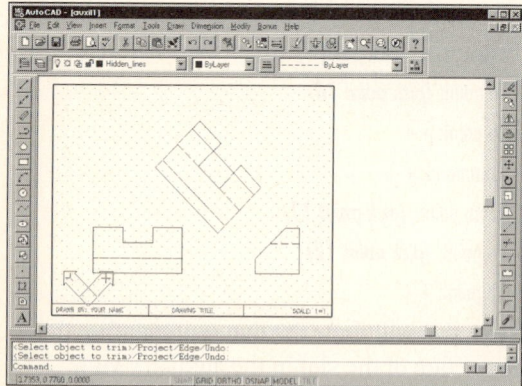

Figure 13.8

Adding Visual Reference Lines

Next, you will add two visual reference lines to your drawing from the extreme outside edges of the surface shown true size in the auxiliary view to the inclined surface in the side view. The lines will provide a visual reference for the angles in your drawing and help show how the auxiliary view aligns with the standard views. You can extend a centerline from the primary view to the auxiliary view or use one or two projection lines for the reference lines. In this case, as there are no centerlines, you will create two reference lines. Leave a gap of about 1/16" between the reference lines and the views.

On your own, thaw layer PROJECTION. Then pick the Properties icon from the Object Properties toolbar and select the two rays that project from the ends of the angled surface to use as visual reference lines and change them to layer THIN. Refer to Figure 13.9 if you are unsure which lines to select.

The rays should change to layer THIN. You can verify this result visually because the lines will turn red, the color for layer THIN.

Linetype Conventions

Two lines of different linetypes should not meet to make one line. You must always leave a small gap on the linetype that has lesser precedence. In the case of the visible and the reference lines, the reference line should have a small gap (about 1/16" on the plotted drawing) where it meets the visible line. In the case of the visible and the hidden lines, the hidden line should have a small gap (about 1/16" on the plotted drawing) where it meets the visible line. You can use the Break command to do this.

On your own, freeze layer PROJECTION again. Then use the Break command to create a gap of about 1/16" between each view and the reference lines. Create a similar gap where the hidden line extends from the visible line, forming the front of the slot in the auxiliary view. Erase the extra portions of the rays where they extend past the side and auxiliary views.

> ■ **TIP** You could also offset the edge of the object by 1/16" in each view and use Trim to create the gap between the lines. ■

Your drawing should look like the one in Figure 13.9.

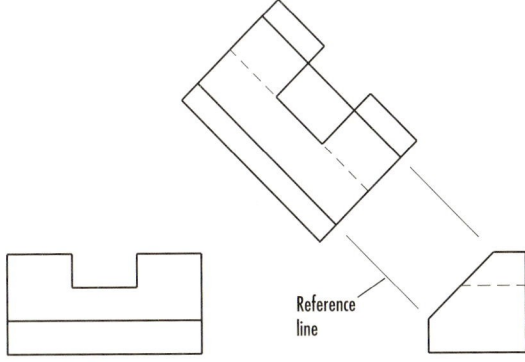

Reference line

Figure 13.9

On your own, save your drawing.

Curved Surfaces

Now you will create an auxiliary view of the slanted surface in a cylinder drawing like the one you created at the end of Tutorial 6. The drawing has been provided for you as *cyl3.dwg* in the data files that came with your software.

On your own, begin a new drawing. Use *cyl3.dwg* as a template and name your new drawing *auxil3.dwg*.

You return to AutoCAD's drawing editor. Figure 13.10 shows the drawing of the orthographic views of the cylinder with the slanted surface that should be on your screen.

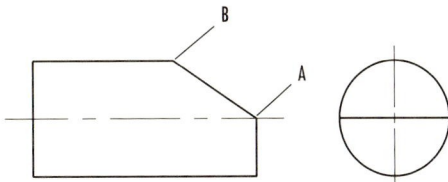

Figure 13.10

The drawing shows the front and side view of the slanted surface but not the true shape of the surface. You will draw a *partial auxiliary view* to show the true shape of the surface. Partial auxiliary views are often used because, in the auxiliary view, the inclined or oblique surface chosen is shown true size and shape, but the other surfaces are all foreshortened. As these other surfaces have already been defined in the basic orthographic views, there is no need to show them in the auxiliary view. A partial auxiliary view shows only the inclined surfaces (leaving out the normal surfaces), thus saving time in projecting the view and giving a clearer appearance to your drawing.

Figure 13.11 shows a front view of an object, along with two partial auxiliary views.

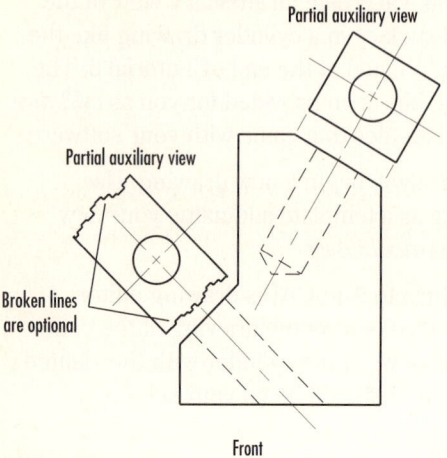

Figure 13.11

Rotating the Snap

This time you will use the Rotate option of the Snap command to rotate the grid and snap so that they align with the inclined surface. Doing so lets you easily project the curved surface to the auxiliary view that you will create.

On your own, be sure that the Intersection running object snap is on, that the grid is on, that Ortho is off, and that layer PROJECTION is the current layer.

Now you will rotate the snap angle with the Snap command. You will pick two points (A and B in Figure 13.10) to define the angle.

Command: **SNAP** ⏎

Snap spacing or ON/OFF/Aspect/Rotate/Style <0.2500>: **R** ⏎

Base point <0.0000,0.0000>: *(pick the intersection labeled A)*

Rotation angle<0>: *(pick the intersection labeled B)*

The message, *Angle adjusted to 326*, will appear. Your screen should be similar to Figure 13.12. Note that the crosshairs and

grid have aligned with the inclined surface, but the UCS icon remains as it was previously, aligned with the World Coordinates.

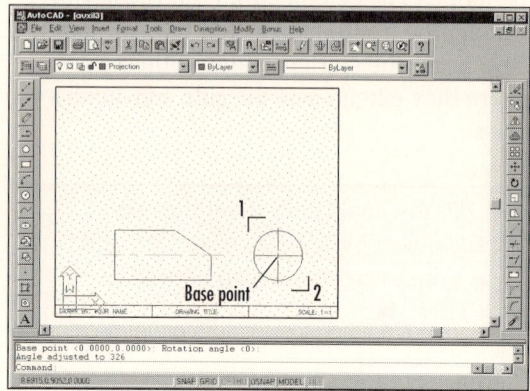

Figure 13.12

Next, you will copy the object in the side view and then rotate it so that it aligns with the rotated grid and snap. Doing so makes projecting the depth of the object from the side view into the auxiliary view easy. Then you do not have to use a reference surface to make depth measurements as you did in the previous drawing. The shapes in this drawing are simple, but these methods will also work for much more complex shapes.

For the next steps you will use Ortho and Snap. Be sure that they are turned on. Refer to Figure 13.12 for the points to select.

Pick: **Copy Object icon**

Select objects: *(pick point 1)*

Other corner: *(pick point 2 to form a window around the side view)*

Select objects: ⏎

<Base point or displacement>/Multiple: *(pick the intersection labeled Base Point)*

Second point of displacement: *(pick a point on the snap above and to the right of the side view)*

A copy of the side view should appear on your screen, as shown in Figure 13.13.

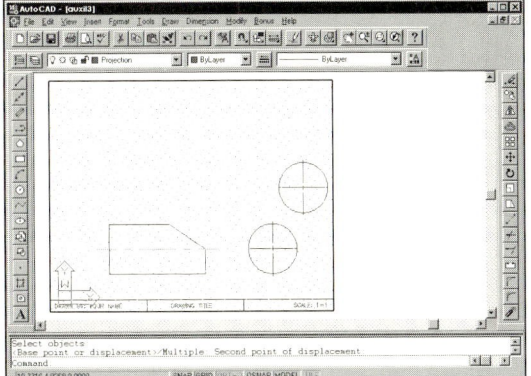

Figure 13.13

Now you will change the copy onto layer PROJECTION because it is part of the construction of the drawing, not something you will leave in the drawing when you are finished. You will freeze layer PROJECTION when you have finished the drawing.

On your own, change the object that you just created to layer PROJECTION.

The objects that you selected are changed onto layer PROJECTION. Their color is now magenta, which is the color of layer PROJECTION.

Now you are ready to rotate the copied objects so that they align with the snap and can be used to project the auxiliary view. You will use the Rotate icon from the Modify toolbar and type *P* for the previous selection set to do so.

Pick: **Rotate icon**

Select objects: **P** ⏎

■ *TIP* Typing *P* gives you the previous selection set, which you just defined when changing properties. ■

Select objects: ⏎

Base point: *(pick the intersection of the centerlines of the copied objects)*

<Rotation angle>/Reference: *(with Ortho turned on, select a point above and to the right of the base point you picked, as shown in Figure 13.14)*

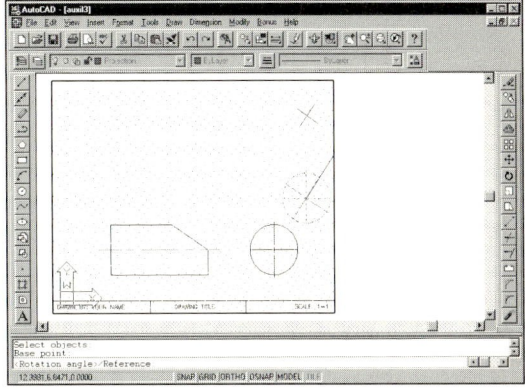

Figure 13.14

The objects that you copied previously should now be rotated into position so that they align with the snap and grid. Your drawing should look like that in Figure 13.15.

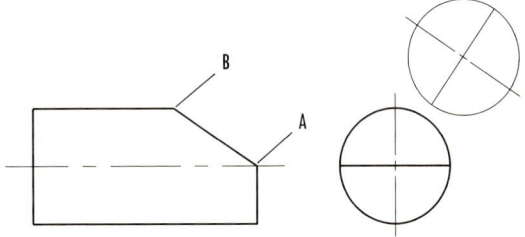

Figure 13.15

Next, you will draw projection lines from the front view to the empty area where the auxiliary view will be located. Start the Ray command by picking it from the Draw menu. (Ortho and the Intersection running object snap should be on.) Refer to Figure 13.15 for the points to select.

Pick: **Draw, Ray**

From point: *(pick the intersection labeled A in Figure 13.15)*

Through point: *(pick a point above the front view, as in Figure 13.16)*

Through point: ⏎

Command: ⏎ *(to restart the Ray command)*

From point: *(pick the intersection labeled B in Figure 13.15)*

Through point: *(pick a point above the front view)*

Through point: ⏎

Figure 13.16 shows an example of these lines.

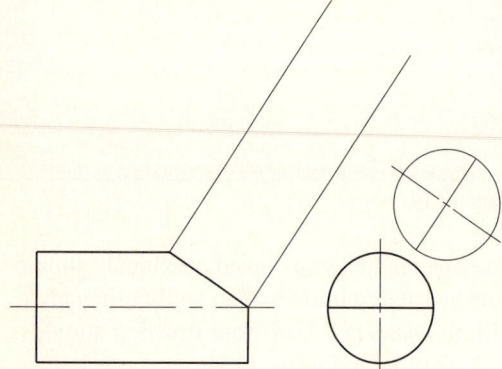

Figure 13.16

On your own, project rays from the intersections in the copied side view so that your drawing looks like the one in Figure 13.17. Then set layer VISIBLE as the current layer.

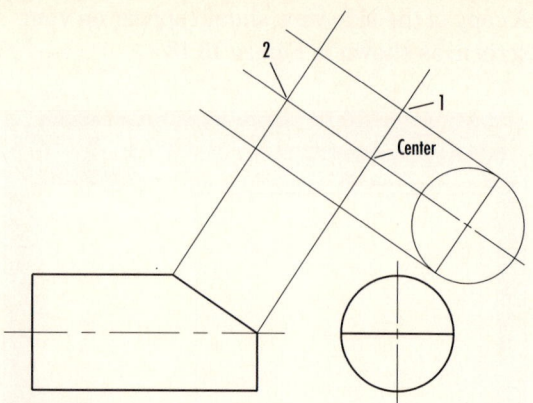

Figure 13.17

Because the object appears as a circle in the side view and is tilted away in the front view, the true shape of the surface must be an ellipse. You will select the Ellipse icon from the Draw toolbar to create the elliptical surface in the auxiliary view. Refer to Figure 13.17 for your selections.

Pick: **Ellipse icon**

Arc/Center/<Axis endpoint 1>: **C** ⏎

Center of ellipse: *(pick the intersection labeled Center)*

Axis endpoint: *(pick intersection 1)*

<Other axis distance>/Rotation: *(pick intersection 2)*

■ *TIP* If the surface in the side view were an irregular curve, you could project a number of points to the auxiliary view and then connect them with a polyline. After connecting them, you could connect them in a smooth curve, using the Fit Curve or Spline option of the Edit Polyline command. Although this method would also produce an ellipse, the Ellipse command is quicker. ■

The ellipse in your drawing should look like the one in Figure 13.18. You will need to trim the lower portion of it to create the final surface.

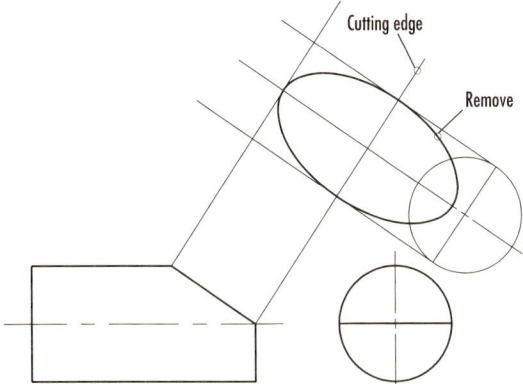

Figure 13.18

You will use the Trim command to eliminate the unnecessary portion of the ellipse in the auxiliary view.

Pick: **Trim icon**

Select cutting edges: **(pick the cutting edge in Figure 13.18)**

Select objects: ⏎

<Select object to trim>/Project/Edge/Undo: **(pick on the ellipse to the right of the cutting edge)**

Select objects: ⏎

The lower portion of the ellipse is now trimmed off.

Change the two projection lines from the front view onto layer THIN on your own. Use the Break command to break them so that they do not touch the views. Draw in the final line across the bottom of the remaining portion of the ellipse on layer VISIBLE. Freeze layer PROJECTION to remove the construction and projection lines from your display.

When you are finished, your screen should look like Figure 13.19. Save your drawing.

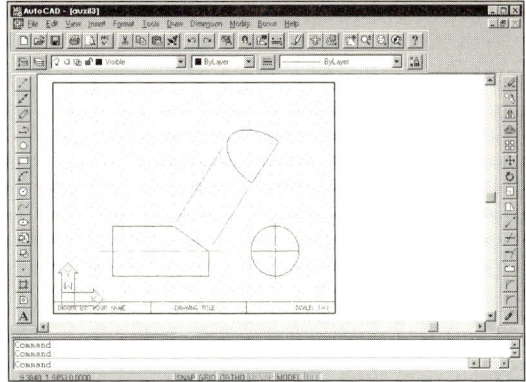

Figure 13.19

Creating an Auxiliary View of a 3D Solid Model

You can easily create auxiliary views of 3D solid models by changing the viewpoint in any viewport to show the desired view. You can view the model from any direction inside any viewport. However, good engineering drawing practice requires that adjacent views align; otherwise, interpreting the drawing can be difficult or impossible. In general, you should show at least two principal orthographic views of the object. Additional views are sometimes placed in other locations on the drawing sheet or on a separate sheet. If you use a separate sheet, you should clearly label on the original sheet the view that is located elsewhere. When placing auxiliary views in other locations, you should show a viewing plane line, similar to a cutting plane line, that indicates the direction of sight for the auxiliary view. You should clearly label the auxiliary view and preserve its correct orientation, rather than rotate it to a different alignment.

On your own, create a new drawing called *aux3d.dwg*. Use the drawing *anglblok.dwg* provided with the data files as the template.

The solid model shown in Figure 13.20 appears on your screen. You will continue working with this drawing, creating an auxiliary view that shows the true size of the inclined surface.

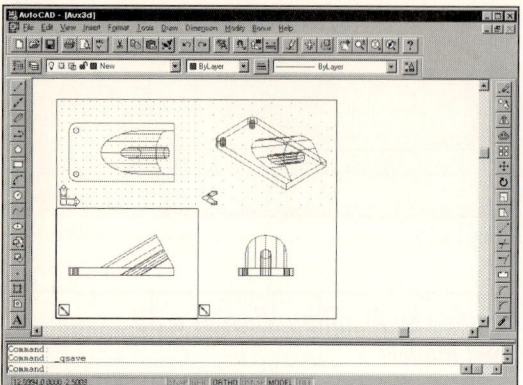

Figure 13.20

You will use hot grips to stretch the viewports to different sizes in order to make room for the auxiliary view. To do so, you will need to thaw the layer containing the viewports because it is turned off in the data file. On your own, thaw layer VPORT but do not make it current. For now leave layer NEW set as the current layer. When layer VPORT is thawed, you should see the outlines of each viewport in your drawing.

Now you are ready to stretch the viewports to different sizes to make room for a new viewport that will show the auxiliary view.

On your own, switch to paper space.

Next, activate the grips for each of the viewports. (Remember, viewports are paper space objects, and you can modify them by using Scale, Rotate, Stretch, and Copy, among other commands, but only when you are in paper space.)

> ■ *TIP* To activate the grip for each of the four viewports, make an implied Crossing box where the viewports come together at the center of the screen. ■

Pick: *(to activate the grips for the viewport borders)*

The grips are activated, as shown in Figure 13.21.

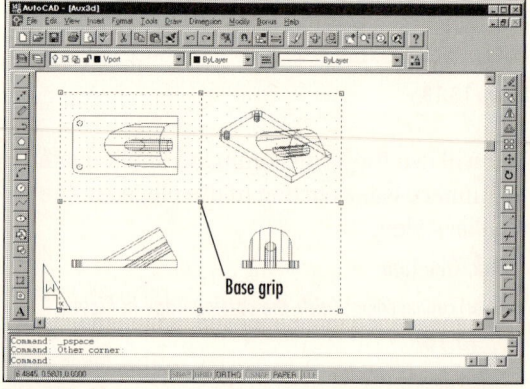

Figure 13.21

Next, pick the grip in the center of the four viewports as the base grip. You will stretch the viewports to the right, making the viewports containing the pictorial view and side view smaller. On your own pick the base grip, as shown in Figure 13.21, and stretch the viewports to the right so that the final result on the screen looks like Figure 13.22.

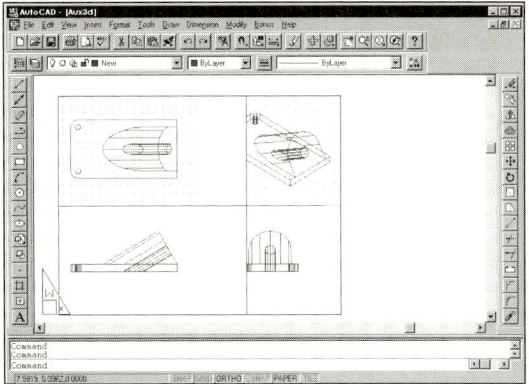

Figure 13.22

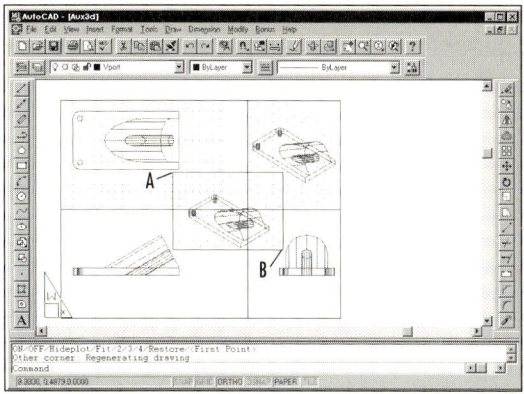

Figure 13.23

Now switch to model space and take the following steps on your own. Make the top right view active and zoom the pictorial view to .75X (so that the view is shown at 3/4 of its current size). Then use the Pan Realtime icon to pan the image until it is centered. Then use Pan Realtime and MVSETUP's Align option to move the views in the lower right and upper left viewports to make room for a new viewport and realign them with the lower left viewport. Before switching to paper space, verify that the upper right viewport is active.

On your own, switch to paper space and set layer VPORT as current. The paper space icon is displayed in the lower left corner of your screen. (The upper right viewport containing the pictorial view should have been the active viewport before you switched to paper space). You will now create your new viewport, using the View menu.

Pick: **View, Floating Viewports, 1 Viewport**

ON/OFF/Hideplot/Fit/2/3/4/Restore/<First point>: *(pick point A in Figure 13.23)*

Other corner: *(pick point B)*

The new viewport appears, as shown in Figure 13.23. The overlapping viewport borders don't matter because you will turn their layer off before plotting your drawing.

Enlarging a Viewport

You will use Zoom Window to enlarge the upper right viewport so that it fills the entire screen. Then, you can switch back to model space to work in the viewport. This method is especially useful if you do not have a large monitor. Sometimes if several small viewports are on the screen at the same time, the object is too small for you to work on effectively. By switching to paper space and zooming in on one viewport, you can enlarge just that viewport on the screen. (Zooming in model space will enlarge only the view within the viewport and retain the size of the viewport on the screen.)

You should still be in paper space.

Pick: **Zoom Window icon**

First corner: *(pick a point above and to the left of the upper right viewport)*

Other corner: *(pick a point below and to the right of the upper right viewport)*

The upper right viewport fills the screen. You are still in paper space, but to work on the model, you must change to model space. Switch to model space on your own. The UCS icon shows inside the enlarged viewport.

Now you are ready to change the viewpoint in the new viewport so that it shows the auxiliary view. You will make this change by aligning the UCS with the angled surface and then using the Plan command to show the view looking straight toward the UCS.

Using 3 Point UCS

The 3 Point option of the UCS command creates a User Coordinate System that aligns with the three points you specify. AutoCAD prompts you to pick the origin, a point in the positive X-direction, and a point in the positive Y-direction. On your own, be sure that OSNAP is active with Endpoint and Quadrant selected. Refer to Figure 13.24 for the points to select.

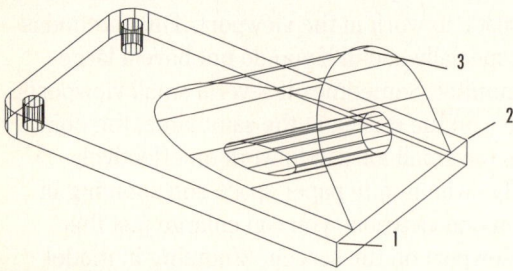

Figure 13.24

Pick: **3 Point UCS icon**

Origin point <0,0,0>: *(pick Endpoint 1)*

Point on positive portion of the X-axis
<14.000,4.5000,0.5000>: *(pick Endpoint 2)*

Point on positive-Y portion of the UCS XY plane
<12.0000,4.5000,0.5000>: *(pick on the upper tessellation line labeled as point 3, using the Quadrant AutoSnap marker)*

Switch back to paper space and use Zoom Previous on your own to restore the original zoom factor, showing the five viewports. When you have restored the previous zoom factor, switch back to model space. Your screen should look like Figure 13.25. The UCS icon

now lines up with the angled surface and is shown as a broken pencil in the lower left viewport. These icons are important because they tell you that the User Coordinate System is correctly aligned with the angled surface. The broken pencil shows when the coordinate system is perpendicular to the viewing direction. Because the UCS is aligned with the angled surface, it is perpendicular to the front view (although along an angled line).

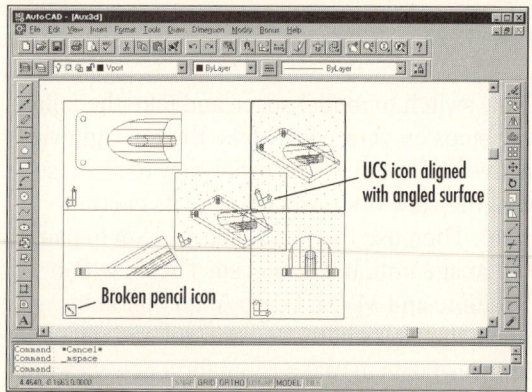

Figure 13.25

Using Plan

The Plan command aligns the viewpoint in the active viewport so that you are viewing the current UCS, a named UCS, or the WCS straight on. You will accept the default, the current UCS, to show the true size view of the angled surface, with which you have already aligned the coordinate system. You will type the Plan command at the command prompt, but it is also available from the View menu under 3D Viewpoint, Plan View.

On your own, activate the center viewport. The crosshairs appear in the center viewport, indicating that it is the active viewport.

■ *TIP* If you have trouble picking inside a viewport because it overlaps other viewports, type Ctrl-*R* to toggle the viewport. ■

Command: **PLAN** ⏎

\<Current UCS>/Ucs/World: ⏎

The view in the center viewport changes so that you are looking straight at the angled surface. On your own, use the Zoom command with a .5XP scaling factor to size the view. (The original views were zoomed to .5XP in the data file.) Next, use the Pan Realtime command to center the view inside the viewport. When you have finished these steps, your screen should look like Figure 13.26.

■ *TIP* If your viewport is not large enough to show the entire view zoomed to .5XP, switch back to paper space and use hot grips to stretch the viewport so that it is slightly larger than before. Switch back to model space when you are finished and use Pan once again to position the view. ■

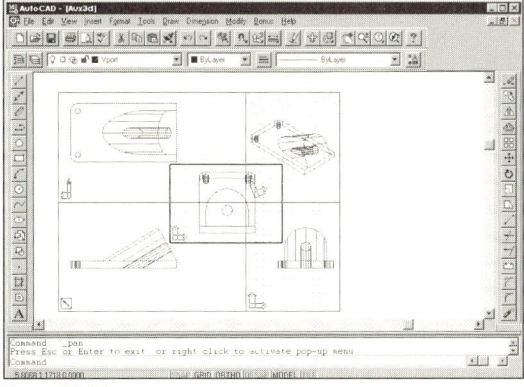

Figure 13.26

■ *TIP* You can set the Ucsfollow variable to generate automatically the plan view of the current UCS. Pick in a viewport to make it active and then set Ucsfollow to 1. AutoCAD automatically generates a plan view of the current UCS in that viewport whenever you change the UCS. ■

Now the view shows the true size of the angled surface, but it is not aligned correctly with the front view. Remember, to show a true-size view of the surface you use a line of sight that is perpendicular to the view in which the surface shows on edge, as in the front view.

Using 3D Dynamic View

The 3D Dynamic View command uses the metaphor of a camera and a target, where your viewing direction is the line between the camera and the target. It is called 3D Dynamic View because while you are in the command options, you can see the view of the selected objects change dynamically on the screen when you move your pointing device. If you do not select any objects for use with the command, you see a special block shaped like a 3D house instead. Use AutoCAD's Help command to discover more information about the options of the 3D Dynamic View command.

In the next steps you will use the 3D Dynamic View command with the Twist option to twist the viewing angle. You will type *DVIEW* at the command prompt, but the 3D Dynamic View command is also available from the View menu. The angled surface is at 60° in the front view. You will type *30* for the twist angle, which will produce a view that is aligned 90° to the front view. On your own, verify that the center viewport is active.

Command: **DVIEW** ⏎

Select objects: *(pick on the solid model)* ⏎

CAmera/TArget/ Distance/POints/PAn/Zoom/TWist/
CLip/Hide/Off/Undo/<eXit>: **TW** ⏎

New view twist <0.00>: **30** ⏎

CAmera/TArget/ Distance/POints/PAn/Zoom/TWist/
CLip/Hide/Off/Undo/<eXit>: ⏎

The view is twisted 30° to align with the edge view of the angled surface. Your screen should look like Figure 13.27.

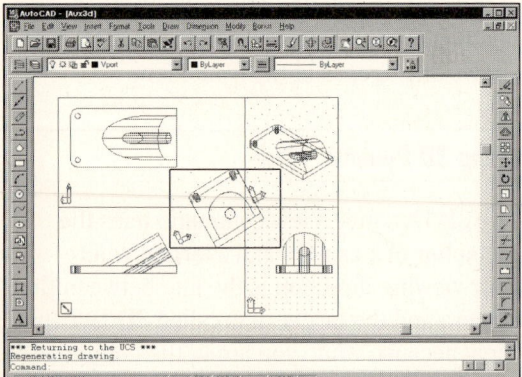

Figure 13.27

On your own, return to paper space and use hot grips to stretch the center viewport so thatit is large enough to show the entire object. When you have it sized as you want, set layer NEW as the current layer and then freeze layer VPORT. When you have finished, your screen should be similar to Figure 13.28.

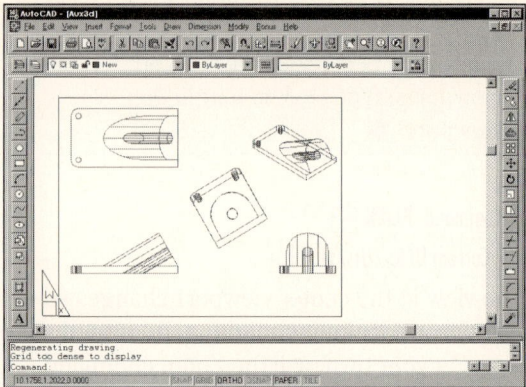

Figure 13.28

Next, you will create an Auxiliary view with the Solview command. On your own, thaw layer VPORT and erase the Auxiliary viewport that you just created.

Using Solview

This time you will use Solview to create your Auxiliary view. After performing the above-mentioned tasks, your screen should look like Figure 13.29.

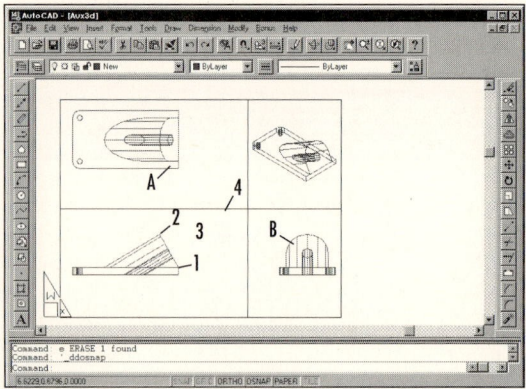

Figure 13.29

Command: **SOLVIEW** ↵

Ucs/Ortho/Auxiliary/Section/<eXit>: **A** ↵

Incline Plane's 1st point: *(select point 1 using Snap to Endpoint)*

Incline Plane's 2nd point: *(select point 2 using Snap to Quadrant)*

Side to view from: *(pick point 3)*

View center: *(pick near point 4)*

View center: ↵

Clip first corner: *(select point A)*

Clip other corner: *(select point B)*

View name: **AUXILIARY** ↵

Ucs/Ortho/Auxiliary/Section/<eXit>: **X** ↵

Your drawing should appear on the screen as shown in Figure 13.30.

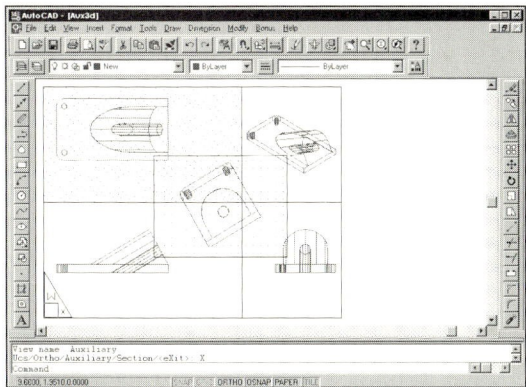

Figure 13.30

On your own, change to model space and turn the grid off in your new viewport. Then change back to paper space and freeze layer VPORT and layer VPORTS created by the Solview command.

When you have finished, your screen should be like Figure 13.31. Plot your drawing from paper space.

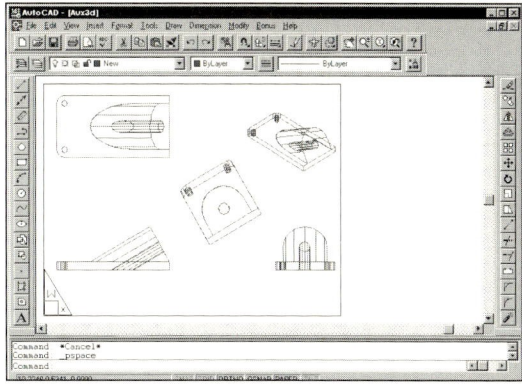

Figure 13.31

■ **TIP** You can add a few projection lines in paper space to show how the auxiliary view aligns. Another acceptable practice is to extend the centerline for the hole in the auxiliary view to the view from which it was projected. You can use paper space to add the centerline if you want. Don't forget that object snap will find locations on model space geometry even when you are in paper space. You can use this to help draw the centerline accurately. ■

Save your drawing, close all but the Draw and Modify toolbars, and exit AutoCAD. You have completed Tutorial 13.

Use the 2D or solid modeling techniques you have learned in the tutorial to create an auxiliary view of the slanted surfaces for the objects below. The letter M after an exercise number means that the given dimensions are in millimeters.

13.1 Stop Block

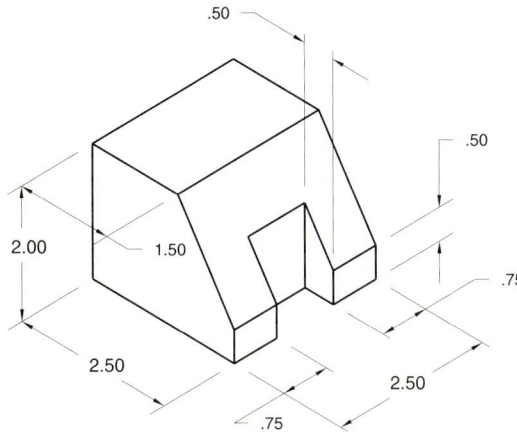

13.2M Lever

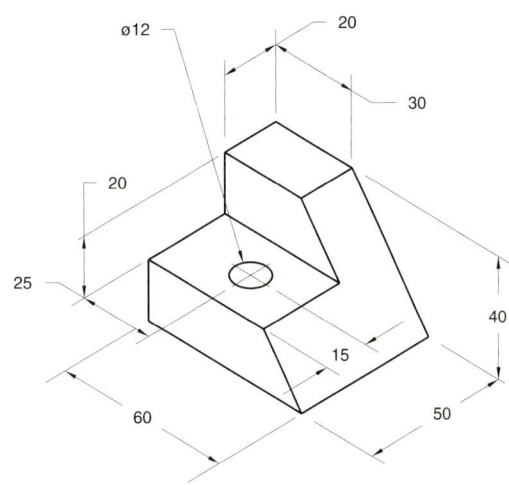

13.3 Bearing

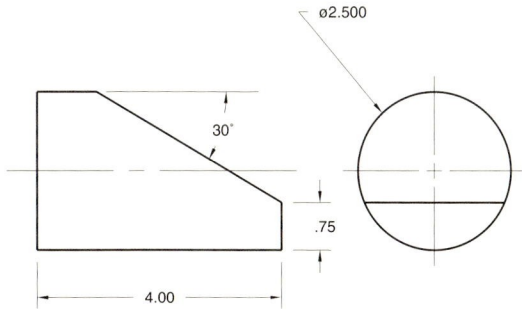

⊙ 13.4 Tooling Support

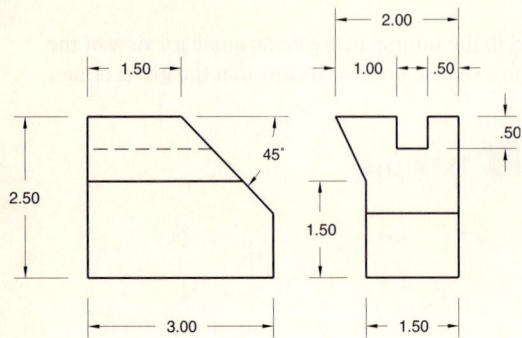

⊙ 13.5 Incline Block

Use solid modeling to construct the object shown. Create an auxiliary view showing the true size of the slanted surface. Use the Vpoint (3,–3,1) for the pictorial view.

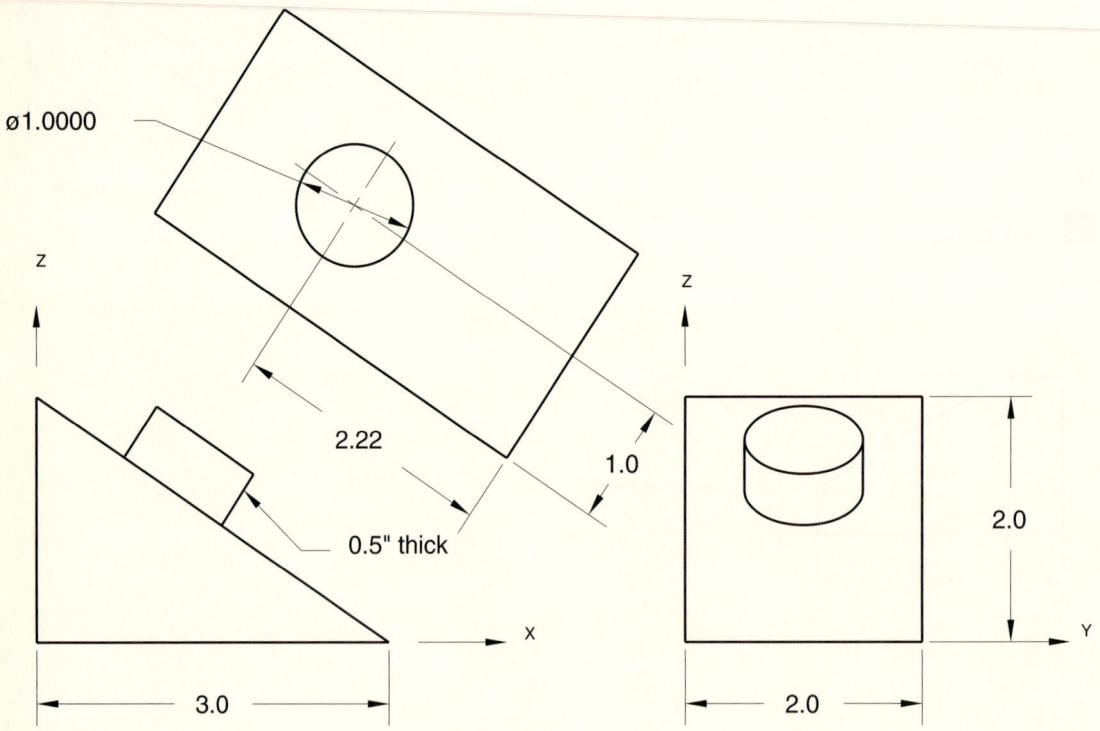

Use solid modeling to create the objects below. Produce the necessary primary and auxiliary views to describe each part.

 **13.6 Column Base**

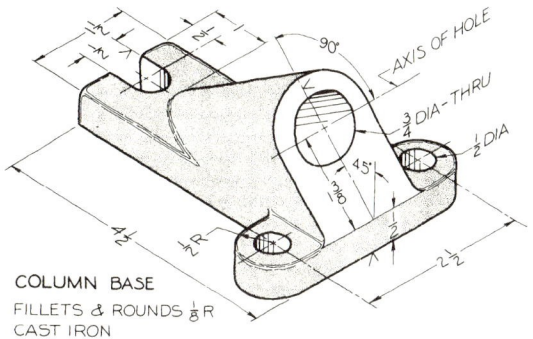

COLUMN BASE
FILLETS & ROUNDS $\frac{1}{8}$ R
CAST IRON

13.8M Hexagon Angle

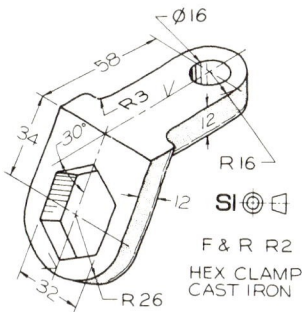

F & R R2
HEX CLAMP
CAST IRON

13.7M Hanger

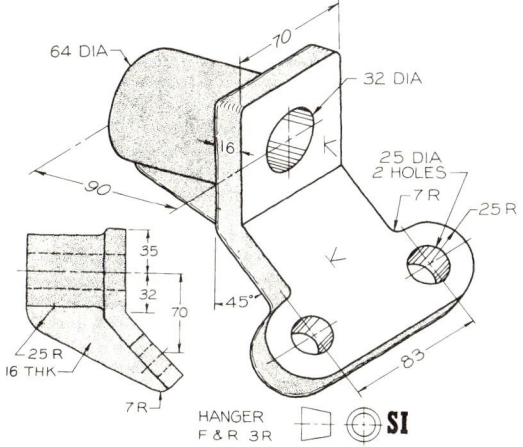

HANGER
F & R 3R   **SI**

13.9M Dovetail Bracket

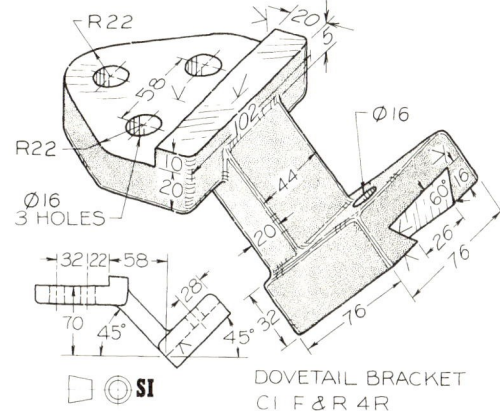

DOVETAIL BRACKET
CI F & R 4R

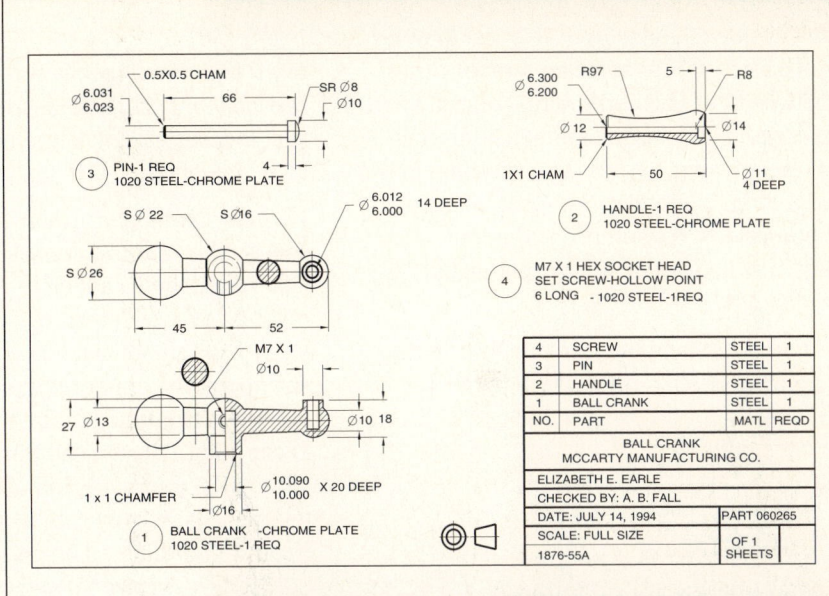

Advanced Dimensioning

Objectives

When you have completed this tutorial, you will be able to

1. Use tolerances in a drawing.

2. Set up the dimension variables to use limit tolerances and variance (plus/minus) tolerances.

3. Add geometric tolerances to your drawing.

4. Use dimension overrides.

Tolerance

No part can be manufactured to exact dimensions. There is always some amount of variation between the size of the actual object when it is measured and an exact dimension specified in the drawing. To take this variation into consideration, tolerances are specified along with dimensions. A tolerance is the total amount of variation that an acceptable part is allowed to have.

To better understand the role that tolerances play in dimensioning, refer to Figure 14.1.

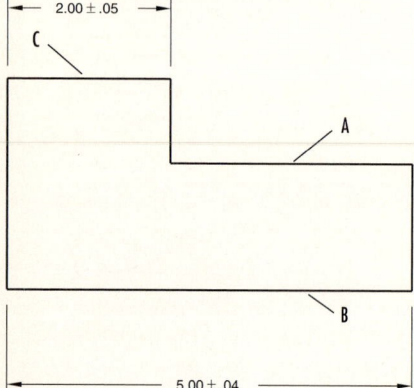

Figure 14.1

The dimensions given in the drawing are used to make and inspect the part. Included with the dimension is the allowable amount of variation (tolerance) that the part can have and still be acceptable. In order to determine whether a part is acceptable, the measurements of the actual part are compared to the toleranced dimensions specified in the drawing. If the part falls within the tolerance range specified, it is an acceptable part.

In Tutorial 7, you learned how to add a general tolerance note to your drawing to specify this allowable variation. AutoCAD provides three ways that you can give a tolerance with the dimension for the feature: limit tolerances, variance tolerances, and geometric tolerancing symbols.

Limit tolerances specify the upper and lower allowable measurements for the part. An actual part measuring anywhere between the two limits is acceptable.

Variance tolerances or *plus/minus tolerances* specify the *nominal dimension* and the allowable range that is added to it. From this information you can determine the upper and lower limits. Add the plus tolerance to the nominal size to get the upper limit; subtract the minus tolerance from the nominal size to get the lower limit. (Or you can think of it as always adding the tolerance, but when the sign of the tolerance is negative, it has the effect of subtracting the value.) The plus and minus values do not always have to be the same. There are two types of variance tolerances: bilateral and unilateral. *Bilateral tolerances* specify a nominal size and both a plus and minus tolerance. *Unilateral tolerances* are a special case where either the plus or the minus value specified for the tolerance is zero. You can create both bilateral and unilateral tolerances with AutoCAD.

Geometric tolerances feature special symbols inside *feature control frames* that describe tolerance zones that relate to the type of feature being controlled. The Tolerance command allows you to create feature control frames quickly and select geometric tolerance symbols for them. You will learn to create these frames later in this tutorial.

To understand tolerances more clearly, consider Figure 14.1. Surface A on the actual object is longest when the 5.00 overall dimension shown for surface B is at its largest acceptable value and when the 2.00 dimension shown for surface C is at its shortest acceptable value. Thus

$$A_{max} = 5.04 - 1.95 = 3.09$$

Surface A is shortest when surface B is at its smallest acceptable value and surface C is at its largest acceptable value. Thus

$$A_{min} = 4.96 - 2.05 = 2.91$$

In other words, the given dimensions with the added tolerances permit surface A to vary between 3.09 and 2.91 units.

Figure 14.2 shows the drawing that you dimensioned in Tutorial 7 with tolerances added to the dimensions. Tolerances A, B, F, and G are examples of bilateral tolerances. Tolerances C, D, and E are limit tolerances.

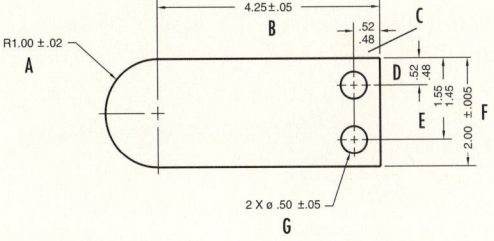

Figure 14.2

Next, you will change the dimensions for the drawing you created in Tutorial 7 to add tolerances.

Starting

Start AutoCAD. On your own, use the drawing *obj-dim.dwg* as the template for your new drawing. Name your new drawing *tolernc.dwg*.

Your screen should be similar to Figure 14.3. Be sure that layer DIM is set as the current layer.

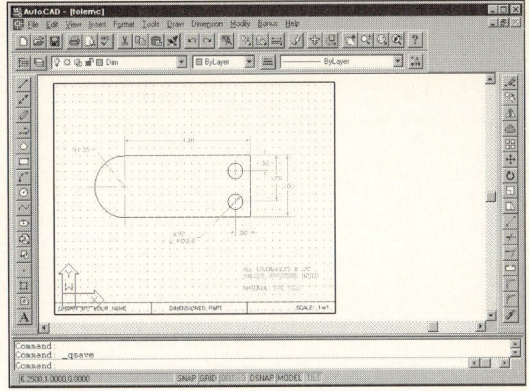

Figure 14.3

In the next steps, you will set up the dimension variables to use tolerances.

Automatic Bilateral Tolerances

You can have AutoCAD automatically add bilateral tolerances by setting the dimensioning variables. You will use the Dimension Styles dialog box to set the dimension variables. First you will show the Dimension toolbar.

Pick: **View, Toolbars, Dimension (then close the Toolbars dialog box)**

The Dimension toolbar appears on your screen. On your own, position it in a location where it will be easily available but not in the way of your drawing. You will use the Dimension toolbar to select the Dimension Styles dialog box to set the variables.

Pick: **Dimension Style icon**

The Dimension Styles dialog box appears on your screen, as shown in Figure 14.4. You will use it to set up the appearance for the tolerances to be added to the dimensions. The dimension style MECHANICAL is current. It is like the style you created in Tutorial 7. You will make changes to it so that the dimensions in the drawing created with that style will show a variance or deviation tolerance.

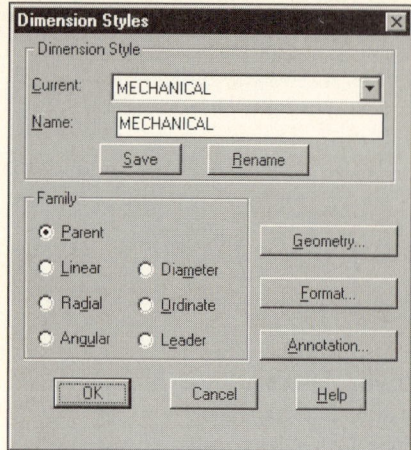

Figure 14.4

Recall from Tutorial 7 that you use the Name text box to select a name for a new dimension style. You could create a new style, but this time you will change style MECHANICAL to include deviation tolerances. Changes that you make to the dimension style affect only the dimensions that you create or update in the future, unless you pick the Save button in the Dimension Style area of the dialog box. Picking the Save button causes AutoCAD to make the changes to the existing dimensions having that style when you exit the Dimension Styles dialog box. When you are done making the following changes to the dimension style, you will save the changes to style MECHANICAL so that the dimensions automatically reflect the changes.

Pick: **Annotation**

The Annotation dialog box appears. On your own, pick in the box to the right of Method in the Tolerance area of the dialog box or pick on the downward-pointing triangle to pull down the list of available tolerance methods shown in Figure 14.5. The choices of tolerance methods are None, Symmetrical, Deviation, Limits, and Basic. None, of course, uses just the dimension and does not add a tolerance, as your current dimensions reflect. *Symmetrical tolerances* are

lateral tolerances with the same upper and lower values; these tolerances generally specify the dimension plus or minus a single value. *Deviation tolerances* are lateral tolerances that have a different upper and lower deviation. Limit tolerances show the maximum value for the dimension preceded by a plus sign and the minimum value preceded by a minus sign. *Basic dimension* is a term used in geometric dimensioning and tolerancing to specify a theoretically exact dimension without a tolerance. Basic dimensions appear in your drawing with a box drawn around them. If you choose any of the tolerance methods, you can preview what your dimensions will look like in the image tiles.

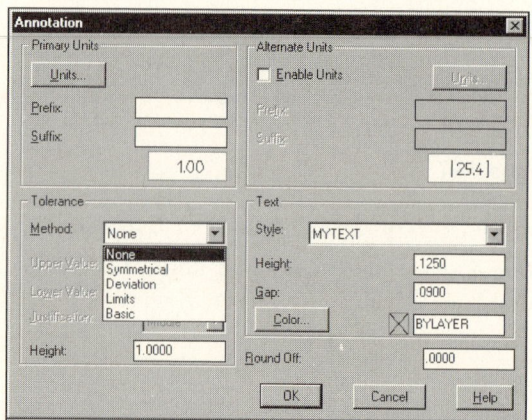

Figure 14.5

Pick: **Deviation**

Note the image tile just above the Tolerance area; it shows a dimension with a deviation style tolerance added. You can also make selections in the dialog box by picking on the image tile until it has the correct appearance; try to do so, as follows.

Pick: **(on the image tile to cycle its appearance to that of limit tolerances)**

Pick: **(again on the image tile to change it to a basic dimension)**

Continue picking the image tile on your own until you cycle through all the options and return to Deviation tolerance, as shown in Figure 14.6.

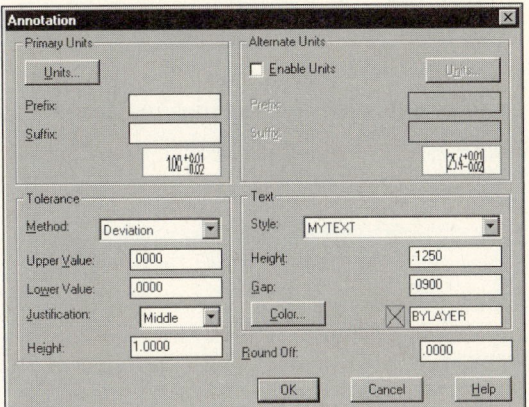

Figure 14.6

Specifying Tolerance Values

The text boxes for setting tolerance values were formerly grayed. Now that you have selected a tolerance method, they are available for input. You will use them to specify the upper and lower values that you want to use with the dimension to specify the allowable deviation. Both values will be positive. The upper value will be added to the dimension value; the lower value will be subtracted from the dimension value. AutoCAD will automatically add the plus or minus sign in front of the value when showing the tolerance. On your own, pick in the text box to the right of Upper Value.

Type: **.05**

On your own, pick in the text box to the right of Lower Value or press TAB.

Type: **.03**

Note that you can also control the justification for the tolerance. For now you will leave it set to Middle, the default.

Setting the Tolerance Text Height

The default setting of 1.0000 appears in the Height text box at the bottom of the Tolerance area. The value set for the tolerance height is a scaling factor. A setting of 1.0000 makes the tolerance values the same height as the standard dimension text. For this tutorial you will set tolerance height to 0.8 (so that the height of the tolerance values will be 8/10 the height of the dimension values). Highlight the value in the box.

Type: **.8**

Setting the Tolerance Precision

The tolerance values can have a different precision than the dimension values. You control the precision by selecting the Units button in the Primary Units area at the upper left of the Annotation dialog box.

Pick: **Units**

The Primary Units dialog box appears on your screen, as shown in Figure 14.7. Recall that you used it in Tutorial 7 to set the number of decimal places for the dimension. Now you will use it to set the precision for the tolerance values. You can also control whether leading and trailing zeros are used. On your own, pick on the Precision value in the Tolerance area (not the Dimension area). A list of precision values pulls down. Pick the two-place decimal 0.00 from the list. Pick OK to exit the Primary Units dialog box.

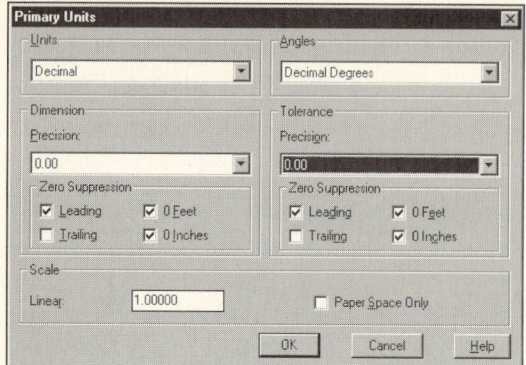

Figure 14.7

The same text style used for the dimension text, MYTEXT, will also be used for the tolerance text. You can control the text style and the dimension text height, color, and gap in the Text area of the Annotation dialog box. You will leave these features set as they are.

Pick: **OK (to exit the Annotation dialog box)**

Pick: **Save (to save your changes to the dimension style MECHANICAL)**

Pick: **OK (to exit the Dimension Styles dialog box)**

You return to your drawing. The dimensions created with style MECHANICAL automatically update to show the deviation tolerance. Your drawing should look like that in Figure 14.8.

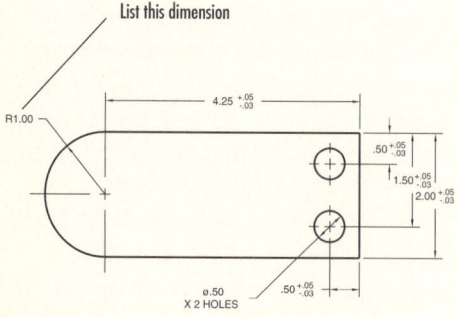

Figure 14.8

Why didn't the radial dimension update? You can find information about the dimension by using the List command. List is on the Inquiry flyout on the Standard toolbar.

Pick: **List icon**

Select objects: **(pick the radial dimension at the left end of the object)** ←

The List command displays the information shown in Figure 14.9 in an AutoCAD Text window.

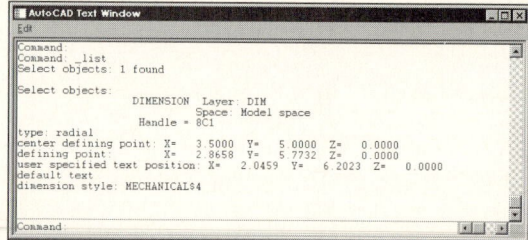

Figure 14.9

Note that the style name for this dimension is MECHANICAL$4. When you list a dimension, the $# code tells you that the dimension uses a child style different from the parent style. Here, it tells you that the dimension was created with a child style that has different settings than the parent style, MECHANICAL. If you need to review parent and child styles, refer to Tutorial 7. The codes for the child types are:

$0	Linear
$2	Angular
$3	Diameter
$4	Radial
$6	Ordinate
$7	Leader (also used for tolerance objects)

You must change the particular child dimension style (for example, radial) in order to change the appearance of the dimension. The reason that the dimension did not update to show the tolerance is because no tolerance was created for the radial dimension child style MECHANICAL$4. To change the child style, you will use the Dimension Styles dialog box and pick the child style for radial and set its properties.

On your own, close the AutoCAD Text window.

Pick: **Dimension Style icon**

The Dimensions Styles dialog box appears on your screen. You will use it to change the radial child style for MECHANICAL. In the Family area of the dialog box,

Pick: **Radial**

Pick: **Annotation**

The Annotation dialog box appears on your screen. On your own, use it to select Deviation for the tolerance method and set the upper value to 0.05 and the lower value to 0.03, as you did for the MECHANICAL parent dimension style. Change the scaling factor for the height in the Tolerance area to 0.8. Use the Units selection to set two decimal places for the tolerance precision. When you have finished making these selections, pick OK to return to the Dimension Styles dialog box. Save the changes you made to the style so that you can apply them to the existing dimensions.

Pick: **Save**

Pick: **OK *(to exit the dialog box)***

Note that the radial dimension now has a tolerance value, as shown in Figure 14.10.

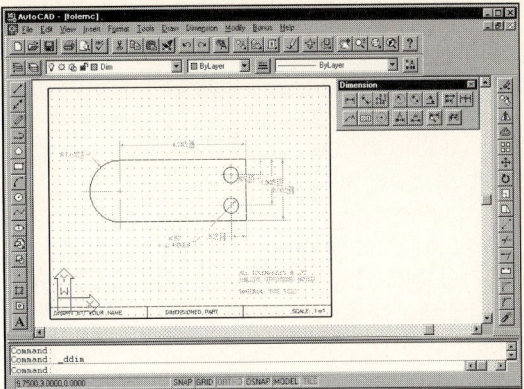

Figure 14.10

Using Dimension Overrides

Dimension overrides let you change the dimension variables controlling a dimension and leave its style as is. A good approach is to change the general appearance of the dimensions by changing the dimension style and to change individual dimensions that need to vary from the overall group of dimensions (to suppress an extension line, for instance) by using overrides.

You can change dimension variables by typing *DIMOVERRIDE* at the command prompt. Dimoverride has the alias Dimover. When you use Dimoverride at the command prompt, AutoCAD prompts you for the exact name of the dimension variable you want to control and its new value. Then you select the dimensions to which you will apply the override. The Clear option of the Dimoverride command lets you remove overrides from the dimensions you select.

You can also use the Properties selection from the Object Properties toolbar to do the same thing.

Pick: **Properties icon**

Select objects: *(pick on the radial dimension for the rounded end)* ⏎

The Modify Dimension dialog box appears on your screen. You will use it to override the dimension style for the radial dimension. Because the 2.00 vertical dimension at the right side of the part controls the part's height, providing the radial dimension is an example of overdimensioning. When you want to call attention to the full rounded end, which is tangent to the horizontal surfaces at the top and bottom of the object, you can show the 1.00 radius value as a *reference dimension*. Or you can simply identify it with the letter R, indicating the radius but leaving its value to be determined by the other surfaces. You will turn off the tolerance and add REF to the end of the radial dimension value to indicate that it is provided as a reference dimension only. You can also note reference dimensions by including the dimension value in square brackets.

Pick: **Annotation**

The Annotation dialog box appears on your screen. On your own, use it to select None as the tolerance method. In the Primary Units area of the dialog box, type REF in the Suffix text box. When you have finished making these selections, pick OK. You return to the Modify Dimension dialog box.

You can also use the Modify Dimension dialog box to override the dimension text by picking the Edit button and using the Edit MText dialog box that appears to change the text. The angle brackets that you see in the Mtext editor are AutoCAD's way of storing the default lengths of the dimensions, which allows them to update. After you have overridden the text, your dimension values no longer update if you stretch or change the object. As usual with the Properties selection, you can change the layer, color, linetype, and linetype scale and thickness. Exit the Modify Dimension dialog box by picking OK.

Your drawing should look like the one in Figure 14.11.

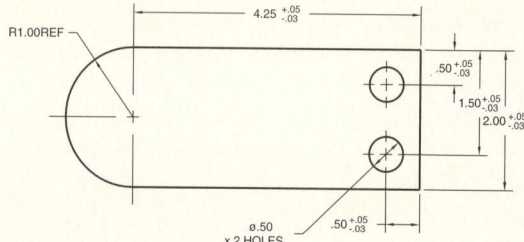

Figure 14.11

Next, use the List icon to list the radial dimension.

Pick: **List icon**

Select objects: *(pick the radial dimension for the rounded end)* ⏎

An AutoCAD Text window appears on your screen, as shown in Figure 14.12, listing the information about the radial dimension. Note that overrides for the dimension variables Dimpost and Dimtol are now listed. Dimpost displays the dimension suffix, which is set to REF. Dimtol is set to Off, so that no deviation tolerance is applied to the dimension.

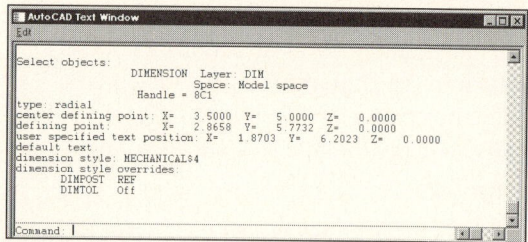

Figure 14.12

When dimensioning, keep your dimension styles organized by creating new dimension styles and controlling the child styles as needed. Use dimension overrides when certain dimensions need a different appearance than the general case for that type of dimension.

You can now close the Text window.

Using Limit Tolerances

Next you will create a new style for dimensions using limit tolerances, and then you will change the object's overall dimension to use that style.

Pick: **Dimension Style icon**

The Dimension Styles dialog box appears on your screen once again. Next you will type in a new style name. On your own, highlight the name MECHANICAL in the Name text box.

Type: **LIMITTOL**

Be sure the Parent radio button is active.

Pick: **Annotation**

On your own, use the image tile in the Annotation dialog box to select the Limits tolerance method. Once it has been selected, the dialog box appears as shown in Figure 14.13.

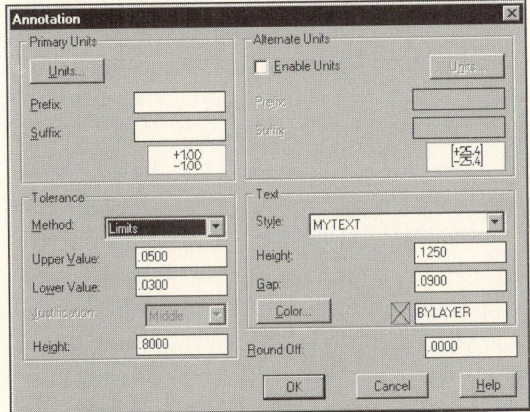

Figure 14.13

You can leave the upper and lower values and text height set as you did for deviation tolerances.

Pick: **OK *(to exit the Annotation dialog box)***

Pick: **Save**

Pick: **OK *(to exit the Dimension Styles dialog box)***

Next, you will change the overall dimension of the object to use the new style, LIMITTOL. To do so you will use the Properties selection from the Object Properties toolbar.

Pick: **Properties icon**

Select objects: **(pick on the longest horizontal dimension)** ⏎

To change the dimension style to LIMITTOL, use the Modify Dimension dialog box that appears. To change the dimension's style, pick on the name MECHANICAL next to Style in the Text area of the dialog box to pull down the list of available styles. On your own, pick LIMITTOL from the list of styles. (If you do not see the selections for changing the dimension style, perhaps you selected more than one dimension. The Properties selection is sensitive to the types of objects selected and displays a different dialog box depending on what you select.) Pick OK to exit the dialog box.

The limit tolerance appears, as shown in Figure 14.14.

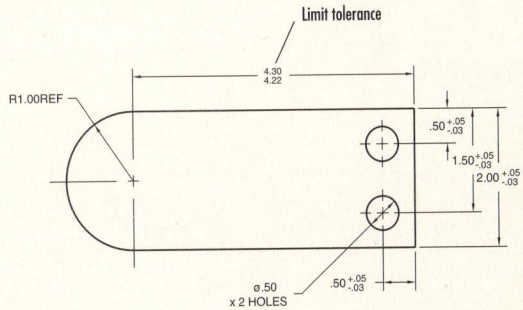

Figure 14.14

Next, you will use the Tolerance command to add a feature control frame specifying a geometric tolerance to the drawing.

Creating Feature Control Frames

Geometric tolerancing allows you to define the shapes of *tolerance zones* in order to control accurately the manufacture and inspection of parts. Consider the location dimensions for the two holes shown in Figure 14.15. The location for the center of hole A is allowed to vary in a square zone 0.1 unit wide, as defined by the location dimensions 1 and 2 and the +/−0.05 tolerance applied to them.

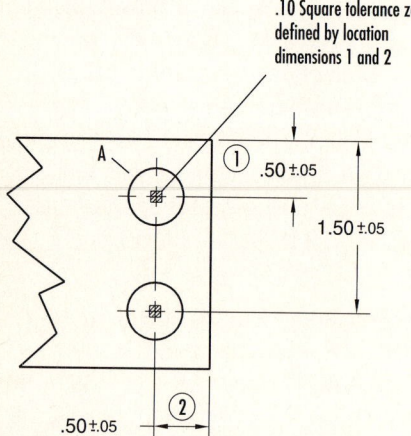

.10 Square tolerance zone defined by location dimensions 1 and 2

Figure 14.15

This tolerance zone is not the same length in all directions from the true center of the hole. The stated tolerances allow the location for the center of the actual hole to be off by a larger amount diagonally, as shown in Figure 14.16.

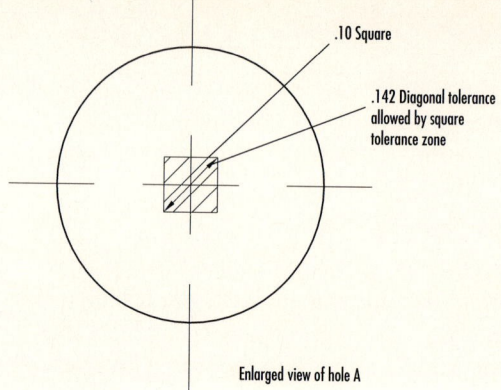

.10 Square

.142 Diagonal tolerance allowed by square tolerance zone

Enlarged view of hole A

Figure 14.16

In other words, some parts that should be rejected would be accepted because the tolerance stated for the location of the hole was considered to be only a horizontal or vertical dimension. Or, if the diagonal was considered when the tolerance was stated, some parts that are off horizontally or vertically, but could have been accepted because of the diagonal dimension, would be rejected. This realization led to the development of specific tolerance symbols for controlling the geometric tolerance characteristics for position, flatness, straightness, circularity, concentricity, runout, total runout, angularity, parallelism, and perpendicularity.

Geometric tolerances allow you to define a *feature control frame* that specifies a tolerance zone shaped appropriately for the feature being controlled (sometimes this method is called feature-based tolerancing). A feature control frame is shown in Figure 14.17.

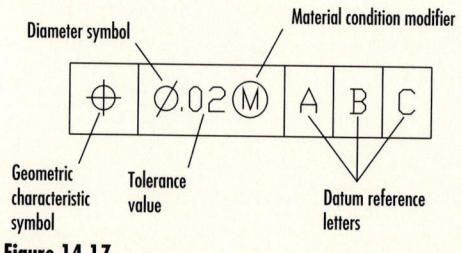

Diameter symbol

Material condition modifier

Geometric characteristic symbol

Tolerance value

Datum reference letters

Figure 14.17

The feature control frame begins with the *geometric characteristic symbol*, which tells the type of geometry being controlled. Figure 14.17 first shows a geometric characteristic symbol for positional tolerance. It next specifies the total allowable tolerance zone; in this case a diametral-shaped zone 0.02 wide around the perfect position for the feature. These two symbols are followed by any modifiers for material condition or projected tolerance zone. The circled M shown is the modifier for maximum material condition, meaning that at maximum material, in this case the smallest hole, the tolerance must apply, but when the hole is larger a greater tolerance zone may be calculated, based on the size of the actual hole measured. The remaining boxes indicate that the position is measured from datum surfaces A, B, and C. Not every geometric tolerance feature control frame requires a datum; for example, flatness can just point to the feature and control it, without respect to another surface. Features such as perpendicularity that are between two different surfaces require a datum.

Among other things, proper application of geometric tolerancing symbols requires that you understand

- the design intent for the part;
- the shapes of the tolerance zones specified with particular geometric characteristic symbols;
- the selection and indication of datum surfaces on the object; and
- the placement of feature control frames in the drawing.

In this tutorial, you will learn how to create feature control frames and add them to your drawing. To understand the topic of geometric tolerancing fully, refer to your engineering design graphics textbook or a specific text on geometric tolerancing. Merely adding feature control frames to the drawing will not result in a drawing that clearly communicates your design intent for the part to the manufacturer.

Next you will create the feature control frame shown in Figure 14.17. To create a feature control frame, you will pick the Tolerance icon from the Dimension toolbar.

Pick: **Tolerance icon**

The Symbol dialog box shown in Figure 14.18 appears on your screen. You will use it to pick the symbol for the geometric characteristic of the geometric tolerance feature control frame that you will create.

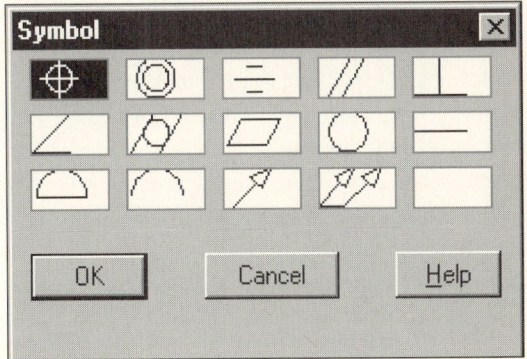

Figure 14.18

Pick: (the positional tolerance symbol in the upper left corner)

Pick: **OK**

> ■ *TIP* In many Windows dialog boxes you can double-click on the desired selection to select it directly and exit the dialog box in one action without having to pick OK. You can double-click the picture of the positional tolerance symbol to select it directly and exit the dialog box. ■

The Geometric Tolerance dialog box appears on your screen. You will make selections so that the dialog box appears as shown in Figure 14.19. The positional tolerance symbol should already be shown at the left.

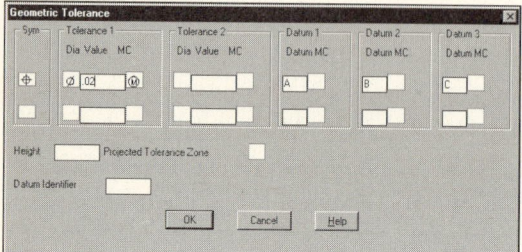

Figure 14.19

The box to the right of the positional tolerance symbol, below the heading Dia, is a toggle. Pick in the empty box to turn on the diameter symbol. When you pick in the empty box, the symbol should appear. Next, click in the banner below the heading Value and type 0.02. To add a modifier, pick on the empty box below the heading MC (for material condition). The Material Condition dialog box appears on your screen, as shown in Figure 14.20.

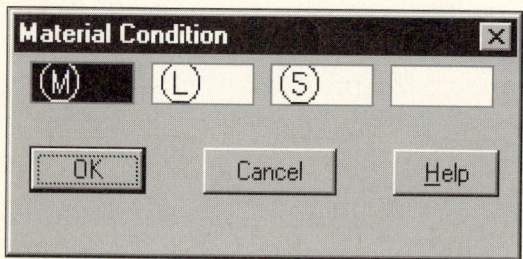

Figure 14.20

Double-click on the circled M to select the modifier for maximum material condition or pick once and then pick OK. You return to the Geometric Tolerance dialog box. The modifier appears for the feature control frame. Next, you will specify datum surfaces A, B, and C. To do so, pick in the empty text box below Datum 1 and type A; repeat this process for datum B and datum C on your own. Note that datums can also have modifiers. If you wanted to, you could pick on the empty box below MC to add a modifier for the datum.

Now you are ready to add the feature control frame to your drawing. On your own, pick OK to exit the Geometric Tolerance dialog box. You will locate the feature control frame below the 0.50 diameter dimension for the two holes.

Enter tolerance location: *(pick the location for the tolerance symbol)*

Your drawing should look like that shown in Figure 14.21.

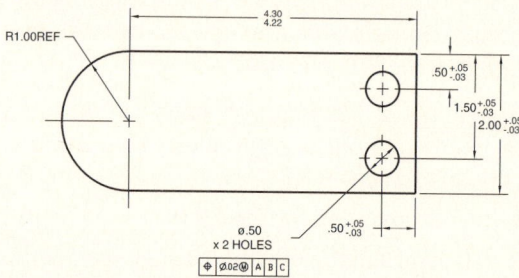

Figure 14.21

Using Leader with Tolerances

 You can also use the Leader command to add a leader with an attached geometric tolerance symbol to your drawing. You will create the next feature control frame by starting with the Leader command. You will pick the Leader icon from the Dimension toolbar. On your own, be sure that OSNAP is on, with Nearest selected.

Pick: **Leader icon**

From point: *(pick at about 2 o'clock on the upper hole, using AutoSnap)*

To point: *(pick above and to the right of the drawing)*

To point (Format/Annotation/Undo)<Annotation>: ⏎

Annotation (or press ENTER for options): ⏎

Tolerance/Copy/Block/None/<Mtext>: **T** ⏎

The Symbol dialog box appears; from there you can start creating the feature control frame. On your own, create another feature control frame like the first one. After you've finished making selections in the Geometric Tolerance dialog box, pick OK. The feature control frame is added to the end of the leader line, as shown in Figure 14.22.

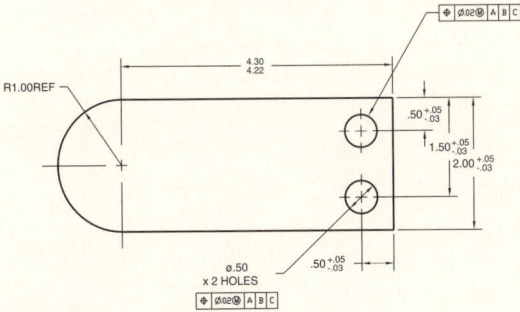

Figure 14.22

Next, you will override the location dimensions for the holes so that they appear as basic dimensions. When you specify a positional tolerance, the true position is usually located with basic dimensions, which theoretically are exact dimensions. The feature control frame for positional tolerance controls the tolerance zone from which the true position can vary. To create basic dimensions by overriding the dimension style,

Pick: **Properties icon**

Select objects: *(pick the 0.50 vertical dimension at the upper right of the object)* ⏎

The Modify Dimension dialog box appears on your screen. On your own, use it to select Annotation and then set the method in the Tolerance area of the dialog box to Basic. When you have finished, pick OK to exit the Annotation dialog box and then pick OK to exit the Modify Dimension dialog box.

Figure 14.23 shows the drawing with the 0.50 basic dimension.

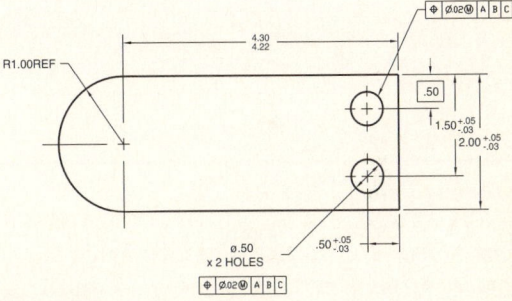

Figure 14.23

On your own, override the dimension style for the 0.50 horizontal dimension below the object, and then for the 1.50 vertical dimension. You must pick only one dimension each time you use the Properties dialog box, or you will get the more generic Change Properties dialog box. When you have finished, save your drawing on your own. Next, you will add datum flags to identify datum surfaces A and B, from which the position feature can vary.

Leaders with associated annotations can cause problems when you save your drawing to Release 13 format. If you were to save your current drawing in Release 13 format and attempt to edit it in Release 13, AutoCAD might crash. Specifically, if you erase a leader with Annotation, any subsequent operation that modifies the leader's associated Mtext, Tolerance, or Block Insert annotation object will cause AutoCAD to crash. A Release 13 application to prevent this problem is available for downloading from the Support & Training area of the Autodesk Internet location (http://www.autodesk.com). If you are unable to obtain this application, do not erase Release 14 leaders saved to Release 13; rather, explode the associated multiline text annotation before editing. This problem occurs only when a drawing is being saved to Release 13, not when it is being saved to Release 12 or when editing or erasing is being done in Release 14. ■

Creating Datum Flags

Datum flags are boxed letters identifying the feature on the object that is being used as a datum. Often they are attached to the extension line that is parallel to the surface being identified as the datum surface. To create a datum flag,

Pick: **Tolerance icon**

From the lower right of the Symbol dialog box,

Pick: **(the blank box)**

Pick: **OK**

The Geometric Tolerance dialog box appears on your screen. You will leave the selections blank, except for the datum identifier. On your own type *–A–* in the Datum Identifier text box. When you have finished, the dialog box should look like Figure 14.24. Pick OK. You will return to your drawing to select the location for the datum flag that you are creating.

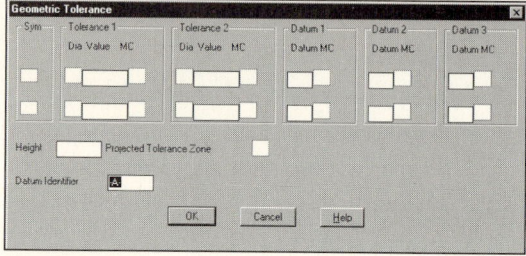

Figure 14.24

Enter tolerance location: **(pick on the extension line)**

The datum flag appears in your drawing, as shown in Figure 14.25. Add datum flag B to the drawing on your own.

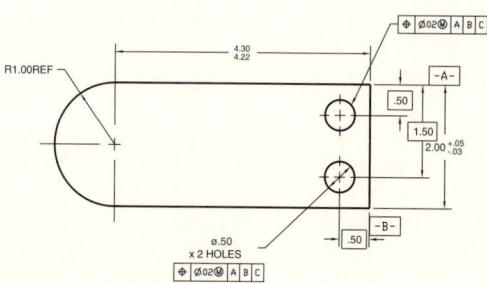

Figure 14.25

Close the Dimension toolbar, save your drawing, and exit AutoCAD. You have completed Tutorial 14.

KEY TERMS

basic dimension
bilateral tolerances
datum flags
deviation tolerances
feature control frame
geometric characteristic
  symbol

geometric tolerances
limit tolerances
nominal dimension
plus/minus tolerances
reference dimension
symmetrical tolerances

tolerance zones
unilateral tolerances
variance tolerances

KEY COMMANDS

Dimoverride
Dimpost

Dimtol
Leader

Tolerance

Draw the necessary views and dimension the following shapes. Include tolerances if indicated. Use either solid modeling techniques or 2D orthographic views.

14.1 Stop Base

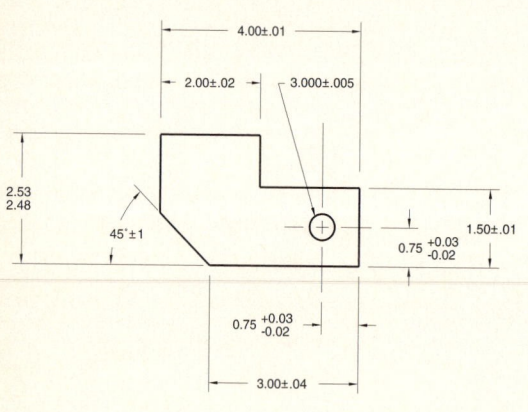

14.2 Tolerance Problem

Calculate the maximum and minimum clearance between Parts A and B and Part C as shown.

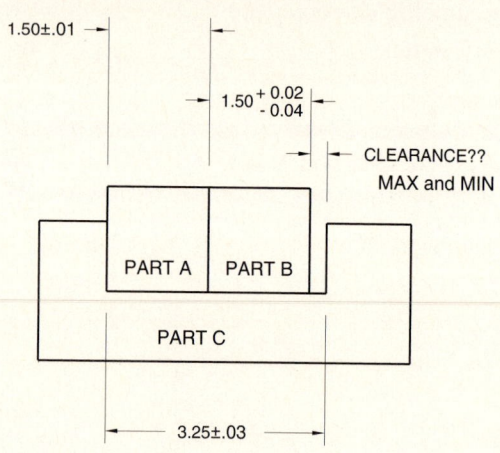

14.3M Lathe Stop

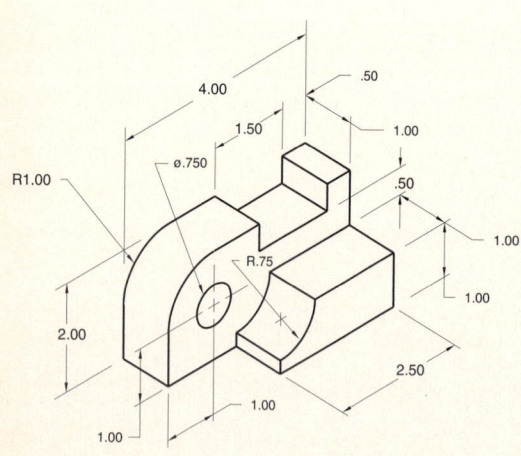

14.4M Mill Tool

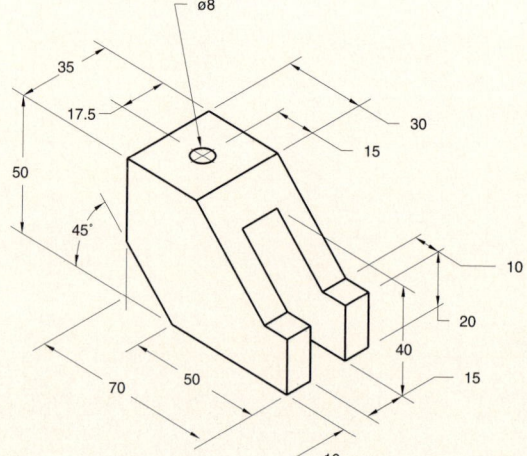

14.5M Shaft Bearing

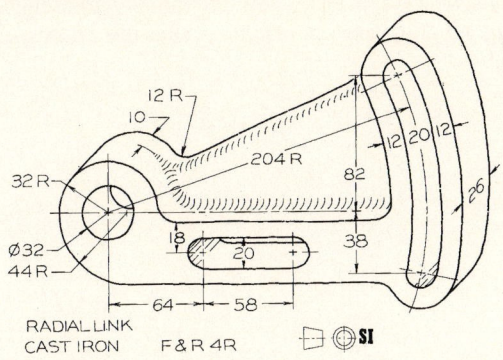

RADIAL LINK
CAST IRON   F & R 4R

14.6 Positional Tolerance

Locate the two holes with an allowable size tolerance of 1.00 mm and a position tolerance of 0.50 mm DIA. Use the proper symbols and dimensions.

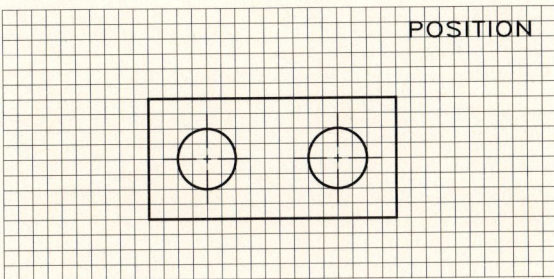

POSITION

14.7 Angular Tolerance

Using a feature control frame and the necessary dimensions, indicate that the allowable variation from the true angle specified for the inclined plane is 0.7mm, using the bottom surface of the object as the datum.

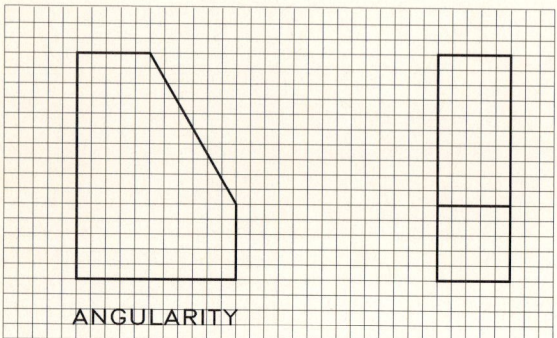

ANGULARITY

14.8 Hub

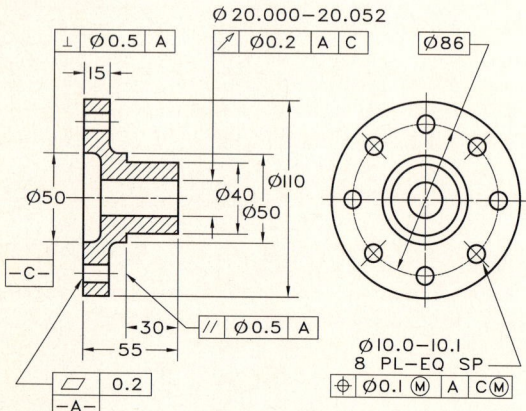

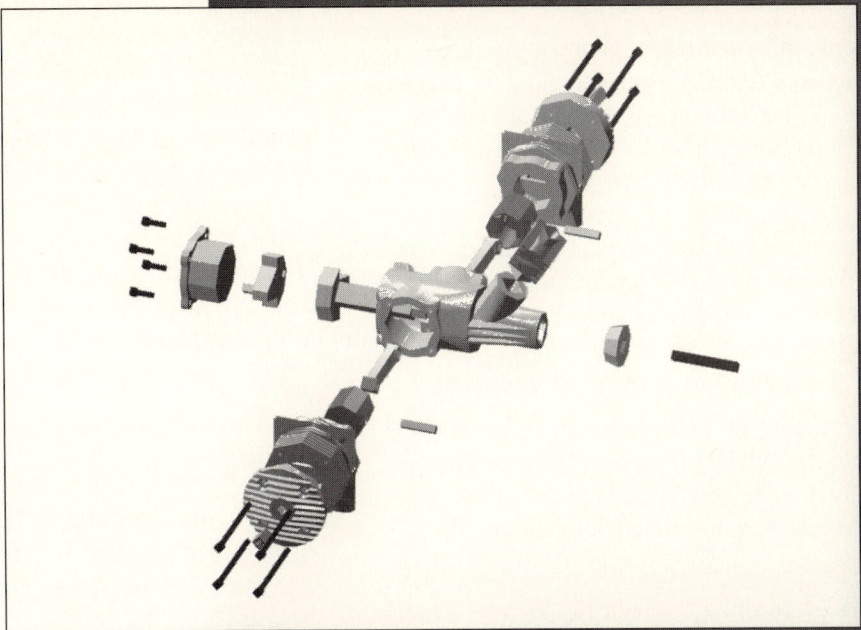

Rendering

Objectives

When you have completed this tutorial, you will be able to

1. Realistically shade a solid model or surface modeled drawing.

2. Set up and use lighting effects.

3. Apply materials and finishes.

4. Save rendered files for use with other programs.

5. Export to 3D Studio format.

Introduction

Render allows quick creation of rendered images of AutoCAD objects that have surfaces. Rendering is the process of calculating the lighting effects, color, and shape of an object in a single two-dimensional (2D) view. A rendered view adds greatly to the appearance of the drawing. It makes the object appear more real and aids in interpretation of the object's shape, especially for those unfamiliar with engineering drawings. Rendered drawings can be used very effectively in creating presentations.

AutoCAD objects that have surfaces can be rendered. Objects that have surfaces include solid models, regions, and objects created with AutoCAD's surface modeling commands.

Rendering is a mathematically complex process. When you are rendering, the capabilities of your hardware, such as speed, amount of memory, and the *resolution* of the display, are particularly noticeable. The complex rendering calculations demand more of your hardware than just running the AutoCAD program, so you may notice that your system seems slower. Display resolution affects not only the speed with which objects are rendered, but also the quality of the rendered appearance. The rendered screens shown in this tutorial were created using Super VGA resolution (800 x 600) and 256 colors. Better appearance can be gained by rendering to a display having higher resolution and more colors; however, increased resolution and colors take longer to render.

In this tutorial you will learn to apply materials, surface maps, and lighting to your drawing and to render views of an object. Applying a material to a surface in your drawing gives the surface its color and shininess; it is similar to painting the surface. Materials can also reflect light like a mirror or be transparent like glass. Adding a surface map is similar to adding wallpaper; *surface maps* are patterns or pictures that you can apply to the surface of an object. Lighting illuminates the surfaces. Lighting is a very important factor in creating the desired effect in your rendered drawing.

Starting

Launch AutoCAD. In this tutorial you will use Render to shade the solid model of the assembly that you created in Tutorial 11. On your own, start a new drawing from the template drawing *assemb1.dwg* provided in the data files for this manual. Save your drawing as *asmshade.dwg*.

The drawing appears on your screen, as shown in Figure 15.1.

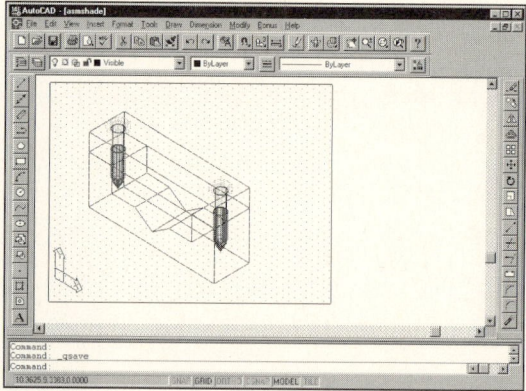

Figure 15.1

Render has its own submenu and toolbar.

Pick: **View, Render**

The Render menu selections pull down from the menu bar, as shown in Figure 15.2.

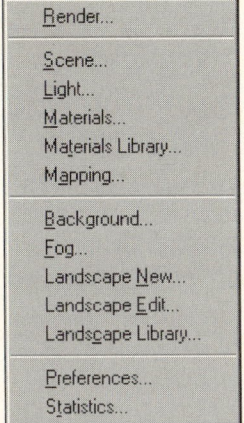

Render...

Scene...
Light...
Materials...
Materials Library...
Mapping...

Background...
Fog...
Landscape New...
Landscape Edit...
Landscape Library...

Preferences...
Statistics...

Figure 15.2

Next, you will turn on the Render toolbar so that you can use it to select commands.

Pick: **View, Toolbars, Render** *(then close the dialog box)*

The Render toolbar appears on your screen, as shown in Figure 15.3. Pick on its title bar and, holding the pick button down, drag it to a location where it will be handy but not in the way of your drawing.

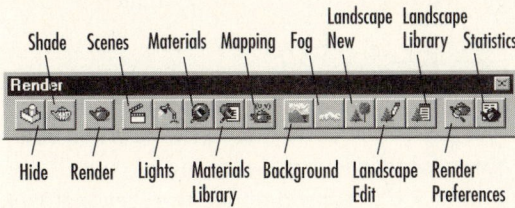

Shade Scenes Materials Mapping Fog Landscape New Landscape Library Statistics

Render

Hide Render Lights Materials Library Background Landscape Edit Render Preferences

Figure 15.3

You will use Render to render the solid model drawing so that it appears realistically shaded on your screen.

The Render Dialog Box

The Render icon opens the Render dialog box to allow you to set rendering options for your drawing.

Pick: **Render icon**

The Render dialog box appears on your screen, as shown in Figure 15.4.

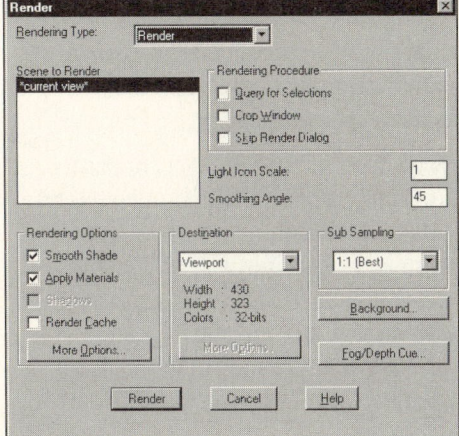

Figure 15.4

Rendering Algorithms

Render should be selected as the Rendering Type. The other rendering types available are Photo Real and Photo Raytrace. The Render rendering type uses either Gourard or Phong shading to add smooth shading and materials, but not surface maps or shadows. *Gourard shading* calculates the shading value for each vertex of a surface and then averages the color across the surface. This method is faster, but less accurate, than *Phong shading*, which calculates the shading value for each pixel on the surface. Phong shading takes longer, but produces more realistic highlights. (When you select Render as the rendering type, you can use the More Options selection to choose between the Gourard and Phong rendering algorithms.)

The Photo Raytrace rendering type uses a ray-tracing algorithm to produce rendered images that have reflections, refractions, and detailed shadows. This method can produce a more realistic result, but because it is more complex, it takes more time to render. The Photo Real rendering algorithm falls somewhere between Render and Photo Raytrace in complexity. It can render objects with surface maps and shadows but not transparent materials or reflections and refractions.

When you are rendering, the best approach often is to start with the least complex method and a lower resolution and progress to more complex renderings. This way, while you are working out details of how to set up the lighting and surfaces for a realistic appearance, you are not having to wait a long time for a complex rendering. For now, you will leave Render selected.

Scene to Render

A *scene* establishes a direction from which you will view the model and the lights you will use for that scene. You make and name scenes with Render's Scenes. (You can select the Scenes command by picking the Scenes icon from the Render toolbar.) To select a view for use in the Scenes command, you must first use AutoCAD's View command and save a named view. Then select the Scenes command and associate that named view with the new scene name. The default selection, current view, always appears and can be used to render the current view on your screen even if you have not created any named scenes.

Background

Selecting the Background button displays the Background dialog box, which lets you select colors, bitmapped images, and gradient backgrounds that blend from one color to the next and merge the rendered image with the current AutoCAD image as the background.

Fog/Depth Cue

Selecting the Fog/Depth Cue button displays the Fog dialog box, which you can use to turn on and adjust this effect. The fog and depth cues provide visual cues about the distance that the object is from the camera, as if you were looking at the object through a fog. As the distance to the object increases, the amount of fog between you and the object increases. Fog uses a white color to obscure more of the object as it gets farther from the camera. The depth cue has the same effect as Fog, but uses black to achieve the visual cue.

Query For Selections

If you check this box, when you choose Render you are returned to your drawing screen to select the objects that you want to render. When you press ⏎ to end the selection, the objects that you have selected are rendered. You do not return to the Render dialog box.

Crop Window

If you select the Crop Window box, when you choose Render you are returned to your drawing to pick a window defining the area that you want to render. Once you've defined the window, AutoCAD renders the area. You do not return to the Render dialog box.

Destination

The Destination selection allows you to send the rendered output to a viewport, the Render Window, or a file. For now you will have the drawing rendered to your viewport. Choosing to send the output to the rendering window uses Render Window, which has its own menu bar selections for copying to the Windows Clipboard, saving the rendering as a bitmap file, and setting the resolution for the rendering window. You can also print the image shown in the rendering window. Selecting rendering to a file allows you to select more options, which let you set different file types, number of colors, and resolutions for the rendered image saved in a file. Be sure that Viewport is selected under the heading Destination.

Rendering Options

Smooth Shading causes rounded surfaces to be shaded so that they appear round. If it is off, rounded surfaces will appear faceted (made up from flat surfaces). Apply Materials tells Render to apply materials to objects. Verify that Smooth Shading and Apply Materials are turned on under Rendering Options. As you have not selected any finishes, the AutoCAD Color Index (ACI) will be used in the scene that you will render, giving your shaded objects a color similar to their original AutoCAD color. Later, when you apply materials to the surfaces of the objects in your drawing, selecting Apply Materials will cause the surfaces to be rendered with the material appearance, not the AutoCAD color.

Selecting Shadows causes shadows to be generated when you use Photo Real or Photo Raytrace as the rendering algorithm. (This option is not available with the Render algorithm.) Using shadows can create drawings that have a more realistic appearance but take

longer to render. Smoothing Angle controls which surfaces are shaded as smoothed rounded surfaces and which are shaded as flat surfaces. The value 45 smoothes all surfaces that meet with a change in angle of less than 45° and considers surfaces with a greater change in angle to be flat intersecting surfaces.

The Render Cache option allows you to Render the drawing to a cache file that it will use again. This option is useful when you will be rendering a drawing a number of times without changing any of the parameters, view positions, and the like. AutoCAD then uses the Rendered cache file, taking less time to calculate the Rendering.

The More Options button displays additional options that you can use in the Render Options dialog box. When you pick it, different dialog boxes showing additional options appear, depending on which rendering type you have selected: Photo Real, Render, or Photo Raytrace.

One of the additional options that you can set is anti-aliasing, which produces a smoother appearance on edges of surfaces by blending additional pixels along the edge with a color between the edge color and the surface color. You can set the level of anti-aliasing used, which is helpful because rendering takes longer when you use high levels of anti-aliasing. Because additional pixels are blended along edge lines, they appear thicker when you utilize anti-aliasing.

Options for Texture Map Sampling control the projection of bitmaps onto objects that are smaller than the bitmaps.

Another of the additional selections lets you control the shading of back surfaces of the object, called *back faces*. Picking Discard Back Faces causes the back surfaces to be ignored. Doing so speeds up rendering time when there are no surfaces in back of the object that you can see. You would not want to select Discard Back Faces when you are looking into an open box and can see the back surface. Back Face Normal is Negative controls the direction for the back surface of the objects. A *normal* is a vector perpendicular to a surface, telling AutoCAD which is the top side of the surface. Selecting Back Face Normal is Negative tells AutoCAD to shade the back surface inside a shape, such as an open box, not the outside back surface.

On your own, pick Cancel to close the Render Options dialog box and return to the Render dialog box.

Help

You can get help for the items in the Render dialog box by picking the Help button at the lower right of the dialog box.

Pick: **Help**

The Render Help window appears on your screen, as shown in Figure 15.5.

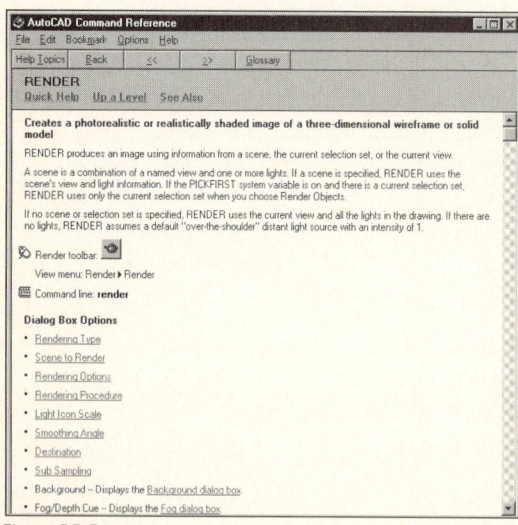

Figure 15.5

Use it as you would the AutoCAD Help window. You can pick an item that is highlighted in the Help window to get help for that topic. Use the buttons below the menu bar to go to the Help index or to move back to the previous help screen. Experiment with Render Help on your own. When you are finished, click the Windows close box to close the Help window on your own. You return to the Render dialog box. If you experimented with any of the settings, return the Render dialog box to the selections shown in Figure 15.4.

Pick: **Render**

Your drawing is rendered to the display. It should appear on your screen as shown in Figure 15.6.

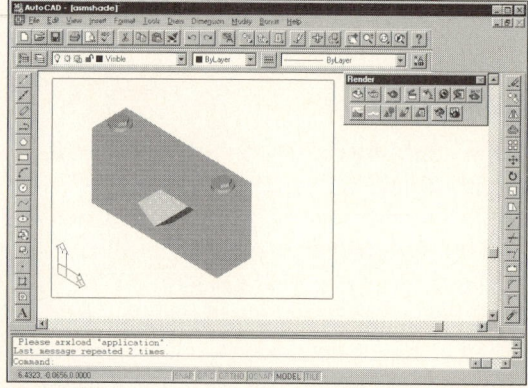

Figure 15.6

Adding Lights

 Note that some of the surfaces in your rendered drawing blend into one another and are hard to see. The reason is that you have not set up any lighting in the drawing. The default lighting is ambient lighting with an intensity of 0.30, which is not very much illumination. *Ambient lighting* illuminates every surface equally, so distinguishing one surface on the object from the next is difficult. The best use for ambient lighting is to add some light to back surfaces of the object so that they do not appear entirely dark. You can improve the look of your rendered drawing by using Render to add additional lights. Lighting is very important in achieving good-looking rendering results.

Pick: **Lights icon**

The Lights dialog box appears on your screen, as shown in Figure 15.7. You will use it to create lighting for your drawing.

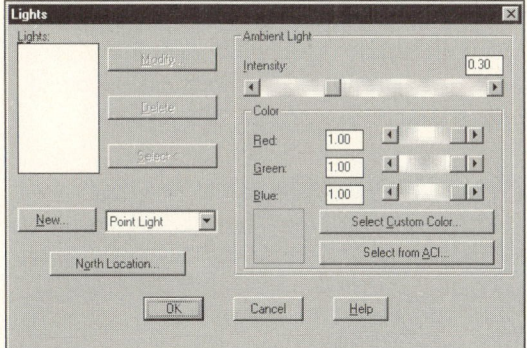

Figure 15.7

In addition to ambient light, Render allows three different types of lights. A *point light* is like a bare light bulb that casts light in every direction (i.e., it is omnidirectional). A *distant light* is like sunlight. It illuminates with equal brightness all the surfaces on which it shines. A *spotlight* highlights certain key areas and casts light only toward the selected target. Point lights and spotlights create indoor lighting effects.

Render provides a sun angle calculator to help you establish the location of the sun or distant light. This feature is particularly useful for architectural applications. You will learn how to use the sun angle calculator later in this tutorial. To see the other Lights options,

Pick: **(the arrow next to Point Light)**

The list of light types appears. Next, you will add a point light to your drawing. Because it is the default, it is already selected in the box to the right of the New button at the left of the dialog box. Leave this selection set to Point Light.

To create a new light for the drawing, you will use the New button.

Pick: **New**

The New Point Light dialog box appears, as shown in Figure 15.8. You will use this dialog box to name the new light and select the features that determine its appearance, such as intensity (brightness), location, color, attenuation (strength), and shadows, in the drawing. You will name the light FILL because you are using it to fill the area where your object is located with more lighting.

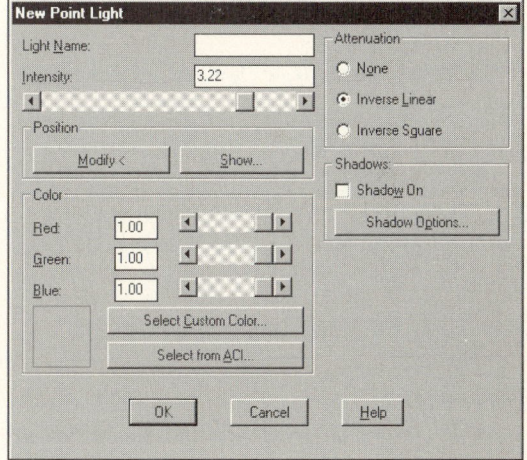

Figure 15.8

Type: **FILL (in the text box to the right of Light Name)**

Next, you will position the light in your drawing. Render places a block as a symbol in your drawing to represent the light.

Pick: **Modify (in the Position area of the dialog box)**

You are returned to your drawing. You can position the light by picking a location from the screen; however, for this example you will type in the coordinates for the light. Positioning the light by picking from the screen is difficult because you are selecting only the X- and Y-coordinates. (You can use point filters and then set the Z-location if you want. Use Help to find out more about point filters.) When specifying the coordinates for the light position, do not position it too close to the object.

Enter light location <current>: **3,5,10** ⏎

You return to the New Point Light dialog box.

Pick: **Show**

The Show selection opens the Show Light Position dialog box and reveals the coordinates of the light's location. This step allows you to determine whether you made an error or to identify the coordinates for lights that you had placed previously or located by picking from the screen.

Pick: **OK** *(to exit the Show Light Position dialog box)*

Next, you will adjust the intensity, or brightness, of the light. You can think of *intensity* as the wattage of a bare light bulb (a 100-watt bulb is brighter than a 60-watt bulb). You will type the value in the box provided. You can also use the slider bar to adjust the intensity of the light.

Select: *(the value in the text box to the right of Intensity)*

Type: **3.94**

You can use the Color area of the dialog box to select a different color for the light. Setting the light to a yellowish color will give a warmer effect, like indoor lamps. For this example you will leave it white.

Pick: **OK** *(to exit the New Point Light dialog box)*

Pick: **OK** *(to exit the Lights dialog box)*

Whenever you make changes that will affect the rendered appearance of your object—adding lights or materials, for instance—you

will need to re-render your drawing. You are now ready to use the Render command to see the effect of adding the point light to your drawing.

Pick: **Render icon**

The Render dialog box that you used earlier appears on your screen. Be sure that the options you set earlier are still current: Rendering Type should be Render, Smooth Shading and Apply Materials should be on, and Destination should be Viewport. When you have verified these selections,

Pick: **Render**

The added lighting should improve the appearance of your rendered drawing on the screen so that it looks like Figure 15.9.

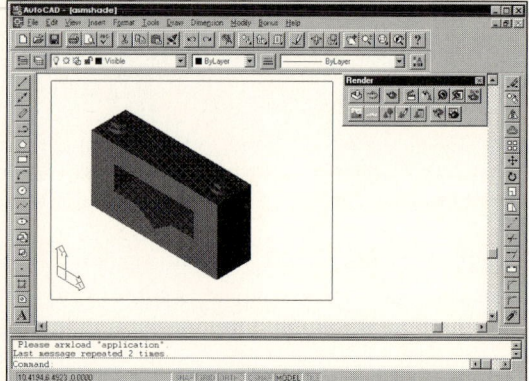

Figure 15.9

The right surface of the object is still not very well lit; not much light is striking that surface. Next, you will add a spotlight behind the object to light that surface and provide highlights in the drawing. You will use the Lights dialog box again to add a new light.

Pick: **Lights icon**

Pick on the name Point Light and from the list of light types that appears,

Pick: **Spotlight**

Pick: **New**

The New Spotlight dialog box appears on your screen. You will use it to name and select the attributes for the spotlight, so that it appears as shown in Figure 15.10.

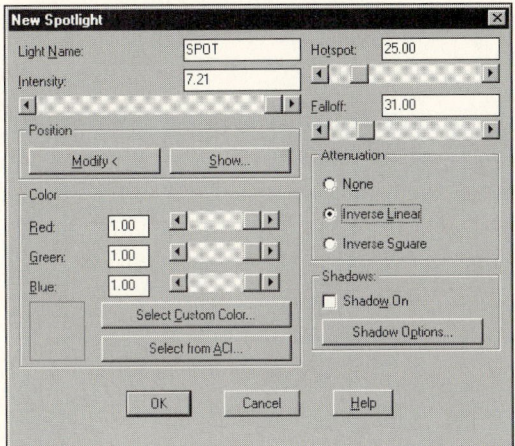

Figure 15.10

Type: **SPOT** *(in the text box to the right of Light Name)*

Set the intensity for the spotlight by moving the scroll bar to the highest value possible or type a value near 7.21 in the box to the right of Intensity.

Now you are ready to position the light in your drawing. When positioning a spotlight, you first specify the target toward which it will point. Then you give the location of the light. In the Position area of the dialog box,

Pick: **Modify**

You will press ↵ to accept the current location for the target (near the center of the object), then type in the location for the light at the next prompt:

Enter light target <current>: ↵

Enter light location <current>: **15,4,−6** ↵

The New Spotlight dialog box returns to your screen. Next, you will set the falloff and hotspot for the spotlight. *Falloff* is the angle of

the cone of light from the spotlight. *Hotspot* is the angle of the cone of light for the brightest area of the beam. The angles for the falloff and hotspot must be between 0° and 160°. Render won't let you set Hotspot to a greater value than Falloff.

On your own, use the slider bar to set the value for Hotspot to 25.00 and the value for Falloff to 31.00. When you are finished, your dialog box should have settings similar to those shown in Figure 15.10. If so, pick OK to exit the New Spotlight dialog box and then pick OK again to exit the Lights dialog box on your own.

You return to the AutoCAD drawing editor. You will use the Render dialog box to render your drawing again to see the effect of adding the spotlight to the drawing.

Pick: **Render icon**

Check the settings in the dialog box to be sure that they are the same ones that appear in Figure 15.4. Then,

Pick: **Render**

The drawing is rendered to your screen and should be similar to Figure 15.11. Note the bright area of the spotlight beam, which you set using Hotspot. The size of the entire circle of light from the spotlight is determined by the setting you entered for Falloff.

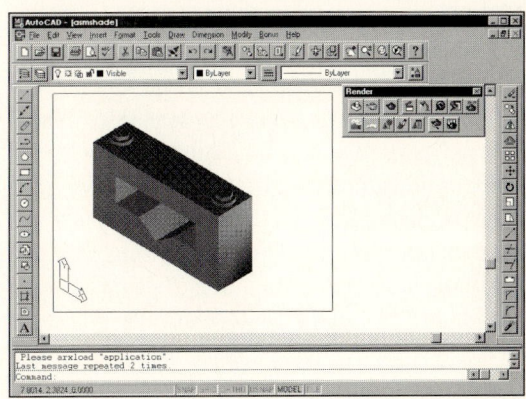

Figure 15.11

Choosing Materials

 You can also improve your rendered drawing by selecting or creating realistic-looking *materials* for the objects in it. You can either use the materials that are provided in the Materials Library or create your own materials with Render. As with lights, a block is added to the drawing to represent materials. To add materials,

Pick: **Materials icon**

The Materials dialog box appears on your screen, as shown in Figure 15.12.

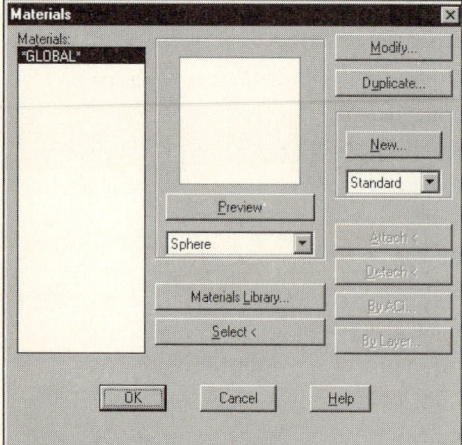

Figure 15.12

You will use the Materials dialog box to select materials from the Materials Library for the objects in your drawing. A single block or solid object in the drawing can have only one material assigned. This limitation usually isn't a problem in engineering drawings because you will typically make each part in an assembly a separate object or block in the drawing. Do not use Boolean operators to join parts to which you want to assign different materials.

To select from the library of premade materials,

Pick: **Materials Library (from the lower center of the dialog box)**

The Materials Library dialog box shown in Figure 15.13 appears on your screen. You will choose materials for your assembly drawing from the list of materials shown at the right in the dialog box.

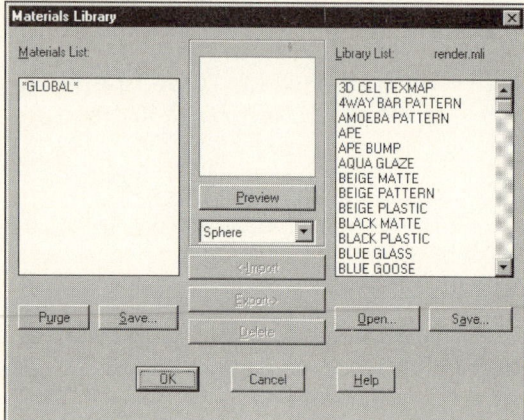

Figure 15.13

■ *Warning:* If you don't see the list of materials shown in Figure 15.13, you need to add the Bonus and Texture files from your AutoCAD CD-ROM. Use the AutoCAD installer to add these files to your AutoCAD program files. ■

Use the scroll bar or pick on the arrows to scroll through the list of materials until you see the selection BRASS VALLEY, which you will use for the washers in the assembly drawing. Position the cursor over the name BRASS VALLEY and press the pick button to select it. The material becomes highlighted in the list.

Many of the materials dialog boxes allow you to preview the selection to see approximately what the material will look like. This preview saves you time if you decide not to use the particular material once you see it or your system is not capable of displaying it. Remember, rendering a complex drawing can be time-consuming. Use the Preview function to see what different materials would look like.

Pick: **Preview**

The material that you selected is applied to a spherical shape in the Preview box, as shown in Figure 15.14.

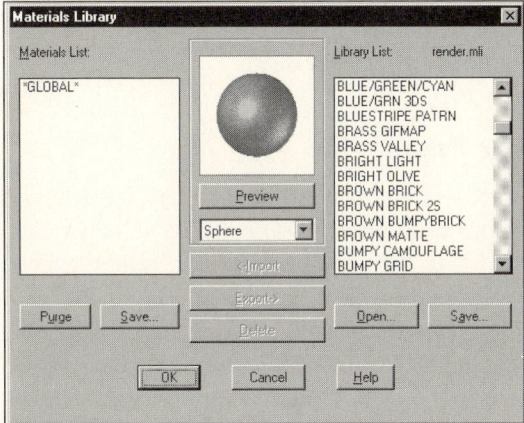

Figure 15.14

On your own, pick on the Sphere selection shown below the Preview button and change it to Cube. Preview the material again. Return the selection to Sphere. Once you have previewed the materials, you will import BRASS VALLEY so that it is available for use in your drawing. Note that you can preview only one material at a time but that you can select several materials to import at the same time.

Pick: **Import**

The selected material is shown in the Materials List at the left of the dialog box. Now you are ready to select another material.

Highlight the materials GRAY MATTE and GRAY SEMIGLOSS on your own from the list of materials at the right in the dialog box. Pick Import to make the materials available for use in the current drawing. The names are added to the list at the left. When you have finished importing the materials,

Pick: **OK (to exit the Materials Library dialog box)**

You return to the Materials dialog box, where you will attach the materials to the objects in your drawing.

Select: **BRASS VALLEY (from the Materials list at the left of the dialog box)**

Pick: **Attach (at the right of the dialog box)**

You are returned to your drawing to select the objects, in this case the two washers. If you accidentally select something else, use the Select Remove option to remove it from the selection set or use Detach later in the dialog box to detach the wrong material. If you have trouble selecting, try to pick on an edge, not in the center of an object, or use implied Windowing.

Pick: **(on the edge of each of the two washers)**

Press: ⏎

The Materials dialog box returns to your screen so that you can continue attaching materials.

Select: **GRAY SEMIGLOSS**

Pick: **Attach**

Pick: **(on the edge of the two screws)**

Press: ⏎

Next, attach the material GRAY MATTE to the base and cover of the assembly.

Select: **GRAY MATTE**

Pick: **Attach**

Pick: **(on the edge of the base and the cover)**

Press: ⏎

Pick: **OK (to exit the Materials dialog box)**

Now that you are finished selecting the materials, you will render the drawing again to view the results of the changes that you made.

Pick: **Render icon**

Accept the same defaults that you used previously.

Pick: **Render**

Your drawing should now show the new materials. Your screen should look like Figure 15.15.

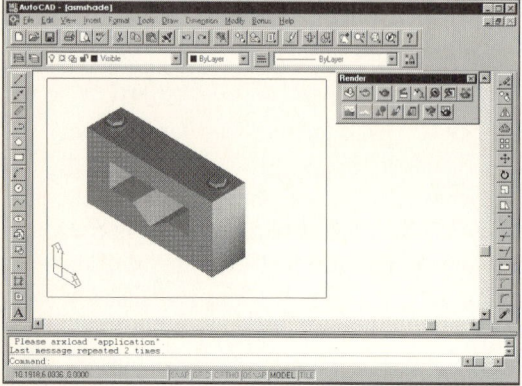

Figure 15.15

You can also use Render to create your own materials. Next, you will create a material that looks like steel to use for the base and cover of the assembly.

Pick: **Materials icon**

When the Materials dialog box appears,

Pick: **New *(from the right-hand side of the dialog box)***

The New Standard Material dialog box shown in Figure 15.16 appears on your screen.

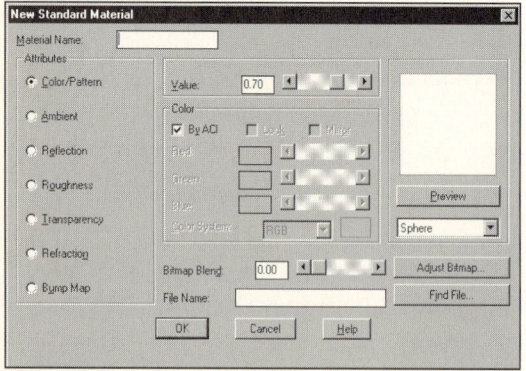

Figure 15.16

Type: **STEEL *(in the text box to the right of Material Name)***

At the left in the New Standard Material dialog box are the various qualities, or material attributes: Color/Pattern, Ambient, Reflection, Roughness, Transparency, Refraction, and Bump Map. These are the attributes that you can control to create materials with different appearances. In general, these qualities indicate how the surface or material responds when light strikes it. To change attributes, pick in the circle to the left of the attribute for which you want to set the values.

Color/Pattern

The Color/Pattern selection determines the color for the material. The default selection for color is By ACI (AutoCAD Color Index), which results in the object being rendered in the color assigned to its lines in the AutoCAD model. You saw this when you rendered the assembly for the first time and the objects were shown in the original layer colors. If you deselect By ACI in the Color area of the dialog box, you can select the other Color options. Near the bottom of the Color area the selection RGB appears, which represents red, green, and blue, the primary colors of light. You can use the slider bars in that section of the dialog box to set the amount of each color of light to use to make up the material color. As you move the slider bars, the resulting color is shown in the empty box at the bottom of the Color area. It is called the *color swatch*.

Keep in mind that colors on your computer screen are made of the primary colors of light (red, green, and blue), not the primary colors of pigment (red, blue, and yellow) that you may be used to using. Changing the selection at the bottom of the Color area from RGB to HLS (hue, lightness, and saturation) lets you use the same slider bars to define the color in terms of these qualities instead of RGB. When you have changed the color slightly, pick Preview.

In addition, you can use a bitmap (a pattern or picture), such as a scanned image or drawing created with a paint-style program, instead of just solid color. To do so, in the lower center area of the dialog box, type the name of a bitmap file in the text box to the right of File Name. You can use bitmaps with the selections Color/Pattern, Reflectivity, Transparency, and Bump Map. The Bitmap Blend value determines the amount of the underlying object color that shows through the bitmap. Values of less than 1.00 allow some of the underlying object color to show through. The lower the value, the more the underlying color shows through the bitmap.

Ambient

The *Ambient* selection allows you to set the color for the object's shadow. You can use the color controls just discussed.

Reflection

The *Reflection* selection allows you to adjust the color of the material's reflective highlight. When you select *ray tracing*, you can specify the value for the material's reflectivity from 0, where it does not reflect any light, to 1, where it is perfectly reflective, like a mirror.

Roughness

The *Roughness* selection lets you specify the surface roughness of the material. Roughness is also related to reflectivity. The less rough the surface is, the more reflective it will be. Change the value and use Preview to see how roughness changes the size of the material's highlight.

Transparency

The settings for *transparency* allow you to make objects that are clear, like glass, or somewhat clear, like colored liquids. Keep in mind that completely transparent objects, which you can't even see in the drawing when they are rendered, still take time to render.

Refraction

The Refraction control is used only with transparent objects. *Refraction* is the amount that light is bent when entering and leaving a transparent object. You must use the Photo Raytrace rendering algorithm in order to see the effect of this setting. Ray tracing uses a more sophisticated rendering algorithm to generate reflections, refractions, and detailed shadows. In ray tracing, the result is calculated for each "ray" of light striking the object. This produces very realistic and detailed effects, but increases the complexity and rendering time.

Bump Map

The Bump Map selection allows you to enter the file name of a bump map. *Bump maps* convert the color intensity or grayscale information to heights to give the appearance that features are raised above the surface, like embossed letters.

Next, you will set the attributes for the material STEEL that you are going to create. On your own, be sure that Color/Pattern is selected and that By ACI is turned off.

Then use the slider bars to adjust the colors for Red, Green, and Blue so that Red and Green are each set to a value of 0.88 and Blue is set to a value of 0.92. Note how the color swatch changes as you move the slider bars. The resulting color should be gray. Preview the result on your own.

Next, you will make the surface somewhat reflective and assign a small value for surface roughness so that the material will look like steel.

Pick: **Reflection**

Use the slider bar to set the value for Reflection to 0.65. Then,

Pick: **Roughness**

Set the value for Roughness to 0.30.

To see the effects of your selections on the material being created,

Pick: **Preview**

Your screen should be similar to Figure 15.17.

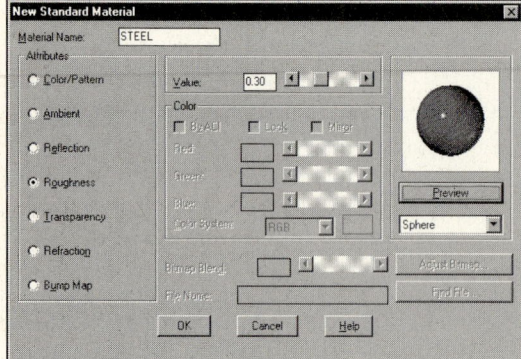

Figure 15.17

Pick: **OK** *(to exit the New Standard Material dialog box)*

You will return to the Materials dialog box. Next, you will attach this material to the base and cover in the assembly drawing in place of GRAY MATTE. To do so, use the dialog box on your own to select the material STEEL that you created and attach it to the base and cover. When you have attached the material, pick OK to exit the dialog box. Then select the Render command and render the drawing to the display on your own.

Adjusting Mapping

 When bitmaps or materials that use bitmapped patterns are applied to surfaces in your drawing, you can control the type of mapping, coordinates, scale, repeat, and other aspects of the maps' appearance. Next, you will apply a bitmapped material to the base of the assembly to see these effects, then use Mapping to adjust them.

Pick: **Materials icon**

Pick: **Materials Library**

Pick: **V Pattern** *(from the list of materials at the right)*

Pick: **Import**

Pick: **OK** *(to exit the Materials Library dialog box)*

Pick: **Attach**

Select objects to attach "V PATTERN" to: *(pick the base)* ⏎

Pick: **OK** *(to exit the Materials dialog box)*

The V Pattern material is attached to the base part of the drawing. Render the current scene to the viewport, but this time choose Photo Real instead of Render as the rendering type.

Pick: **Render icon**

Pick: **Photo Real** *(from within the pull down menu to the right of Rendering Type)*

Pick: **Render**

Your screen should look like Figure 15.18.

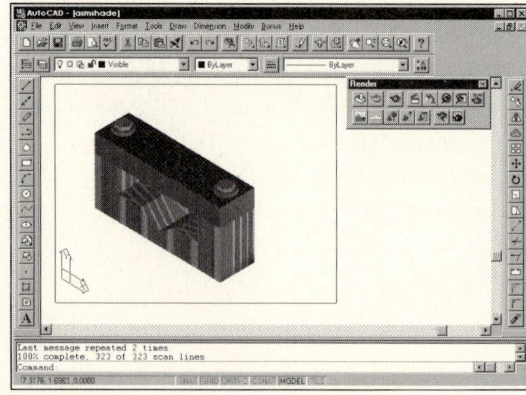

Figure 15.18

Pick: **Mapping icon**

The Setuv command is echoed at the command prompt, followed by the prompt,

Select objects: *(pick the base)*

Select objects: ⏎

The Mapping dialog box appears on your screen, as shown in Figure 15.19. To preview the mapping for the Planar method of projection, which is currently selected,

Pick: **Preview**

The preview of the bitmapped material in planar projection is shown in the upper right of the dialog box in Figure 15.19.

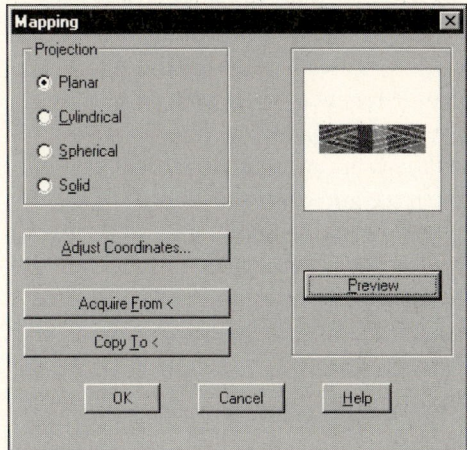

Figure 15.19

To adjust the mapping for a cylindrical or spherical object, you can pick the Cylindrical or Spherical radio button from the left of the dialog box. Try each one on your own and then use Preview to see the effect of that projection method. In general, you would use cylindrical mapping for cylindrical objects, spherical mapping for spherical objects, and so on. As the base is composed of plane surfaces, return the

selection to Planar on your own. Solid mapping is a special type that you should use to represent granite, marble, and wood solid materials. Figure 15.20 shows the Materials dialog box with the list of these materials pulled down. Picking one of these materials and then the New button from the Materials dialog box displays a dialog box that you can use to create that type of custom materials.

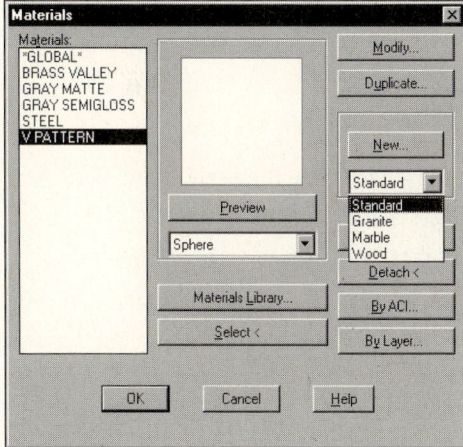

Figure 15.20

If you experimented with the solid materials, return to the Mapping dialog box on your own. Next, you will adjust the mapping coordinates to change the direction for the pattern.

Pick: **Adjust Coordinates**

The Adjust Planar Coordinates dialog box appears on your screen.

Pick: **WCS XZ Plane**

Pick: **Preview**

The preview of the object with the pattern applied in the XZ-direction appears, as shown in Figure 15.21.

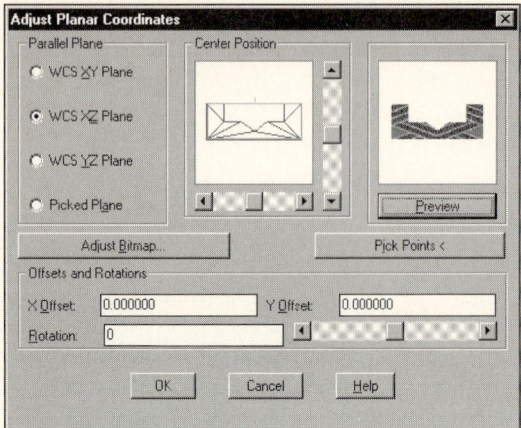

Figure 15.21

Next, you will change the scale of the pattern.

Pick: **Adjust Bitmap**

The Adjust Object Bitmap Placement dialog box appears on your screen. On your own, use it to change the scale to 2.5 and preview the result, shown in Figure 15.22.

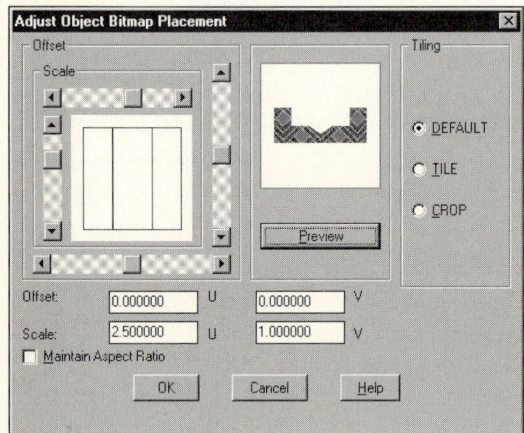

Figure 15.22

In addition to changing its scale, you can select the radio buttons at the right of the dialog box to tile or crop the bitmapped image. Tiling causes it to repeat its pattern across the object. Crop uses the bitmapped image like a decal: It appears only at the single location specified by the offsets. When you create materials that use bitmaps, you can either tile or crop them. The DEFAULT selection uses whatever you set for the material as a whole when you created it. Mapped coordinates are given relative to the U- and V-directions. The U- and V-coordinates are equivalent to X- and Y-coordinates for the map. The letters U and V are used to indicate that they are part of a local coordinate system for the map that is independent of the X- and Y-coordinate system of the drawing. You specify offset and scale using these U- and V-directions. When you select Maintain Aspect Ratio, the U and V sliders work as a unit to maintain the proportion of the bitmapped pattern. The scale and offsets set in this dialog box are additive to the ones set for the material as a whole. The offset is added to the materials offset, and the scale is multiplied by the scale for the material. To exit the dialog box,

Pick: **OK**

Pick: **OK (to exit the Adjust Planar Coordinates dialog box)**

Pick: **OK (to exit the Mapping dialog box)**

Render the drawing on your own.

Using mapped materials is a good way to add detail to your rendered drawing without creating complex models. For example, you could model every brick in a wall as a separate solid and create a different solid for the mortar between them. Doing so would create a very complex solid model with a lot of vertices.

Instead, you can create the entire wall as a single flat surface and then apply the appearance of brick to it as a bitmapped pattern. The drawing will look very detailed but will have far fewer vertices and take less file space than the previous example. You can also combine bitmaps and bump maps in the same drawing to give a raised appearance to parts of the material.

Rendering to a File

Rendering your drawing to a file allows you to use it in other programs. For example, you can render your drawing to a Windows metafile format and paste it into various Windows applications.

Pick: **Render icon**

Use the Destination area of the Render dialog box to select File on your own. (Pick on the word Viewport and use the list that appears to make your selection.)

Pick: **More Options (below Destination)**

The File Output Configuration dialog box appears on your screen, as shown in Figure 15.23. You can use the selection at its upper left to save your drawing in many different file formats, among them TGA (TARGA), TIFF, Postscript, PCX, and BMP. If you have a paint-style program that uses one of these formats, you may be able to edit and print the saved image file. You can often also use these formats to insert the drawing in a word processing program. You can set the

resolution for the output file by using the selection below File Type, where you see 640 x 480 (VGA) and the number of colors for the resulting file. You do not need to change the resolution, but change the file type to TIFF to save your file in that format, then

Pick: **OK**

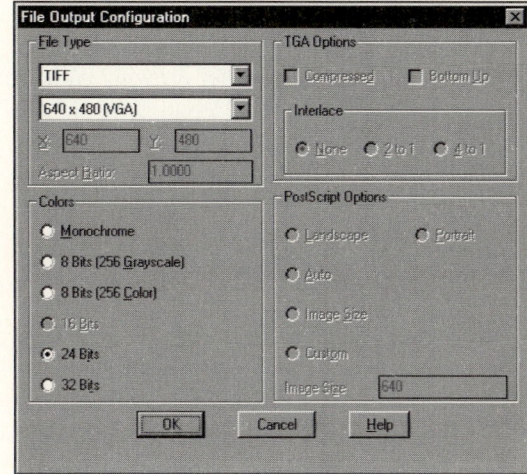

Figure 15.23

The file will be saved with the extension *.tif* to your working directory. To render to the file,

Pick: **Render**

Use the Rendering File dialog box that appears to specify the name and directory in which to save the file. On your own, try importing your rendered drawing as a graphic into your Windows word processor. Then return to your AutoCAD drawing session.

Using the Sun Angle Calculator

The sun angle calculator is particularly useful in architectural drawings for showing the different effects that sunlight has on the rendered scene during different times of the day and year. You can use the sun angle calculator to obtain these effects easily, instead of going through the laborious process of looking up the information in tables and adding the correct lighting effects to your drawing. You can even set the geographic location by picking it from the map. The sun will be added as a distant light in your rendered drawing, so

Pick: **Lights icon**

Pick: **Distant Light *(from the list of light types that pulls down to the right of New)***

Pick: **New**

The New Distant Light dialog box shown in Figure 15.24 appears on your screen.

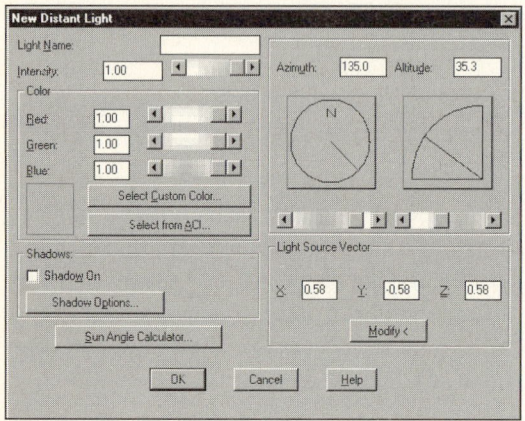

Figure 15.24

Type: **SUN *(in the text box to the right of Name)***

You can use the slider bars or the color wheel to set the color for the distant light. Leave it

white for now, as the sun usually casts white light, which is composed of the full spectrum of light colors. The area at the top right of the dialog box allows you to specify the azimuth and altitude for the sun if you want to do so. You can use the Sun Angle Calculator button at the bottom left of the dialog box and let it determine these values for you. You will do that now.

Pick: **Sun Angle Calculator**

The Sun Angle Calculator dialog box appears on your screen. You can use it to specify the date, time of day, time zone, latitude, and longitude to determine the angle of the sun.

On your own, set the date to March 20 (3/20) and the clock time to 6:00 hours (6 A.M.) MST. Turn Daylight Savings on by picking in the box to its left so that a check is displayed. Note that the azimuth and altitude settings change as you make selections. When you are finished, the dialog box should look like Figure 15.25.

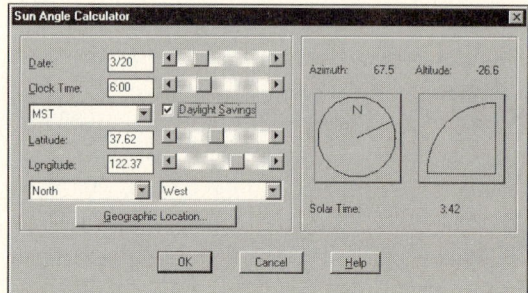

Figure 15.25

Next, you will specify the geographic location because it works with the date and time of day to determine the angle of the sun.

Pick: **Geographic Location**

The Geographic Location dialog box shown in Figure 15.26 appears on your screen.

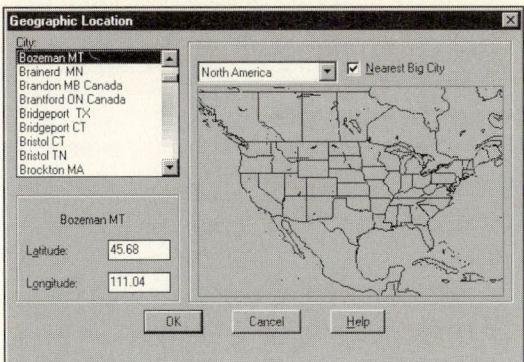

Figure 15.26

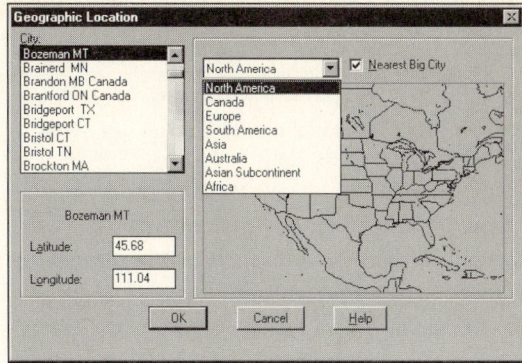

Figure 15.27

Move the cursor over the map until it is positioned as shown in Figure 15.26 and press the pick button. Note that Bozeman, MT (cultural Mecca and center of the universe) appears as the Nearest Big City in the list at the left. Below, the latitude and longitude for Bozeman are displayed. You can also scroll down the list of cities to make a selection. Note the words North America at the upper left of the map area.

Pick: **(on the words North America)**

A list of the other continents pulls down, as shown in Figure 15.27. You can use the scroll bar to select from the other continents. The major cities of other continents are stored in the database, and you can select them by picking from the map or by scrolling down the list of cities.

Do not change the continent selection at this time. Leave it set to North America.

Pick: **OK** *(to exit the Geographic Location dialog box)*

Pick: **OK** *(to exit the Sun Angle Calculator dialog box)*

Pick: **OK** *(to exit the New Distant Light dialog box)*

Pick: **OK** *(to exit the Lights dialog box)*

Your drawing returns to the screen. Now, you will render the drawing again to see the part as though the sun were shining on it at 6 A.M. on March 20 in Bozeman, Montana.

On your own, use the Render dialog box and set the Destination back to Viewport. Render the drawing and note the effect of the sunlight. Then, you will modify the light, SUN, in the Lights dialog box. On your own, select the Lights command and pick SUN as the light you want to modify. Use the sun angle calculator to change the time of day to 4 P.M. (The time-of-day clock is a 24-hour clock.) Render the drawing again and note the change. Although this assembly is not an architectural drawing, you can imagine what a great advantage the sun angle calculator is for architectural uses. You can also use it to evaluate the suitability of a site for solar power and other types of heat transfer analysis.

Adding Backgrounds

 You can add a background to your rendering by using the Background icon on the Render toolbar. Backgrounds can be solid colors, gradient colors (which blend one to three colors), images (bitmaps), or you can use Merge to select the current AutoCAD image as the background.

Pick: **Background icon**

You will set up the Background dialog box so that it looks like Figure 15.28. You will use it to select an image as the background. Using the radio buttons across the top of the dialog box,

Pick: **Image**

Pick: **Find File**

Select the file *c:\AutoCADR14\Textures\ cloud.tga* from the dialog box that appears (be sure to change the file type to **.tga*, since it defaults to **.bmp*). Then pick OK on your own to return to the Background dialog box. You can adjust the appearance of the bitmap, as you did earlier in this tutorial, by picking Adjust Bitmap. Environment lets you use an image that Render maps onto a sphere surrounding your drawing. This environment is then reflected onto other objects when you use reflective and refractive materials in your drawing. When you use the Photo Real renderer, reflective objects mirror the background image. When you use the Raytrace renderer, the environment is raytraced.

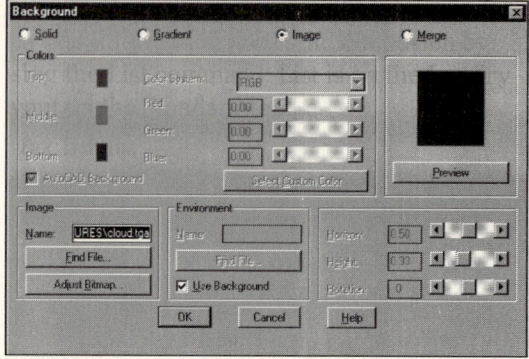

Figure 15.28

Pick: **OK *(to exit the Background dialog box)***

On your own, render your drawing to show the cloud background.

Next, you will restore the world coordinate system so that the landscape objects you add will have the correct orientation.

Command: World UCS

The UCS icon changes back to the WCS.

Using Landscape Objects

 Landscape objects are bitmapped images that are applied to extended entity objects and added to your drawing. Landscape objects have grips at the base, top, and corners that you can use to move, scale, rotate, and otherwise edit them. You can create your own landscape objects and add them to the landscape library. To add a landscape object to the drawing,

Pick: **Landscape New icon**

The Landscape New dialog box appears on your screen, as shown in Figure 15.29. On your own, select People #2. Change Height to 6.0 and pick Preview to preview the person that you will add to your drawing.

Figure 15.29

To locate the landscape object in your drawing,

Pick: **Position**

Choose the location of the base of the landscape
object: **5,5,0** ⏎

You return to the Landscape New dialog box.

Pick: **OK** *(to add the landscape object and exit the
dialog box)*

The landscape object is added to your drawing.
It appears as a triangular shape. On your own,
use Zoom All to zoom out so that you can see
the entire landscape object. If necessary, use
Zoom Window afterward to size your view of
the landscape object and the assembly. Render
the drawing on your own. Although the effects
of the background and person are not particularly applicable to this engineering assembly
drawing, you can easily imagine creating
realistic effects in the rendering by using
backgrounds and landscape objects.

Exporting to 3D Studio

You can use the materials and lights that you
have created with Render in Autodesk's 3D
Studio animation program. Once you've
exported your drawing for use in 3D Studio,
you can import it and add 3D animation effects
in 3D Studio.

> ■ *TIP* You can't export solid materials
> such as granite, marble, and wood to
> 3D Studio. ■

To export to 3D Studio, use the File selection
at the left of the menu bar to pick the Export
command.

Pick: **File, Export**

The Export Data dialog box shown in
Figure 15.30 appears on your screen.

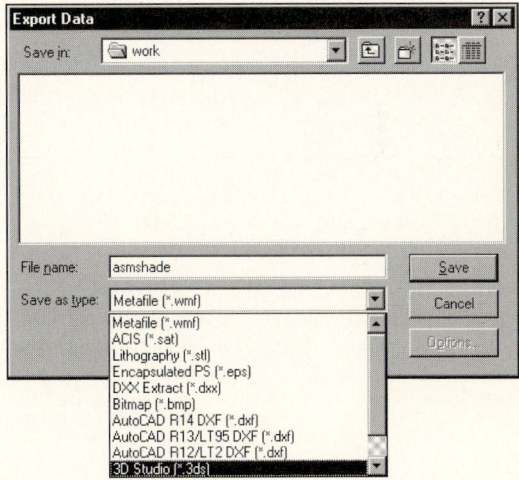

Figure 15.30

Use the selection under the heading Save as Type to select 3D Studio (*.*3ds*) as the type of file to export. Use the dialog box to select the *c:\work* directory and name your drawing file *asmshade.3ds* on your own. (AutoCAD will automatically assign the file extension .*3ds* to your file when you pick the 3D Studio export type.) When you are finished naming the file, pick Save on your own to exit the dialog box.

Select objects: **ALL** ⏎

Select objects: ⏎

After you exit the file naming dialog box and have selected the objects to be exported, the 3D Studio File Export Options dialog box shown in Figure 15.31 appears on your screen. You will use it to set the method that 3D Studio should use to determine individual objects and the smoothing and welding to be done.

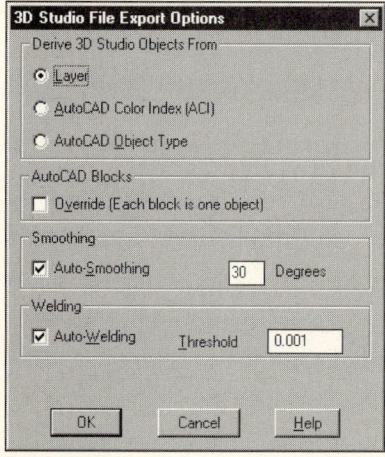

Figure 15.31

3D Studio can determine which objects are individual items and can therefore be moved and animated separately in three ways. If you select the Layer option, each object that you want to have as a separate 3D Studio object must be on its own layer in the drawing. This method works quite well if the objects are on separate layers. (You can easily change the layer of an object using AutoCAD's Properties

icon if the objects are not on separate layers.) Choosing AutoCAD Color Index creates the object from objects that have the same AutoCAD color. Choosing Object Type creates individual objects from the same type of object in the drawing. Arcs, lines, and polylines must have a nonzero thickness in order to be exported as 3D Studio objects.

Pick: **AutoCAD Object Type** *(from the selections available in the dialog box)*

Pick: **Override** *(so that each block is one object)*

Auto-Smoothing makes faceted surfaces appear smooth in the drawing. The smooth appearance of contoured objects, such as spheres, is controlled by the Facetres variable in AutoCAD. Setting Facetres to a higher value causes the object to appear to be made of more facets and thus more closely approximating the surface. Even when a lower value for Facetres is used, Render and 3D Studio have the ability to make such an object appear smooth and round, like a ball. When Auto-Smoothing is turned on, the setting in the Degrees area specifies the maximum number of degrees for smoothing; in other words, surfaces that meet at an angle greater than the number of degrees specified won't be smoothed. The default is 30°.

Leave Auto-Smoothing turned on and the angle set to 30°.

Auto-Welding joins vertices that are no farther apart than the specified *threshold value* into one vertex. This feature simplifies complex geometry. As in Render, the complexity of the objects and the demands of the lighting and shading can slow rendering and animating. Simplifying the drawing in this way can speed up rendering and animating in 3D Studio. For the assembly drawing, multiple vertices are not close together, so this will not have a large effect.

Leave Auto-Welding turned on and Threshold set to 0.001.

Pick: **OK (to export the assembly drawing to 3D Studio format)**

Your computer will take a moment to process the file and export it into 3D Studio format. When it has finished, close the Render toolbar and save your drawing.

> ■ **TIP** You can insert many Raster file formats into an AutoCAD drawing. You can manipulate them in a variety of ways, including a new AutoCAD function called transparency. The Transparency command affects whether an AutoCAD drawing underneath the Raster file image can show through. When exporting an AutoCAD drawing that contains a Raster file, only the following file types are exportable: BMP, TGA, PCX, TIFF, and GIF. ■

Exporting Other File Types

AutoCAD supports the export of many other file types in addition to that for 3D Studio.

Pick: **File, Export**

The Export Data dialog box returns to your screen, as shown in Figure 15.30. In the file types you will find:

Metafile	**.wmf*
ACIS	**.sat*
Lithography	**.stl*
Encapsulated PS	**.eps*
DXX Extract	**.dxx*
Bitmap	**.bmp*
AutoCAD R14 DXF	**.dxf*
AutoCAD R13/LT 95 DXF	**.dxf*
AutoCAD R12/LT 2 DXF	**.dxf*
3D Studio	**.3ds*
Block	**.dwg*
Drawing Web Format	**.dwf*

The Drawing Web Format is a new tool in AutoCAD Release 14. It allows you to convert an AutoCAD drawing so that it can be viewed on the Web. Along with this feature is an Internet toolbar.

Pick: **View, Toolbars**

Pick: **Inet (from within the Menu Group at the bottom of the dialog box)**

Pick: **Internet Utilities**

Pick: **Close**

The Internet Utilities toolbar appears, as shown in Figure 15.32.

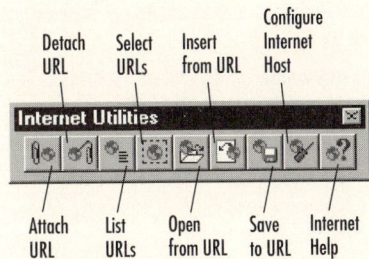

Figure 15.32

The toolbar offers a variety of Internet tools for attaching AutoCAD drawings to Web sites. You can also attach a Web link to a particular part of a drawing so that, when you select it, you are automatically sent to that Web site (if your computer system is configured properly). For more information on properly using the Internet Utilities, consult AutoCAD Help.

You have completed Tutorial 15 and this tutorial guide to AutoCAD. Exit AutoCad.

KEY TERMS

ambient	hotspot	refraction
ambient lighting	intensity	resolution
back faces	materials	roughness
bump maps	normal	scene
color swatch	Phong shading	spotlight
distant light	point light	surface maps
falloff	ray tracing	threshold value
Gourard shading	reflection	transparency

KEY COMMANDS

Background	Mapping	Scenes
Landscape New	Materials	
Lights	Render	

Use solid modeling to create the object shown below or retrieve your file from the exercises you did for Tutorial 11. Produce three different renderings of the object, varying the type, location, intensity, and color of the light. Experiment with different materials and backgrounds.

15.1 Clamping Block

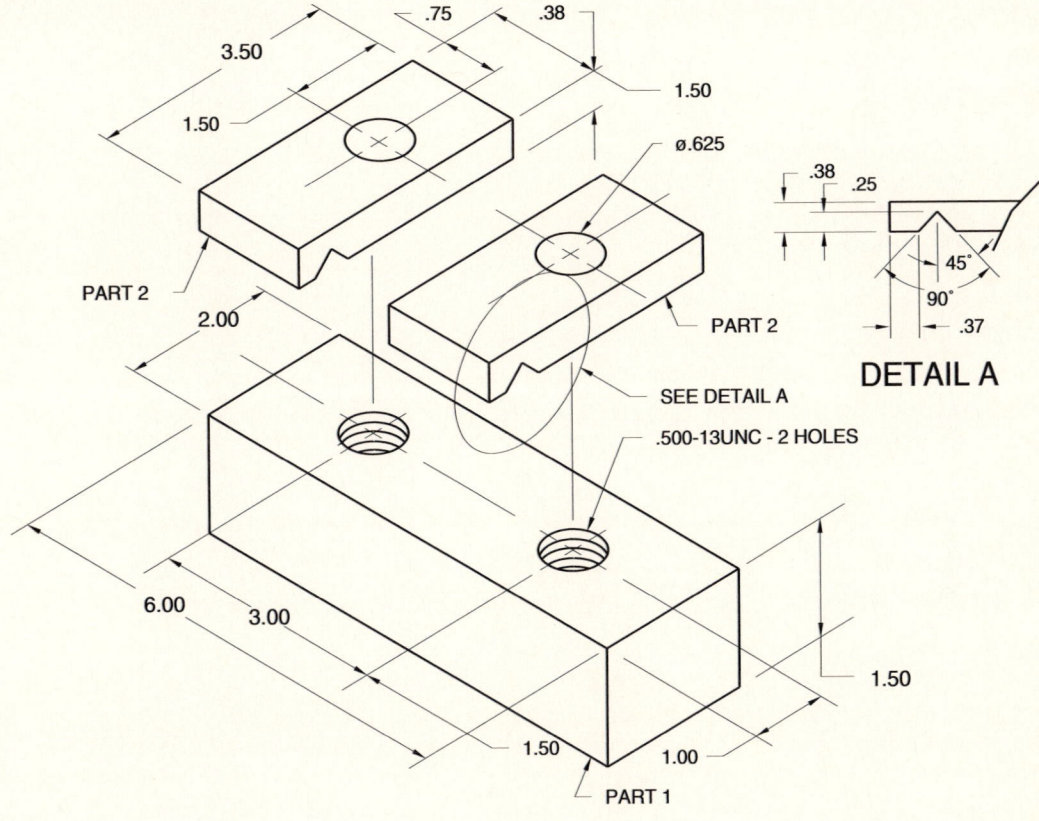

15.2 Pressure Assembly

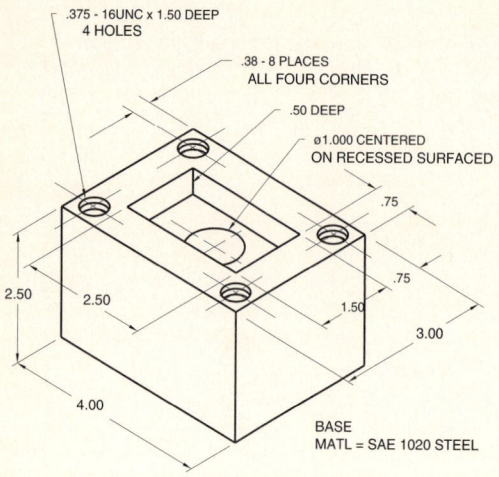

.375 - 16UNC x 1.50 DEEP
4 HOLES

.38 - 8 PLACES
ALL FOUR CORNERS

.50 DEEP

ø1.000 CENTERED
ON RECESSED SURFACED

.75

.75

.75

1.50

2.50

2.50

1.50

3.00

4.00

BASE
MATL = SAE 1020 STEEL

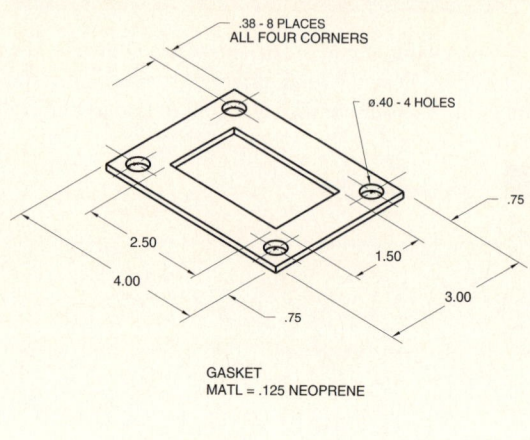

.38 - 8 PLACES
ALL FOUR CORNERS

ø.40 - 4 HOLES

.75

2.50

1.50

4.00

.75

3.00

GASKET
MATL = .125 NEOPRENE

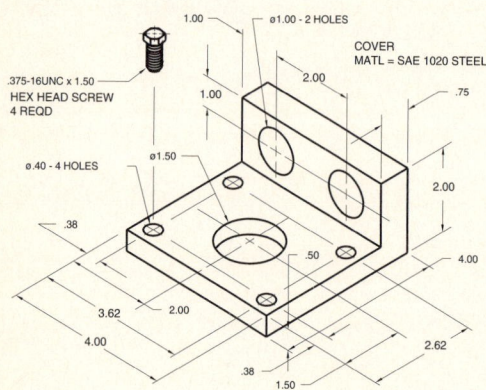

1.00

ø1.00 - 2 HOLES

COVER
MATL = SAE 1020 STEEL

.375-16UNC x 1.50
HEX HEAD SCREW
4 REQD

2.00

.75

1.00

ø1.50

2.00

ø.40 - 4 HOLES

.38

.50

4.00

3.62

2.00

2.62

.38

1.50

15.3 Hub Assembly

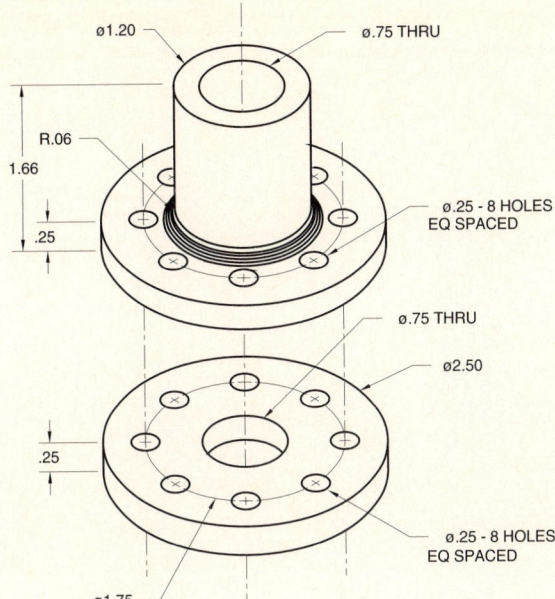

ø1.20

ø.75 THRU

R.06

1.66

.25

ø.25 - 8 HOLES
EQ SPACED

ø.75 THRU

ø2.50

.25

ø.25 - 8 HOLES
EQ SPACED

ø1.75

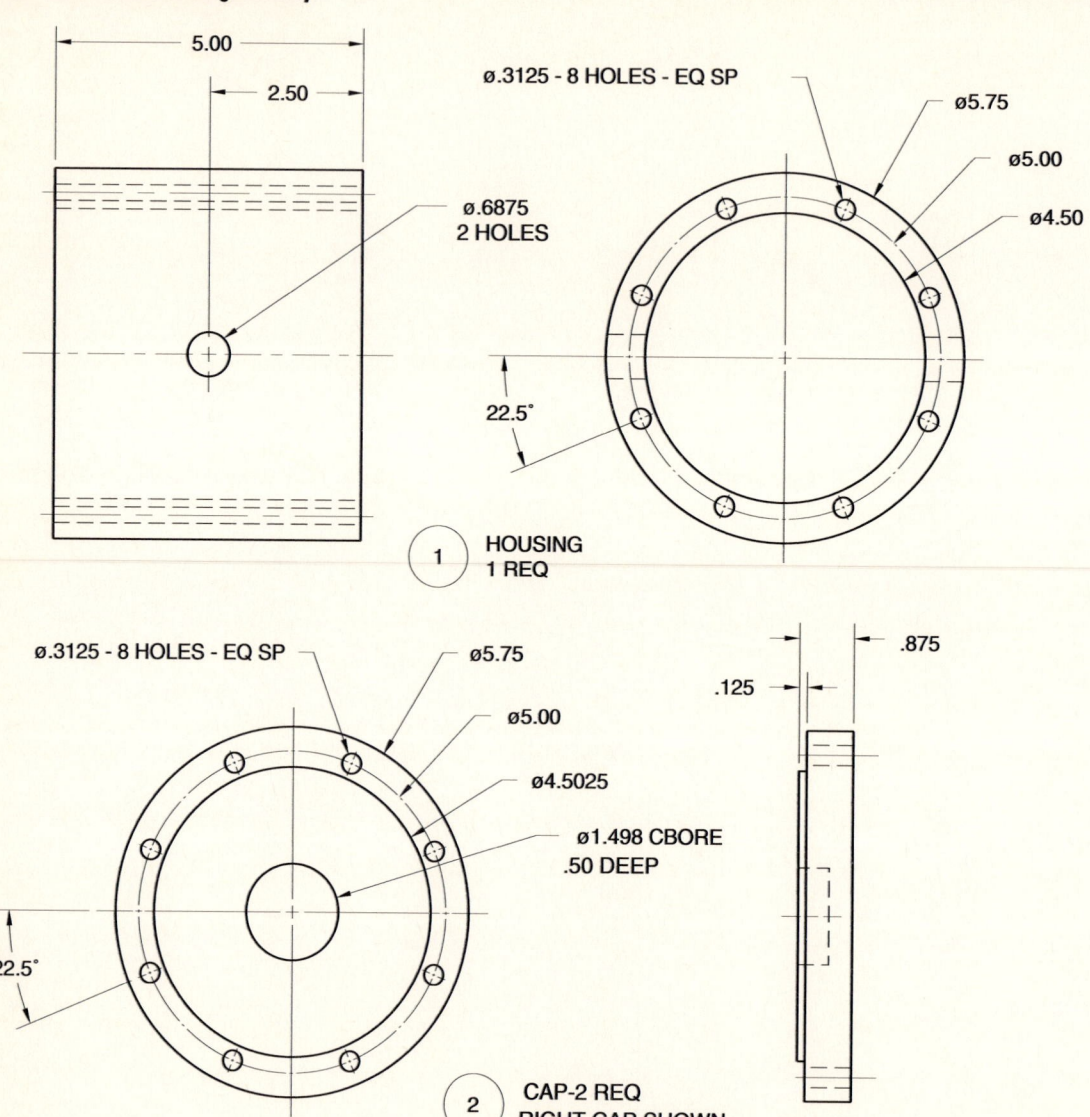

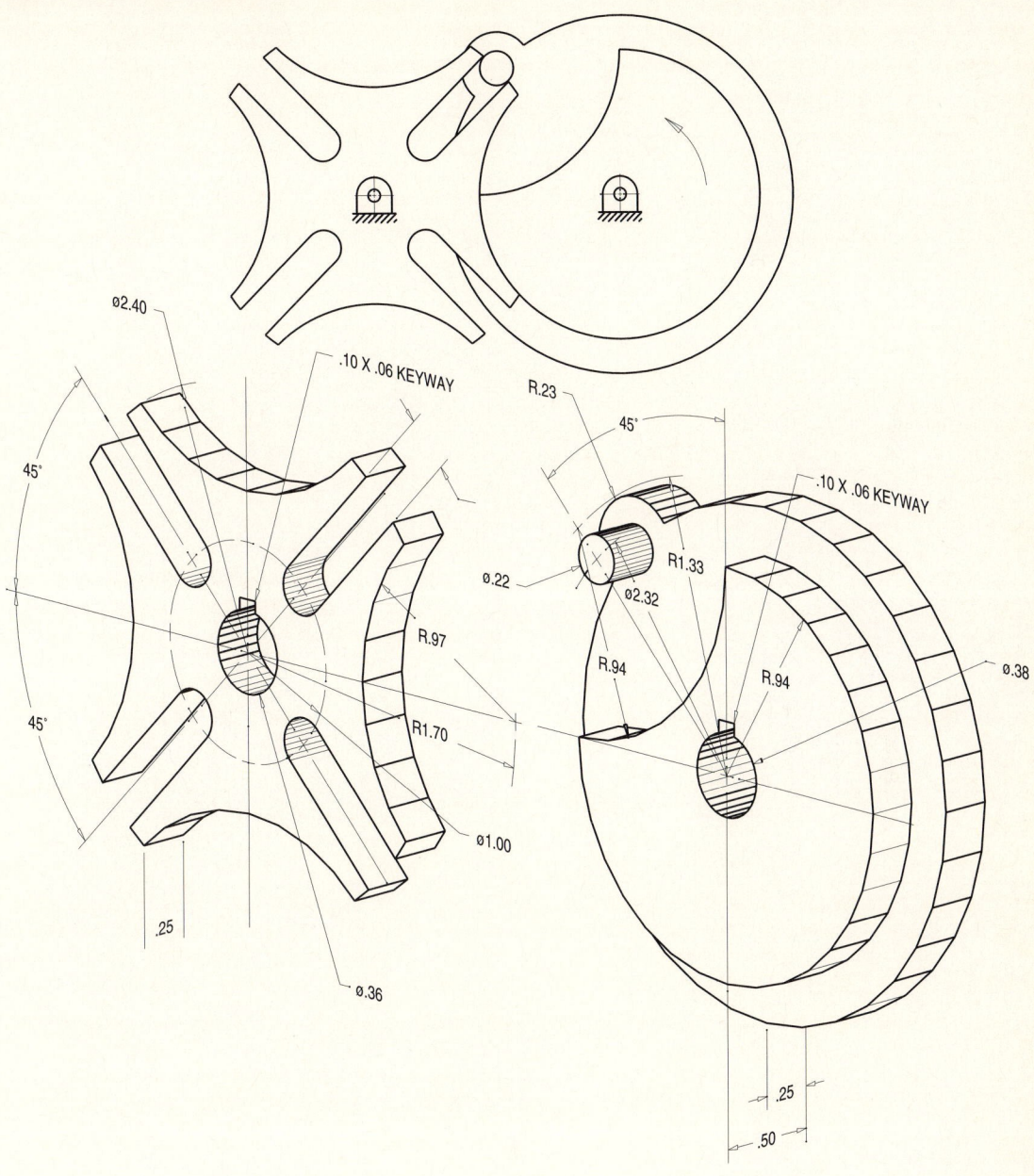

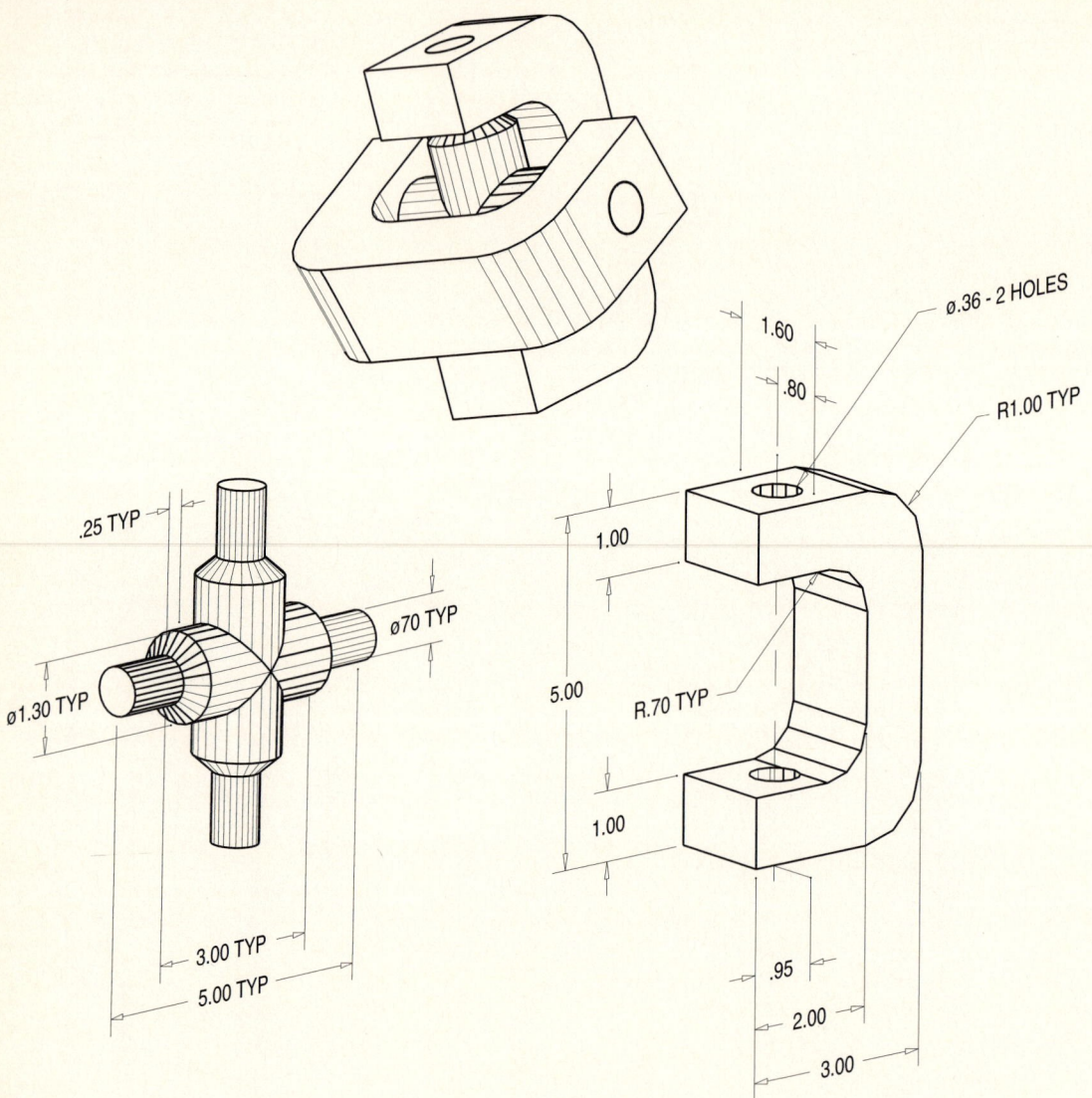

.25 TYP

ø70 TYP

ø1.30 TYP

3.00 TYP

5.00 TYP

1.60

.80

ø.36 - 2 HOLES

R1.00 TYP

1.00

5.00

R.70 TYP

1.00

.95

2.00

3.00

15.7 Staircase

Retrieve your file from Exercise 11.6 and render the stairs using wood materials and a distant light source.

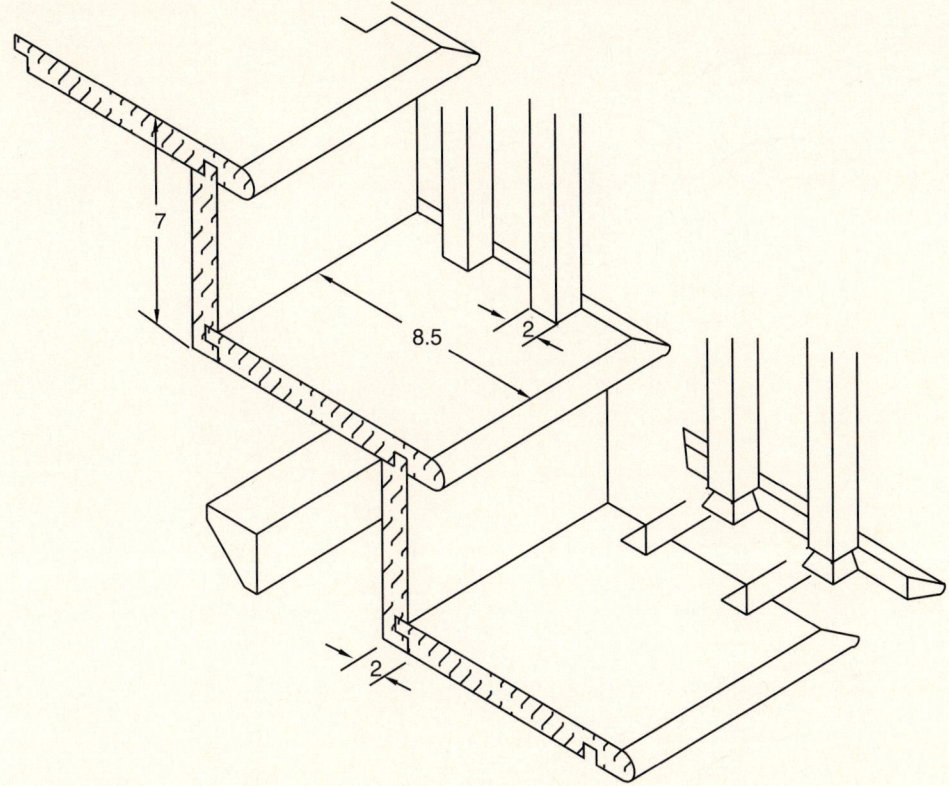

15.8M Pulley Arm

Use solid modeling to create a pictorial view like the one shown. Render the model to create a pictorial section.

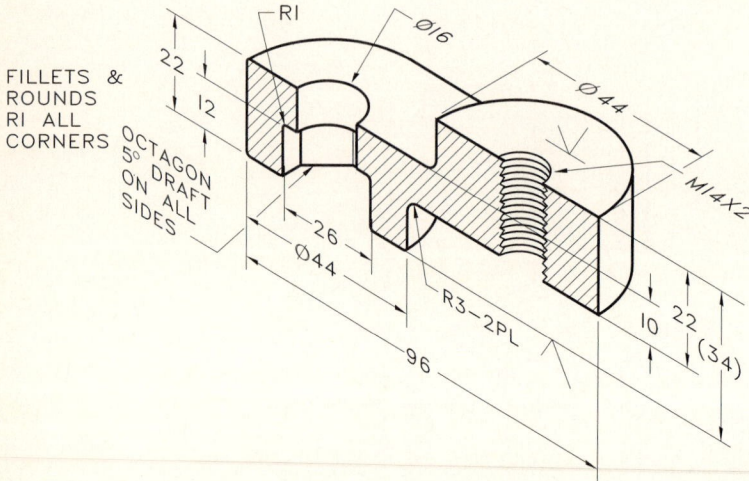

FILLETS &
ROUNDS
R1 ALL
CORNERS

OCTAGON
5° DRAFT
ON ALL
SIDES

R1

Ø16

Ø44

M14X2

22

12

26

Ø44

R3-2PL

96

22

10

(34)

Retrieve your file from Exercise 11.8M or 11.9M. Assign materials, create lighting and render your drawing to a file. Import the rendered drawing into your Windows word processor and add ball tags and a parts list to create a rendered isometric assembly drawing.

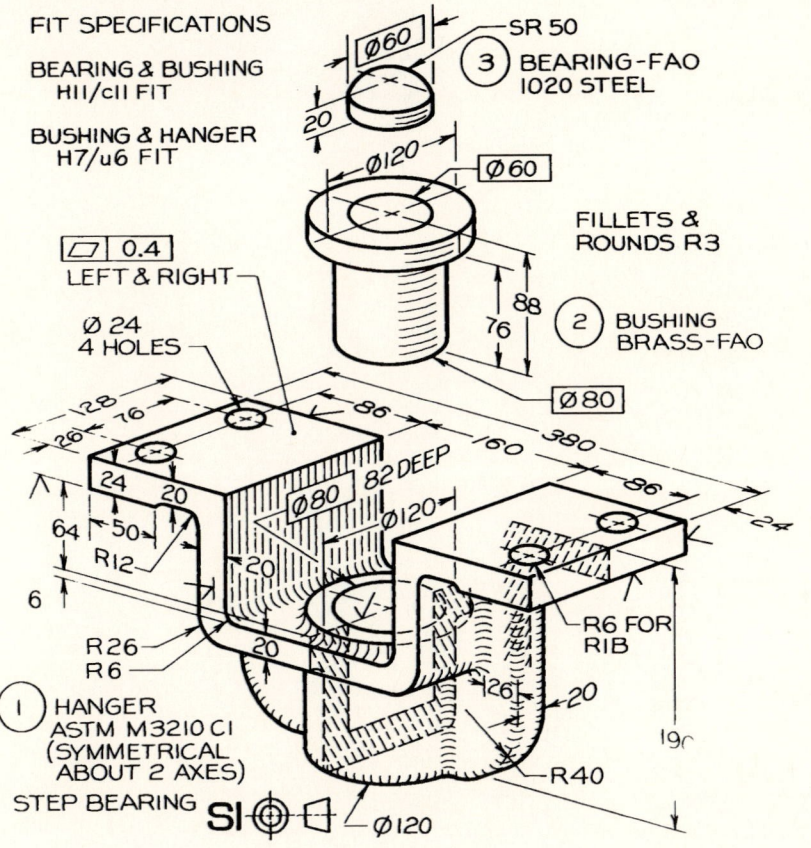

FIT SPECIFICATIONS

BEARING & BUSHING
H11/c11 FIT

BUSHING & HANGER
H7/u6 FIT

15.10M Tensioner

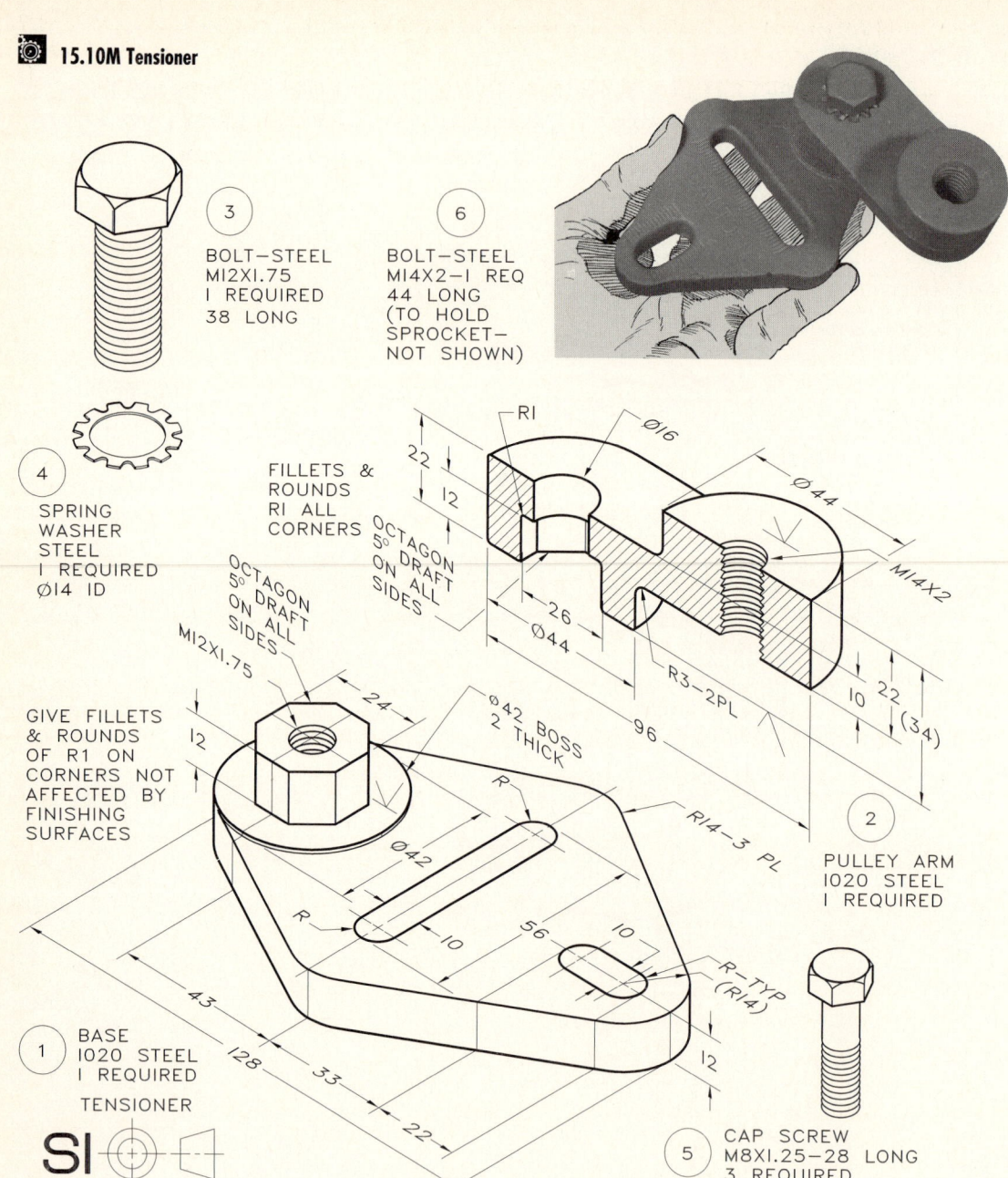

③ BOLT—STEEL
M12X1.75
1 REQUIRED
38 LONG

⑥ BOLT—STEEL
M14X2—1 REQ
44 LONG
(TO HOLD
SPROCKET—
NOT SHOWN)

④ SPRING
WASHER
STEEL
1 REQUIRED
Ø14 ID

FILLETS &
ROUNDS
R1 ALL
CORNERS

R1

Ø16

Ø44

M14X2

22

12

OCTAGON
5° DRAFT
ON ALL
SIDES

OCTAGON
5° DRAFT
ON ALL
SIDES

26

Ø44

R3—2PL

96

10

22

(34)

② PULLEY ARM
1020 STEEL
1 REQUIRED

OCTAGON
5° DRAFT
ON ALL
SIDES

M12X1.75

GIVE FILLETS
& ROUNDS
OF R1 ON
CORNERS NOT
AFFECTED BY
FINISHING
SURFACES

24

12

Ø42 BOSS
2 THICK

R

R14—3 PL

Ø42

R

10

56

10

R—TYP
(R14)

12

43

128

33

22

① BASE
1020 STEEL
1 REQUIRED

TENSIONER

SI

⑤ CAP SCREW
M8X1.25—28 LONG
3 REQUIRED

This grid includes the commands available for use with AutoCAD Release 14 for Windows. You should use AutoCAD's on-line help feature for complete information about the commands and their options.

Command or Icon Name	Description	Toolbar	Icon	Menu Bar	Command or Alias
2D Solid	Creates solid-filled polygons.	Surfaces		Draw, Surfaces, 2D Solid	SOLID, SO
3 Point UCS	Selects a user coordinate system based on any three points in the graphics window.	Standard (UCS flyout) or UCS		Tools, UCS, 3 Point	UCS
3D	Creates three-dimensional mesh objects	N/A		N/A	3D
3D Array	Creates a three-dimensional array.	N/A		Modify, 3D Operation, 3D Array	3DARRAY, 3A
3D Dynamic View	Changes the 3D view of a point.	N/A		View, 3D Dynamic View	DVIEW, DV
3D Face	Creates a three-dimensional surface.	Surfaces		Draw, Surfaces, 3D Face	3DFACE, 3F
3D Mesh	Creates a free-form polygon mesh.	Surfaces		Draw, Surfaces, 3D Mesh	3DMESH
3D Mirror	Creates a mirror image of objects about a plane.	N/A		Modify, 3D Operation, Mirror 3D	MIRROR3D
3D Polyface Mesh	Creates a 3D polyface mesh by vertices.	N/A		N/A	PFACE
3D Polyline	Creates a 3D polyline of straight line segments.	N/A		Draw, 3D Polyline	3DPOLY, 3P
3D Rotate	Rotates an object in a 3D coordinate system.	N/A		Modify, 3D Operation, Rotate 3D	ROTATE3D

Command or Icon Name	Description	Toolbar	Icon	Menu Bar	Command or Alias
3D Studio In	Imports a 3D Studio (*.3ds) file into an AutoCAD drawing.	N/A		Insert, 3D Studio	3DSIN
3D Studio Out	Saves a file formatted for use in 3D Studio (*.3ds).	N/A		N/A	3DSOUT
About	Displays information about AutoCAD.	N/A		N/A	'ABOUT, ABOUT
Acadprefix	System variable that stores the directory path, if any, specified by the ACAD environment variable. (read only)	N/A		N/A	ACADPREFIX
Acadver	System variable that stores the AutoCAD version number. (read only)	N/A		N/A	ACADVER
ACIS In	Imports an ACIS (*.sat) file into an AutoCAD drawing.	N/A		N/A	ACISIN
ACIS Out	Saves a file formatted in ACIS (*.sat).	N/A		N/A	ACISOUT
Administration	Performs administrative functions with external database commands.	External Database		Tools, External Database, Administration	ASEADMIN, AAD
Aerial View	Opens the Aerial View window.	Standard		View, Aerial View	DSVIEWER, AV
Align	Moves and rotates objects to align with other objects.	N/A		Modify, 3D Operation, Align	ALIGN, AL
Dimension Text Edit	Aligns dimension text to Home, Left, Right or Angle position.	Dimension		Dimension, Align Text	DIMTEDIT, DIMTED
Aligned Dimension	Creates an aligned linear dimension.	Dimension		Dimension, Aligned	DIMALIGNED, DAL, DIMALI

Command or Icon Name	Description	Toolbar	Icon	Menu Bar	Command or Alias
AME Convert	Converts AME solid models to AutoCAD solid objects.	N/A		N/A	AMECON-VERT
Angular Dimension	Dimensions an angle.	Dimension		Dimension, Angular	DIMANGU-LAR, DAN, DIMANG
Aperture	Controls the size of the object snap target box.	N/A		N/A	'APERTURE, APERTURE
Application Load	Loads AutoLISP (*.ads), and (*.arx) files.	N/A		Tools, Load Application	'APPLOAD, APPLOAD, AP
Arc	Draws an arc based on three points.	Draw		Draw, Arc, 3 Points	ARC, A
Arc Center Start Angle	Draws an arc of a specified angle based on the center and start points.	N/A		Draw, Arc, Center Start Angle	ARC, A
Arc Center Start End	Draws an arc based on the center, start, and endpoints.	N/A		Draw, Arc, Center Start End	ARC, A
Arc Center Start Length	Draws an arc of a specified length based on the center and start points.	N/A		Draw, Arc, Center Start Length	ARC, A
Arc Continue	Draws an arc from the last point of the previous line or arc drawn.	N/A		Draw, Arc, Continue	ARC, A
Arc Start Center Angle	Draws an arc of a specified angle based on the start and center points.	N/A		Draw, Arc, Start Center Angle	ARC, A
Arc Start Center End	Draws an arc based on the start, center, and endpoints.	N/A		Draw, Arc, Start Center End	ARC, A
Arc Start Center Length	Draws an arc of a specified length based on the start and center points.	N/A		Draw, Arc, Start Center Length	ARC, A
Arc Start End Angle	Draws an arc of a specified angle based on the start and endpoints.	N/A		Draw, Arc, Start End Angle	ARC, A
Arc Start End Direction	Draws an arc in a specified direction based on the start and endpoints.	N/A		Draw, Arc, Start End Direction	ARC, A

Command or Icon Name	Description	Toolbar	Icon	Menu Bar	Command or Alias
Arc Start End Radius	Draws an arc of a specified radius based on the start and endpoints.	N/A		Draw, Arc, Start End Radius	ARC, A
Area	Finds the area and perimeter of objects or defined areas.	Standard (Inquiry flyout) or Inquiry		Tools, Inquiry, Area	AREA, AA
Array	Creates multiple copies of objects in a regularly spaced circular or rectangular pattern.	Modify		Modify, Array	ARRAY, AR
Attdia	System variable that controls whether the Insert command uses a dialog box for attribute value entry.	N/A		N/A	ATTDIA
Attribute Display	Globally controls attribute visibility.	N/A		View, Display, Attribute Display	'ATTDISP, ATTDISP
Audit	Evaluates the integrity of a drawing.	N/A		File, Drawing Utilities, Audit	AUDIT
AutoCAD Runtime Extension	Provides information, loads/unloads ARX applications.	N/A		N/A	ARX
Back View	Sets the viewing direction for a three-dimensional visualization of the drawing to the back.	Standard (Viewpoint flyout) or Viewpoint		View, 3D Viewpoint, Back	VPOINT, -VP, DDVPOINT, VP
Background	Sets a background for your rendered scene.	Render		View, Render, Background	BACK-GROUND
Base	Specifies the insertion point for a block or drawing.	N/A		Draw, Block, Base	'BASE, BASE
Baseline Dimension	Continues a linear, angular, or ordinate dimension from the baseline of the previous or selected dimension.	Dimension		Dimension, Baseline	DIMBASE-LINE, DBA, DIMBASE

Command or Icon Name	Description	Toolbar	Icon	Menu Bar	Command or Alias
Bitmap Out	Saves AutoCAD objects in a bitmap (*.bmp) format.	N/A		N/A	BMPOUT
Blipmode	System variable that controls the display of marker blips.	N/A		Tools, Drawing Aids	'BLIPMODE, BLIPMODE
Bottom View	Sets the viewing direction for a three-dimensional visualization of the drawing to the bottom.	Standard (Viewpoint flyout) or Viewpoint		View, 3D Viewpoint, Bottom	VPOINT, -VP, DDVPOINT, VP
Boundary	Creates a region or polyline of a closed boundary.	N/A		Draw, Boundary	BOUNDARY, BO, -BOUND-ARY, -BO
Boundary Hatch	Automatically fills the area you select with a pattern.	Draw		Draw, Hatch	BHATCH, BH, H
Box	Creates a three-dimensional rectangular-shaped solid.	Solids		Draw, Solids, Box	BOX
Break	Removes part of an object or splits it in two.	Modify		Modify, Break	BREAK, BR
Browser	Launches the Web browser defined in the system registry.	Standard		N/A	BROWSER
Calculator	Calculates mathematical and geometrical expressions.	N/A		N/A	'CAL, CAL
Cancel	Exits current command without performing operation.	N/A		N/A	Esc
Celtype	System variable that sets the linetype of new objects.	N/A		N/A	CELTYPE
Center Mark	Draws center marks and centerlines of arcs and circles.	Dimension		Dimension, Center Mark	DIMCENTER, DCE
Chamfer	Draws a straight line segment (called a chamfer) between two given lines.	Modify		Modify, Chamfer	CHAMFER, CHA

Command or Icon Name	Description	Toolbar	Icon	Menu Bar	Command or Alias
Change	Changes the properties of existing objects.	N/A		N/A	CHANGE, -CH
Circle	Draws circles based upon center point and radius.	Draw		Draw, Circle, Center Radius	CIRCLE, C
Circle Center Diameter	Draws circles based upon center point and diameter.	N/A		Draw, Circle, Center Diameter	
Circle Tan Tan Radius	Draws circles based upon two tangent points and a radius.	N/A		Draw, Circle, Tan Tan Radius	
Circle Tan Tan Tan	Draws circles based upon three tangent points.	N/A		Draw, Circle, Tan Tan Tan	
Circle 3 Points	Draws circles based upon 3 points.	N/A		Draw, Circle, 3 Points	
Circle 2 Points	Draws circles based upon 2 points.	N/A		Draw, Circle, 2 Points	
Cmddia	System variable controls whether dialog boxes are enabled for more than just Plot and external database commands.	N/A		N/A	CMDDIA
Color Control	Sets the color for new objects.	N/A		Format, Color	'COLOR, COLOR, 'DDCOLOR, DDCOLOR, COL
Compile	Compiles PostScript font files and shape files.	N/A		N/A	COMPILE
Cone	Creates a three-dimensional solid primitive with a circular base tapering to a point perpendicular to its base.	Solids		Draw, Solids, Cone	CONE

Command or Icon Name	Description	Toolbar	Icon	Menu Bar	Command or Alias
Configure	Reconfigures AutoCAD.	N/A		Tools, Preferences	CONFIG
Construction Line	Creates an infinite line.	Draw		Draw, Construction Line	XLINE
Continue Dimension	Adds the next chained dimension, measured from the last dimension.	Dimension		Dimension, Continue	DIMCONTINUE, DIMCONT, DCO
Convert	Converts associative hatches and 2D polylines to Release 14 format.	N/A		N/A	CONVERT
Copy	Copies an existing shape from one area on the drawing to another area.	Modify		Modify, Copy	COPY, CP, CO
Copy Link	Copies the current view to the Windows Clipboard for linking to other OLE applications.	N/A		Edit, Copy Link	COPYLINK
Copyclip	Copies objects to the Windows Clipboard.	Standard		Edit, Copy	COPYCLIP
Copyhist	Copies all text in the command line history to the Clipboard.	N/A		N/A	COPYHIST
Cutclip	Copies objects to the clipboard and erases them from the drawing.	Standard		Edit, Cut	CUTCLIP
Cylinder	Creates a three-dimensional solid primitive similar to an extruded circle, but without a taper.	Solids		Draw, Solids, Cylinder	CYLINDER
Database List	Lists database information for drawing objects.	N/A		N/A	DBLIST
Define Attribute	Creates attribute text.	N/A		Draw, Block, Define Attributes	ATTDEF, -AT, DDATTDEF, AT
Delay	Adds a timed pause (33 seconds maximum) within a script.	N/A		N/A	'DELAY, DELAY

Command or Icon Name	Description	Toolbar	Icon	Menu Bar	Command or Alias
Delobj	Controls whether objects used to create other objects are retained or deleted from the drawing database.	N/A		N/A	DELOBJ
Detach	Removes an externally referenced drawing.	N/A		N/A	XREF, XR, -XREF, -XR
Diameter Dimension	Adds center marks to a circle when providing its diameter dimension.	Dimension		Dimension, Diameter	DIMDIAME-TER, DIMDIA, DDI
Diastat	System variable that stores the exit method of the most recently used dialog box. (read only)	N/A		N/A	DIASTAT
Dimalt	System variable that enables alternate units.	N/A		N/A	DIMALT
Dimaltd	System variable that sets alternate unit precision.	N/A		N/A	DIMALTD
Dimaltf	System variable that sets alternate unit scale.	N/A		N/A	DIMALTF
Dimalttd	System variable that stores the number of decimal places for the tolerance values of an alternate units dimension.	N/A		N/A	DIMALTTD
Dimalttz	System variable that toggles the suppression of zeros for tolerance values.	N/A		N/A	DIMALTTZ
Dimaltu	System variable that sets the units format for alternate units of all dimension style family members, except angular.	N/A		N/A	DIMALTU
Dimaltz	System variable that toggles suppression of zeros for alternate units dimension values.	N/A		N/A	DIMALTZ

Command or Icon Name	Description	Toolbar	Icon	Menu Bar	Command or Alias
Dimapost	System variable that specifies the text prefix and/or suffix to the alternate dimension measurement for alternate types of dimensions except angular.	N/A		N/A	DIMAPOST
Dimasz	System variable that controls the size of dimension line and leader line arrowheads.	N/A		N/A	DIMASZ
Dimaso	System variable that controls the creation of associative dimension objects.	N/A		N/A	DIMASO
Dimaunit	System variable that sets the angle format for angular dimensions.	N/A		N/A	DIMAUNIT
Dimblk	System variable that sets the name of a block to be drawn instead of the normal arrowhead at the end of the dimension line or leader line.	N/A		N/A	DIMBLK
Dimblk1	System variable that, if Dimsah is on, specifies the user-defined arrowhead block for the first end of the dimension line.	N/A		N/A	DIMBLK1
Dimblk2	System variable that, if Dimsah is on, specifies the user-defined arrowhead block for the second end of the dimension line.	N/A		N/A	DIMBLK2
Dimcen	System variable that controls the drawing of circle or arc centermarks or centerlines.	N/A		N/A	DIMCEN
Dimclrd	System variable that assigns colors to dimension lines, arrowheads, and dimension leader lines.	N/A		N/A	DIMCLRD

Command or Icon Name	Description	Toolbar	Icon	Menu Bar	Command or Alias
Dimclre	System variable that assigns colors to dimension extension lines.	N/A		N/A	DIMCLRE
Dimclrt	System variable that assigns color to dimension text.	N/A		N/A	DIMCLRT
Dimdec	System variable that sets the number of decimal places for the tolerance values of a primary units dimension.	N/A		N/A	DIMDEC
Dimdle	System variable that extends the dimension line beyond the extension line when oblique strokes are drawn instead of arrowheads.	N/A		N/A	DIMDLE
Dimdli	System variable that controls the dimension line spacing for baseline dimensions.	N/A		N/A	DIMDLI
Dimension Edit	Edits dimensions.	Dimension		Dimension, Oblique	DIMEDIT, DIMED, DED
Dimension Status	Shows the dimension variables and their current settings.	N/A		N/A	DIM
Dimension Style	Controls dimension styles.	Dimension		Dimension, Style	DDIM, D, DIMSTYLE, DST, DIMSTY
Dimension Update	Updates the dimensions based on changes to the dimension styles and variables.	Dimension		Dimension, Update	DIM (w/ update)
Dimexe	System variable that determines how far to extend the extension line beyond the dimension line.	N/A		N/A	DIMEXE
Dimexo	System variable that determines how far extension lines are offset from origin points.	N/A		N/A	DIMEXO

Command or Icon Name	Description	Toolbar	Icon	Menu Bar	Command or Alias
Dimfit	System variable that controls the placement of text and attributes inside or outside extension lines.	N/A		N/A	DIMFIT
Dimgap	System variable that sets the distance around the dimension text when you break the dimension line to accommodate dimension text.	N/A		N/A	DIMGAP
Dimjust	System variable that controls the horizontal dimension text position.	N/A		N/A	DIMJUST
Dimlfac	System variable that sets the global scale factor for linear dimensioning measurements.	N/A		N/A	DIMLFAC
Dimlim	System variable that, when on, generates dimension limits as the default text.	N/A		N/A	DIMLIM
Dimoverride	Allows you to override dimensioning system variable settings associated with a dimension object, but will not affect the current dimension style.	N/A		N/A	DIMOVER-RIDE, DIMOVER, DOV
Dimpost	System variable that specifies a text prefix and/or suffix to the dimension measurement.	N/A		N/A	DIMPOST
Dimrnd	System variable that rounds all dimensioning distances to the specified value.	N/A		N/A	DIMRND
Dimsah	System variable that controls the use of user-defined attribute blocks at the ends of the dimension lines.	N/A		N/A	DIMSAH

Command or Icon Name	Description	Toolbar	Icon	Menu Bar	Command or Alias
Dimscale	System variable that sets overall scale factor applied to dimensioning variables that specify sizes, distances, or offsets.	N/A		N/A	DIMSCALE
Dimsd1	System variable that, when on, suppresses drawing of the first dimension line.	N/A		N/A	DIMSD1
Dimsd2	System variable that, when on, suppresses drawing of the second dimension line.	N/A		N/A	DIMSD2
Dimse1	System variable that, when on, suppresses drawing of the first extension line.	N/A		N/A	DIMSE1
Dimse2	System variable that, when on, suppresses drawing of the second extension line.	N/A		N/A	DIMSE2
Dimsho	System variable that, when on, controls redefinition of dimension objects while dragging.	N/A		N/A	DIMSHO
Dimsoxd	System variable that, when on, suppresses drawing of dimension lines outside the extension lines.	N/A		N/A	DIMSOXD
Dimtad	System variable that controls the vertical position of text in relation to the dimension line.	N/A		N/A	DIMTAD
Dimtdec	System variable that sets the number of decimal places to display the tolerance values for a dimension.	N/A		N/A	DIMTDEC

Command or Icon Name	Description	Toolbar	Icon	Menu Bar	Command or Alias
Dimtfac	System variable that specifies a scale factor for the text height of tolerance values relative to the dimension text height.	N/A		N/A	DIMTFAC
Dimtih	System variable that controls the position of dimension text inside the extension lines for all dimension types except ordinate dimensions.	N/A		N/A	DIMTIH
Dimtix	System variable that draws text between extension lines.	N/A		N/A	DIMTIX
Dimtm	System variable that, when Dimtol or Dimlim is on, sets the minimum (or lower) tolerance limit for dimension text.	N/A		N/A	DIMTM
Dimtofl	System variable that, when on, draws a dimension line between the extension lines even when the text is placed outside the extension lines.	N/A		N/A	DIMTOFL
Dimtoh	System variable that, when turned on, controls the position of dimension text outside the extension lines.	N/A		N/A	DIMTOH
Dimtol	System variable that generates plus/minus tolerances.	N/A		N/A	DIMTOL
Dimtolj	System variable that sets the vertical justification for tolerance values relative to the nominal dimension text.	N/A		N/A	DIMTOLJ

Command or Icon Name	Description	Toolbar	Icon	Menu Bar	Command or Alias
Dimtp	System variable that, when Dimtol or Dimlim is on, sets the maximum (or upper) tolerance limit for dimension text.	N/A		N/A	DIMTP
Dimtsz	System variable that specifies the size of oblique strokes drawn instead of arrowheads for linear, radius, and diameter dimensioning.	N/A		N/A	DIMTSZ
Dimtvp	System variable that adjusts text placement relative to text height.	N/A		N/A	DIMTVP
Dimtxsty	System variable that specifies the text style of a dimension.	N/A		N/A	DIMTXSTY
Dimtxt	System variable that specifies the height of the dimension text, unless the current text style has a fixedheight.	N/A		N/A	DIMTXT
Dimtzin	System variable that controls the suppression of zeros for tolerance values.	N/A		N/A	DIMTZIN
Dimunit	System variable that sets the units format for all dimension family members, except angular.	N/A		N/A	DIMUNIT
Dimupt	System variable that controls the cursor functionality for User Positioned Text.	N/A		N/A	DIMUPT
Dimzin	System variable controlling the suppression of the inch portion of a feet-and-inches dimension when distance is an integral number of feet, or the feet portion when distance is less than a foot.	N/A		N/A	DIMZIN

Command or Icon Name	Description	Toolbar	Icon	Menu Bar	Command or Alias
Dispsilh	System variable that controls the display of silhouette curves of body objects in wire-frame mode.	N/A		N/A	DISPSILH
Distance	Reads out the value for the distance between two selected points.	Standard, (Inquiry flyout) or Inquiry		Tools, Inquiry, Distance	'DIST, DIST, DI
Divide	Places points or blocks along an object to create segments.	N/A		Draw, Point, Divide	DIVIDE, DIV
Donut	Draws concentric filled circles or filled circles.	N/A		Draw, Donut	DONUT
Dragmode	Controls the way drag-ged objects are displayed	N/A		N/A	'DRAGMODE, DRAGMODE
Drawing Aids	Sets the drawing aids for the drawing.	N/A		Tools, Drawing Aids	'DDRMODES, RM, DDR-MODES
Draworder	Changes the display order of images and objects.	Modify II		Tools, Display Order	DRAWORDER, DR
DWF Out	Saves drawing objects in Drawing Web format (*.dwf).	N/A		N/A	DWFOUT
DXB In	Imports coded binary files (*.dxb).	N/A		N/A	DXBIN
DXF In	Imports drawing saved in Drawing Interchange (*.dxf) format.	N/A		N/A	DXFIN
DXF Out	Saves drawing objects in Drawing Interchange (*.dxf) Format.	N/A		N/A	DXFOUT
Edge	Changes the visibility of three-dimensional face edges.	Surfaces		Draw, Surfaces, Edge	EDGE
Edge Surface	Creates a three-dimensional polygon mesh.	Surfaces		N/A	EDGESURF

Command or Icon Name	Description	Toolbar	Icon	Menu Bar	Command or Alias
Edit Attribute	Edits attribute values.	Modify II		Modify, Object, Attribute, Single	ATTEDIT, -ATE
Edit Attribute Globally	Edits attribute values.	N/A		Modify, Object, Attribute, Global	DDATTE, ATE
Edit Hatch	Modifies an existing associative hatch block.	Modify II		Modify, Object, Hatch	HATCHEDIT, HE
Edit Links	Updates, changes, and cancels existing links.	N/A		Edit, OLE Links	OLELINKS
Edit Multiline	Edits multiple parallel lines.	Modify II		Modify, Object, Multiline	MLEDIT
Edit Polyline	Changes various features of polylines.	Modify II		Modify, Object, Polyline	PEDIT, PE
Edit Spline	Edits a spline object.	Modify II		Modify, Object, Spline	SPLINEDIT, SPE
Edit Text	Allows you to edit text inside a dialogue box.	Modify II		Modify, Object, Text	DDEDIT, ED
Elev	Sets elevation and extrusion thickness of new objects.	N/A		N/A	'ELEV, ELEV
Elevation	System variable that stores the current three-dimensional elevation relative to the current UCS for the current space.	N/A		N/A	ELEVATION
Ellipse	Used to draw ellipses.	Draw		Draw, Ellipse, Center	ELLIPSE, EL
Ellipse Axis End	Used to draw ellipses.	N/A		Draw, Ellipse, Axis End	

Command or Icon Name	Description	Toolbar	Icon	Menu Bar	Command or Alias
Ellipse Arc	Used to draw ellipses.	N/A		Draw, Ellipse, Arc	
Erase	Erases objects from drawing.	Modify		Modify, Erase	ERASE, E
Explode	Separates blocks or dimensions into component objects so that they are no longer one group and can be edited.	Modify		Modify, Explode	EXPLODE, X
Export	Exports a drawing to a file of type DXF so that another application can use it.	N/A		File, Export	EXPORT, EXP
Export Links	Exports link information for selected objects.	External Database		Tools, External Database, Export Links	ASEEXPORT, AEX
Extend	Extends the length of an existing object to meet a boundary.	Modify		Modify, Extend	EXTEND, EX
Extract Attributes	Extracts attribute data from a drawing.	N/A		N/A	ATTEXT, DDATTEXT
External Reference	Attaches, Detaches, Reloads, Unloads, Binds, etc. other drawings as external references, without really adding them to your current drawing.	Draw (Insert flyout) or Insert or Reference		Insert, External Reference	XREF, XR, -XREF, -XR
External Reference Attach	Uses another drawing as an external reference, without really adding it to your current drawing.	Reference		N/A	XATTACH, XA
External Reference Bind	Binds symbols of an xref to a drawing.	Reference		Modify, Object, External Reference, Bind	XBIND, XB, -XBIND, -XB
External Reference Clip	Defines an xref clipping boundary.	Reference		N/A	XCLIP, XC

Command or Icon Name	Description	Toolbar	Icon	Menu Bar	Command or Alias
External Reference Clip Frame	System Variable that controls visibility of clipping boundaries.	Reference		Modify, Object, External Reference, Frame	XCLIPFRAME
Extrude	Extrudes a two-dimensional shape into a three-dimensional object.	Solids		Draw, Solids, Extrude	EXTRUDE, EXT
Facetres	System variable that adjusts the smoothness of shaded and hidden line-removed objects.	N/A		N/A	FACETRES
Filedia	System variable that suppresses the display of the file dialog boxes.	N/A		N/A	FILEDIA
Fill	Controls the filling of multilines, traces, solids, and wide polylines.	N/A		N/A	'FILL, FILL
Fillet	Connects lines, arcs, or circles with a smoothly fitted arc.	Modify		N/A	FILLET, F
Fog	Provides visual information about the distance of objects from the camera.	Render		View, Render, Fog	FOG
Front View	Sets the viewing direction for a three-dimensional visualization of the drawing to the front.	Standard (Viewpoint flyout) or Viewpoint		View, 3D Viewpoint, Front	VPOINT, VP, DDVPOINT, -VP
Global Linetype	System variable that sets the current global linetype scale factor for objects.	N/A		N/A	CELTSCALE
Graphics Screen	Switches from the text window back to the graphics window.	N/A		N/A	'GRAPHSCR, GRAPHSCR
Grid	Displays a grid of dots at desired spacing on the screen.	Status bar		N/A	'GRID, GRID, or toggle off and on with F7

Command or Icon Name	Description	Toolbar	Icon	Menu Bar	Command or Alias
Grips	Enables grips and sets their color.	N/A		Tools, Grips	'GRIPS, GRIPS, 'DDGRIPS, DDGRIPS, GR
Handles	System variable that controls the use of handles in solid models.	N/A		N/A	HANDLES
Hatch	Draws an unassociative hatch pattern.	N/A		N/A	HATCH, -H
Help	Displays information explaining commands and procedures.	Standard	[?]	Help	'HELP, '?, HELP, ?, or toggle on with F1
Hide	Hides an object.	Render		View, Hide	HIDE, HI
Highlight	System variable that controls object highlighting.	N/A		N/A	HIGHLIGHT
Image	Inserts a Raster Image into the current drawing.	Draw (Insert flyout) or Reference		Insert, Raster Image	IMAGE, IM, -IMAGE, -IM
Image Adjust	Controls the fade, contrast, and brightness of a selected image.	Reference		Modify, Object, Image, Adjust	IMAGEAD-JUST, IAD
Image Attach	Attaches a new image to the drawing.	Reference		N/A	IMAGEAT-TACH, IAT
Image Clip	Creates clipping boundaries for image objects.	Reference		Modify, Object, Image Clip	IMAGECLIP, ICL
Image Frame	Controls whether an image frame is displayed or not.	Reference		Modify, Object, Image, Frame	IMAGE-FRAME
Image Quality	Controls the quality of images.	Reference		Modify, Object, Image, Quality	IMAGE-QUALITY
Image Transparency	Controls whether the background of an image is opaque or clear.	Reference		Modify, Object, Image, Transparency	TRANS-PARENCY

Command or Icon Name	Description	Toolbar	Icon	Menu Bar	Command or Alias
Import	Imports many different file formats into AutoCAD.	Draw (Insert flyout) or Insert		Insert	IMPORT, IMP
Insert Block	Places blocks or other drawings into a drawing via the command line or via a dialog box.	Draw (Insert flyout) or Insert		N/A	INSERT, -I, DDINSERT, I
Insert Multiple Blocks	Inserts multiple instances of a block in a rectangular array.	N/A		Insert, Block	MINSERT
Insert Objects	Inserts a linked or embedded object.	N/A		Insert, OLE Objects	INSERTOBJ, IO
Interfere	Shows where two solid objects overlap.	Solids		Draw, Solids, Interference	INTERFERE, INF
Intersect	Forms a new solid from the area common to two or more solids or regions.	Modify II		Modify, Boolean, Intersect	INTERSECT, IN
Isolines	System variable that specifies the number of isolines per surface on objects.	N/A		N/A	ISOLINES
Isometric Plane	Controls which isometric drawing plane is active: left, top, or right.	N/A		N/A	'ISOPLANE, ISOPLANE, or toggle with Ctrl-E
Landscape Edit	Allows you to edit a landscape object.	Render		View, Render, Landscape Edit	LSEDIT
Landscape Library	Maintains libraries of landscape objects.	Render		View, Render, Landscape Library	LSLIB
Landscape New	Lets you add realistic landscape items, such as trees and bushes, to your drawings.	Render		View, Render, Landscape New	LSNEW

Command or Icon Name	Description	Toolbar	Icon	Menu Bar	Command or Alias
Layers	Creates layers and controls layer color, linetype, and visibility.	Object Properties		Format, Layer	'LAYER, LAYER, 'LA, LA, 'DDLMODES, DDLMODES, -LAYER, -LA
Leader	Creates leader lines for identifying lines and dimensions.	Dimension		Dimension, Leader	LEADER, LEAD, LE
Left View	Sets the viewing direction for a three-dimensional visualization of the drawing to the left side.	Standard (Viewpoint flyout) or Viewpoint		View, 3D Viewpoint, Left	VPOINT, -VP, DDVPOINT, VP
Lengthen	Lengthens an object.	Modify		Modify, Lengthen	LENGTHEN, LEN
Lights	Controls the light sources in rendered scenes.	Render		View, Render, Lights	LIGHT
Limits	Sets up the size of the drawing.	N/A		Format, Drawing Limits	'LIMITS, LIMITS
Line	Draws straight lines of any length.	Draw		Draw, Line	LINE, L
Linear Dimension	Creates linear dimensions.	Dimension		Dimension, Linear	DIMLINEAR, DIMLIN, DLI
Linetype	Changes the pattern used for new lines in the drawing.	Object Properties		Format, Select Linetype	'LINETYPE, LINETYPE, -LT 'DDLTYPE, DDLTYPE, LT
Linetype Scale	System variable that changes the scale of the linetypes in the drawing.	N/A		Options, Linetypes, Global Linetype Scale	'LTSCALE, LTSCALE, LTS
Links	Manipulates links between objects and an external database.	External Database		Tools, External Database, Links	ASELINKS, ALI

Command or Icon Name	Description	Toolbar	Icon	Menu Bar	Command or Alias
List	Provides information on an object, such as the length of a line.	Standard (Inquiry fly-out) or Inquiry		Tools, Inquiry, List	LIST, LI, LS
List Variables	Lists the values of system variables.	N/A		Tools, Inquiry, System Variable	'SETVAR, SETVAR, SET
Load	Makes shapes available for use by the Shape command.	N/A		N/A	LOAD
Locate Point	Finds the coordinates of any point on the screen.	Standard (Inquiry flyout) or Inquiry		Tools, Inquiry, ID Point	'ID, ID
Log File Off	Closes the log file.	N/A		Tools, Preferences	LOGFILEOFF
Log File On	Writes the text window contents to a file.	N/A		Tools, Preferences	LOGFILEON
Match Properties	Copies the properties of one object to other selected objects.	Standard		Modify, Match Properties	'MATCHPROP, MATCHPROP, MA, PAINTER
Make Block	Groups a set of objects together so that they are treated as one object.	Draw		Draw, Block, Make	BMAKE, B, BLOCK, -B
Mapping	Lets you map materials onto selected geometry.	Render		View, Render, Mapping	SETUV
Materials	Controls the material a rendered object appears to be made of, and allows the creation of new materials.	Render		View, Render, Materials	RMAT
Materials Library	Manages all the materials available for your drawing.	Render		View, Render, Materials Library	MATLIB
Mass Properties	Calculates and displays the mass properties of regions and solids.	Standard (Inquiry flyout) or Inquiry		Tools, Inquiry, Mass Properties	MASSPROP

Command or Icon Name	Description	Toolbar	Icon	Menu Bar	Command or Alias
Measure	Marks segments of a chosen length on an object by inserting points or blocks.	N/A		Draw, Point, Measure	MEASURE, ME
Menu	Loads a menu file.	N/A		N/A	MENU
Menu Load	Loads partial menu files.	N/A		Tools, Customize Menus	MENULOAD
Menu Unload	Unloads partial menu files.	N/A		N/A	MENUUNLOAD
Mirror	Creates mirror images of shapes.	Modify		Modify, Mirror	MIRROR, MI
Model Space	Enters model space.	Status bar		View, Model Space (Tiled)	MSPACE, MS
Modify Properties	Controls properties of existing objects.	N/A		Modify, Properties	DDMODIFY, MO
Move	Moves an existing shape from one area on the drawing to another area.	Modify		Modify, Move	MOVE, M
Mtprop	Changes paragraph text properties.	N/A		N/A	MTPROP
Multiline	Creates multiple, parallel lines.	Draw		Draw, Multiline	MLINE, ML
Multiline Style	Defines a style for multiple, parallel lines.	N/A		Format, Multiline Style	MLSTYLE
Multiline Text	Creates a paragraph that fits within a nonprinting text boundary.	Draw		Draw, Text, Multiline Text	MTEXT, MT, T, -MTEXT, -T
Multiple	Repeats the next command until canceled.	N/A		N/A	MULTIPLE
MV Setup	Helps set up drawing views using a LISP program.	N/A		N/A	MVSETUP

Command or Icon Name	Description	Toolbar	Icon	Menu Bar	Command or Alias
Mview	Creates viewports in paperspace (when Tilemode = 0).	N/A		View, Floating Viewports (all options)	MVIEW, MV
Named UCS	Selects a named and saved user coordinate system.	Standard (UCS flyout) or UCS		Tools, UCS, Named UCS	DDUCS, UC
Named Views	Sets the viewing direction to a named view.	Standard (Viewpoint flyout) or Viewpoint		View, Named Views	'VIEW, VIEW, -V, 'DDVIEW, DDVIEW, V
NE Isometric View	Sets the viewing direction for a three-dimensional visualization of the drawing.	Standard (Viewpoint flyout) or Viewpoint		View, 3D Viewpoint, NE Isometric	VPOINT, -VP, DDVPOINT, VP
New Drawing	Sets up the screen to create a new drawing.	Standard		File, New	NEW
NW Isometric View	Sets the viewing direction for a three-dimensional visualization of the drawing.	Standard (Viewpoint flyout) or Viewpoint		View, 3D Viewpoint, NW Isometric	VPOINT, -VP, DDVPOINT, VP
Object Group	Creates a named selection set of objects.	N/A		Tools, Object Group	GROUP, G, -GROUP, -G
Object Selection Filters	Creates lists to select objects based on properties.	N/A		N/A	'FILTER, FILTER, FI
Object Selection Settings	Defines Selection Mode.	N/A		Tools, Selection	'DDSELECT, DDSELECT, SE
Object Snap Settings (Running Object Snap)	Sets running objects snap modes and changes target box size.	Standard (Object Snap flyout) or Object Snap or Status Bar		Tools, Object Snap Settings	'OSNAP, OSNAP, 'DDOSNAP, DDOSNAP, -OSNAP, -OS, OS
Object UCS	Selects a user coordinate system based on the object.	Standard (UCS flyout) or UCS		Tools, UCS, Object	UCS

Command or Icon Name	Description	Toolbar	Icon	Menu Bar	Command or Alias
Offset	Draws objects parallel to a given object.	Modify		Modify, Offset	OFFSET, O
Oops!	Restores erased objects.	N/A		N/A	OOPS
Open Drawing	Loads a previously saved drawing.	Standard		File, Open	OPEN
Ordinate Dimension	Creates ordinate point dimensions.	Dimension		N/A	DIMORDI-NATE, DIMORD, DOR
Origin UCS	Creates a user-defined coordinate system at the origin of the view.	Standard (UCS flyout) or UCS		Tools, UCS, Origin	UCS
Ortho	Restricts movement to only horizontal and vertical.	Status bar		N/A	'ORTHO, ORTHO, or toggle on and off with F8
Pan Down	Moves the drawing around on the screen without changing the zoom factor.	N/A		View, Pan, Down	
Pan Left	Moves the drawing around on the screen without changing the zoom factor.	N/A		View, Pan, Left	
Pan Point	Moves the drawing around on the screen without changing the zoom factor.	N/A		View, Pan, Point	'-PAN, -PAN, '-P, -P
Pan Right	Moves the drawing around on the screen without changing the zoom factor.	N/A		View, Pan, Right	
Pan Realtime	Moves the drawing around on the screen without changing the zoom factor.	Standard		View, Pan, Real Time	'PAN, PAN, 'P, P,
Pan Up	Moves the drawing around on the screen without changing the zoom factor.	N/A		View, Pan, Up	

Command or Icon Name	Description	Toolbar	Icon	Menu Bar	Command or Alias
Paper Space	Enters paper space when TILEMODE is set to 0.	Status bar		View, Paper Space	PSPACE, PS
Paper Space Linetype Scale	System variable that causes the paper space LTSCALE factor to be used for every viewport.	N/A		N/A	PSLTSCALE
Paste From Clipboard	Places objects from the Windows Clipboard into the drawing using the Insert command.	Standard		Edit, Paste	PASTECLIP
Paste Special	Inserts and controls data from the Clipboard.	N/A		Edit, Paste Special	PASTESPEC, PA
Perimeter	System variable that stores the last perimeter value computed by the Area or List commands. (read only)	N/A		N/A	PERIMETER
Plan	Creates a view looking straight down along the Z axis to see the XY coordinates.	N/A		View, 3D Viewpoint, Plan View	PLAN
Point	Creates a point object.	Draw		Draw, Point, Single Point	POINT, PO
Point Filters (.X, .Y, .Z, .XY, .XZ, .YZ)	Lets you selectively find single coordinates out of a point's three-dimensional coordinate set.	N/A		N/A	(to be used within other commands).X or .Y or .Z or .XY or .XZ or .YZ
Point Style	Specifies the display mode and size of point objects.	N/A		Format, Point Style	'DDPTYPE, DDPTYPE
Polygon	Draws regular polygons with 3 to 1024 sides.	Draw		Draw, Polygon	POLYGON, POL
Polyline	Draws a series of connected objects (lines or arcs) that are treated as a single object called a polyline.	Draw		Draw, Polyline	PLINE, PL

Command or Icon Name	Description	Toolbar	Icon	Menu Bar	Command or Alias
Preferences	Customizes the AutoCAD settings.	N/A		Tools, Preferences	PREFER-ENCES, PR
Preset UCS	Selects a preset UCS.	Standard (UCS flyout) or UCS		Tools, UCS, Preset UCS	DDUCSP, UCP
Previous UCS	Returns you to the previously used user coordinate system.	Standard (UCS flyout) or UCS		Tools, UCS, Previous	UCS
Print/Plot	Prints or plots a drawing	Standard		File, Print	PLOT, PRINT
Print Preview	Previews the drawing before printing or plotting.	Standard		File, Print Preview	PREVIEW, PRE
Projmode	System variable that sets the current projec-tion mode for trim or extend operations.	N/A		N/A	PROJMODE
Properties	Used to change the properties of an object.	Object Properties		Modify, Properties	CHPROP, DDCHPROP, CH
PS Drag	Controls the appearance of a PostScript image as it is dragged into place.	N/A		N/A	PSDRAG
PS Fill	Fills a 2D outline with a PostScript pattern.	N/A		N/A	PSFILL
PS In	Imports a file in PostScript (*.eps) format.	N/A		N/A	PSIN
PS Out	Saves drawing objects in PostScript (*.eps) format.	N/A		N/A	PSOUT
Purge	Removes unused items from the drawing database.	N/A		File, Draw-ing Utilities, Purge	PURGE, PU
Qtext	Controls the display and plotting of text and attribute text.	N/A		N/A	'QTEXT, QTEXT
Quit	Exits AutoCAD with the option not to save your changes, and returns you to the operating system.	N/A		File, Exit	QUIT, EXIT

Command or Icon Name	Description	Toolbar	Icon	Menu Bar	Command or Alias
Radius Dimension	Adds centermarks to an arc when providing its radius dimension.	Dimension		Dimension, Radius	DIMRADIUS, DIMRAD, DRA
Ray	Creates a semi-infinite line.	N/A		Draw, Ray	RAY
Recover	Repairs a damaged drawing.	N/A		File, Drawing Utilities, Recover	RECOVER
Rectangle	Creates rectangular shapes.	Draw		Draw, Rectangle	RECTANG, REC
Redefine	Restores AutoCAD internal commands.	N/A		N/A	'REDEFINE, REDEFINE
Redefine Attribute	Redefines a block and updates associated attributes.	N/A		N/A	ATTREDEF
Redo	Reverses the effect of the most recent UNDO command.	Standard		Edit, Redo	REDO, Ctrl-Y
Redraw All	Redraws all viewports at once.	Standard		View, Redraw	'REDRAWALL, REDRAWALL, RA
Redraw View	Removes excess blipmarks added to the drawing screen and restores objects partially erased while you are editing other objects.	N/A		N/A	'REDRAW, REDRAW, 'R, R
Regenauto	Controls the automatic regeneration of a drawing.	N/A		N/A	'REGENAUTO, REGENAUTO
Regenerate	Recalculates the display file in order to show changes.	N/A		View, Regen	REGEN, RE
Regenerate All	Regenerates all viewports at once.	N/A		View, Regen All	REGENALL, REA

Command or Icon Name	Description	Toolbar	Icon	Menu Bar	Command or Alias
Region	Creates a two-dimensional area object from a selection set of existing objects.	Draw		Draw, Region	REGION, REG
Reinit	Reinitializes the digitizer, input/output port and program parameters.	N/A		N/A	REINIT
Rename	Changes the names of objects.	N/A		Format, Rename	RENAME, DDRENAME, REN, -REN
Render	Produces a realistically shaded 2D view of a 3D model.	Render		View, Render, Render	RENDER, RR
Render File Options	Sets the render to file options for rendering.	N/A		N/A	RFILEOPT
Render Preferences	Lets you control how images are rendered and how you use the Render commands.	Render		View, Render, Preferences	RPREF, RPR
Render Screen	Renders to a full-screen display only.	N/A		N/A	RENDSCR
Replay	Displays a BMP, TGA, or TIFF image	N/A		Tools, Display Image, View	REPLAY
Restore UCS	Restores a saved user coordinate system.	N/A		Tools, UCS, Restore	UCS
Resume	Continues an interrupted script.	N/A		N/A	'RESUME, RESUME
Revolve	Creates a three-dimensional object by sweeping a two-dimensional polyline, circle or region about a circular path to create a symmetrical solid.	Solids		Draw, Solids, Revolve	REVOLVE, REV
Revolved Surface	Creates a rotated surface about a selected axis.	Surfaces		Draw, Surfaces, Revolved Surface	REVSURF

Command or Icon Name	Description	Toolbar	Icon	Menu Bar	Command or Alias
Right View	Sets the viewing direction for a three-dimensional visualization of the drawing to the right side.	Standard (Viewpoint flyout) or Viewpoint		View, 3D Viewpoint, Right	VPOINT, DDVPOINT, VP, -VP
Rotate	Rotates all or part of an object.	Modify		Modify, Rotate	ROTATE, RO
Rows	Displays, edits, creates links and creates selection sets of table data.	External Database		Tools, External Database, Rows	ASEROWS, ARO
Rscript	Creates a continually repeating script.	N/A		N/A	RSCRIPT
Ruled Surface	Creates a polygon mesh between two curves.	Surfaces		Draw, Surfaces, Ruled Surface	RULESURF
Save	Saves a drawing using Quick Save.	Standard		File, Save	QSAVE
SAVE	Saves a drawing to a new file name, without changing the name of the drawing on your screen.	N/A		N/A	SAVE
Save As	Saves a drawing, allowing you to change its name or location.	N/A		File, Save As	SAVEAS
Save As R12/LT 2	Saves the current drawing in AutoCAD Release 12 and LT 2 format.	N/A		File, Save As	SAVEASR12
Save As R13/LT 95	Saves the current drawing in AutoCAD Release 13 and LT 95 format.	N/A		File, Save As	
Save Image	Saves a rendered image to a file.	N/A		Tools, Display Image, Save	SAVEIMG
Save Slide	Creates a slide file of the current viewport.	N/A		N/A	MSLIDE
Save Time	System variable that controls how often a file is automatically saved.	N/A		N/A	SAVETIME

Command or Icon Name	Description	Toolbar	Icon	Menu Bar	Command or Alias
Save UCS	Saves a user coordinate system.	N/A		Tools, UCS, Save	UCS
Scale	Changes the size of an object in the drawing database.	Modify		Modify, Scale	SCALE, SC
Scenes	Saves and recalls named scenes, making it easier to quickly change to a different viewpoint and lighting.	Render		View, Render, Scene	SCENE
Script	Executes a sequence of commands from a script.	N/A		Tools, Run Script	SCRIPT, SCR
SE Isometric View	Sets the viewing direction for a three-dimensional visualization of the drawing.	Standard (Viewpoint flyout) or Viewpoint		View, 3D Viewpoint, SE Isometric	VPOINT, DDVPOINT, VP, -VP
Section	Uses the intersection of a plane and solids to create a region.	Solids		Draw, Solids, Section	SECTION, SEC
Select	Selects objects with a selection method listed below. Commands must be typed following the Select command.	N/A		N/A	'SELECT, SELECT
Select Add	Places selected objects in the previous selection set (default method).	N/A		N/A	ADD (at Select Objects: prompt)
Select All	Places all objects in the selection set.	N/A		N/A	ALL (at Select Objects: prompt)
Select Auto	Selects objects individu-ally, objects within a left-to-right window, and objects crossing and within a right-to-left window (default selection method).	N/A		N/A	AUTO (at Select Objects: prompt)
Select Box	Selects objects within a left-to-right window, and/or objects crossing and within a right-to-left window (default selection method).	N/A		N/A	BOX (at Select Objects: prompt)

Command or Icon Name	Description	Toolbar	Icon	Menu Bar	Command or Alias
Select Crossing	Selects all objects within and crossing the selection box (default selection from right to left).	N/A		N/A	CROSSING (at Select Objects: prompt)
Select Crossing Polygon	Selects all objects within and crossing a user-defined polygon boundary.	N/A		N/A	CPOLYGON (at Select Objects: prompt)
Select Fence	Selects objects crossing a user-defined polyline.	N/A		N/A	FENCE (at Select Objects: prompt)
Select Group	Selects all objects within a specified group.	N/A		N/A	GROUP (at Select Objects: prompt)
Select Last	Re-highlights last selection.	N/A		N/A	LAST (at Select Objects: prompt)
Select Multiple	Selects objects without highlighting them.	N/A		N/A	MULTIPLE (at Select Objects:)
Select Previous	Places previously selected objects in the selection set.	N/A		N/A	PREVIOUS (at Select Objects: prompt)
Select Remove	Removes selected objects in the previous selection set.	N/A		N/A	REMOVE (at Select Objects: prompt)
Select Single	Selects a single.	N/A		N/A	SINGLE (at Select Objects:)
Select Undo	Cancels the selection of the most recently selected object.	N/A		N/A	UNDO (at Select Objects: prompt)
Select Window	Selects all objects within a user-defined window.	N/A		N/A	WINDOW (at Select Objects:)
Select Window Polygon	Selects all objects within a user-defined polygon.	N/A		N/A	WPOLYGON (at Select Objects:)
Select Objects	Makes a selection set from rows linked to text sets and graphic sets.	External Database		Tools, External Database, Select Objects	ASESELECT, ASE

Command or Icon Name	Description	Toolbar	Icon	Menu Bar	Command or Alias
Set Variables	Changes the values of system variables.	N/A		Tools, Inquiry, Set Variable	'SETVAR, SETVAR, SET
Setup Drawing	Generates Profiles and sections in viewports created with SOLVIEW.	Solids		Draw, Solids, Setup, Drawing	SOLDRAW
Setup Profile	Creates profile images of 3D solids.	Solids		Draw, Solids, Setup, Profile	SOLPROF
Setup View	Creates floating viewports for orthographic projection, section views, and multi-views of 3D solids.	Solids		Draw, Solids, Setup, View	SOLVIEW
Shade	Creates a shaded view of a model.	Render		View, Shade	SHADE, SHA
Shape	Inserts a shape into the current drawing.	N/A		N/A	SHAPE
Shell	Accesses operating system commands.	N/A		N/A	SHELL
Show Materials	Shows you the material and attachment method for the selected object.	N/A		N/A	SHOWMAT
Single Line Text	Adds text to a drawing.	N/A		Draw, Text, Single Line Text	DTEXT, DT
Sketch	Creates a series of free-hand line segments.	N/A		N/A	SKETCH
Slice	Slices a set of solids with a plane.	Solids		Draw, Solids, Slice	SLICE, SL
Snap	Limits cursor movement on the screen to set interval so objects can be placed at precise locations easily.	Status bar		N/A	SNAP, SN, or toggle off and on with F9
Snap From	Establishes a temporary reference point as a basis for specifying subsequent points.	Standard (Object Snap flyout) or Object Snap		N/A	'OSNAP, OSNAP, 'DDOSNAP, DDOSNAP, OS, -OS, -OSNAP

Command or Icon Name	Description	Toolbar	Icon	Menu Bar	Command or Alias
Snap to Apparent Intersection	Snaps to a real or visual three-dimensional inter-section formed by objects you select or by an exten-sion of those objects.	Standard (Object Snap flyout) or Object Snap		N/A	'APP, APP, EAPPINT, APPINT
Snap to Center	Snaps to the center of an arc, circle, or ellipse.	Standard (Object Snap flyout) or Object Snap		N/A	'CEN, CEN
Snap to Endpoint	Snaps to the closest end-point of an arc, elliptical arc, ray, multiline, or line, or to the closest corner of a trace, solid, or three-dimensional face.	Standard (Object Snap flyout) or Object Snap		N/A	'ENDP, ENDP, END
Snap to Insertion	Snaps to the insertion point of text, a block, a shape, or an attribute.	Standard (Object Snap flyout) or Object Snap		N/A	'INS, INS
Snap to Intersection	Snaps to the intersection of a line, arc, spline, ellip-tical arc, ellipse, ray, con-struction line, multiline, or circle.	Standard (Object Snap flyout) or Object Snap		N/A	'INT, INT
Snap to Midpoint	Snaps to the midpoint of an arc, elliptical arc spline, ray, solid, construction line, multiline, or line.	Standard (Object Snap flyout) or Object Snap		N/A	'MID, MID
Snap to Nearest	Snaps to the nearest point of an arc, elliptical arc, ellipse, spline, ray, multi-line, line, circle, or point.	Standard (Object Snap flyout) or Object Snap		N/A	'NEA, NEA
Snap to Node	Snaps to a point object.	Standard (Object Snap flyout) or Object Snap		N/A	'NOD, NOD
Snap to None	Turns object snap mode off.	Standard (Object Snap flyout) or Object Snap		N/A	'NONE, NONE

Command or Icon Name	Description	Toolbar	Icon	Menu Bar	Command or Alias
Snap to Perpendicular	Snaps to a point perpendicular to an arc, elliptical arc, ellipse, spline, ray, construction line, multiline, line, solid, or circle.	Standard (Object Snap flyout) or Object Snap		N/A	'PER, PER
Snap to Quadrant	Snaps to a quadrant point of an arc, elliptical arc, ellipse, solid, or circle.	Standard (Object Snap flyout) or Object Snap		N/A	'QUA, QUA
Snap to Quick	Snaps to the first object snap point found.	Standard (Object Snap flyout) or Object Snap		N/A	'QUI, QUI
Snap to Tangent	Snaps to the tangent of an arc, elliptical arc, ellipse, or circle.	Standard (Object Snap flyout) or Object Snap		N/A	'TAN, TAN
Spelling	Checks spelling in a drawing.	Standard		Tools, Spelling	'SPELL, SPELL, SP
Sphere	Creates a three-dimensional solid sphere.	Solids		Draw, Solids, Sphere	SPHERE
Spline	Fits a smooth curve to a sequence of points within a specified tolerance.	Draw		Draw, Spline	SPLINE, SPL
SQL Editor	Executes Structured Query Language (SQL) statements.	External Database		Tools, External Database, SQL Editor	ASESQLED, ASQ
Statistics	Displays information about your last rendering.	Render		View, Render, Statistics	STATS
Status	Displays drawing statistics, modes, and extents.	N/A		N/A	'STATUS, STATUS
STL Out	Stores solids in an ASCII or binary file.	N/A		N/A	STLOUT
Stretch	Stretches the objects selected.	Modify		Modify, Stretch	STRETCH, S

Command or Icon Name	Description	Toolbar	Icon	Menu Bar	Command or Alias
Style	Lets you create new text styles and modify existing ones.	N/A		Format, Text Style	'STYLE, STYLE, ST
Subtract	Removes the set of a second solid or region from the first set.	Modify II		Modify, Boolean, Subtract	SUBTRACT, SU
Surftab1	System variable that sets the number of tabulations to be generated for Rulesurf and Tabsurf.	N/A		N/A	SURFTAB1
Surftab2	System variable that sets mesh density in the N direction for Revsur and Edgesurf.	N/A		N/A	SURFTAB2
SW Isometric View	Sets the viewing direction for a three-dimensional visualization of the drawing.	Standard (Viewpoint flyout) or Viewpoint		View, 3D Viewpoint, SW Isometric	VPOINT, DDVPOINT, VP, -VP
Syswindows	Arranges windows.	N/A		N/A	SYSWINDOWS
Tablet	Calibrates the tablet with a paper drawing's coordinate system.	N/A		Tools, Tablet	TABLET, TA
Tabmode	System variable that controls the use of tablet mode.	N/A		N/A	TABMODE
Tabulated Surface	Creates a tabulated surface from a path curve and direction vector.	Surfaces		Draw, Surfaces, Tabulated	TABSURF
Target	System variable that stores location (in UCS coordinates) of the target point for the current viewport (read only).	N/A		N/A	TARGET
Text	Creates a single line of user-entered text.	N/A		N/A	TEXT

Command or Icon Name	Description	Toolbar	Icon	Menu Bar	Command or Alias
Text Screen	Switches from the graphics screen to the text screen.	N/A		View, Display, Text Window	'TEXTSCR, TEXTSCR, or toggle back and forth with [F2]
Thickness	Sets the current 3D solid thickness.	N/A		N/A	THICKNESS, TH
Tiled Viewports	Divides the graphics window into multiple tiled viewports and manages them.	N/A		View, Tiled Viewports	VPORTS
Tilemode	When set to 1, allows creation of tiled viewports in model space; when set to 0, enables use of paper space.	Status bar		View, Model Space (Tiled)	TILEMODE, TI
Time	Displays the date and time statistics of a drawing.	N/A		Tools, Inquiry, Time	'TIME, TIME
Tolerance	Creates geometric tolerances and adds them to a drawing in feature control frames.	Dimension		Dimension, Tolerance	TOLERANCE, TOL
Toolbar	Displays, hides, and positions all toolbars.	N/A		View, Toolbars	TOOLBAR, TO
Tooltips	System variable that controls the display of ToolTips.	N/A		N/A	TOOLTIPS
Top View	Sets the viewing direction for a three-dimensional visualization of the drawing to the top.	Standard (Viewpoint flyout) or Viewpoint		View, 3D Viewpoint, Top	VPOINT, DDVPOINT, VP, -VP
Torus	Creates a donut-shaped solid.	Solids		Draw, Solids, Torus	TORUS, TOR
Trace	Creates solid lines.	N/A		N/A	TRACE
Treestat	Displays information on the drawing's current spatial index.	N/A		N/A	TREESTAT

Command or Icon Name	Description	Toolbar	Icon	Menu Bar	Command or Alias
Trim	Removes part of an object at its intersection with another object.	Modify		Modify, Trim	TRIM, TR
UCS	Manages User Coordinate Systems.	Standard (UCS Flyout) or UCS		Tools, UCS	UCS, DDUCS, UC
UCS Follow	System variable that allows you to generate a plan view of the current UCS in a viewport whenever you change the UCS.	N/A		N/A	UCSFOLLOW
UCS Icon	System variable that repositions and turns on and off the display of the UCS icon.	N/A		N/A	UCSICON
Undefine	Allows an application-defined command to override an AutoCAD command.	N/A		N/A	UNDEFINE
Undo	Reverses the effect of previous commands.	Standard		Edit, Undo	UNDO, U
Union	Adds two separate sets of solids or regions together to create one solid model.	Modify II		Modify, Boolean, Union	UNION, UNI
Units	Controls the type and precision of the values used in dimensions.	N/A		Format, Units	'UNITS, UNITS, -UN 'DDUNITS, DDUNITS, UN
View	Saves views of an object to be plotted, displayed, or printed.	N/A		N/A	'VIEW, VIEW, 'DDVIEW, DDVIEW, -V
View Resolution	Sets the resolution for object generation in the current viewport	N/A		N/A	VIEWRES
View Slide	Displays a raster-image slide file in the current viewport.	N/A		N/A	VSLIDE
View UCS	Selects a user coordinate system aligned to a specific view.	Standard (UCS flyout) or UCS		Tools, UCS, View	UCS

Command or Icon Name	Description	Toolbar	Icon	Menu Bar	Command or Alias
Viewport Layer	Controls the visibility of layers within specific viewports.	N/A		N/A	VPLAYER
Wedge	Creates a three-dimensional solid with a closed face tapering along the X axis by specifying the center.	Solids		Draw, Solids, Wedge	WEDGE, WE
WMF In	Imports a file saved in Windows Metafile (*.wmf) format.	N/A		Insert, Windows Metafile	WMFIN
WMF Options	Sets options for importing a Windows Metafile.	N/A		N/A	WMFOPTS
WMF Out	Saves drawing objects in Windows Metafile (*.wmf) format.	N/A		File, Save As, (set file as type *.wmf)	WMFOUT
World UCS	Selects the world coordinate system as the user coordinate system.	Standard (UCS flyout) or UCS		Tools, UCS, World	UCS
Write Block	Saves a block as a separate file so that it may be used in other drawings.	N/A		N/A	WBLOCK, W
X Axis Rotate UCS	Selects a user coordinate system by rotating the X axis.	Standard (UCS flyout) or UCS		Tools, UCS, X Axis Rotate	UCS
Xplode	Breaks a component object into its component objects; see also Explode.	N/A		N/A	XPLODE
Y Axis Rotate UCS	Selects a user coordinate system by rotating the Y axis.	Standard (UCS flyout) or UCS		Tools, UCS, Y Axis Rotate	UCS
Z Axis Rotate UCS	Selects a user coordinate system by rotating the Z axis.	Standard (UCS flyout) or UCS		Tools, UCS, Z Axis Rotate	UCS
Z Axis Vector UCS	Selects a user coordinate system defined by a vector in the Z direction.	Standard (UCS flyout) or UCS		Tools, UCS, Z Axis Vector	UCS

Command or Icon Name	Description	Toolbar	Icon	Menu Bar	Command or Alias
Zoom All	Resizes to display the entire drawing in the current viewport.	Standard (Zoom flyout) or Zoom		View, Zoom, All	'ZOOM, ZOOM, 'Z, Z
Zoom Center	Resizes to display a window by entering a center point, then a magnification value or height.	Standard (Zoom flyout) or Zoom		View, Zoom, Center	'ZOOM, ZOOM, 'Z, Z
Zoom Dynamic	Resizes to display the generated portion of the drawing with a view box.	Standard (Zoom flyout) or Zoom		View, Zoom, Dynamic	'ZOOM, ZOOM, 'Z, Z
Zoom Extents	Resizes to display the drawing extents.	Standard (Zoom flyout) or Zoom		View, Zoom, Extents	'ZOOM, ZOOM, 'Z, Z
Zoom In	Resizes areas of the drawing in the screen to make it appear closer or larger.	Standard (Zoom flyout) or Zoom		View, Zoom, In	'ZOOM, ZOOM, 'Z, Z
Zoom Out	Resizes areas of the drawing in the screen so that more of the drawing fits in the screen.	Standard (Zoom flyout) or Zoom		View, Zoom, Out	'ZOOM, ZOOM, 'Z, Z
Zoom Previous	Resizes to the previous view.	Standard		View, Zoom, Previous	'ZOOM, ZOOM, 'Z, Z
Zoom Realtime	Resizes the view in a real-time fashion.	Standard		View, Zoom, Realtime	'ZOOM, ZOOM, 'Z, Z
Zoom Scale	Resizes to display at a specified scale factor.	Standard (Zoom flyout) or Zoom		View, Zoom, Scale	'ZOOM, ZOOM, 'Z, Z
Zoom Vmax	Resizes so that the drawing is viewed from as far out as possible on the current viewports virtual screen without forcing a complete regeneration of the drawing.	N/A		N/A	'ZOOM, ZOOM, 'Z, Z
Zoom Window	Resizes to display an area specified by two opposite corner points of a rectangular window.	Standard (Zoom flyout) or Zoom		View, Zoom, Window	'ZOOM, ZOOM, 'Z, Z

absolute coordinates The exact location of a specific point in terms of x, y, and z from the fixed point of origin.

absolute value The numerical value or magnitude of the quantity, without regard to its positive or negative sign.

alias A short name that can be used to activate a command; you can customize command aliases by editing the file *acad.pgp*.

ambient The overall amount of light that exists in the environment of a rendered scene.

And gate An electronic logic symbol; if either input is zero, the output is zero.

angle brackets A value that appears in angle brackets < > is the default option for that command, which will be executed unless it is changed.

aperture A type of cursor resembling a small box placed on top of the crosshairs; used to select in the object snap mode.

architectural units Drawings made with these units are drawn in feet and fractional inches.

aspect ratio The relationship of two dimensions to each other.

associative dimensioning Dimensioning where each dimension is inserted as a group of drawing objects relative to the points selected in the drawing. If the drawing is scaled or stretched, the dimension values automatically update.

associative hatching The practice of filling an area with a pattern which automatically updates when the boundary is modified.

attribute Text information associated with a block.

attribute prompt The prompt, which you define, that appears in the command area when you insert the block into a drawing.

attribute tag A variable name that is replaced with the value that you type when prompted as you insert the attribute block.

AutoSnap A feature which displays a marker and description to indicate which object snap location will be selected.

auxiliary view An orthographic view of an object using a direction of sight other than one of the six basic views (front, top, right-side, rear, bottom, left-side).

ball tag A circled number identifying each part shown in an assembly drawing; also called a balloon number.

base feature In AutoCAD Designer, the main feature of a drawing, from which other features are defined based on the constraints put on the model.

base grip The selected grip, used as the base point for hot grip commands.

baseline dimensioning A dimensioning method in which each successive dimension is measured from one extension line or baseline.

basic view One of the six standard views of an object: front, top, right side, rear, bottom, or left side.

bearing The angle to turn from the first direction stated toward the second direction stated.

bicubic surface A sculptured plane, mathematically describing a sculptured surface between three-dimensional curves.

bidirectional associativity Describes the link between the drawing and the model in AutoCAD Designer. If a change is made to the model, the drawing is automatically updated to reflect the change; if a change is made to the parametric dimensions in the drawing, both the drawing and the model update automatically.

bilateral tolerances Tolerances specified by defining a nominal dimension and the allowable range of deviation from that dimension, both plus and minus.

blipmarks Little crosses that appear on the screen, indicating where a location was selected.

block A set of objects that have been grouped together to act as one, and can be saved and used in the current drawing and in other drawings.

block name Identifies a particular named group of objects.

block reference A particular insertion of a block into a drawing (blocks can be inserted more than once).

Boolean operators Find the union (addition), difference (subtraction), and intersection (common area) of two or more sets.

Buffer An electronic logic symbol.

bump map Converts color intensity or grayscale information to heights to give the appearance that features are raised above the surface, like embossed letters.

buttons A method of selecting options by picking in a defined area of the screen resembling a box or push button.

Cartesian coordinate system A rectangular coordinate system created by three mutually perpendicular coordinate axes, commonly labeled x, y, and z.

chained (continued) dimensioning A dimensioning method in which each successive dimension is measured from the last extension line of the previous dimension.

chamfer A straight line segment connecting two otherwise intersecting surfaces.

chord length The straight line distance between the start point and the endpoint of an arc.

circular view A view of a cylinder in which it appears as a circle (looking into the hole).

circumscribed Drawn around the outside of a base circle.

clearance fit The space available between two mating parts, where the greatest shaft size always is smaller than the smallest hole size, thus producing an open space.

colinear constraint Lying on the same straight line.

command aliasing The creation and use of alternative short names for commands, such as LA for Layer.

command prompt The word or words in the command window that ask for the next piece of information.

command window The lines of text below the graphics window that indicate the status of commands and prompt for user input.

construction plane A plane that is temporarily defined during commands for creating new drawing objects.

context sensitive Recognizes when you are in a command, and displays information for that command.

Coons patch A bicubic surface interpolated between four edges.

coordinate system locator An icon to help you visually refer to the current coordinate system.

coordinate values Used to identify the location of a point, using the Cartesian coordinate system; x represents the horizontal position on the X axis, and y represents the vertical position on the Y axis. In a three-dimensional drawing, z represents the depth position on the Z axis.

crosshatching The practice of filling an area with a pattern to differentiate it from other components of a drawing.

current layer The layer you are working on. New drawing objects are always created in the layer that is current.

cursor (crosshairs) A mark that shows the location of the pointing device in the graphics window of the screen; used to draw, select, or pick icons, menu items, or objects. The appearance of the cursor may change, depending on the command or option selected.

custom hatch pattern A design that you have previously created and stored in the file *acad.pat* or another *.pat* file of your own making to use to fill an area.

customize To change the toolbars, menus, and other aspects of the program to show those commands and functions that you want to use.

cutting edges Objects used to define the portions to be removed when trimming an object.

cutting plane line Defines the location on the object where the sectional view is taken.

datum surface A theoretically exact geometric reference used to establish the tolerance zone for a feature.

default The value that AutoCAD will use unless you specify otherwise; appears in angle brackets < > after a prompt.

default directory The DOS directory to which AutoCAD will save all drawing files unless instructed otherwise.

delta angle The included angular value from the start point to the endpoint.

diameter symbol ⌀ Indicates that the value is a diameter.

difference The area formed by subtracting one region or solid from another.

dimension line Drawn between extension lines with an arrowhead at each end; indicates how the stated dimension relates to the feature on the object.

dimension style A group of dimension features saved as a set.

dimension value The value of the dimension being described (how long, how far across, etc.); placed near the midpoint of the dimension line.

dimension variables Features of dimensions that can be altered by the user; you control the features by setting the variables using the Dimension Style dialog box.

dimensions Describe the sizes and locations of a part or object so that it can be manufactured.

distance across the flats A measurement of the size of a hexagon from one flat side to the side opposite it.

docked Refers to the toolbar's ability to be attached to any edge of the graphics window.

draft angle The taper on a molded part that makes it possible to easily remove the part from the mold.

drag To move an object on the screen and see it at the same time, in order to specify the new size or location.

edge view A line representing a plane surface shown on its end.

elements   Multilines comprising up to 16 lines.

engineering units   Drawings made with these units are drawn in feet and decimal inches.

export   To save a file from one application as a different file type for use by another application.

extension line   Relates a dimension to the feature it refers to.

extension line offset   Specifies a distance for the gap between the end of the extension line and the point that defines the dimension.

extrusion   Creates a long three-dimensional strip with the shape of a closed two-dimensional shape, as if material had been forced through a shaped opening.

falloff   The angle of the cone of light from a spotlight.

feature   Any definable aspect of an object—a hole, a surface, etc.

file extension   The part of a file name that is composed of a period, followed by one to three characters, and that helps you to identify the file.

fillet   An arc of specified radius that connects two lines, arcs, or circles, or a rounded interior corner on a machined part.

floating   Refers to the toolbar's or command window's ability to be moved to any location on the screen.

floating viewports   A window, created in paper space, through which you can see your model space drawing. They are very useful for plotting the drawing, adding drawing details, showing an enlarged view of the object, or showing multiple views of the object.

flyout   A sub-toolbar that becomes visible when its representative icon on the main toolbar is chosen.

font   A character pattern defined by the text style.

foreshortened   Appears smaller than actual size, due to being tipped away from the viewing plane.

fractional units   Drawings made with these units express lengths less than 1 as fractions (e.g., 15 1/4).

freeze   The practice of making a layer invisible and excluding it from regeneration and plotting.

generated layer   A layer created automatically (instead of by the user) during commands such as SOLPROF.

global linetype scaling factor   The factor of the original size at which all lines in the drawing are displayed.

globe   Used to select the direction for viewing a three-dimensional model.

graphics window   The central part of the screen, which is used to create and display drawings.

grid   Regularly spaced dots covering an area of the graphics window to aid in drawing.

group   A named selection set of objects.

handle number   A unique identifier assigned to each AutoCAD drawing object, including viewports.

heads-up   You can enter the command while looking at the screen; by picking an icon, for example, you do not need to look down to the keyboard to enter the command.

hidden line   Represents an edge that is not directly visible because it is behind or beneath another surface.

highlight   A change of color around a particular command or object, indicating that it has been selected and is ready to be executed or worked on in some way.

hot grips A method for editing an already drawn object using only the pointing device, without needing to use the menus, toolbars, or keyboard.

hotspot The angle of the cone of light for the brightest area of the beam of a spotlight.

image tile An active picture which displays dialog box choices as graphic images rather than words.

implied Crossing mode When this mode is activated, the first corner of a box is started when you select a point on the screen that is not on an object in the drawing. If the box is drawn from right to left, everything that partially crosses as well as items fully enclosed by the box are selected.

implied Windowing mode When this mode is activated, the first corner of a box is started when you select a point on the screen that is not on an object in the drawing. If the box is drawn from the left to right, a window is formed that selects everything that is entirely enclosed in the box.

import To open and use a file created by a different application than the one being used.

inclined surface Slanted at an angle, a surface that is perpendicular to one of the three principal views and tipped away from the other principal views.

included angle The angular measurement along a circular path.

inscribed Drawn inside a base circle.

inspection The act of examining the part against its specifications.

instance Each single action of a command.

intensity The brightness level of a light source.

intersection The point where two lines or surfaces meet, or the area shared by overlapping regions or solid models.

Inverter gate A symbol used to draw an electronic logic circuit; for inputs of 1, the output is 0, for inputs of 0, the output is 1.

layer A method of separating drawing objects so that they can be viewed individually or stacked like transparent acetates, allowing all layers to show. Used to set color and linetype properties for groups of objects.

leader line A line from a note or radial dimension that ends in an arrowhead pointing at the feature.

limiting element The outer edge of a curved surface.

limiting tolerances Tolerances specified by defining an upper and lower allowable measurement.

lock To prohibit changes to a layer, although the layer is still visible on the screen.

major arc An arc which comprises more than $180°$ of a circle.

major axis The long axis of symmetry across an ellipse.

manufacture To create a part according to its specifications.

mass properties Data about the real-world object being drawn, such as its mass, volume, and moments of inertia.

matrix An array of vertices used to generate a surface.

menu bar The strip across the top of the screen showing the names of the pull-down menus, such as File, Edit, etc.

minor axis The short axis of symmetry across an ellipse.

mirror line A line that defines the angle and distance at which a reversed image of a selected object will be created.

miter line A line drawn above the side view and to the right of the top view, often drawn from the top right corner of the front view, used to project features from the side view onto the top view of an object.

model space The AutoCAD drawing database, when an object exists as a three-dimensional object.

Nand gate A symbol used to draw an electronic logic circuit; when both inputs are 1, output is 0, when any input is 0, output is 1.

Nor gate A symbol used to draw an electronic logic circuit; when both inputs are 0, output is 1, any input of 1, output is 0.

normal surface A surface that is perpendicular to two of the three principal orthographic views and appears at the correct size and shape in a basic view.

noun/verb selection A method for selecting an object first and then the command to be used on it.

nurb A cubic spline curve.

Object Properties toolbar Contains the icons whose commands control the appearance of the objects in your drawing.

offset distance Controls the distance from an existing object at which a new object will be created.

options The choices associated with a particular command or instruction.

Or gate A symbol used to draw an electronic logic circuit; any input of 1 results in output of 1, when both inputs are 0, output is 0.

orthographic view A two-dimensional drawing used in representing a three-dimensional object. A minimum of two orthographic views are necessary to show the shape of a three-dimensional object, because each view shows only two dimensions.

output file The file to which attribute data is extracted.

overall dimensions The widest measurements of a part; needed by manufacturers to determine how much material to start with.

override mode When the desired object snap mode is selected during each command it is to be used for.

paper space A mode that allows you to arrange, annotate, and plot different views of a model in a single drawing, as you would arrange views on a piece of paper.

parametric A modeling method that allows you to input sizes for and relationships between the features of parts. Changing a size or relationship will result in the model being updated to the new appearance.

partial auxiliary view An auxiliary view that shows only the desired surfaces.

parts list Provides information about the parts in an assembly drawing, including the item number, description, quantity, material, and part number.

perspective icon A cube drawn in perspective that replaces the UCS icon when perspective viewing, activated by the 3D Dynamic View command, is in effect.

pictorial full section A section view showing an object cut in half along its vertical center line.

pin registry A process in which a series of transparent sheets are punched with a special hole pattern along one edge, allowing the sheets to be fitted onto a metal pin bar. The metal pin bar keeps the drawings aligned from one sheet to the next. Each sheet is used to show different map information.

plan view The top view or view looking straight down the Z axis toward the XY plane.

plus/minus tolerances Another name for variance tolerances.

point filter Option that lets you selectively find single coordinates out of a point's three-dimensional coordinate set.

point light A light source similar to a bare light bulb that casts light in every direction.

pointer An indicator established from the current drawing to the original drawing, or external reference.

polar array A pattern created by repeating objects over and over in a circular fashion.

polar coordinates In AutoCAD, the location of a point, defined as a distance and an angle from another specified point, using the input format @DISTANCE<ANGLE

polygon A closed shape with sides of equal length.

polyline A series of connected objects (lines or arcs) that are treated as a single object.

precedence When two lines occupy the same space, precedence determines which one is drawn. Continuous lines take precedence over hidden lines, and hidden lines take precedence over center lines.

primitives Basic shapes that can be joined together or subtracted from each other to form more complex shapes.

profile A two-dimensional view of a three-dimensional solid object created using the Solprof (Setup Profile) command.

project To transfer information from one view of an object to another by aligning them and using projection lines.

projection lines Horizontal lines that stretch from one view of an object to another to show which line or surface is which.

prototype drawing A drawing saved with certain settings that can be used repeatedly as the basis for starting new drawings.

quadrant point Any of the four points that mark the division of a circle or arc into four segments: 0°, 90°, 180°, and 270°.

ray tracing The method of calculating shading by tracking reflected light rays hitting the object. Both AutoVision and 3D Studio use the Phong algorithm to perform ray tracing to produce realistic shading.

real-world units Units in which drawings in the database are drawn; objects should always be drawn at the size that the real object would be.

rectangular array A pattern created by repeating objects over and over in columns and rows.

rectangular view A view of a cylinder in which it appears as a rectangle (looking from the side).

reference surface A surface from which measurements are made when creating another view of the object.

reflection The degree to which a surface bounces back light.

refraction The degree to which an object changes the angle of light passing through it.

regenerate To recalculate from the drawing database a drawing display that has just been zoomed or had changes made to it.

region A two-dimensional area created from a closed shape or a loop.

relative coordinates The location of a point in terms of the X and Y distance, or distance and angle, from a previously specified point @X,Y.

rendered Lighting effects, color, and shape applied to an object in a single two-dimensional view.

resolution Refers to how sharp and clear an image appears to be and how much detail can be seen; the higher the resolution, the better the quality of the image. This is determined by the number of colors and pixels the computer monitor can display.

revolution Creating a three-dimensional object by sweeping a two-dimensional polyline, circle, or region about a circular path to create a symmetrical solid that is basically circular in cross-section.

romans A font.

root page The original standard screen menu column along the right side of the screen, accessed by picking the word AutoCAD at the top of the screen.

roughness The apparent texture of a surface. The lower the surface roughness, the more reflective the surface will be.

round A convex arc, or a rounded external corner on a machined part.

running mode When the current object snap mode is automatically used any time a command calls for the input of a point or selection.

scale factor The multiple or fraction of the original size of a drawing object or linetype at which it is displayed.

section lines Show surfaces that were cut in a section view.

section view A type of orthographic view that shows the internal structure of an object; also called a cutaway view.

selectable group A named selection set in which picking on one of its members selects all members that are in the current space, and not on locked or frozen layers. An object can belong to more than one group.

selection filters A list of properties required of an object for it to be selected.

selection set All of the objects chosen to be affected by a particular command.

session Each use, from loading to exiting, of AutoCAD.

shape-compiled font Character shapes drawn in AutoCAD using vectors.

sign Positive or negative indicator that precedes a centermark value.

slicing plane A plane that is defined for use in slicing a solid object. During the Slice command, the object can be removed on either side of the slicing plane or the selected object can be parted along the slicing plane and both sides can be retained.

solid modeling A type of three-dimensional modeling that represents the volume of an object, not just its lines and surfaces; this allows for analysis of the object's mass properties.

spotlight A light source used to highlight certain key areas and cast light only toward the selected target.

Standard toolbar Contains the icons which control file operation commands.

status bar The rectangular area at bottom of screen that displays time and some command and file operation information.

style Controls the name, font file, height, width factor, obliquing angle, and generation of the text.

submenu A list of available options or commands relating to an item chosen from a menu.

surface modeling A type of three-dimensional modeling that defines only surfaces and edges so that the model can be shaded and have hidden lines removed; resembles an empty shell in the shape of the object.

swatch An area that shows a sample of the color or pattern currently selected.

sweeping Extrusion along a non-linear path rather than a straight line.

target area The square area on an object snap aperture, in which at least part of the object to be selected must fit.

template drawing A drawing with the extension *.dwt* having certain settings that can be used repeatedly as the basis for starting new drawings. Any drawing can be renamed and used as a template.

template file Specifies the file format for the Extract Attribute command as *.cdf* of *.sdf*.

tessellation lines To cover a surface with a grid or lines, like a mosaic. Tessellation lines are displayed on a curved surface to help you visualize it; the number of tessellation lines determines the accuracy of surface area calculations.

threshold value Determines how close vertices must be in order to be automatically welded during the rendering process.

through point The point through which the offset object is to be drawn.

tiled viewports If more than one viewport is created, they cannot overlap. They must line up next to each other, like tiles on a floor.

tilemode An AutoCAD system variable that controls the use of tiled or floating viewports or paper space.

title bar The rectangular area at the top of the screen which displays the application and file names.

toggle A switch that turns certain settings on and off.

tolerance The amount that a manufactured part can vary from the specifications laid out in the plans and still be acceptable.

tool tip The name of the icon that appears when you hold the cursor over the icon.

toolbar A strip containing icons for certain commands.

transparency A quality that makes objects appear clear, like glass, or somewhat clear, like colored liquids.

transparent command A command that can be selected while another command is operating.

typing cursor A special cursor used in dialog boxes for entering text from the keyboard.

UCS icon Indicates the current coordinate system in use and the direction in which the coordinates are being viewed in 3D drawings.

unidirectional dimensioning The standard alignment for dimension text, which is to orient the text horizontally.

union The area formed by adding two regions or solid models together.

units AutoCAD can draw figures using metric, decimal, scientific, engineering, or architectural measurement scales. These are set with the Units command.

update To regenerate the active part or drawing using any new dimension values or changed sketches.

User Coordinate System (UCS) A set of x, y, and z coordinates whose origin, rotation, and tilt are defined by the user. You can create and name any number of User Coordinate Systems.

user-defined hatch pattern A predefined hatch pattern for which you have specified the angle, spacing, and/or linetype.

variance tolerances Tolerances specified by defining a nominal dimension and the allowable range of deviation from that dimension; includes bilateral and unilateral tolerances.

vector A directional line. Also, a way of storing a graphic image as a set of mathematical formulas.

Venn diagram Pictorial description named after John Venn, an English logician, where circles are used to represent set operations.

viewpoint The direction from which you are viewing a three-dimensional object.

viewport A "window" showing a particular view of a three-dimensional object.

virtual screen A file containing only the information displayed on the screen. AutoCAD uses this to allow fast zooming without having to regenerate the original drawing file.

Windows close box A box that appears in the upper right corner of a window that closes the window.

wireframe modeling A type of three-dimensional modeling that uses lines, arcs, circles, and other objects to represent the edges and other features of an object; so called because it looks like a sculpture made from wires.

World Coordinate System (WCS) AutoCAD's system for defining three-dimesional model geometry using x, y, and z coordinate values; the default orientation is a horizontal X axis with positive values to the right, a vertical Y axis with positive values above the X axis, and a Z axis that is perpendicular to the screen and has positive values in front of the screen.

World Coordinates The basis for all user coordinate systems.

Chapter 1 and 2 openers, and Tutorial 1, 2, 3, 4, 5 openers courtesy of Autodesk, Inc.

Exercises 1.2, 1.3, 1.6, 1.6, 2.2,2.3,2.5, 3.4, 4.1, 4.2, 4.4, 5.1, 6.4, 6.5, 6.6, 6.7, 7.1, 7.2, 7.3, 7.6, 7.7, 8.1, 8.3, 8.4, 8.8, 9.5, 9.6, 10.3, 10.4, 10.6, 10.7, 11.3,11.4, 11.5, 11.7, 15.3, 15.4, 15.5, 15.6 courtesy of the Spocad Centers of Gonzaga University School of Engineering.

Exercises 1.9, 3.5, 3.6, 8.6, 8.7, 10.2, 12.4, 12.5, 12.6, and 13.6 courtesy of Karen L. Coen-Brown, Engineering Mechanics Department, University of Nebraska-Lincoln.

Exercises 1.10, 2.7, 2.8, 2.9, 2.10, 3.8, 3.9, 3.10, 4.7, 4.8, 4.9, 4.10, 5.2M, 5.3, 5.4, 5.5M, 5.10M, 6.8M, 6.9, 6.10, 7.9, 7.10, 8.9, 8.10, 9.9, 11.8M, 11.9M, 11.10, 12.10, 13.7, 13.8, 13.9, 13.10, 14.5, 14.6, 14.7, 14.8, 15.8, 15.9, 15.10, and Tutorial 7, 13, and 14 openers courtesy of James H. Earle.

Exercise 4.3 courtesy of D. Krall, Norfolk State University.

Exercises 4.6, 7.4, 7.5, 9.1, 9.2, 9.3, 12.7, and 12.9 courtesy of Tom Bryson, University of Missouri-Rolla.

Exercise 5.8 courtesy of Kyle Tage.

Tutorial 6 opener courtesy of Kim manner, Department of mechanical Engineering, University of Wisconsin-Madison.

Tutorial 8 opener courtesy of Torian Roesch.

Tutorial 9 opener courtesy of Craig Bradley.

Exercise 9.4 courtesy of Mary Ann Koen, University of Missouri-Rolla.

Tutorial 10 opener courtesy of Doug Baese.

Exercise 10.1 courtesy of Kevin Berisso.

Tutorial 11 opener courtesy of Wendy Warren.

Tutorial 15 opener courtesy of Randy Harris.

connecting with fillets, 77–79
continuing, 55–56
fillets. *See* Fillet(s)
major, 55
specifying start, center, and chord length,
    54–55
specifying start, center, and endpoints, 53–54
specifying start, center, and included angles,
    54
specifying start, end, and radius, 55
through three points, 52–53
Arc command, 11, 52–56
Architectural units, 25
Area(s), finding, 84
Area card, Quick Setup Wizard, 99
Area command, 84
Array command, 121–122
Arrow(s)
   oblique stroke, 201
   size and style, 202
Arrow option, Leader command, 354
ASSIST item, screen menus, 10
Associative dimensioning, 197, 207
   automatically updating values, 219–222
Associative hatching, 356
   changing hatches, 363–365
Asterisk (*)
   wildcard character, G–31
   Write Block command, 235
At symbol (@), relative coordinates, 19
Attribute(s), 327–330
   attribute insertion point, 329–330
   attribute mode, 328
   attribute prompt, 328
   attribute tag, 328
   attribute text options, 328
   changing attribute values, 332
   default attribute value, 328
   defining attributes, 327
Attach Xref dialog box, 320
Attdia system variable, G–15, 99
Attribute Extraction dialog box, 334–335

AutoCAD
   exiting, G–31, 36
   loading, G–18
Autosave feature, 361
Auto-Smoothing, 440
AutoSnap feature, 50–51
AutoSnap index card, Osnap Settings dialog
   box, 51
Auto-Welding, 440
Auxiliary views, 379–395
   curved surfaces, 385–389
   drawing using 2D, 381–385
   enlarging viewports, 391–392
   partial, 385–386
   Plan command, 392–393
   solid modeling, 389–391
   Solview command, 394–395
   3D dynamic view, 393–394
   3D solid models, 389–391

Back face(s), 423
Back Face Normal is Negative option, Render
   dialog box, 423
Back option, Undo command, 25
Background(s), renderings, 438
Background dialog box, 422, 438
Backing out, of commands, G–24
Backing up
   commands, G–24
   customizable files, G–7–8
Backslash, programming toolbar buttons, 239
Backslash-n, programming toolbar buttons,
   239
Ball tags
   creating blocks, 327
   defining blocks, 330–331
   inserting, 331–332
Base grip, 123
Base point option, Mirror command, 125
Baseline Dimension command, 213, 216
Baseline dimensioning, 213–214
Basic dimension, 404

Extend command, 113–114
Extending lines, 113–114
Extension line(s), 196
Extension Line area, Geometry dialog box, 202
Extension line offset, 196
Extents option, Print/Plot Configuration dialog box, 88
External Reference command, 319–321
External Reference dialog box, 320, 322
Extrusion, creating solid models, 277–278, 279

Facetres variable, 440
Falloff, 427
FATAL ERRORs, 13
Feature control frames, 410–412
File(s). *See also* Drawings; *specific files*
  accessing from Web browsers, G–9
  Comma Delimited, 334
  customizable, backing up, G–7–8
  data, for tutorials, installing, G–9–10
  downloading via FTP, G–9
  Drawing Interchange, 334
  exporting. *See* Exporting
  locating and managing, G–31–32
  menu, editing, 242–243
  naming, G–32, 12–13. *See also* Filename extensions
  opening, 185
  output, 334
  rendering to, 435
  saving, G–31–32, 12–13, 21, 45–46
  Space Delimited, 334
  template, 334
  transferring, 36
File menu, 6, 36
File Output Configuration dialog box, 435
Filedia system variable, G–15, 99
Filename extensions, G–7, G–8, G–10, G–31, 12, 441
Files card, Preferences dialog box, G–11, 9
Fillet(s), 77–79
  solid models, 299

Fillet command, 77–79, 299
Fillet option, Rectangle command, 29
Filters, selection, 115
Find tab, Help window, G–28
Finding. *See* Locating
Fit area, Format dialog box, 203–204
Fit option, Edit Polyline command, 81
Fit options
  Edit Polyline command, 80
  Mview command, 150
Floating toolbars, G–23, 5
Floating viewports, 144, 150–151
Flyouts, G–23, 5. *See also specific flyouts*
  adding to toolbars, 242
  creating, 241–242
Fog dialog box, 422
Font(s)
  default, selecting, 142
  shape, 209–210
  shape-compiled, 33
Font Style option, Style command, 34
fontmap.ps file, G–8
Format dialog box, 203–204, 209
Format menu, 6
Format option, Leader command, 354
Fraction(s), stacked, 35
Fractional units, 25
Freeze/Thaw in Current Viewport command, 310–311
Freezing, layers, 245
Freezing layers, 49
From object snap, 112
FTP, downloading data files for tutorials, G–9

General card, Preferences dialog box, G–11, G–14, 8
Geographic Location dialog box, 437
Geometric characteristic symbol, 411
Geometric tolerance(s), 402
Geometric tolerance characteristics, 410
Geometric Tolerance dialog box, 412
Geometry calculator, 304

Geometry dialog box, G–26, 200–203

Geometry option, Dimension Styles dialog box, 221

Global linetype scaling factor, 102–103
  setting, 183–184

Gourard shading, 421, 422

Graphics cursor, 4

Graphics window, G–10, G–11, G–19, 4

GRID button, toggling, 14

Grid command, 13–14, 18

Grips dialog box, 122

Grouping objects, 246–247

Hatch command, 365

Hatchedit dialog box, 364

Hatching, 350, 351, 356–363
  associative, 356

Help. *See* On-line help

Help button, dialog boxes, G–26

Help dialog box, 15–16

Help menu, 7

Help window, G–28, G–29

Hexagons, 104–105
  size, 104

Hidden lines, 165, 170–172
  duplicate, eliminating, 308–309
  plotting from, 305–311

HIDDEN linetype, 138, 352

Hide command, 371

Hide Lines option, Print/Plot Configuration
    dialog box, 88

Hideplot option, Mview command, 150

Highlighted buttons, 6

Holes, centerlines, 178

Horizontal Justification area, Format dialog
    box, 204

Horizontal option, Construction line command, 167

Hot grips, 122–126
  activating, 122–123
  Mirror command, 125
  Move command, 123–124

noun/verb selection, 125–126
  Rotate command, 124
  Scale command, 124–125
  Stretch command, 123, 124

Hotspots, 427

Icons
  creating, G–8
  renaming, G–8

Image tiles, G–27, 202

Implied Windowing mode, 22

Inclined surfaces. *See* Slanted surfaces

Included angle, drawing arcs, 54

Index tab, Help window, G–28

Infinite lines, 118

Input buttons, G–27

Insert Block command, 148, 233, 236

Insert command, 322

Insert dialog box, 148, 233, 234, 323

Insert menu, 6

Inserting
  ball tags, 331–332
  blocks, 233–234
  drawings as blocks, 322–324
  existing drawings, 148–149
  flyouts in toolbars, 242
  title blocks and text in paper space, 151–152

Insertion object snap, 112

Installing data files for tutorials, G–9–10

Intensity, lighting, 426

Interfere command, 324–325

Interference, checking for, 324–325

Internet Utilities toolbar, G–10

Intersection object snap, 109, 382

Intersection operator, 281–283

Invisible mode, attributes, 328

Island Detection, 358

ISOLINES command, 270–271

Isolines variable, setting, 270–271

Isometric views, exploded, 325–326

Joining arcs to arcs or lines, 55–56

files, 185
Ortho command, 167
Orthographic views. *See* 2D orthographic drawings
OSNAP button, 52
Osnap Settings dialog box, 51
Output devices, determining limits, 143
Output files, 334
Overall dimensions, 196
Overall Scale, Geometry dialog box, 201
Overrides
  dimensioning, 407–408
  object snaps, 105–106

Pan command, 11, 59–61, 150
Pan Realtime command, 280, 360
Paper
  orientation, G–14, 88–89
  size, specifying, 88–89
Paper space, 144–155
  adding title blocks and text, 151–152
  creating viewports, 144, 150–151
  dragging model space drawing in viewport, 150
  enabling, 144–146
  inserting existing drawings, 148–149
  model space compared, 144
  plotting from, 154–155
  plotting solid models from, 276
  scaling, 201
  setting limits, 147
  XP scaling factor, 149–150
Parent styles, saving, 206–207
Partial auxiliary views, 385–386
Parts lists, 333–337
  extracting attribute information, 334–335
  object linking and embedding, 336–337
  word processing programs, 335–336
Pasting, parts list creation, 335–336
.pat filename extension, G–7
Pattern Properties area, Boundary Hatch dialog box, 357

Pattern Type area, Boundary Hatch dialog box, 357
.pcp filename extension, G–7
Percent sign, double (%%), degree symbol, 32
Performance card, Preferences dialog box, G–11, G–12–13, 8
Perpendicular object snap, 109–110
Perspective icon, 261
PHANTOM linetype, 352
Phong shading, 421–422
Photo Raytrace option, Render dialog box, 422
Photo Real option, Render dialog box, 422
Pick Point button, 329
Pick Points method, hatching patterns, 358
Picking, G–4. *See also* Selecting
  commands, G–20, G–21, 7
  menu options, G–20, G–21
  objects, G–20
Picking option, selecting objects, 115
Pictorial full sections, 350
Pin registry drafting, 135
PLAN command, 303
Plan command, 392–393
Pline command, 11
Plot command, 143, 154–155
Plot dialog box, G–27
Plotter(s)
  determining limits, 143
  selecting, 88
Plotter configuration, G–15
Plotting, 311
  batch, 154
  hidden lines, 305–311
  from paper space, 154–155
  specifying areas of drawing, 88
Plus sign (+), programming toolbar buttons, 239
Plus/minus tolerances, 402
Point(s)
  entering, G–20
  locating, 84
Point lights, 425

Point Style dialog box, 153
Pointer(s), 332
Pointer card, Preferences dialog box, G–11, G–14, 9
Pointing, G–4, G–20–21
Polar arrays, 121
Polar coordinates, 19–20
Polygon(s), 103–105
    hexagons, 104–105
    inscribed and circumscribed, 103–104
Polygon command, 103–105
Polygon option, selecting objects, 115
Polyline(s), 79–82, 279
    editing, 80–81
Polyline command, 79–82
Precedence, lines, 172–173
Precision
    dimensions, 205
    tolerance, 405–407
Preferences dialog box, G–11–15, G–27, 7–9
Preset mode, attributes, 328
Previewing drawings, 89–90
Previous option, Zoom command, 58
Primary Units area, Annotation dialog box, 204–205
Primary Units dialog box, 205, 405–406
Primitives, 44, 262
Printer(s)
    determining limits, 143
    selecting, 88
Printer card, Preferences dialog box, G–11, G–14–15, 9
Printing
    paper orientation, G–14, 88–89
    specifying areas of drawing, 88
Print/Plot Configuration dialog box, 87–90, 143, 145
Profiles card, Preferences dialog box, G–11, G–12, G–15, 9
Programmed commands, 239–241
Project option
    Extend command, 113

Trim command, 71–72, 72
Projection lines, 165
Prompts, attribute, 328
Properties
    linetypes, 102
    objects, changing, 76–77
Properties card, Multiline Text Editor dialog box, 35
Properties command, 355–356
Pspace command, 11
Pull-down menus, G–25
Purge command, 155

Quadrant object snap, 107
Quadrant point, 107
Question mark (?)
    Help command icon, 15
    wildcard character, G–31
Quick object snap, 112
Quick Setup Wizard, 98–99
QUIT command, 36

Radio buttons, G–27
Radius Dimension command, 212
Ray casting, 358
Ray command, 174, 382
Realtime option
    Pan command, 59–61
    Zoom command, 58
Real-world units, 27
Rectangle(s), drawing with Box command, 262
Rectangle command, 29–30
Rectangular arrays, 121
Rectangular view, 178
Redo command, 25
Redraw All command, 22, 298
Redraw command, 11, 298
Reference dimensions, 408
Reference lines, 382, 384
Reference surfaces, 380
Reflection option, New Standard Material dialog box, 431

virtual, 60

Screen menus, 9–10

Screen resolution, G–6

Scroll bars, G–29–30

dialog boxes, G–27

SDF (Space Delimited Files), 334

Section command, 369–370

Section views, 350

conventions, 350–351

creating from solid models, 365–366

2D, creating, 350

Select Add option, selecting objects, 115

Select All option, selecting objects, 115

Select Color dialog box, 48, 136–137, 140, 201, 202

Select Crossing option, selecting objects, 115

Select Drawing dialog box, 322

Select Drawing File dialog box, 148

Select Fence option, selecting objects, 115

Select File dialog box, 44, 45

Select file to attach dialog box, 320

Select Group option, selecting objects, 115

Select Last option, selecting objects, 115

Select Layer dialog box, 356

Select Linetype dialog box, 137, 138, 140

Select Previous option, selecting objects, 115

Select Remove option, selecting objects, 115

Select Window option, selecting objects, 115

Selecting. *See also* Picking

active viewports, 264

default fonts, 142

objects, 115–116

plotters and printers, 88

Selection filters, 115

Selection sets, G–20, 21, 115

Semicolon (;), programming toolbar buttons, 239

Setup Profile command, 305

Setuv command, 433

Shading

Gourard, 421, 422

Phong, 421–422

Shadows option, Render dialog box, 423

Shape(s). *See also specific shapes*

dimensioning, 197

Shape fonts, 209–210

Shape-compiled fonts, 33

Show Light Position dialog box, 426

.shp filename extension, G–8

Size

arrows, 202

centermarks, 202–203

enlarging viewports, 391–392

hexagons, 104

paper, selecting, 88–89

presetting for drawings, 27

Sizing

drawings, 27–29

viewports, 145–146

Slanted surfaces, 173–174

aligning UCS with, 381–384

projecting on cylinders, 184–186

Slash (/)

programming toolbar buttons, 239

stacked fractions and text, 35

Slice command, 300–304, 371

Slider boxes, G–29

Smooth Shading option, Render dialog box, 423

SNAP button, toggling, 14

Snap command, 14, 18, 386–388

Snap From icon, 304

SOLDRAW command, 366–367

Solid fill, hatching, 362

Solid modeling, 257–337

attaching models, 318–319

auxiliary views, 389–391

ball tags. *See* Atrribute(s); Ball tags

chamfers, 299

changing externally referenced drawings, 332–333

complex solid models, 271–276

converting AME solid models, 283

converting to ACIS solid models, 283